Manufacturing Systems Engineering

Manufacturing Systems Engineering

Edited by
Jeff Hansen

Willford Press
www.willfordpress.com

Published by Willford Press,
118-35 Queens Blvd., Suite 400,
Forest Hills, NY 11375, USA

ISBN: 978-1-68285-333-7

Cataloging-in-publication Data

Manufacturing systems engineering / edited by Jeff Hansen.
p. cm.
Includes bibliographical references and index.
ISBN 978-1-68285-333-7
1. Production engineering. 2. Systems engineering. 3. Mechanical engineering. I. Hansen, Jeff.
TS176 .M38 2017
658.5--dc23

For information on all Willford Press publications
visit our website at www.willfordpress.com

Willford Press

Printed in China.

Contents

Preface

Manufacturing Engineering studies in detail the tools and techniques along with the processes involved in the production of any product. This book outlines the various branches and processes of manufacturing with the use of multiple case studies from across the globe. Different approaches, evaluations, methodologies and advanced studies on manufacturing engineering have been presented in this book. A number of latest researches have been included to keep the readers up-to-date with the global concepts and systems in this area of study. It strives to serve as a resource guide for students and experts alike and contribute to the growth of this discipline.

This book is a comprehensive compilation of works of different researchers from varied parts of the world. It includes valuable experiences of the researchers with the sole objective of providing the readers (learners) with a proper knowledge of the concerned field. This book will be beneficial in evoking inspiration and enhancing the knowledge of the interested readers.

In the end, I would like to extend my heartiest thanks to the authors who worked with great determination on their chapters. I also appreciate the publisher's support in the course of the book. I would also like to deeply acknowledge my family who stood by me as a source of inspiration during the project.

Editor

1

An electromagnetism-inspired method for a generalized flowshop problem

Majid Khalili*

Department of Industrial Engineering, Islamic Azad University, Karaj Branch, Karaj, Iran

Abstract – There are two common simplifying assumptions made in the scheduling of operations. One is that machines are available continuously, and the other is that once a job completes one process it is immediately available for its next. Moreover, in flowshops, it is commonly assumed that jobs must visit all machines. In practice, however, these assumptions may not be justifiable in some applications. This paper deals with a generalized flow shop (GFS) problem with machine availability constraints, transportation times between stages, and machine skipping. The electromagnetism-like method (EM) has been successfully applied to some NP-hard problems and this has motivated us to apply and assess the effectiveness of an EM algorithm in the GFS scenarios. Simulated annealing (SA) and a number of other well-recognized heuristics to the given GFS scheduling problem that minimizes two independent objective functions, namely the total tardiness and the total completion time also has been applied. In order to evaluate the performance of the proposed EM and SA, a set of practical instances has been considered. The related results are analyzed in two parts; in term of the objective functions, and the observed effects on variables in our instances. Extensive experiments and statistical analyses demonstrate that our proposed EM is more efficient than SA and other heuristics with regards to the objective functions considered in this paper.

Key words: Generalized flowshop scheduling, Transportation times, Machine availability, Electromagnetism-like method

1. Introduction

Production scheduling is defined as: determining the process sequence by which a given set of parts are processed on a certain number of machines in order to satisfy some performance measures under the production constraints. Through scheduling, all parts in a manufacturing system can be processed by passing through each machine according to a predetermined sequence. Production scheduling develops into a very active and relevant field of research after the first methodical study by Johnson [1]. Since, many papers with a variety of practical and impractical assumptions have been published; however, there always exists a gap between the theory and practice in the literature for this field.

One of the most thoroughly studied scheduling problems is the flow shop (FS) problem. In regular FS, we have a set of n jobs $\{J_1, J_2, \ldots, J_n\}$ and a set of m machines $\{M_1, M_2, \ldots, M_m\}$. Each job has a set of m operations and each operation j can be done by one machine. All the jobs have the same processing routes, starting from machine 1 until finishing at machine m. Each job has a known and fixed processing time for operation j. P_{ij} denotes the processing time of job i on machine j. By removing the restriction that all jobs need to be processed on all machines, the regular FS converts to a more realistic problem [2]. In this problem, we still require jobs to move from the first to the last machine.

Three types of decision-making goals are prevalent in scheduling: (1) Efficient utilization of resources: minimizing completion time or makespan; (2) rapid response to demands, or minimizing the work-in-progress: mean completion time, mean flow time, or mean waiting time; and (3) close conformance to prescribed deadlines: total tardiness, maximum tardiness, and the number of tardy jobs. The tardiness of each job is equivalent to the amount of time when the job is completed after its due date. The first and second types of decision-making goals are process oriented and the third one, as we show in Figure 1, is considered as a customer-oriented objective. In generalized flow shop (GFS), minimizing the total completion time (TCT) is synonymous to maximizing the throughput.

Since the best sequence with respect to the TCT minimization may lead to a large number of jobs being completed after their due dates, we also consider the total tardiness (TT) minimization. The TT minimization is vital prominence in make-to-order or just-in-time environments [3]. The main aim of

*e-mail: Khalili.mj@gmail.com

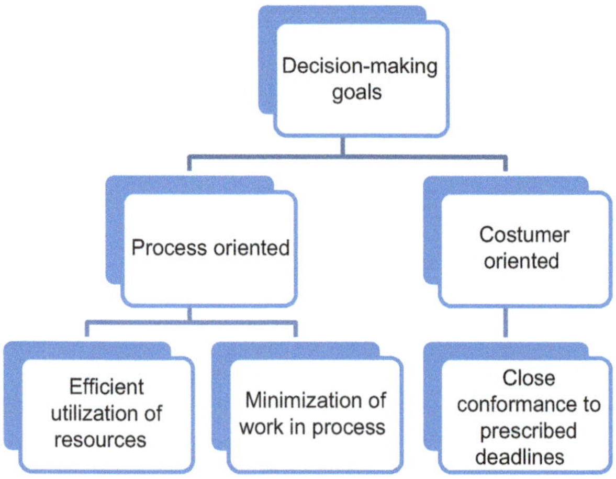

Figure 1. Classification of decision-making goals in scheduling problems.

considering two independent objectives is to gauge the robustness of our proposed EM by two essentially different objectives and study as to whether the high performance of EM is transferable to other objective functions. In the other words, we consider independently the total tardiness and total completion time in order to analyze the efficiency and robustness of our proposed EM on each of these objectives against some existing methods.

In real-world situations machines are not continuously available due to many reasons. For example it is possible that certain important jobs may have been promised during the previous planning horizon. Another common reason for this unavailability could be breakdown (a stochastic event) or preventive maintenance (a deterministic event). A breakdown makes the shop behavior hard to predict, and thereby reduces the efficiency of the production system. Therefore, scheduling maintenance to reduce the breakdown rate has commonly been recognized by the decision makers. This paper investigates the deterministic case of machine unavailability before making decisions for job sequence starts.

In most studies on scheduling it is assumed that processing of a job i on machine j commences without delay after finishing its process on machine $j-1$. In practice, in most manufacturing and distribution systems, semi-finished jobs are transferred from one processing facility to another by transporters such as robots and conveyors, and finished jobs are delivered to warehouses or customers by vehicles such as trucks. The transportation time from machine $j-1$ to machine j is T_f and transportation time from machine j to machine $j-1$ is T_b. The time to load and unload the transporter is included in the transportation time. When transporter reaches machine j, it delivers the job to the machine if the setup time has been completed and machine is ready to receive the job. Otherwise, it places the job in a waiting line and immediately starts its return to machine $j-1$. In other words, once the transporter leaves the first machine, it always returns in time $T_f + T_b$ to take the next job. This transportation time could be either job-dependent or job in-dependent. We assume that all transportations are job-independent; and transportations between two machines are handled by one transporter. This transporter can only carry one job at a time; therefore, a job may have to wait for the transporter before it is transported.

In regular flowshop, the earliest starting time of job j processed on machine i is computed by:

$$s_{ji} = \max\left\{C_{ji-1}, C_{j'i}\right\}, \tag{1}$$

where C_{ji-1} is the completion time of job j on machine $i-1$ (i.e., $C_{ji-1} = s_{ji-1} + p_{ji-1}$) and j' is the predecessor of job j in the sequence.

In this paper, flowshop with transportation times is considered, on which the earliest starting time, s_{ji}, is calculated by:

$$s_{ji} = \max\left\{C_{ji-1} + T_{f_j}, C_{j'i-1} + 2T_{f_j} + T_{b'_j}, C_{j'i}\right\} \tag{2}$$

where T_{f_j} and T_{b_j} are transportation time from machine $i-1$ to machine i and return time of a transporter from machine i to machine $i-1$. Precisely, the term of $C_{ji-1} + T_{f_j}$ determines the s_{ji} when job j does not wait for the transporter while the term of $C_{j'i-1} + 2T_{f_j} + Tb_j$ is used when job j has to wait for the transporter. The term of $C_{j'i}$ shows the availability time of machine i.

Electromagnetism-inspired algorithms have recently been classified as a meta-heuristic approach to tackle complex optimization problems. The motivation behind this meta-heuristic approach has risen from the attraction-repulsion mechanism of electromagnetic theories and this basic idea that in meta-heuristics we desire to bring our search closer to a region with the superior objective function and at same time, go away from the region with the inferior objective function to move the solution gradually toward the optimality. The EM shows very high performance than other meta-heuristics in NP-hard problems [4, 5]. In our case, machine availability constraints and transportation times as well as skipping probability of some jobs from some stages are assumed. To judge the performance of the proposed EM, we go through the corresponding problem under minimizing two separate objectives, total tardiness from make-to-order environments, and total completion times from make-to-stock environments.

The GFS is categorized as a combinatorial optimization problem known to be strongly NP-hard requiring an effective metaheuristic algorithm to be utilized to overcome the complexity of the problem considered and solve it. Therefore, in this paper we have employed EM which is one of the recently introduced metaheuristic for the first time. This is the first time that the EM algorithm is being employed for the problem considered GFS. Therefore, there is no similar EM in the literature to compare against. We have compared the most relevant and effective algorithms in the literature of the problem at hand to prove the superiority of our algorithm. Meanwhile, based on your comments, we mentioned our pioneering effort to employ EM in our problem in the paper.

The purpose of this paper is to investigate a realistic flowshop case with three practical assumptions, transportation time, machine availability constraints and machine skipping. Doing so, we aim to establish a simple criterion to integrate GFS production scheduling with preventive maintenance (PM) activities to yield a high level of machine availability noting that PM operations have priority over production operations. So PM

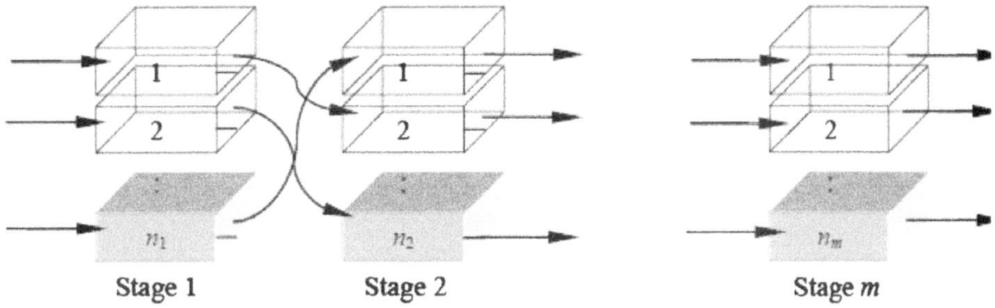

Figure 2. 3D flowshop scheduling problem.

operations must be scheduled first, and then jobs sequenced along with them. To solve such an NP-hard problem [5], we propose to apply the EM. This proposed algorithm tackles two independent objectives and we examine its robustness in different situations. Finally, we compare these results with the results of a number of other related algorithms. The following assumptions are usually characterized to GFS: each machine in each stage can process at most one job at a time, each job can be processed by at most one machine at a time in each stage, the process of a job on a machine cannot be interrupted, machines aren't available continuously and once a job completes one process it isn't immediately available for its next (Figure 2).

The problem under consideration generalizes the problem studied by the other paper. This generalization comes from assuming machine availability constraints. This paper considers total completion minimization as well as total tardiness, while the other paper minimizes makespan and total tardiness.

The rest of this paper is organized as follows: Section 2 gives literature review on the problem. We introduce the preventive maintenance and how to integrate it with production operations sequencing in Section 3. In Section 4, the proposed algorithm is presented. In Section 5 we evaluate the algorithms by a benchmark. Finally, Section 6 concludes the paper.

2. Literature review

The FS has been a very dynamic research domain in meta-heuristic and heuristic techniques, principally because of the difficulty encountered by exact methods to solve large or even medium instances.

Salvador [6] was the first who defined the FS problem. A detailed survey for the FS problem has been given by Linn and Zhang [7] and Wang [8]. Moursli and Pochet [9] presented a branch-and-bound algorithm to minimize the makespan in FS. Exact algorithm is applied to large FS problems; such an approach can take hours or days to derive a solution so Sriskandarajah and Sethi [10] proposed a heuristic algorithm for solving a special case of the FS problem. Guinet et al. [11] proposes heuristic techniques for a simplified FS makespan problem with two stages and only one machine at stage two. Guinet et al. [11] also proposed a heuristic for the makespan problem in a two-stage FS based on Johnson's rule [1]. To obtain a near optimal solution, meta-heuristic algorithms have also been proposed. Nowicki and Smutnicki [12] proposed a tabu search (TS) algorithm for the FS makespan problem. Gourgand et al. [13] and Zhang and Wu [14] presented several simulated annealing (SA)-based algorithms for the FS problem. A genetic algorithm has been widely used in many previous works for the FS makespan problem, see e.g., Reeves [15].

Tavakkoli-Moghaddam et al. [16] proposed a memetic algorithm (MA) combined with a new local search engine, namely, nested variable neighborhood search (NVNS), to solve a FS scheduling problem with processor blocking. The performance of the proposed EM was verified by a number of instances and then compared with genetic algorithms. Cheng et al. [17] addressed the earliness/tardiness scheduling problem with identical parallel machines, and they apply a genetic algorithm to solve this problem. Yang [18] was the first to use EM to train a neural network. The results show a great saving on the computation memories and time, and indicate that EM performed much better than genetic algorithm in finding the global optimum. For its merit of simple concept and economic computational cost, EM has been used in the areas of function optimization, fuzzy neural network training, project scheduling, and other combinatorial optimization fields [5] but seldom have used in scheduling problems, so we were motivated to solve our problem with this method. Khalili and Tavakkoli-Moghadam [19] proposed a new multi-objective electromagnetism algorithm for a bi-objective flowshop scheduling problem.

Because of its significance both in industrial and theory applications, the GFS has attracted the attention by many researchers. To obtain a near-optimal solution, meta-heuristic algorithms have also been proposed. Different genetic algorithms (GA) are applied by Chen and Neppalli [20], Aldowaisan and Allahverdi [21]. Among the other metaheuristics, one could refer the reader to particle swarm optimization (PSO) by Pan et al. [22], simulated annealing (SA) by Fink and Voß [23], ant colony optimization (ACO) by Shyu et al. [24] and tabu search (TS) by Grabowski and Pempera [25]. Khalili [26] proposed an iterated local search algorithm for flexible flow lines with sequence dependent setup times to minimize total weighted completion and also studied multi-objective no-wait hybrid flowshop scheduling problems to minimise both makespan and total tardiness [27].

Next, briefly review the related research on availability constraint and the traveling time between stages, which are other characteristics of the problem at hand. Traveling times between stages in most investigations have been considered as processing time. Hurink [28] assumed that unlimited buffer spaces exist between the machines, and all transportation is accomplished through a single robot. Soukhal et al. [29] analyzed two-machine FS scheduling problems with transportation and assumed that finished jobs are transported from the processing facility and delivered to customers by trucks. In the other words, both transportation capacity and transportation times were explicitly taken into account.

Scheduling problems with availability constraints have been extensively examined by Ruiz et al. [30]. For more details, a survey of existing methods for solving scheduling problems under availability constraints as well as complexity results can be found in Schmidt [31]. Lee [32] handled the preemptive FS problem with two machines and one unavailability period first imposed on machine 1 and then on machine 2 with the makespan objective. Both problems are proved to be NP-hard in the ordinary sense, and heuristics with error bounding analysis are proposed. Blazewicz [33] investigated the two-machine problem with arbitrary number of unavailability periods on one machine, and proved that the makespan minimization problem is strongly NP-hard. Breit [34] investigated the problem of scheduling n preemptable jobs in a two-machine FS where the first machine is unavailable for processing during a given time interval. A more complex hybrid FS problem is tackled in Allaoui and Artiba [35].

3. Preventive maintenance

Machine scheduling is concerned with the problem of optimally scheduling available machines. However, the majority of the scheduling literature carries a common assumption that machines are available at all times. This availability assumption may not be true in real industrial settings, since a machine may become unavailable during certain periods of time. For instance, a machine may not be available at the beginning of a planning horizon as it may continue to process the late jobs from the previous horizon. Similarly, when a machine breakdown has occurred or a preventive maintenance activity has been scheduled. A breakdown causes the shop behavior difficult to predict, and thereby reduces the efficiency of the production system. Therefore, scheduling maintenance to reduce the breakdown rate has commonly been recognized by the decision makers. It is known that maintenance plays an important role in many industries because more reliable production systems with higher serviceable performance is what factories are mostly intending to reach; on the other hand, maintenance systems are designed to make sure that production facilities are serviceable and reliable to operate to achieve target production levels [30].

Maintenance policies influence the machine availability and the machine utilization ratio. Maintenance actions usually can be classified into two major categories: corrective maintenance (CM) and preventive maintenance (PM). CM is the actions carried out when a failure has already occurred. PM is the action taken on a system while it is still in service, but is carried out in order to keep the system at the desired level of performance. In PM, activities are conducted at fixed time intervals determined for machines and facilities before a failure or breakdown occurs. Here, the main question is whether PM decisions and production sequencing should be jointly scheduled. Ruiz et al. [30] investigated this issue and concluded that the optimal decision is case dependent.

Table 1. Processing times (P_i) for single machine with n = 4.

	Jobs (i)			
	1	2	3	4
P_i	25	30	25	20

According to different criteria, various PM policies are defined. The commonly used PM policy in industry is "preventive maintenance at fixed predefined time intervals" [30]. PM operations are scheduled in advance at pre-determined time intervals without taking into account a probabilistic behavior for time to failure. The intervals between PM operations can be weekly, monthly or other periods depending on the system. In this policy, fixed time intervals (T_{PM}) are determined and PM operations are carried out exactly at those times. Our criterion works as follows: whenever a new job is to be processed on each machine, the completion time is computed. If this time exceeds the pre-defined T_{PM}, then the process of the next job is postponed and the PM is carried out first because PM operations have priority over processing. It should be pointed out that has been assumed that the processing of a job cannot be interrupted (or preemption is not allowed or the jobs are non-resumable), and availability time of any machine is after finishing the process of the last job on that machine. The following example shows how PM operations scheduling and single machine scheduling are jointly considered. Suppose a shop decides to carry out PM at every 50 time units. The durations of these PM operations (D_{PM}) for the machine are 15 time units. The processing times are shown in Table 1. The durations of these PM operations (D_{PM}) for the machine are 15 time units. The processing times (P_i) are also shown in Table 1. In fact, will been scheduled the jobs in the production horizon shown in Figure 3.

Figure 4 depicts a sequence of all the jobs as {4, 1, 3, 2}. As it can be seen in this figure, the first maintenance is carried out after processing jobs 4 and 1, the accumulated total processing time is 25 + 20 = 45 and it is not possible to process the third job of sequence (Job 3), because it has a processing time of 25 time units, which would result in an accumulated total processing time of 70 units, which is longer than T_{PM} = 50. So, machine will be idle for 5 time units, and then PM operation begins at time 50 units, and lasts for 15 time units. After carrying out PM, Job 3 is processed. Then, once more it is not possible to process the next job (Job 2) since a processing time of 90 + 30 = 120 will be accumulated. So, in the same manner, second PM operation is carried out from 100 to 115, then Job 2 is processed.

4. Proposed algorithms

4.1. Heuristics

In general, the heuristics applied to each decision making goal, process or customer oriented, are different. These two

	1	2	3	4
P_i	25	30	25	20

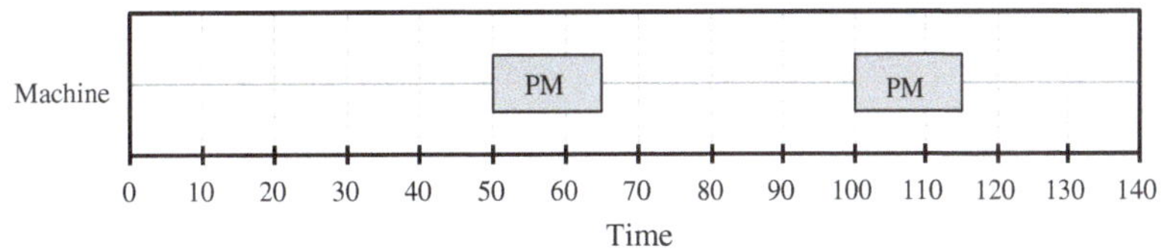

Figure 3. Gantt chart of the production horizon after applying PM.

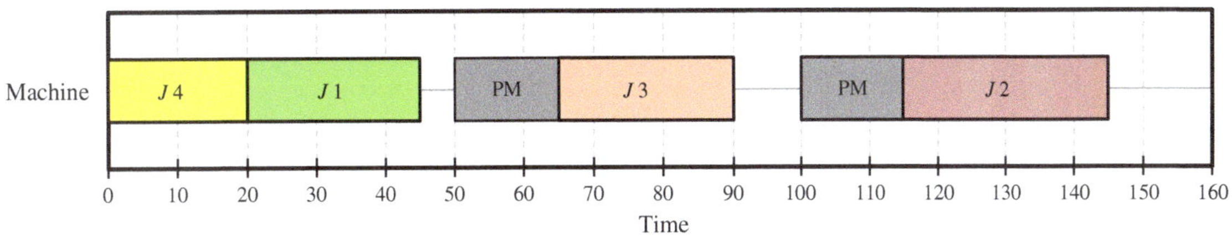

Figure 4. Gantt chart of the solution for the given example.

groups have been introduced and explain the heuristics, which we have used in this study:

1. Heuristics to solve scheduling problem with emphasis on completion times such as SPT, LPT, NEH (Nawaz, Enscore and Ham), and (g/2, g/2) Johnson' rule [1].
2. Heuristics to solve scheduling problem with emphasis on tardiness such as EDD, SLACK and NEH_EDD.

The above seven heuristics are briefly described below.

- Shortest Processing time (or SPT): jobs are processed on machine 1 in increasing order of the processing times on machine 1. That is, the process of jobs with shortest processing times is first started. At subsequent machines, jobs are sorted in earliest ready time order.
- LPT arranges jobs on machine 1 in descending order of processing times of jobs on machine 1.
- (g/2, g/2) Johnson's rule, the sum of processing time of jobs on machine 1 to [m/2] and the sum over machine [m/2] + 1 to m are calculated to order jobs on machine 1 [1].
- NEH, This heuristic can be divided into three simple steps:

  1. The total processing times for all the jobs on m machines are computed as follows:

  $$p_i = \sum_{j=1}^{m} P_{ij}; \quad i = 1, \ldots, n$$

  2. Jobs are sorted in descending order of P_i. Then, the first two jobs (those with higher P_i) are taken and the two possible schedules containing them are evaluated.
  3. Take job i (i = 3, …, n) and find the best schedule by placing it in all the possible i positions in the sequence of jobs that are already scheduled. For example, if i = 4, the already built sequence would contain the first three jobs of the sorted list calculated in step 2, then, the fourth job could be placed either in the first, second, third or the last position of the sequence. The best sequence of the four would be selected for the next iteration.

- EDD is a well known dispatching rule and orders the jobs according to imminent due dates.
- SLACK, another name of this method is the minimum slack because in this method at time t, the job with the minimum value of $d_j - C_j(s)$ is selected, where $C_j(s)$ will be the completion time of job $j \notin s$ if it is scheduled at the end of sequence s.
- In NEH_EDD we consider the due dates for defining an initial order in which the jobs are considered for insertion. The initial order in NEH_EDD is based on the earliest due date dispatching rule that arranges jobs in ascending order of their due dates.

4.2. Simulated annealing

Simulated annealing (SA) was first presented as a search algorithm for combinational optimizations (CO) problems in Cerny [36]. SA is modeled after physical annealing of solid metal. In annealing, a metal is first heated to a high temperature and then cooled down with a very slow rate to the room temperature. Sometimes, if cooling is not fast enough, quenching is done. In SA, solutions are randomly generated from a set of feasible solutions. This process accepts not only those solutions that improve the objective function, but also those solutions, which do not improve objective function on the basis of transition probability (TP). Transition probability depends on the change in objective function and the annealing temperature. The main features of SA that make this algorithm more sophisticated are perturbation annealing schedule and the

```
Procedure Simulated_annealing
t = T_0
x = initialization                      % Initial solution by another algorithm
x_best = x
while stopping criterion is not met do
    for iter = 1 to max do
        s = move x by an operator       % generating a neighbor solution from x
        if f(s) < f(x) then             % Acceptance criterion
            x = s
            if f(s) < f(x_best) then    % check with the best solution
                x_best = s
            endif
        else
            if random < exp{-(f(s) - f(x))/t} then
                x = s
            endif
        endif
    endfor
    t = α·t                             % temperature decrease
endwhile
```

Figure 5. General outline of a simulated annealing algorithm.

transition probability. Perturbation generates a new solution, and the annealing schedule controls the initial temperature, final temperature and the rate of cooling, while transition probability help heuristic to escape local optima. SA is commonly said to be the oldest among the meta-heuristics and one of the first algorithms that had an explicit strategy to avoid local optima.

The fundamental idea is to generate a new job sequence s by a random rule from the neighborhood of present sequence x. This new sequence is accepted or rejected by another random rule. A parameter t, called the temperature, controls the acceptance rule. The variation between objective values of two candidate solutions is computed $\Delta C = obj(s) - obj(x)$, where obj is the value of the objective function. If ($\Delta C \leq 0$, sequence s is accepted. Otherwise, sequence s is accepted with probability equal to $\exp(\Delta C/t_i)$. The algorithm proceeds by trying a fixed number of neighborhood moves (*max*) at temperature t_i, while temperature is gradually decreased. The procedure is repeated until a stopping criterion is met.

Moves resulting in solutions of worse quality (uphill move) than the current solution may be accepted to escape from local minimum. SA starts at a high temperature (T_0), so most of the moves are accepted at first steps of the procedure. The probability of doing such a move is decreased during the search. Figure 5 shows the general outline of SA algorithm.

The commonly used encoding scheme for FS problem is permutation of jobs which shows the job sequence on machine 1. It is known that initial solution can influence the quality of the solutions. A good initial solution can result in a better and faster subsequent result. There are some alternative choices to consider different parameters and adjust them by tuning and settings. The proposed SA algorithm checks 100 neighbors at temperature t_i (i.e., $max = 100$).

Moving operator generates a neighbor solution from current candidate solution by making a slight change in it. This operator must work in such way to avoid infeasible solutions. In this research, two different move operators have been considered:

- Swap operator (SO): the positions of two randomly selected jobs are swapped (Figure 6).
- Single point operator (SPO): the position of one randomly selected job is randomly changed (Figure 7).

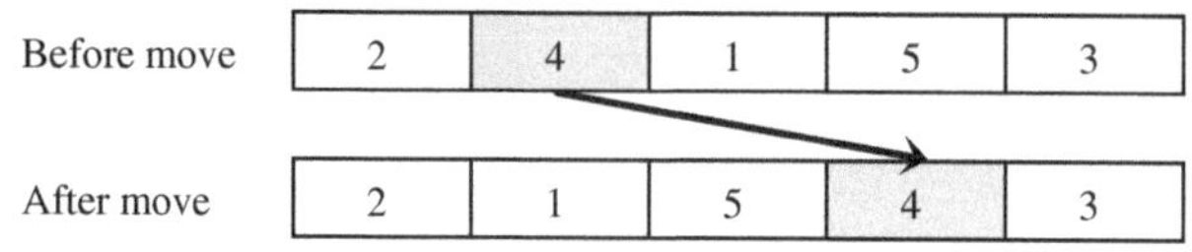

Figure 6. Single point operator.

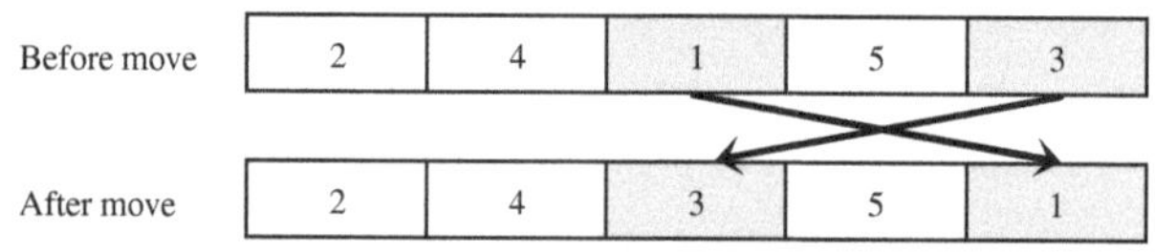

Figure 7. Swap operator.

The algorithm developed in this paper uses different operators from the other paper. As mentioned earlier, to avoid a local minimum, solutions with worse objective values are probabilistically accepted depending on the value of the temperature. When the procedure proceed, the temperature is slightly lowered under a mechanism which is called cooling schedule. Here, exponential cooling schedule has been used, $T_i = \alpha \times T_{i-1}$ (where $\alpha \in (0, 1)$ is temperature reducing rate), which is often believed to be an excellent cooling recipe.

4.3. Electromagnetism-like method

Birbil and Fang [37] proposed the electromagnetism-like method (EM), which is a flexible and effective population-based algorithm to search for the optimal solution of global optimization problems. EM originates from the electromagnetism theory of physics by considering potential solutions as electrically charged particles spread around the solution space. This meta-heuristic method utilizes an attraction-repulsion mechanism to move the particles towards optimality. Debels et al. [38] applied successfully a hybridization of the EM with scatter search for resource-constrained project scheduling problem. EM is useable for particular set of optimization problems with bounded variables.

Each candidate solution as a charged particle has been considered. The charge of each candidate solution is related to the objective function value. The size of attraction or repulsion over candidate solutions in the population is calculated by this charge. The direction of this charge for candidate solution i is determined by vectorially adding the forces from all other solutions on the candidate solution i. In this mechanism, a candidate solution with good objective function value attracts the other ones; candidate solutions with unfavourable objective function repel the other population members; and better the result of the objective function value the higher the size of attraction.

EM has four phases including *Initialization* of algorithm, computation of *total force* exerted on each particle, *movement* along the direction of the force, and the *local search*.

4.3.1. Encoding scheme and initialization

The most frequently used encoding for the flowshop problem is a simple permutation of the jobs. The relative order of the jobs in the permutation indicates the processing order of

Procedure *Local_search*
$k = 1$
while $k < n + 1$ **do**
v = Shift x_{ik} to a new random position
if $f(v) < f(x_i)$ **then**
$x_i = v$
$k = n$
endif
$k = k + 1$
endwhile

Figure 8. Procedure of the local search.

the jobs on the first machine in the shop. It is necessary to mention that a drawback of the algorithms proposed for generalized flowshop problems is that they only order jobs according to earliest ready time of jobs at the beginning of each stage. However in GFS problems, it is very likely to have some jobs with the same ready time at beginning of each stage. For example, all the jobs which skip the first stage would have the ready time of zero at the beginning of the second stage. So we need to establish a clear criterion to order these jobs. In this paper, if some jobs have the same ready time at the beginning of stage t ($t = 2, 3, \ldots, g$), the same as their order at stage $t - 1$ has been arranged. We subject all the compared algorithms to this criterion.

Traditionally, in an EM, the initial population is generated randomly. However, it is known that the initial solutions can affect on the quality of the results obtained by the algorithms. The initialization procedure in such a complex combinatorial problem has to be made with great care, to ensure convergence to desirable objective function values in a reasonable amount of time. Because of this, initial solutions for proposed EM are generated by EDD and NEH_EDD for total tardiness objective and SPT and NEH for the completion time objective. The population size (*popsize*-2) is randomly generated. The objective function values of solutions are calculated and the best one is recorded as x_{best}.

4.3.2. Local search

The proposed EM is hybridized with a local search in order to improve the performance of the algorithm. The procedure for this local search can be described as follows: the first job in the sequence of candidate solution i (x_{i1}) is relocated to a new random position in the sequence. If this new sequence (v) results in a better makespan, the current solution (x_i) is replaced by the new sequence (v). This procedure iterates at most for all the subsequent jobs in sequence. If there is any improvement in the kth $< n$, the local search for the current solution terminates. Subsequently, the best solution is updated. Procedure for the local search is shown in Figure 8.

4.3.3. Computation of total forces

The charge of each solution candidate i is calculated in relation to its objective function by:

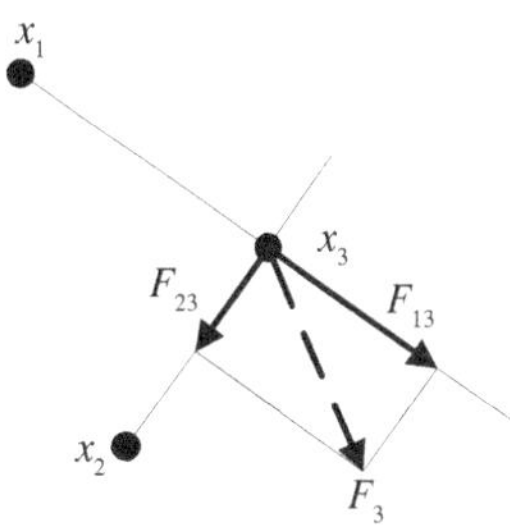

Figure 9. Example of exertion of forces.

$$q_i = \exp\left(-n \cdot \frac{f(x_i) - f(x_{\text{best}})}{\sum_{j=1}^{\text{popsize}} \left(f(x_j) - f(x_{\text{best}})\right)}\right), \quad \forall i\; i = 1, 2, \ldots, \text{popsize}. \tag{3}$$

This formula ensure that the better objective values are assigned higher charges. The *total force* F_i exerted on candidate solution i is also calculated by the following formula:

$$F_i = \begin{cases} \sum_{j \neq i}^{P} (x_j - x_i) \dfrac{q_i \cdot q_j}{\|x_j - x_i\|^2}; & f(x_j) < f(x_i), \\ \sum_{j \neq i}^{P} (x_i - x_j) \dfrac{q_i \cdot q_j}{\|x_j - x_i\|^2}; & f(x_i) < f(x_j). \end{cases} \tag{4}$$

A two-dimensional example *total force vector* F_i exerted on candidate solutions is shown in Figure 9. The force exerted by x_1 on x_3 is F_{13} (repulsion: the objective function of x_1 is worse than that of x_3) and the force exerted by x_2 on x_3 is F_{23} (attraction: the objective function of x_2 is better than that of x_3). F_3 is the *total force* exerted on x_3 by x_1 and x_2.

4.3.4. Moving by total forces

All the candidate solutions are moved with the exception of the current best solution. The move for each candidate solution is in the direction of the *total force* exerted on it by a random step length. This length is generated from a uniform distribution between [0, 1]. By selecting random length that candidate solutions have a nonzero probability to move to the unvisited solution space along this direction has been guaranteed. Moreover, by normalizing the *total force* exerted on each candidate solution, we are able to avoid producing infeasible solutions. Figure 10 shows a pseudo code of our proposed EM.

5. Experimental evaluation

In this section, we apply a benchmark to evaluate the performance of our proposed algorithms. They are implemented in MATLAB 7.0 and run on a Pentium IV PC with an Intel processor running at 3 GHz and 1 GB of RAM memory. Relative percentage deviation (*RPD*) for total completion time as a common performance measure to compare the methods has been used. The best solutions, Min_{sol}, obtained for each instance

Procedure *Electromagnetism-like method*

```
Initialization
While termination criterion are not satisfied do
    for i = 1 to popsize do
        x_i = Local search
    endfor
    Update x_best

    for i = 1 to popsize do                                   %Procedure computation_of_total_forces
        calculate (q_i)
        F_i = 0
    endfor
    for i = 1 to popsize do
        for j = 1 to popsize do
            if f(x_j) < f(x_i) then
                F_i = F_i + (x_j − x_i) (q_i·q_j)/||x_j − x_i||^2   (attraction)
            else
                F_i = F_i − (x_j − x_i) (q_i·q_j)/||x_j − x_i||^2   (repulsion)
            endif
        endfor
    endfor

    for i = 1 to popsize do                                   %Procedure Moving_by_total_forces
        if i ≠ best then
            β = random [0, 1]
            F^i = F_i/||F_i||
            for k = 1 to n do
                if F_ik > 0 then
                    x_ik = x_ik + β · F_ik(1 − x_ik)
                else
                    x_ik = x_ik + β · F_ik(x_ik)
                endif
            endfor
        endif
    endfor
Endwhile
```

Figure 10. Pseudo code of the proposed EM.

are computed by any of the five algorithms. *RPD* is obtained by the following equation:

$$\mathrm{RPD} = \frac{\mathrm{Alg_{sol}} - \mathrm{Min_{sol}}}{\mathrm{Min_{sol}}} \times 100 \tag{5}$$

where $\mathrm{Alg_{sol}}$ is the objective function value obtained for a given algorithm and instance. Clearly, lower values of RPD are preferred. In the case of the total tardiness GFS problem, the best solution could be zero (and therefore optimal), then in the above equation we will have division by zero. Moreover, if the best solution is a small value, the performance measure underestimates an algorithm which obtains a solution slightly worse than the best. Therefore, a different performance ratio is usually used in tardiness cases to avoid these problems, which has been termed the relative deviation index (RDI). The RDI is obtained by:

$$\mathrm{RDI} = \frac{\mathrm{Alg_{sol}} - \mathrm{Min_{sol}}}{\mathrm{max_{sol}} - \mathrm{Min_{sol}}} \times 100. \tag{6}$$

With this measure, an index between 0 and 100 is obtained for each method such that a good solution will have an index very close to 0. Note that if the worst and the best solutions take the same value, all the methods provide the best (or same) solution and hence, the index value will be 0 (i.e., the best index value) for all methods.

Indeed, the above formulas are two different ways of normalization. They calculate the deviation from the best solution, which is always a positive number.

5.1. Data generation

Data required for the given problem consist of the number of jobs (n), number of machines (m), range of processing times (p) and transportation times (T_f and T_b), the ready times (r), skipping probability, time interval between two consecutive PM (T_{PM}) and duration of PM operations (D_{PM}). Our instances based on benchmark values [39] that are shown in Table 2 has been guaranteed.

We have $n = \{20, 50, 100, 200, 500\}$ and $m = \{5, 10, 20\}$, which results in 15 combinations of $n \times m$. The processing time in Taillard's instances [39] are generated from a uniform distribution over the range [1, 99]. GFS is considered by allowing

Table 2. Factor levels.

Factor	Levels
Number of jobs	20 50 100 200 500
Number of machines (m)	5 10 20
Range of processing times (p)	U [1, 99]
Range of transportation times ($T_f + T_b$)	U [1, 30]
Skipping probabilities	0.1 0.4
Range of time interval between two consecutive PM (T_{PM})	U [200, 300]
Range of duration of PM operations (T_{PM})	U [1, 50] U [1, 99] U [1, 150]

some jobs to skip some stages. The probability of skipping a stage is set at 0.1 or 0.4. The transportation times (T_f and T_b) come from a uniform distribution in the range [1, 30], where the running average will work out about 30% of the processing time. T_{PM} for each machine are obtained from a uniform distribution in the range [200, 300]. D_{PM} of each machine are distributed uniformly over three ranges [1, 50], [1, 99] and [1, 150], where the running average will be about 50%, 100% and 150% of the processing times, respectively. The different levels of factors result in $5 \times 3 \times 1 \times 1 \times 2 \times 1 \times 3 = 90$ different scenarios. Ten instances for each scenario, similar to a Taillard's benchmark have been produced. Therefore, $90 \times 10 = 900$ instances have been.

To generate due dates for all n jobs an approach similar to [40] have been used. The following steps are applied to produce the due dates [41]:

1. Compute the total processing time of each job on all g stages.

$$p_i = \sum_{t=1}^{g} p_{it}, \quad \forall i \in N. \tag{7}$$

2. Determine a due date for each job:

$$d_i = (p_i) \times (1 + \text{random} \times 3), \quad \forall i \in N, \tag{8}$$

where *random* is a random number from a uniform distribution over the range [0, 1].

5.2. Parameter tuning

The quality of algorithms is significantly influenced by the values of parameters. In this section, the behavior of different operators and parameters of SA and the proposed EM have been studied. In order to tune the algorithms, we apply a full factorial design in the design of experiment (DOE) approach [42]. Five instances for each combination of n and m, which results in a total of 75 instances have been randomly generated. The stopping criterion is $n \times m \times 0.2$ s of the computational time. This criterion allows for more time as the number of jobs or machines increases.

5.2.1. Simulated annealing

The proposed SA has three parameters, initial solution, {initial temperature, cooling rate} and the move operator. The considered levels of the parameters are:

- Initial solution (*IS*): two levels (SPT, NEH).
- {Initial temperature (T_0), cooling rate (α)}: three levels ({50, 0.985}, {100, 0.98}, {200, 0.97}).
- Move operator (*MO*): two levels (SO, SPO).

So, $2 \times 3 \times 2 = 12$ different SAs are obtained by these levels and all the 75 instances are solved by them. The results are analyzed by means of the analysis of variance (ANOVA) technique. Three main hypotheses, including normality, homogeneity of variance and independence of residuals have been examined and no bias have been found to question the validity of the experiments. The means plot and least significant differences (LSD) intervals (at the 95% confidence level) for the levels of *IS*, {T_0, α} and *MO* parameter are shown in Figures 11–13, respectively.

Figure 10 illustrates that the initial solution of NEH provides statistically better results than SPT. In Figure 11 can be seen that there is statistically significant difference between two move operators and that SPO is superior and also in Figure 12 can be seen $T_0 = 50$ and $\alpha = 0.985$ results in statistically better output than either of the other two sets. As the results of the analysis, the parameters *IS* = NEH, *MO* = SPO and (T_0, α) = (50, 0.985) have been set.

5.2.2. Electromagnetism-like method

One of the advantages of the proposed EM is that it has only one parameter, *popsize* (number of population). The considered levels for *popsize* are 2, 4, 6 and 8. All 75 instances are solved by the EM algorithm obtained by the above values of *popsize*. The results are assessed by means of Analysis of Variance (ANOVA) technique. The means plot and least significant differences (LSD) intervals for various levels of *popsize* parameter factor are shown in Figure 14. This figure reveals that the number of population of four provides statistically better results than other values of *popsize* = 2, 6, 8.

5.3. Experimental results

In this subsection, the authors intend to compare our proposed EM with other existing methods. As described before, this phase includes two subsections each of which considers one objective function. In each subsection, the proposed EM

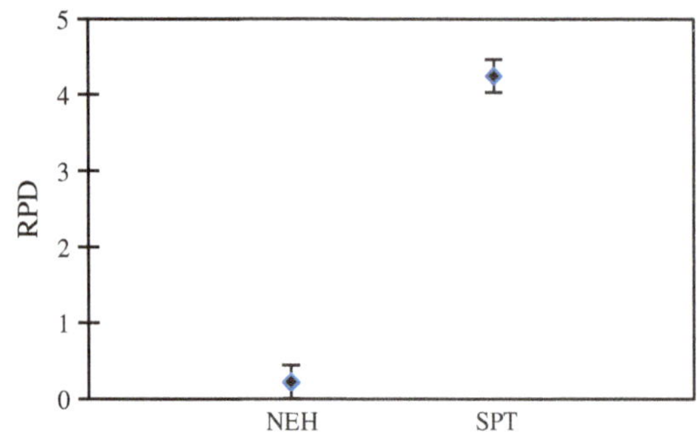

Figure 11. Means plot and LSD intervals for the initial solution of SA.

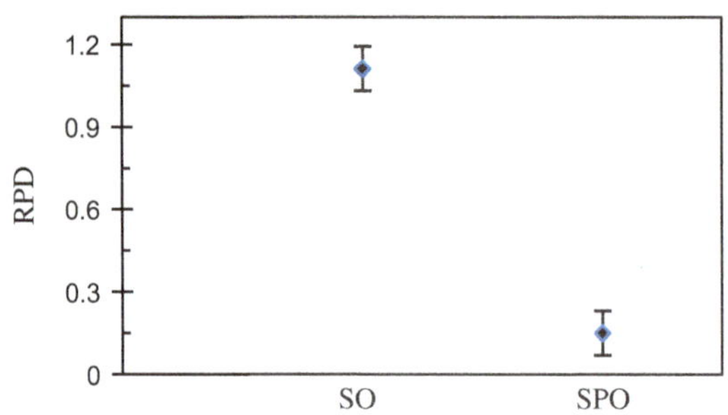

Figure 12. Means plot and LSD intervals for the move operator of SA.

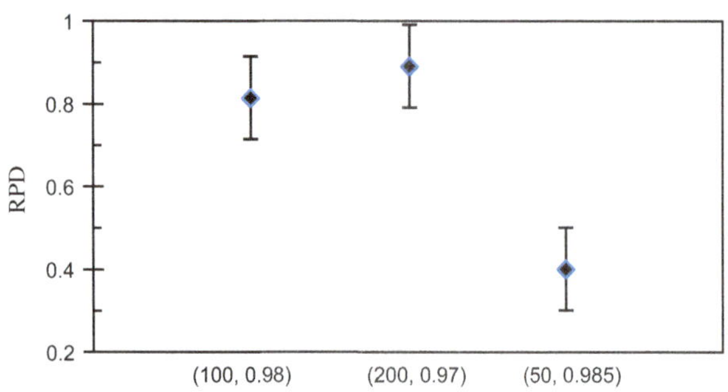

Figure 13. Means plot and LSD intervals for the initial parameters (temperature, cooling rate) in SA.

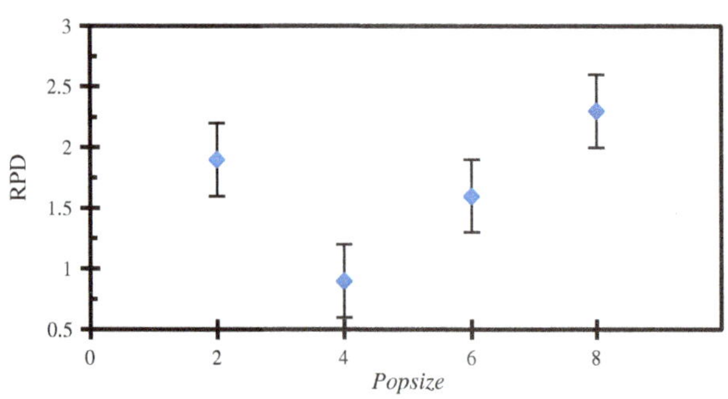

Figure 14. Means plot and LSD intervals for number of population in the proposed EM.

Table 3. Average relative percentage index ($\overline{\text{RPI}}$) for the algorithms grouped by n and m.

Instances	EDD	SLACK	NEH_EDD	SA	EM
20 × 5	97.50	93.25	36.43	6.42	**3.14**
20 × 10	80.00	76.40	29.87	3.36	**1.97**
20 × 20	60.20	66.08	26.54	4.41	**3.57**
50 × 5	91.05	96.50	39.18	18.08	**6.53**
50 × 10	89.00	92.63	36.67	17.83	**2.81**
50 × 20	91.00	95.18	19.20	15.94	**1.69**
100 × 5	90.53	94.46	48.24	26.57	**8.88**
100 × 10	95.47	96.28	46.02	28.89	**2.18**
100 × 20	92.63	98.05	35.96	19.80	**3.03**
200 × 5	90.34	92.47	38.76	17.94	**5.63**
200 × 10	93.00	98.00	26.02	18.67	**7.63**
200 × 20	92.02	98.74	25.27	18.54	**6.60**
500 × 5	93.00	98.00	24.68	15.46	**8.46**
500 × 10	92.63	98.03	18.02	14.51	**6.78**
500 × 20	91.36	97.79	13.29	11.40	**9.31**
Average	89.32	92.79	30.94	15.85	**5.21**

with some well-known heuristics as well as SA for associated objective functions has been compared. In the first subsection, the EM is compared with SA, EDD, SLACK and NEH_EDD, and in the next stage it is compared with SA, SPT, ($g/2$, $g/2$) Johnson's rule, and NEH [1]. In each subsection, the effects of variables such as problem size and numbers of machines on the performance of the algorithms are investigated. The stopping criterion is $n \times m$ 0.2-s computation time.

As mentioned earlier, the total completion time (TCT) belonging to the process oriented goals, and the total tardiness (TT) belonging to the costumer oriented goals independently to analyze the efficiency and robustness of our proposed EM on each of these objectives against some existing methods has been considered.

5.3.1. Analysis of the total tardiness

In this section, the authors compare our proposed EM algorithm with SA, NEH_EDD, SLACK and EDD with the aim of minimizing the total tardiness. The results of the experiments, averaged for each combination of n and m are shown in Table 3. As expected, the meta-heuristics perform better than the heuristics, and the worst performing algorithms in almost all the instances are EDD and SLACK with RDI of 89.32% and 92.79%, respectively. The proposed EM provides best results among these algorithms with the RDI of 5.21%.

For further analysis, the ANOVA has been carried out. The means plot for the different algorithms with the least significant difference (LSD) intervals are shown in Figure 15. As it is seen, the proposed EM provides statistically better results than the other methods. NEH_EDD supersedes the traditional EDD and SLACK. EDD and SLACK are statistically the same.

The results also show that EM statistically outperforms the other algorithms. It is interesting to see the performance of NEH in comparison with two other heuristics. NEH_EDD significantly supersedes EDD and SLACK. In addition, we carry out a two-way ANOVA for each variable to find the interaction between them and performance of the algorithms. Figure 16 shows the trend of the quality of the methods evaluated as

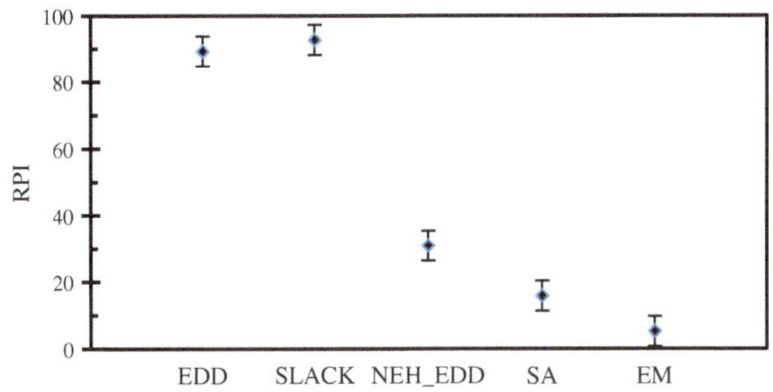

Figure 15. Means plot and LSD intervals (at 95% confidence level) for the type of the algorithm factor.

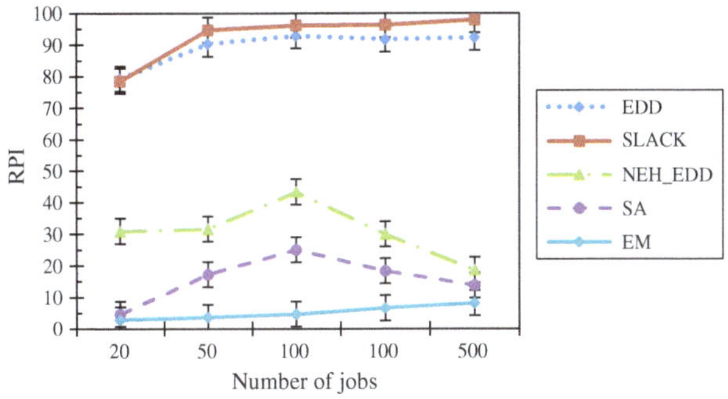

Figure 16. Means plot and LSD intervals for the interaction between the algorithm type and the number of jobs.

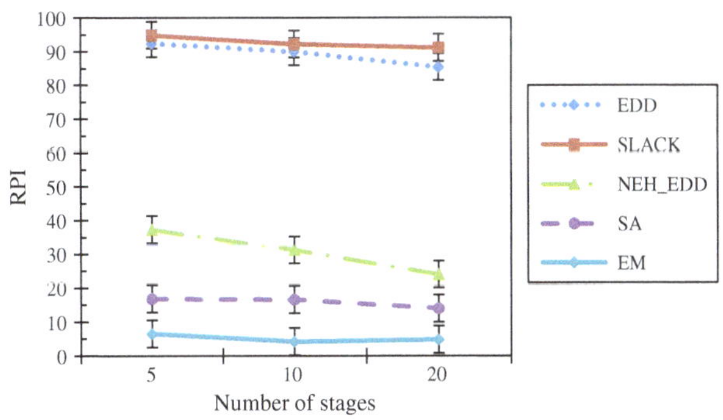

Figure 17. Means plot and LSD intervals for the interaction between the algorithm type and the number of stages.

the number of jobs increases. EM and SA are the same in the case of $n = 200$ and the greatest difference between SA and EM is shown when $n = 100$ and again the same in the case of $n = 500$. As illustrated in Figure 17, the higher the number of stages, the better the heuristics performs, while there is no change of the metrics for EM and SA.

Figure 16 depicts the means plot and LSD intervals for the interaction between the various algorithm as a functions of the number of stages. It is interesting to see in this figure that the proposed EM shows a robust performance in almost all cases against the number of stages.

Table 4. Average relative percentage deviation ($\overline{\text{RPI}}$) for the algorithms grouped by n and m.

Instances	SPT	Johnson's	NEH	SA	EM
20 × 5	21.57	22.33	3.25	**0.81**	0.83
20 × 10	22.51	23.18	4.63	0.75	**1.68**
20 × 20	14.13	14.76	2.58	0.92	**0.90**
50 × 5	25.44	29.70	7.44	5.12	**0.01**
50 × 10	22.01	24.26	6.19	3.34	**0.28**
50 × 20	16.82	16.01	3.85	1.07	**0.89**
100 × 5	17.66	24.33	6.74	4.11	**0.23**
100 × 10	20.07	23.07	5.83	3.94	**0.08**
100 × 20	15.51	16.60	4.53	2.71	**0.34**
200 × 5	19.45	29.28	6.42	4.42	**0.27**
200 × 10	19.22	24.02	6.75	5.84	**0.07**
200 × 20	19.35	20.11	6.45	5.70	**0.01**
500 × 5	12.82	27.92	4.56	3.35	**1.57**
500 × 10	11.55	25.07	5.01	3.68	**1.84**
500 × 20	11.80	20.45	4.28	3.07	**1.21**
Average	17.99	22.74	5.23	3.26	**0.68**

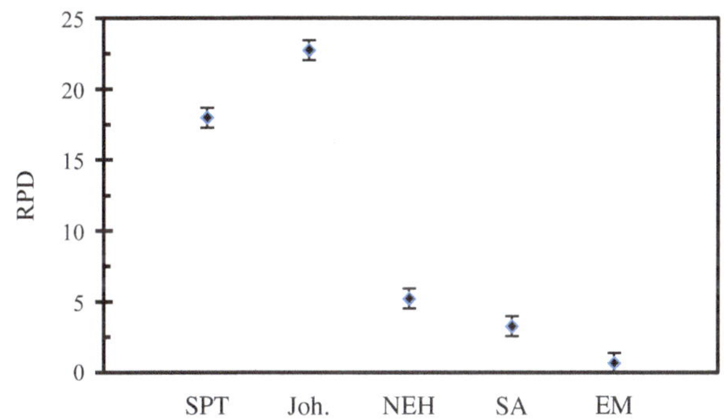

Figure 18. Means plot and LSD intervals (at 95% confidence level) for the type of the algorithm factor.

5.3.2. Analysis of the total completion times

In the total completion time objective, the authors compare the results of SA, NEH, and EM. First, the algorithms in terms of the selected objective function (i.e., total completion time) have been analyzed, and then effects of variable, such as problem size and numbers of machines, on the performance of the algorithms are examined. RPD measure to compare the algorithms has been used. The stopping criterion is again $n \times m \times 0.2$ s of the computational time. EM outperforms the other algorithms again, which support its robustness (Table 4).

Here an analysis similar to the previous section has been applied. The means plot for the different algorithms with the LSD intervals are shown in Figure 18. The Johnson's rule and SPT have a poor performance while the proposed EM gives excellent results than the heuristics and the SA.

Figure 19 shows the quality trend of the evaluated methods as the number of jobs increases. As seen, there is almost no effect on the performance of all five algorithms with regards to the problem size. Only SPT in comparison with the (g/2, g/2) Johnson's rule improves when the number of jobs increases. As shown in Figure 20, increasing the number of

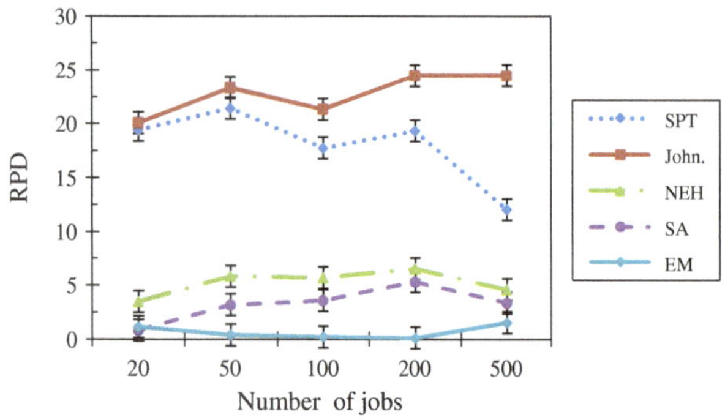

Figure 19. Means plot and LSD intervals for the interaction between the algorithm type and the number of jobs.

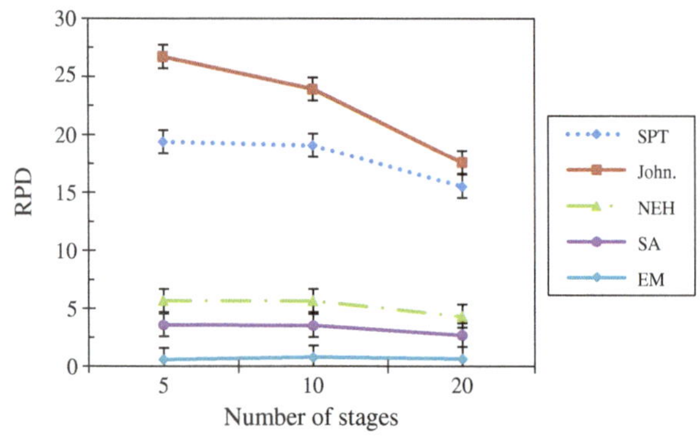

Figure 20. Means plot and LSD intervals for the interaction between the algorithm type and the number of stages.

stages results in better performance for the heuristics. It is noteworthy to see that EM again shows a robust performance in almost all cases with the number of jobs as the variable.

It is concluded that the algorithms perform differently in various situations. In particular, the number of jobs, as a factor, has noticeable impact on the performance quality of the algorithms.

6. Conclusion

In this paper, an EM for a generalized flow shop (GFS) problem with machine availability constraints and transportation times between stages to minimize two independent objectives has been proposed. EM and also SA as well as a number of other well-known heuristics to minimize total tardiness and total completion times, independently. The results were analyzed in terms of objective functions and the effects of variable factors on the instances considered in this paper. The results show that the proposed EM performs very well. This is the first time that an EM has been applied to GFS problems, and the computational results show it holds a good promise. In fact, the overall performance of our proposed method can be regarded as a very good meta-heuristic algorithm since it does not make the use of problem specific knowledge such as the critical paths concept or extensive speed-ups as used in other high-performing algorithms.

In modern manufacturing systems such as agile and flexible manufacturing systems, there is an increasing demand for customized products, which are produced in smaller lot sizes than before. Therefore, there appear to be an increased focus on finding new methods that have the most of the strategic advantages of a GFS but also can provide some of the operational advantages of an assembly line without flexibility limitations of Cellular Manufacturing Systems. Virtual Cellular Manufacturing Systems (VCMSs) are a new manufacturing technology generated from the changing and dynamic marketing environment, which has gained momentum during the last decade. Therefore, extensions of the proposed EM to VCMS considering other objectives or features, such as sequence-dependent setup times, are possible. For further insight, it will be interesting to work on other iterative algorithms or on hybridization of the proposed EM with other algorithms to achieve even better results.

References

1. S.M. Johnson, Optimal two and three-stage production schedules with set up times included, Naval Research Logistics Quarterly 1 (1954) 61–68.
2. W.H. Yang, A study on the intelligent neural network training using the electromagnetism algorithm, Unpublished Master Thesis, Dept. of Industrial Engineering and Management, I-Shou University, Kaohsiung County, Taiwan, 2002.
3. R. Ruiz, T. Stützle, An Iterated Greedy heuristic for the sequence dependent setup times flowshop problem with makespan and weighted tardiness objectives, European Journal of Operational Research 187, 3 (2008) 1143–1159.
4. M. Khalili, M.J. Tarokh, B. Naderi, Using electromagnetism algorithm for determining the number of kanbans in a multistage supply chain system, Journal of Industrial Engineering 6 (2010) 63–72.
5. P. Wu, W.-H. Yang, N.-C. Wei, An electromagnetism algorithm of neural network analysis – an application to textile retail operation, Journal of the Chinese Institute of Industrial Engineers 21 (2004) 59–67.
6. M.S. Salvador, A solution to a special case of flow shop scheduling problems, in: S.E. Elmaghraby (Ed.), Symposium of the Theory of Scheduling and its Applications, Springer, New York, 1973, pp. 83–91.
7. R. Linn, W. Zhang, Hybrid flow shop scheduling: a survey, Computers & Industrial Engineering 37, 1–2 (1999) 57–61.
8. H. Wang, Flexible flowshop scheduling: optimum, heuristics, and artificial intelligence solutions, Expert Systems 22, 2 (2005) 78–85.
9. O. Moursli, Y. Pochet, A branch-and-bound algorithm for the hybrid flowshop, International Journal of Production Economics 64, 1–3 (2000) 113–125.
10. C. Sriskandarajah, S.P. Sethi, Scheduling algorithms for flexible flowshops: worst and average case performance, European Journal of Operational Research 43, 2 (1989) 143–160.
11. A. Guinet, M.M. Solomon, P.K. Kedia, A. Dussauchoy, A computational study of heuristics for two-stage flexible flowshops, International Journal of Production Research 34, 5 (1996) 1399–1415.
12. E. Nowicki, C. Smutnicki, The flow shop with parallel machines: a tabu search approach, European Journal of Operational Research 106, 2–3 (1998) 226–253.

13. M. Gourgand, N. Grangeon, S. Norre, Metaheuristics for the deterministic hybrid flow shop problem, Proceeding of the International Conference on Industrial Engineering and Production Management (IEPM'99), Glasgow, United Kingdom, July 12–15 (1999), pp. 136–145.
14. R. Zhang, C. Wu, A simulated annealing algorithm based on block properties for the job shop scheduling problem with total weighted tardiness objective, Computers & Operations Research 385 (2011) 854–867.
15. C.R. Reeves, A genetic algorithm for flowshop sequencing, Computers & Operations Research 22, 1 (1995) 5–13.
16. R. Tavakkoli-Moghaddam, N. Safaei, F. Sassani, A memetic algorithm for the flexible flow line scheduling problem with processor blocking, Computers & Operations Research 36 (2009) 402–414.
17. R. Cheng, M. Gen, M. Tozawa, Minmax earliness/tardiness scheduling in identical parallel machine system using genetic algorithms, Computers & Industrial Engineering 29, 1–4 (1995) 513–517.
18. J. Yang, Minimizing total completion time in two-stage hybrid flow shop with dedicated machines, Computers & Operations Research 38, 7 (2011) 1045–1053.
19. M. Khalili, R. Tavakoli-Moghadam, A multi-objective electromagnetism algorithm for a bi-objective flowshop scheduling problem, Journal of Manufacturing Systems 31 (2012) 232–239.
20. C. Chen, R. Neppalli, Genetic algorithms applied to the continuous flow shop problem, Computers & Industrial Engineering 30, 4 (1996) 919–929.
21. T. Aldowaisan, A. Allahverdi, New heuristics for m-machine no-wait flowshop to minimize total completion time, Omega 32, 5 (2004) 345–352.
22. Q.K. Pan, M.F. Tasgetiren, Y.C. Liang, A discrete particle swarm optimization algorithm for the no-wait flowshop scheduling problem, Computers & Operations Research 35 (2008) 2807–2839.
23. A. Fink, S. Voß, Solving the continuous flow-shop scheduling problem by metaheuristics, European Journal of Operational Research 151 (2003) 400–414.
24. S.J. Shyu, B.M.T. Lin, P.Y. Yin, Application of ant colony optimization for no-wait flowshop scheduling problem to minimize the total completion time, Computers & Industrial Engineering 47 (2004) 181–193.
25. J. Grabowski, J. Pempera, Some local search algorithms for no-wait flow-shop problem with makespan criterion, Computers & Operations Research 32 (2005) 2197–2212.
26. M. Khalili, An iterated local search algorithm for flexible flow lines with sequence dependent setup times to minimize total weighted completion, International Journal of Management Science and Engineering Management 7, 1 (2012) 63–66.
27. M. Khalili, Multi-objective no-wait hybrid flowshop scheduling problem with transportation times, Journal International Journal of Computational Science and Engineering 7, 2 (2012) 147–153.
28. J. Hurink, S. Knust, Makespan minimization for flow-shop problems with transportation times and a single robot, Discrete Applied Mathematics 112 (2001) 199–216.
29. A. Soukhal, A. Oulamara, P. Martineau, Complexity of flow shop scheduling problems with transportation constraints, European Journal of Operational Research 161 (2005) 32–41.
30. R. Ruiz, C.J. Garica-Diaz, C. Maroto, Considering scheduling and preventive maintenance in the flowshop sequencing problem, Computers & Operations Research 34 (2007) 3314–3330.
31. G. Schmidt, Scheduling with limited machine availability, European Journal of Operational Research 121 (2000) 1–15.
32. C.Y. Lee, Minimizing the makespan in the two-machine flowshop scheduling problem with an availability constraint, Operations Research Letters 20 (1997) 129–139.
33. J. Blazewicz, J. Breit, P. Formanowicz, W. Kubiak, G. Schmidt, Heuristic algorithms for the two-machine flowshop problem with limited machine availability, Omega 29 (2001) 599–608.
34. J. Breit, A polynomial-time approximation scheme for the two-machine flow shop scheduling problem with an availability constraint, Computers & Operations Research 33 (2006) 2143–2153.
35. H. Allaoui, A. Artiba, Integrating simulation and optimization to scheduling a hybrid flow shop with maintenance constraints, Computers & Industrial Engineering 47 (2004) 431–450.
36. V. Cerny, Thermodynamical approach to the travelling salesman problem: an efficient simulation algorithm, JOTA 45 (1985) 41–51.
37. I. Birbil, S.C. Fang, An electromagnetism-like mechanism for global optimization, Journal of Global Optimization 25 (2003) 263–282.
38. D. Debels, B.D. Reyck, R. Leus, M. Vanhoucke, A hybrid scatter search/electromagnetism meta-heuristic for project scheduling, European Journal of Operational Research 169 (2006) 638–653.
39. E. Taillard, Benchmarks for basic scheduling problems, European Journal of Operational Research 64, 2 (1993) 278–285.
40. R. Ruiz, A. Allahverdi, Some effective heuristics for no-wait flowshops with setup times to minimize total completion time, Annals of Operation Research 156 (2007) 143–171.
41. B. Naderi, M. Mousakhani, M. Khalili, Scheduling multi-objective open shop scheduling using a hybrid immune algorithm, The International Journal of Advanced Manufacturing Technology 66, 5–8 (2013) 895–905.
42. D.C. Montgomery, Design and Analysis of Experiments, Fifth edition, John Wiley & Sons, 2000.

2

Comparative analysis of aluminium surface roughness in end-milling under dry and minimum quantity lubrication (MQL) conditions

Ugochukwu C. Okonkwo[1,*], Imhade P. Okokpujie[1], Jude E. Sinebe[1], and Chinedu A.K. Ezugwu[2]

[1] Department of Mechanical Engineering, Nnamdi Azikiwe University, PMB 5025 Awka, Nigeria
[2] Department of Mechanical Engineering, Landmark University, PMB 1001 Omu-Aran, Kwara, Nigeria

Abstract – In this study an experimental investigation of effects of cutting parameters on surface roughness during end milling of aluminium 6061 under dry condition and minimum quantity lubrication (MQL) condition were carried out. Spindle speed (*N*), feed rate (*f*), axial depth of cut (*a*) and radial depth of cut (*r*) were cutting parameters chosen as input variables in the investigation of the surface roughness quality. The experimental design adopted for this study was the central composite design (CCD) of response surface methodology. Thirty samples were run in a CNC milling machine for each condition and the surface roughness measured using Mitutoyo surface tester. A comparison showing the effects of cutting parameters on the surface roughness for dry and MQL conditions in end-milling of aluminium were evaluated. Surface roughness values for MQL condition were lower with up to 20% reduction when compared to dry conditions. MQL cutting condition was found to be better and more reliable because it is environmentally friendly and gives better surface finish. With the obtained optimum input parameters for surface roughness, production operations will be enhanced.

Key words: Surface roughness, Minimum quantity lubrication (MQL), CNC end milling, Response surface methodology

1. Introduction

The quality of machined surface is characterized by the accuracy of its manufacture with respect to the dimensions specified by the designer. Every machining operation leaves characteristic evidence on the machined surface. This evidence in the form of finely spaced micro irregularities left by the cutting tool. Each type of cutting tool leaves its own individual pattern that therefore can be identified. This pattern is known as surface finish or surface roughness. The surface roughness of a machined surface depends on many factors and can be grouped as follows:

- Geometric factors which are the type of machining operation, the cutting geometry and the feed.
- Work material factors which includes the built up edges effect, damaged to the surface caused by chip curling back into work, tearing of work surface during chip formation when machining ductile materials, friction between the tool flank and newly generated work surface etc.
- Vibration and machine tool factors which are ones related to the machine tool, tooling and setup in operation.

The characteristic feature of the milling process is that each milling cutter tooth removes its share of the stock in the form of small individual chips. It is of three types which are: peripheral milling, face milling and end milling. End milling is one of the most common metal removal operation encountered in industrial process. It is widely used in the manufacturing industries which include the automotive and aerospace sectors, where quality is an important factor in the production of slots, pockets, precision molds, and dies. In end milling, the cutter generally rotates on an axis vertical to the work-piece. It can be tilted to machine tapered surfaces. Cutting teeth are located on both the end face of the cutter and the periphery of the cutter body.

The primary function of the MQL in metal machining operations is to serve as a coolant, also as a lubricant thereby reducing friction and tool wear. It is generally agreed that the application of MQL can improve the tool life and results in

*e-mail: ugocoons@yahoo.com

good surface finish by reducing thermal distortion and flushing away of machined chips. What is perhaps even more important is to ensure proper filtration of the fluid as suspended swarf can cause random deep scratches on the job. Obtaining a good surface quality is very important in every engineering component design or fabrication. Good surface finish influences mechanical properties such as fatigue behavior, wear, corrosion, lubrication, and electrical conductivity. Therefore measurement of surface finish and characterization also plays an important part in the prediction of a machining performance. One way to check if a machined material has good quality is through the measurement of its surface roughness (arithmetic average, R_a).

The knowledge of tool and cutting parameters can stem from either pure experimental analysis or hybrid of experimental and numerical/theoretical analysis [1]. Matsubara et al. [2] developed a theoretical model for accuracy of end-milling by investigating the transfer matrix and instantaneous chip thickness leading respectively to derivation of the static stiffness of the end mill and the instantaneous cutting forces. Budak and Alintas [3] presented a model that highlighted dependence of surface accuracy on the cutting parameters. Insperger et al. [4] generated both the stability diagram and the surface location error diagram and discussed the selection of optimal spindle speed considering both diagrams. Surface roughness is the inherent irregularities left by a single-point tool like turning tool or milling tool on a machined surface. Surface roughness is noted by Field et al. [5] that surface roughness is predominantly considered as the most important feature of practical engineering surfaces due to its crucial influence on the mechanical and physical properties of a machined part.

The roughness of a machined surface is an indication of relative vibration between the tool and work piece during a machining operation as the work of Peigne et al. [6] in which they studied the effects of the cutting vibratory phenomena and their impacts on the surface roughness of the machined surface suggests. The parameters of machining process are expected to affect this relative vibration thus have effects on component surface roughness. The obvious machining parameters of a machining process such as end-milling are the cutting speed, the axial depth of cut, the radial depth of cut and the feed rate. These are the most easily controlled parameters of the machining process being at the disposal of the operator to choose or to vary continuously in process. Other parameters include tool geometry (given in terms of tool angles like rake angle, flank or tool relief angle, notch angle), tool and work piece material and tool wear. Tool wear being a tribological phenomenon develops with progression of machining and then causes progressive increase in surface roughness.

Surface roughness has been attributed to cutting conditions, tool geometry and mechanical stiffness. Various other studies have considered the behavior of surface roughness under different tool-work-piece material combinations and experiments. Kishawy et al. [7] studied the effect of flood coolant, and dry cutting, on tool wear, surface roughness and cutting forces. A study of the surface integrity produced by end mill tool using a Taguchi orthogonal array has been presented by Mantle and Aspinwall [8], Wang and Chang [9] analyzed the influence of cutting conditions and tool geometry on surface roughness of slot end milling operation. Feasibility study and development of an in-process based recognition system to predict the surface roughness of machined parts in the end milling process has been presented by Tsai et al. [10]. Similarly, Ertekin et al. [11] has identified the most influential and common sensory features for dimensional accuracy and surface roughness in CNC milling operations.

Ginta et al. [12] developed an effective methodology to determine the performance of uncoated WC-Co inserts in predicting minimum surface roughness in end milling of titanium alloys Ti-6Al-4V under dry conditions. Central composite design (CCD) of response surface methodology (RSM) was employed to create an efficient analytical model for surface roughness in terms of cutting parameters: cutting speed, axial depth of cut, and feed per tooth. End milling tests were conducted on Vertical Machining Center (VMC ZPS, Model: MLR 542) with full immersion cutting and under dry condition. They concluded with CCD being a successful technique to predict the surface roughness produced in end-milling of titanium alloy Ti-6Al-4V using uncoated inserts under dry conditions. Linear CCD model proved inadequate while quadratic CCD model was adequate with 95% accuracy. The two developed models indicated that feed was the most predominant cutting condition followed by cutting speed and depth of cut. Interaction effect between cutting speed and feed will also give a high effect on surface roughness values.

Arokiadass et al. [13] also studied the influence of four machining parameters including spindle speed (N), feed rate (f), depth of cut (d), and various percentage weight of silicon carbide (S) on surface roughness (R_a). The RSM was employed to establish the mathematical relationship between the response and the various process parameters. The result they obtained showed that the quadratic model was statistically significant for analysis of surface roughness. Their model indicated that the feed rate was the most dominant parameter on surface roughness followed by spindle speed and %weight of SiC. Depth of cut has less influence on surface roughness. The effect of cutting parameters on surface roughness during end milling of aluminium under (MQL) was carried out by Okokpujie and Okonkwo [14]. The mathematical model developed by using least square approximation method shows accuracy of 89.5% which is reasonably reliable for surface roughness prediction.

2. Materials and methods

The work piece material used for the study was a rectangular 6061 aluminium blocks of 2000 mm × 50 mm × 5 mm. Method used for the experimental investigations is explained thus:

- Preparation of the vertical CNC milling machine system ready for performing the machining operation, cutting of the work piece of the aluminium 6061 rectangle plate into different sizes of 10, 15, 20, 25 and 30 mm. A total of 60 pieces, 30 pieces for the dry cutting condition and 30 pieces for MQL condition.
- Fixing of the high speed steel (HSS) end milling cutter of 12 mm diameter on the spindle taper of the machine.

Figure 1. Experimental setup for MQL end milling operation.

Figure 2. Experimental setup for dry end milling operation.

- Mounting the work piece, clamped on a vice mounted on top of the table of the machine as shown in Figures 1 and 2, respectively.
- Creating CNC part programs on CNC professional software for tool paths, with specific commands using different factor levels of spindle speed, feed rate, axial depth of cut and radial depth of cut, taking reference for *Y* axis, and *Z* axis then performing end milling operation.
- After each machining the surface roughness of the work piece was measured with the press-o-firm and Mitutoyo surface tester.

Detailed information on chemical composition of the 6061 aluminium is provided in Table 1, and details of the experimental outlay in Table 2, only up-milling cutting mode was investigated.

The experiment was performed on SIEG 3/10/0016 and SIEG 3/10/0010 table top CNC machine vertical milling centre. The vertical milling centre has three (3) planes namely *x*, *y* and *z* planes. The experimental set up is shown in Figures 1 and 2.

Second-order rotatable central composite design of response surface methodology (RSM) was employed in the experimental design. RSM is a collection of mathematical and statistical techniques for empirical model building. By careful design of experiments, the objective is to optimize a response (output variable) which is influenced by several independent variables (input variables). An important aspect of RSM is the design of experiments. A second-order model can be constructed efficiently with central composite designs (CCD). CCD are first-order (2^N) designs augmented by additional centre and axial points to allow estimation of the tuning parameters of a second-order model. By considering all the factorial corner points, some of the central replicates and all the axial points second-order rotatable central composite design requires between 25 and 33 experimental runs depending on the number of the central replicates considered while a full factorial design will require $5^4 = 625$ experimental runs. This explains the choice of second-order rotatable central composite design which tremendously reduces needed number of experimental runs for the MQL cutting conditions, which doubles the calculated number of experimental runs. The design expert 9.0.1 was used in analysis and presentation of results.

Table 1. Chemical composition of Al-6061.

Element	Mg	Fe	Si	Cu	Mn	V	Ti	AL
Weight %	1.08	0.17	0.63	0.32	0.52	0.01	0.02	Remainder

The response surface methodology (RSM) is the procedure for determining the relationship between the independent process parameters with the desired response and exploring the effect of these parameters on responses, including six steps. These are in the order:

- Define the independent input variables and the desired responses with the design constants.
- Adopt an experimental design plan.
- Perform regression analysis with the quadratic model of RSM.
- Calculate the statistical analysis of variance (ANOVA) for the independent input variables in order to find which parameter significantly affects the desired response.
- Determine the situation of the quadratic model of RSM and decide whether the model of RSM needs screening variables or not.
- Optimize and conduct confirmation experiment and verify the predicted performance characteristics.

The experimental plan was developed to assess the influence of spindle speed (*N*), feed rate (*f*), axial depth of cut (*a*) and radial depth of cut (*r*) on the surface roughness parameters (R_a). Five levels were allocated for each cutting variable as given in Table 3. The variable levels were chosen within the intervals recommended by cutting tool manufacturer. Four cutting variables at five levels led to a total of 30 tests for each condition.

The required number of experimental points for four-factor in the CCD with one replication of factorial and axial parts having, factorial design is $= 2^f = 2^4 = 16$, the axial point or

Table 2. Details of the experimental outlay.

Exp. runs	Material	MQL cutting condition	Cutting tool	Input parameters	Response parameters
1 to 30	Al-6061 alloy	10% boric acid + base oil SAE 40	High speed steel	Cutting speed, feed rate, axial depth of cut and radial depth of cut	Surface roughness

Table 3. Factor levels to be used in the experimental design.

Variable	Levels				
	−2	−1	0	1	2
Spindle speed [rpm]	1000	1500	2000	2500	3000
Feed rate [mm/min]	100	150	200	300	500
Radial depth of cut [mm]	0.5	1	1.5	2.0	2.5
Axial depth of cut [mm]	10	15	20	25	30

Table 4. Standard experimental matrix for the factorial design, star point and center points for CCD for both dry and MQL conditions.

Exp. no	Coded value			
	x_1	x_2	x_3	x_4
1	−1	−1	−1	−1
2	1	−1	−1	−1
3	−1	1	−1	−1
4	1	1	−1	−1
5	−1	−1	1	−1
6	1	−1	1	−1
7	−1	1	1	−1
8	1	1	1	−1
9	−1	−1	−1	1
10	1	−1	−1	1
11	−1	1	−1	1
12	1	1	−1	1
13	−1	−1	1	1
14	1	−1	1	1
15	−1	1	1	1
16	1	1	1	1
17	−2	0	0	0
18	2	0	0	0
19	0	−2	0	0
20	0	2	0	0
21	0	0	−2	0
22	0	0	2	0
23	0	0	0	−2
24	0	0	0	2
25	0	0	0	0
26	0	0	0	0
27	0	0	0	0
28	0	0	0	0
29	0	0	0	0
30	0	0	0	0

where x_1 = spindle speed (rpm), x_2 = feed rate (mm/min), x_3 = axial depth of cut (mm), x_4 = radial depth of cut (mm).

star point is $= 2 \times f = 2 \times 4 = 8$, where f = number of factors, the center point chosen for this experiment is 6, which is $= 16 + 8 + 6 = 30$. Therefore the thirty experiments are carried out according to the blocked central composite design (CCD).

The experimental plan gave rise to the second order polynomial model by applying multiple regression analysis. The experimental matrix for the factorial design, star point and center points are shown in Table 4.

3. Results

3.1. Comparisons between surface roughness for dry and MQL condition

Table 5 shows the results of the experiments for the various cutting parameters and measured Ra values for the end milling of aluminium under dry cutting condition (column 7), MQL condition (column 8) and percentage reduction in surface roughness (column 9). A total of 60 experiments were carried out 30 for dry condition and 30 for MQL condition. The coded values of Table 4 have been replaced with the actual values.

The concept of MQL is a precise solution in achieving reduced tool wear and good surface finish as well as long tool life while maintaining cutting force and power at reasonable level. MQL not only reduces tool wear, tool life and surface finish, but reduces the waste involved in conventional application of cutting fluid. Table 5 shows the comparison between the surface finish obtained in dry and MQL condition. It can be seen that there was a significant improvement in the surface finish of about a maximum of 20% which is shown in the reduction in the average surface roughness value of the testing data.

The results in Table 5 which were clearly depicted in Figure 3 for surface roughness generally show that application of MQL greatly improves the surface finish when compared with dry cutting condition, all other advantages notwithstanding. It can be seen that a reduction of up to 20% surface roughness value can be obtained in the machining of aluminium in end milling processes by the application of MQL. These findings were in line with observations made by Abhang and Hameedullah [15] in a related study. The comparison between dry and MQL machining environment is show in Figure 13. Besides, there is more efficient penetration of fluid into the cutting region, which is environmentally friendly, and cause less pollution. This means that machining in MQL environment leads to a better surface finish and less hazardous environment.

3.2. Effects of cutting parameters on surface roughness in dry and MQL conditions

The effects of cutting parameters on surface roughness in end milling of aluminium were investigated using contour plots of the results obtained in dry and MQL conditions. The graphical evaluation was obtained by plotting surface roughness values against the various cutting parameters (axial depth

Table 5. Experimental results of surface roughness of dry and MQL condition.

Sth	Run	x_1	x_2	x_3	x_4	Dry surface roughness (R_a)	MQL surface roughness (R_a)	% reduction in surface roughness (R_a)
13	1	1500	150	25	2	1.12	0.94	16.07
14	2	2500	150	25	2	0.95	0.85	10.53
8	3	2500	300	25	1	1.17	1.02	12.82
11	4	1500	300	15	2	1.27	1.11	12.60
9	5	1500	150	15	2	1.10	0.92	16.36
24	6	2000	200	20	2.5	1.21	1.10	9.09
1	7	1500	150	15	1	1.08	0.90	16.67
25	8	2000	200	20	1.5	1.20	1.01	15.83
5	9	1500	150	25	1	1.04	0.88	15.38
18	10	3000	200	20	1.5	0.61	0.51	16.39
20	11	2000	500	20	1.5	1.31	1.16	11.45
16	12	2500	300	25	2	1.26	1.10	12.70
19	13	2000	100	20	1.5	0.58	0.50	13.79
4	14	2500	300	15	1	1.13	0.98	13.27
22	15	2000	200	30	1.5	1.16	1.00	8.62
23	16	2000	200	20	0.5	1.03	0.88	−11.96
26	17	2000	200	20	1.5	1.17	1.08	7.69
10	18	2500	150	15	2	1.05	0.93	11.43
2	19	2500	150	15	1	0.84	0.74	11.90
27	20	2000	200	20	1.5	1.18	1.08	8.47
17	21	1000	200	20	1.5	1.28	1.12	12.50
12	22	2500	300	15	2	1.22	1.07	12.30
15	23	1500	300	25	2	1.29	1.14	11.63
21	24	2000	200	10	1.5	1.15	0.92	20.00
30	25	2000	200	20	1.5	1.19	0.96	19.33
3	26	1500	300	15	1	1.26	1.06	15.87
7	27	1500	300	25	1	1.24	1.04	16.13
6	28	2500	150	25	1	0.75	0.60	20.00
29	29	2000	200	20	1.5	1.13	1.01	10.62
28	30	2000	200	20	1.5	1.15	1.01	12.17

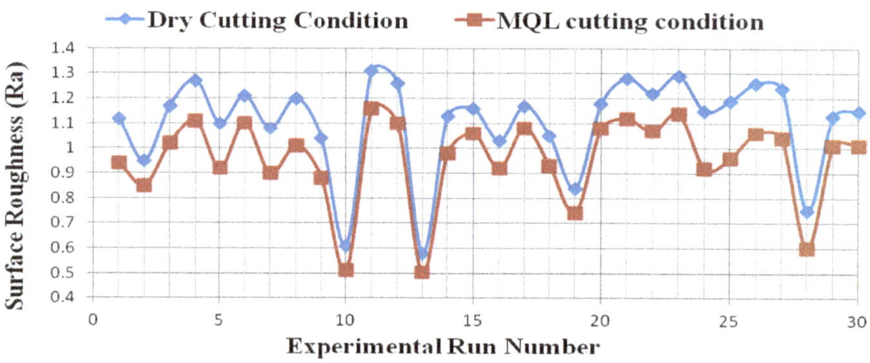

Figure 3. Comparisons between surface roughness for dry and MQL condition.

of cut, radial depth of cut, spindle speed and feed rate). Surface roughness values were simultaneously plotted against two cutting parameters while keeping the other two constant. Figures 4–13 show the experimental results obtained from the effect of cutting parameters on surface roughness.

Effects of spindle speed and feed rate on surface roughness

Figures 4 and 5 show the surface roughness contour plots of spindle speed versus the feed rate in dry and MQL, respectively. At each instance, the axial depth of cut and radial depth of cut were kept at 25 mm and 2 mm.

It can be seen from the Figures 4 and 5 that there is indeed an interaction which has a nonlinear relationship on the surface roughness. Specifically, at spindle speed of 1500 rpm and feed rate of 150 mm/min, the surface roughness obtained during dry condition was 1.12 μm while that for the MQL condition was 0.94 μm. The highest surface roughness of the dry condition revolves around 1.31 μm while that of the MQL condition revolves around 1.16 μm as shown in Figures 4 and 5, respectively. A surface roughness as low as 0.58 and 0.50 can be obtained from dry and MQL condition and these region is at the down-left side of the two figures. On a general note, increasing the spindle speed reduces the surface roughness.

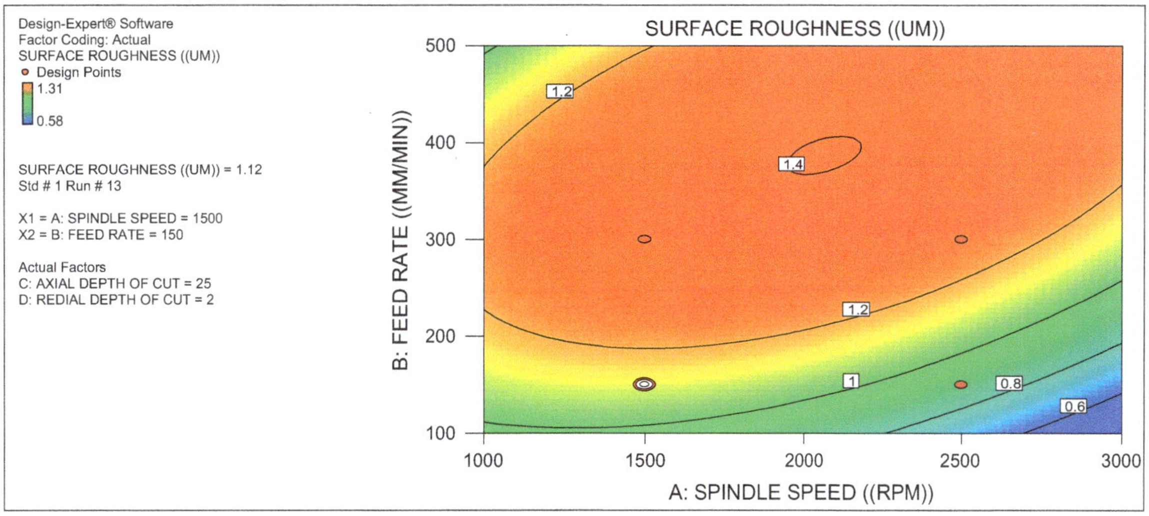

Figure 4. Surface roughness contour plot for spindle speed vs. feed rate in dry condition.

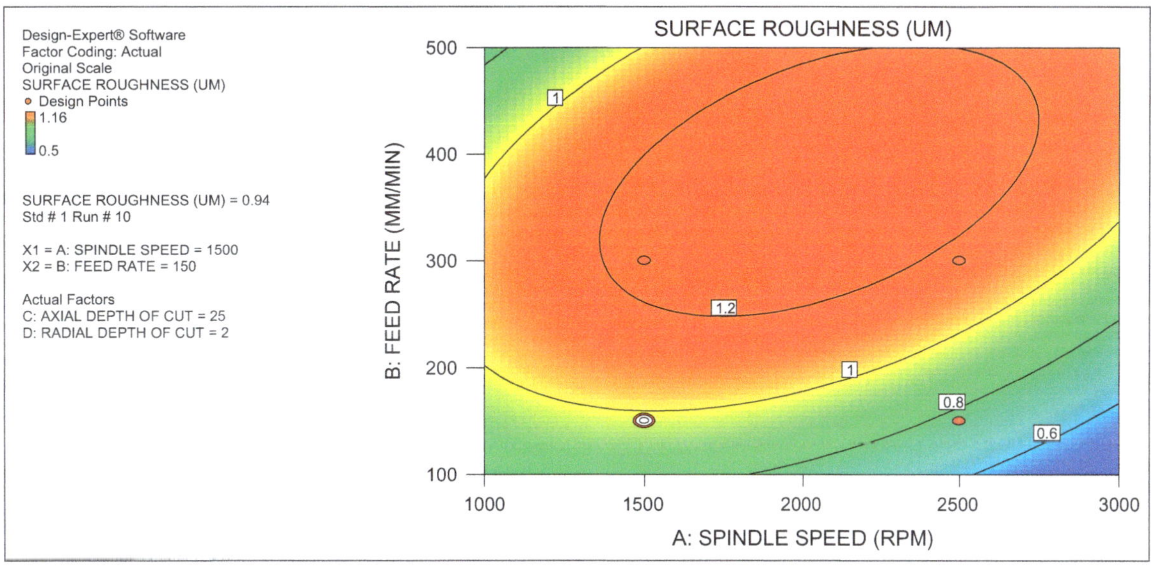

Figure 5. Surface roughness contour plot for spindle speed vs. feed rate in MQL condition.

Apparently, increasing the spindle speed increases the cutting force and eliminates the built-up edge (BUE) tendency. At low spindle speed (rpm), the unstable larger BUE is formed and also the chips fracture readily producing the rough surface. As the spindle speed (rpm) increases, the BUE vanishes, chip fracture decreases, and hence, the roughness decreases. These findings were in line with observations made by Tosun and Huseyinoglu [16]; Korkut and Donertas [17] in related studies.

In this investigation it was also observe that an increase in feed rate significantly increases the surface roughness. Increasing feed rate increases vibration and heat generated, which courses an increase in surface roughness. As the feed rates were increased, chips become discontinuous and were deposited between work piece and tool leading to increased coefficient of friction and more interruption resulting in poor surface finish. In MQL environment, due to the lubricating effect at cutting zone during machining, friction between work piece and tool was reduced, consequently, the surface roughness values were reduced when compared with dry condition. This results corresponds to the result obtained by Arokiadass et al. [13] in his research on predictive modeling of surface roughness in end milling of Al/SiC metal matrix composite, where roughness increases as cut feed rate increased. Tracing the values of the contour plots against the values of feed rate and the surface roughness, one can note that the MQL environment has a finer surface than that of the dry condition.

Effects of spindle speed and axial depth of cut on surface roughness

Figures 6 and 7 show the surface roughness contour plot for spindle speed versus the axial depth of cut for dry and

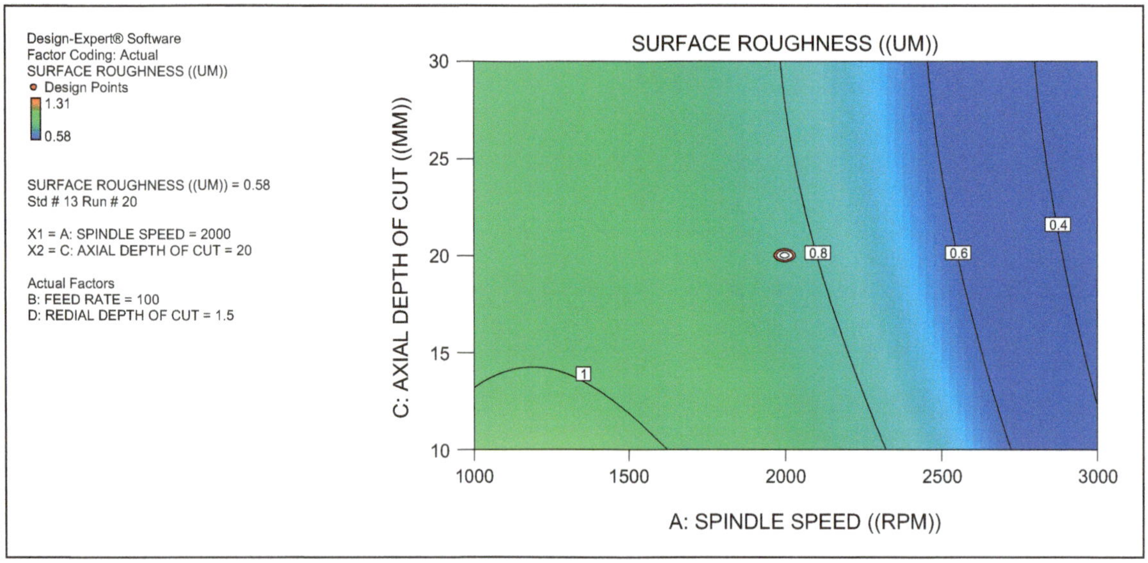

Figure 6. Surface roughness contour plot for spindle speed vs. axial depth of cut in dry condition.

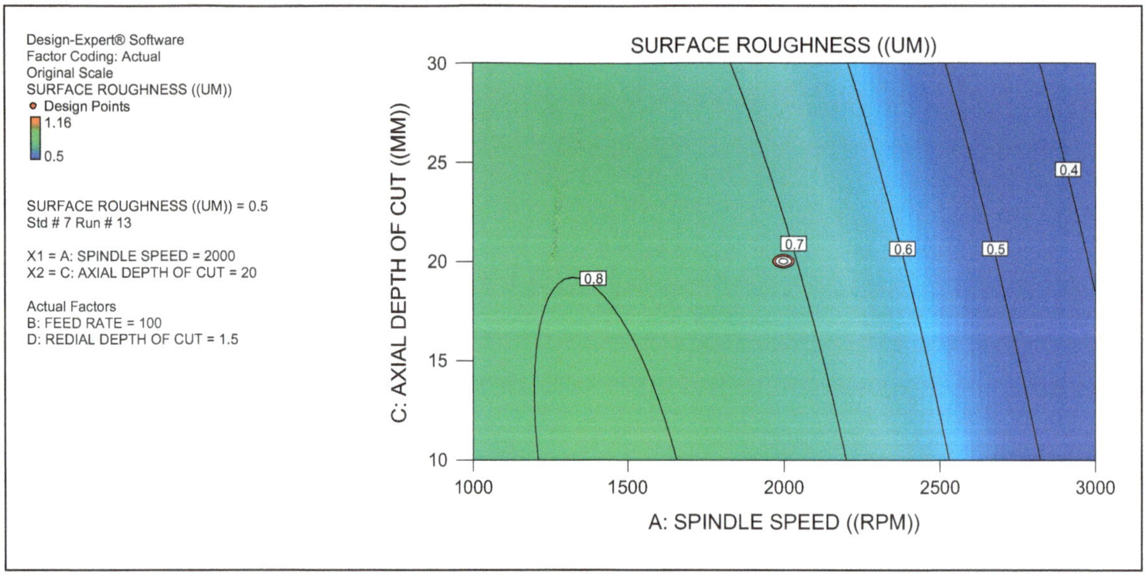

Figure 7. Surface roughness contour plot for spindle speed vs. axial depth of cut in MQL condition.

MQL conditions, respectively. At each instance, the feed rate and radial depth of cut are kept at 100 mm and 1.5 mm, respectively. It can be clearly seen that the spindle speed has more impact when compared with the axial depth of cut when the color variation and contour lines are studied. At higher spindle speed, the surface roughness was greatly reduced even when the axial depth of cut was increased.

On the other hand, it was observed that the surface roughness witnessed under dry condition was far more especially using low spindle speed compared with that of the MQL condition. Specifically, at spindle speed of 2000 rpm and feed rate of 20 mm/min, the surface roughness obtained during dry condition was 0.58 μm while that for the MQL condition was 0.50 μm.

Effects of the feed rate and axial depth of cut on surface roughness

Figures 8 and 9 show the surface roughness contour plot for the feed rate and axial depth of cut for dry and MQL conditions, respectively. At each instance, the spindle speed and radial depth of cut are kept at 2000 rpm and 1.5 mm, respectively.

Studying the contour lines of Figures 8 and 9, one can clearly notice that the feed rate has more impact than the axial depth of cut. At low feed rate, the surface roughness was very small notwithstanding the variation of the axial depth of cut. However, as the feed rate increases, it deteriorates the surface finish. This result was also in line with the work done by Adeel et al. [18] where their result shows that surface roughness

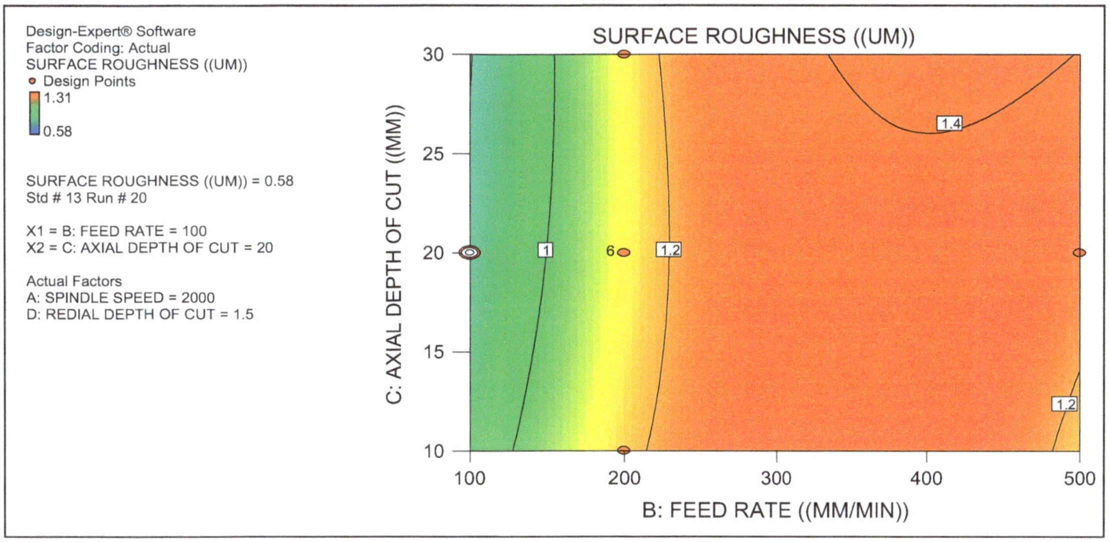

Figure 8. Surface roughness contour plot for feed rate vs. axial depth of cut in dry condition.

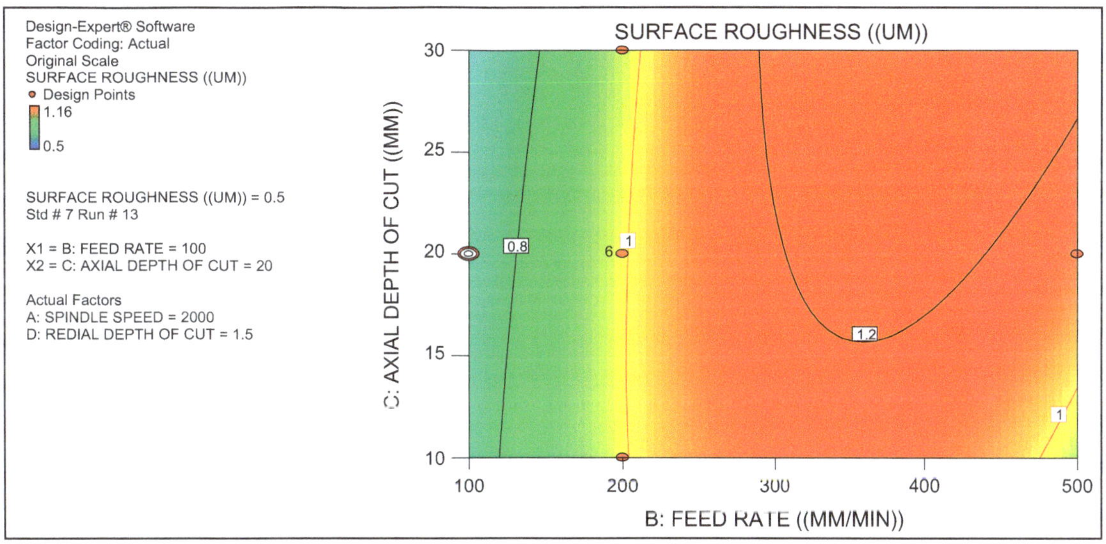

Figure 9. Surface roughness contour plot for feed rate vs. axial depth of cut in MQL condition.

values increase as the feed rate is increased. The above scenario occurs for both MQL and dry conditions. On the other hand, at a higher feed rate and lower axial speed, the surface roughness reduces, which perhaps may be as a result of reduction in vibration. However as the axial depth of cut is increased, maintaining the same feed rate, the roughness surfaces again.

Effects of the feed rate and radial depth of cut on surface roughness

Figures 10 and 11 show the surface roughness contour plot for the feed rate and radial depth of cut for dry and MQL conditions, respectively. At each instance, the spindle speed and radial depth of cut are kept at 2000 rpm and 20 mm, respectively.

It is essentially noted that just like in the case of the feed rate versus the axial depth of cut, in which the depth of cut has more impact on the surface roughness. In the case of radial depth of cut, the feed rate also has more impact on the surface roughness.

Effect of axial depth of cut and radial depth of cut on surface roughness

Figures 12 and 13 show the surface roughness contour plot for axial depth of cut versus radial depth of cut for dry condition and MQL condition, respectively. At each instance, the spindle speed and the feed rate were kept constant at 2000 rpm and 100 mm, respectively.

Comparing the impacts caused by the spindle speed and feed rate, it was found that both the axial depth of cut and radial depth of cut have less impact, with the axial depth of cut being the least. This trend corresponds with the one obtained by Arokiadass et al. [13] where he also concluded that

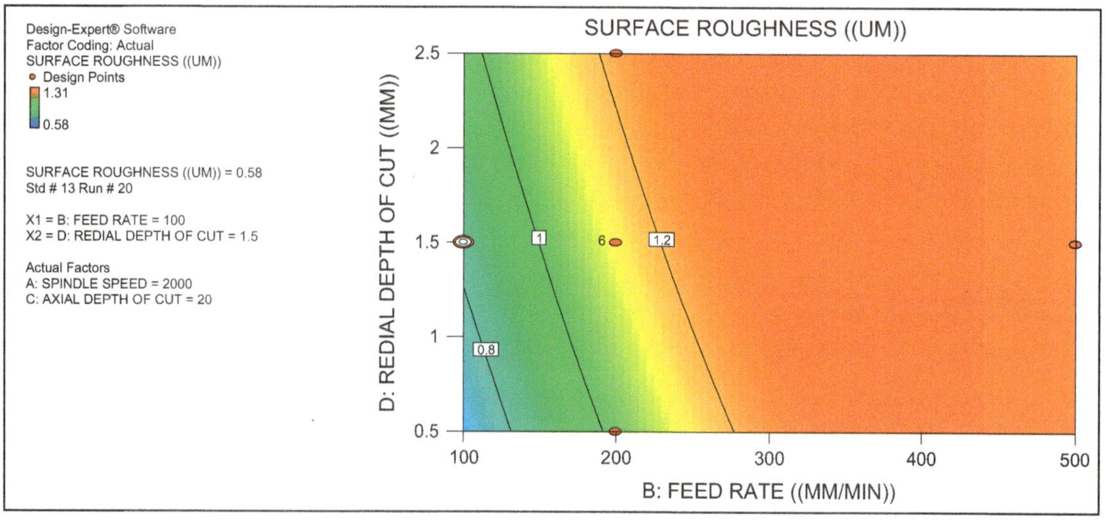

Figure 10. Surface roughness contour plot for feed rate vs. radial depth of cut in dry condition.

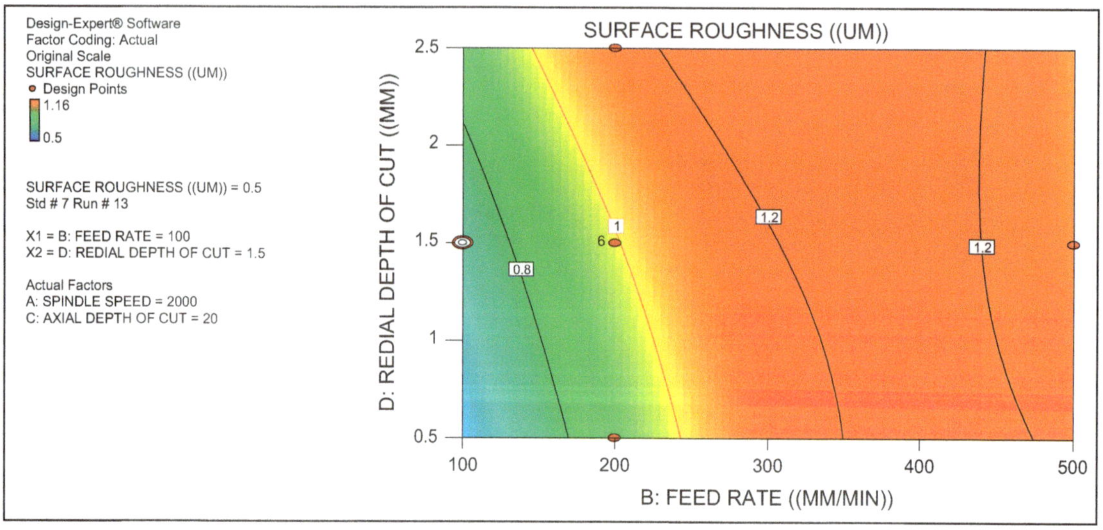

Figure 11. Surface roughness contour plot for feed rate vs. radial depth of cut in MQL condition.

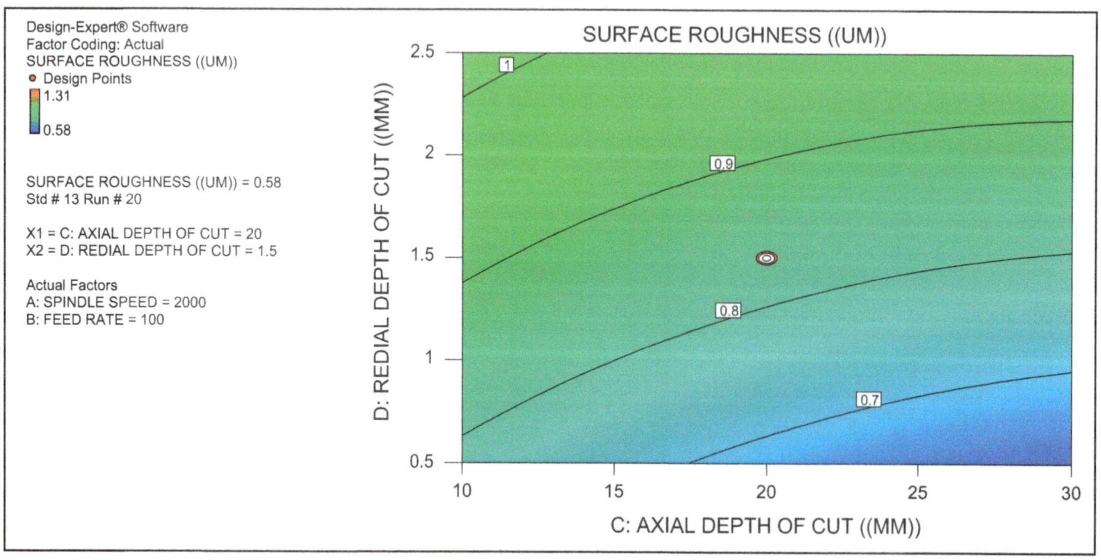

Figure 12. Surface roughness contour plot for axial depth of cut vs. radial depth of cut in dry condition.

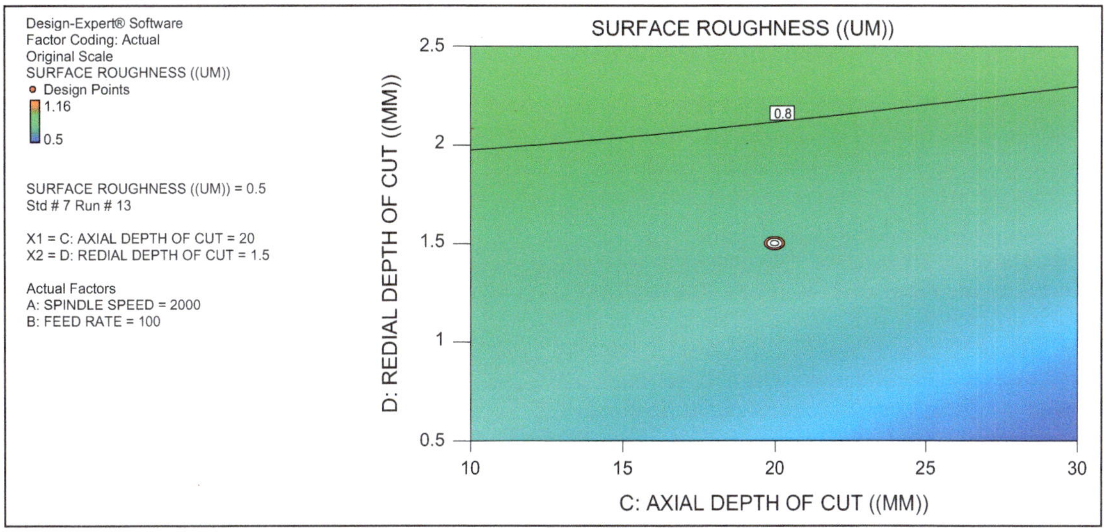

Figure 13. Surface roughness contour plot for axial depth of cut vs. radial depth of cut in MQL condition.

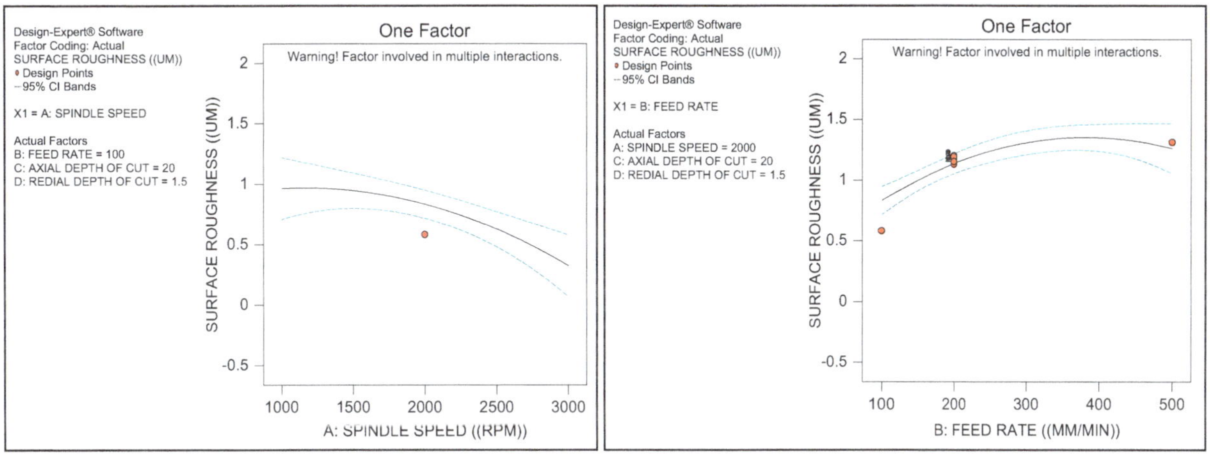

Figure 14. Effects of spindle speed on surface roughness as one factor and feed rate on surface roughness as one factor.

the depth of cut has minimal effect on surface roughness, even though he did not distinguish the depth of cut in question.

The trend of this variation in surface roughness is the same in both dry and MQL conditions but of course deterioration of surface roughness for dry condition than it was with the MQL. This is because the frictional effect is reduced due to the lubrication at the work piece and tool interface. This is in line with model created for both dry and MQL condition where depth of cut has least influence on surface roughness. This is in line with model created for both dry and MQL condition where radial depth of cut has less influence on surface roughness.

3.3. One factor effects of spindle speed and feed rate

It was noted that the spindle speed and feed rate have more impact than the rest. The effects that each has based on the levels adopted for each parameter in this research were shown in Figure 14. It clearly shows that increasing the spindle speed reduces the surface roughness while increasing the feed rate does the opposite (i.e. increases the surface roughness).

4. Conclusions

Experimental works have been carried out on aluminium 6061 in two different environments, namely; dry and MQL environments. A comparison of the effects of cutting parameters in dry and MQL environments on the surface roughness was made. Surface roughness values for MQL condition were lower with up to 20% reduction when compared to dry conditions. MQL cutting condition was found to be best and more reliable condition because it is environmentally friendly and gives better surface finish. The important conclusions drawn from the research are summarized as follows:

- From the experimental values of Table 5, the optimum or minimum surface roughness during cutting process occurs at spindle speed of 2000 rpm, feed rate of 100 mm/min, axial depth of cut 20 mm and radial depth of cut 1.5 mm. For this condition, the minimum surface roughness was 0.58 μm for dry and 0.50 μm for MQL environments.

- Spindle speed and feed rate as independent factors have the most influence on the surface roughness. However, while increase in the spindle speed increases the surface roughness, that of the feed rate reduces it. This was clearly depicted in Figure 14. Similarly radial depth of cut has little effect on the surface roughness and axial depth of cut has no significant effect on the surface roughness.
- Interaction effects between spindle speed and feed rate also possesses a major effect over the surface roughness, followed by axial depth of cut and radial depth of cut.

References

1. U.C. Okonkwo, C.O. Osueke, C.A.K. Ezugwu, Minimizing machining time in pocket milling based on optimal combination of the three basic prescription parameters, International Research Journal of Innovative Engineering 1 (2015) 12–34.
2. T. Matsubara, H. Yamamoto, H. Mizumoto, Study on accuracy in end mill operations, Bulletin of the Japan Society of Precision Engineering 21 (1987) 95–100.
3. E. Budak, Y. Altintas, Peripheral milling conditions for improved dimensional accuracy, International Journal of Machine Tools & Manufacture 34 (1994) 907–918.
4. T. Insperger, J. Gradišek, M. Kalveram, G. Stépán, E. Govekar, Machine tool chatter and surface location error in milling processes, Journal of Manufacturing Science and Engineering 128 (2006) 913.
5. M. Field, J.F. Kahles, W.P. Koster, Surface Finish and Surface Integrity, ASM Handbook, vol. 16, Machining, 9th edn., ASM Publication, ASM, Metal Park, Ohio, 1989.
6. G. Peigne, H. Paris, D. Brissaud, A. Gouskov, Impact of the cutting dynamics of small radial immersion milling operations on machined surface roughness, International Journal of Machine Tools & Manufacture 44 (2004) 1133–1142.
7. H.A. Kishawy, M.E. Dumitrescu, M.A. Elbestawi, Effect of coolant strategy on tool performance chip morphology and surface quality during high-speed machining of A356 aluminum alloy, International Journal of Machine Tools & Manufacture 45 (2005) 219–227.
8. A.L. Mantle, D.K. Aspinwall, Surface integrity of a high speed milled gamma titanium aluminide, Journal of Materials Processing Technology 118 (2001) 143–150.
9. M.Y. Wang, H.Y. Chang, Experimental study of surface roughness in slot end milling, International Journal of Machine Tools & Manufacture 44 (2004) 51–57.
10. Y.H. Tsai, J.C. Chen, S.J. Lou, An in-process surface recognition system based on neural networks in end milling cutting operations, International Journal of Machine Tools & Manufacture 39 (1999) 583–605.
11. Y.M. Ertekin, Y. Kwon, T.L. Tseng, Identification of common sensory features for the control of CNC milling operations under varying cutting conditions, International Journal of Machine Tools & Manufacture 43 (2003) 897–904.
12. T.L. Ginta, A.K.M. Nurul Amin, H.C.D. Mohd Radzi, M.A. Lajis, Development of surface roughness models in end milling titanium alloy Ti-6Al-4V using uncoated tungsten carbide inserts, European Journal of Scientific Research 28 (2009) 542–551.
13. R. Arokiadass, K. Palaniradja, N. Alagumoorthi, Predictive modeling of surface roughness in end milling of Al/SiCp metal matrix composite, Archives of Applied Science Research 3 (2011) 228–236.
14. I.P. Okokpujie, U.C. Okonkwo, Effects of cutting parameters on surface roughness during end milling of aluminium under minimum quantity lubrication (MQL), International Journal of Science and Research 4 (2015) 2937–2942.
15. L.B. Abhang, M. Hameedullah, Experimental investigation of minimum quantity lubricants in alloy steel turning, International Journal of Engineering Science and Technology 2 (2010) 3045–3053.
16. N. Tosun, M. Huseyinoglu, Effect of MQL on surface roughness in milling of AA7075-T6, Materials and Manufacturing Processes 25 (2010) 793–798.
17. I. Korkut, M.A. Donertas, The influence of feed rate and cutting speed on the cutting forces, surface roughness and tool-chip contact length during face milling, Materials & Design 28 (2007) 308–312.
18. A.H. Suhail, N. Ismail, S.V. Wong, N.A. Abdul Jalil, Optimization of cutting parameters based on surface roughness and assistance of workpiece surface temperature in turning process, American Journal of Engineering and Applied Sciences 3 (2010) 102–108.

3

Numerical Modelling of Multilayered Coatings – Latest Developments and Applications

Deqiang Yin[1,2,*], Zhenhai Xu[2], Jiling Feng[3], and Yi Qin[2]

[1] School of Manufacturing Science and Engineering, Sichuan University, Chengdu, P.R. China
[2] Department of Design, Manufacture and Engineering Management, University of Strathclyde, Glasgow G1 1XQ, Scotland, UK
[3] School of Engineering, Manchester Metropolitan University, Manchester, M15 6BH, UK

Abstract – The remarkable mechanical properties of multilayered coatings, such as super-hardness, excellent resistance against cracking, low wear-rate and high thermal-stability, are due to their unique interfacial structures and deformation mechanisms at the nanometer scale. The multilayered coating process itself is a typical multi-scale phenomenon. The modelling of multilayered coatings has become an important topic in research recently, largely due to the recent progress that has been made in the numerical modelling of materials and structures at different length-scales as well as the improved effectiveness achieved in linking such progress in numerical modelling to enable multi-scale modelling. In this paper, numerical modelling for the analysis of multilayered coatings at individual length-scales: Continuum, Molecular and Nano-scale, is reviewed, along with that at multi-scale modelling. Examples are presented showing numerical models obtained using: the Finite Element Method (FEM), Molecular Dynamics (MD), First-principles calculations and Multi-scale modelling, are presented. Their relative limitations are discussed and challenges to their future work highlighted.

Key words: Numerical modelling, Multilayered coatings, Multiscale modelling, Transition metal nitrides

1. Introduction

Multilayered coatings, especially with transition metal nitrides at nanoscale, are currently subjects to intensive research because of their potential for a wide range of technological applications [1, 2] and also because they are, fundamentally, of scientific importance [3, 4]. Figure 1 shows the representative structures of multilayered coatings (a) TiN/CrN and (b) TiAlN/VN. Indeed, the interest in nitride coatings is two-fold: transition metal nitrides, which are well classified as refractory compounds, exhibit unusual physical and mechanical properties such as extreme hardness, high thermal- and chemical-stability, good corrosion resistance, and low, stable friction under even dry contact conditions [5–9]. These intriguing properties render them applicable for many uses, for instance, in relation to: engineering tools, magnetic and electric components, superconducting devices, etc., whilst periodic coating of these nitrides in the nano-scale dimension [10, 11] onto such tools components and devices can enhance most, if not all, of their functions to a level that none of their individual constituents can ever match [12]. A representative example is that the multilayered TiN/VN superlattice, produced by alternating depositions of its two host mononitrides, which reaches a hardness value as high as 56 GPa [13–15], this being much higher than that of either of its host coating alone. Moreover, multilayered coatings enable the design of constituted materials with desired properties. For example, of all the transition metal nitrides, TiN is the most commonly used nitride. It is of interest largely because TiN is hard and wear-resistant, has a bright golden color, and can be coated onto many engineering materials. However although TiN/VN multilayered coatings exhibit the highest hardness, their widespread applications are currently limited by their propensity to be oxidized, especially at a high temperature, which degrades mechanical properties of TiN/VN multilayered coatings [16–19]. To avoid this problem, multilayered coatings TiN/CrN or TiN/AlN lead to improved thermal stability and excellent oxidation resistance with no notable loss in hardness [20–26]. Transmission electron microscopy (TEM) studies [27] have attributed the property improvement to a thin Cr_2O_3 layer that can suppress the formation of oxides and voids.

Apart from the sophisticated applications in tooling industry and other engineered components, due to their excellent mechanical properties, especially the hardness, origin from

*e-mail: deqiang.yin@scu.edu.cn

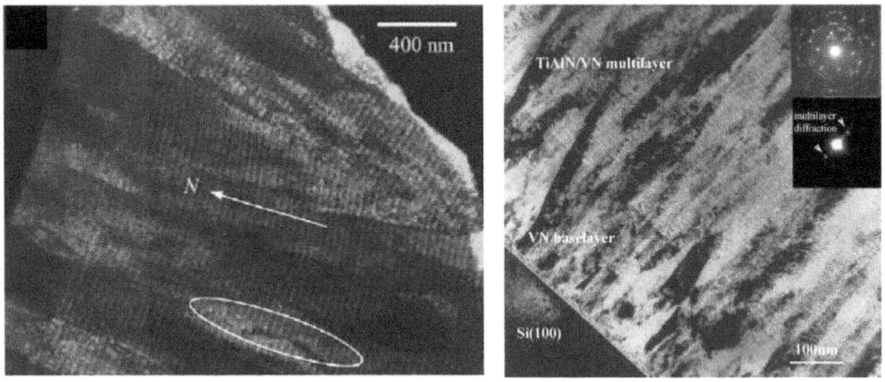

(a) (b)

Figure 1. Transmission electron microscopy (TEM) image of (a) TiN/CrN [28–30]; and (b) TiAlN/VN [2] multilayered coatings.

the nature of transition metal nitrides, multilayered coatings, composed of other functional constituents, are also used in the fields such as optical filters [31–40], magnetic storage [41, 42], sensors and microelectronics [43–46]. For instances, Wang et al. [32] fabricated a TiO_2/SiO_2 multilayered optical filters with graded refractive index profiles, deposited on glass (BK7) and Si(1 0 0) substrates simultaneously and sequentially, by helicon plasma sputtering at room temperature, and found that the measured transmittance spectrum exhibited a reflectance of 99.8% at a central wavelength of 730 nm and high transmittance over the wavelength region outside of the reflected band, as a result of the suppression of the sidelobes. Furthermore, Chhajed et al. [35] designed and produced a nanostructured multilayered tailored-refractive-index antireflection coatings on a glass substrate by using low-refractive-index silica, and reported that the measured average optical transmittance between 1 000 and 2 000 nm is improved from 92.6 to 99.3% at normal incidence. Moreover, Raut et al. [39] described different types of the Anti-Reflective Coatings (ARCs) in greater detail and illustrated the state-of-the-art fabrication techniques. Louis et al. [40] also gave an overview on the progress in the thin film and surface physics involved in multilayered systems with nanometer scale periodicity, wherein a special attention was given to the development of thin diffusion barrier layers between the materials in the multilayered coatings to enhance the optical contrast and to reduce the inter-diffusion.

From the above brief description, it can be seen that the multilayered coatings could be tailored/designed based on the choice of individual constituents for a particular multilayered systems, and these intriguing advantages make them applicable for many uses.

Improvements in mechanical properties due to introducing multilayered coatings, encompassing post-processing behavior, have attracted many studies from both experimental and theoretical. Experimentally, the studies have covered the effects of: (i) preliminary treatment of the substrates; (ii) various kinds of material combination, as well as the thickness ratio of neighboring layers; (iii) post-processing, and evaluation of the mechanical, thermal, and magnetic properties via bending, buckling, nano-indentation, scratching, tensile test, acidic corrosion, high-temperature test, etc. [47, 48]. For instance, understanding of the origin of property improvement through multilayered coatings is twofold: on one hand, it is attributed to the interfaces in multilayered coatings. Nordin et al. [49] reported a significant slowdown of corrosion for the TiN/CrN multilayered coatings because the oxidation is found to be almost inversely proportional to the number of interfaces in the multilayered coatings, as is confirmed in other previous works [50, 51]. On the other hand, it is ascribed to the residual stress present in the multilayered coatings [52]. Mendibide et al. [28–30] successfully deposited TiN/CrN onto the tool steel and investigated the effect of residual stress [53], finding, by transmission electron microscopy (TEM) and synchrotron analysis, that the wear resistance of TiN/CrN is improved due to the mode shift of crack propagation induced by the fluctuating residual stress field across the multilayered coatings. In addition, they claimed that interfaces in the multilayered coatings also play a crucial role in improving the refractory properties and wear resistance. More detailed information of multilayered coatings could be found in the literatures, such as references [54–57]. Correspondingly, various theoretical models have also been developed to explore the mechanisms for the enhanced chemical and mechanical properties of multilayered coatings. However, a better understanding of the underlying mechanisms of multilayered coatings still requires a more novel and more applicable model. In this paper, recent advances in the numerical modelling of multilayered coatings are critically reviewed with a reference to modelling at length-scales and the numerical methods used.

Evidently, one of the notable characteristics of multilayered coatings is that the thickness of individual layer is in the nanoscale, that is to say, it is generally from several nanometers to tens of nanometers while the thickness of the whole system may be in micrometers to millimeters, inducing an abundance of numerical simulations at different scales. Based on the scale applied, the existing numerical models can be generally classified into three categories: continuum models, molecular dynamics models, and nano-scale models, as illustrated in Figure 2. There are also some coupled models or methods combining two of them, such as the quasi-continuum (QC) method, which is a mixed continuum and atomistic approach for simulating the mechanical response of polycrystalline materials. Importantly, there are two basic requirements for building a theoretical

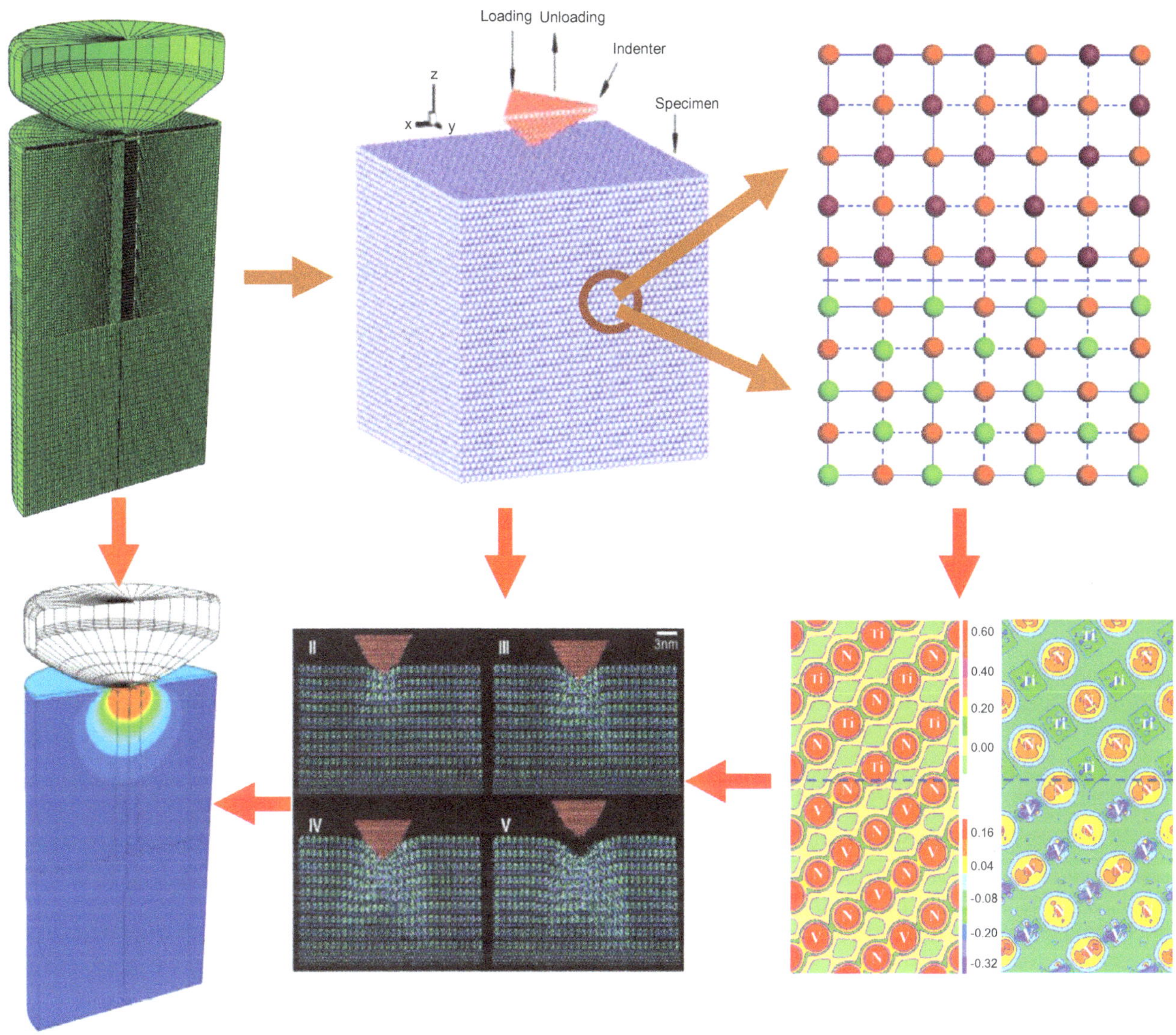

Figure 2. Schematic numerical simulation process from Macro- to Micro- to Nano-scale.

model. One is that the model has to be "big" enough to represent the behaviours/properties of the multilayered coating system studied, and another is that the size of model is limited by the computational power. Consequently, a theoretical model may only be able to consider a thin layer containing the coatings-substrate system, composed of single, duplex or a limited number of phases.

2. Continuum models

Intuitively, the structure of super-hard multilayered coatings at the nanometer scale is still a continuum [58]. Thus, a natural attempt to model multilayered coatings is based on conventional fracture mechanics and continuum mechanics scaled down to the dimension of nano-layered. A representative volume element (RVE) for a constitutive model is shown in Figure 3. It is worth noting that for RVE, the layers of two phases are parallel to each other, and arranged alternatively with perfect bonds, and with a constant interlayer spacing. The x_1x_2

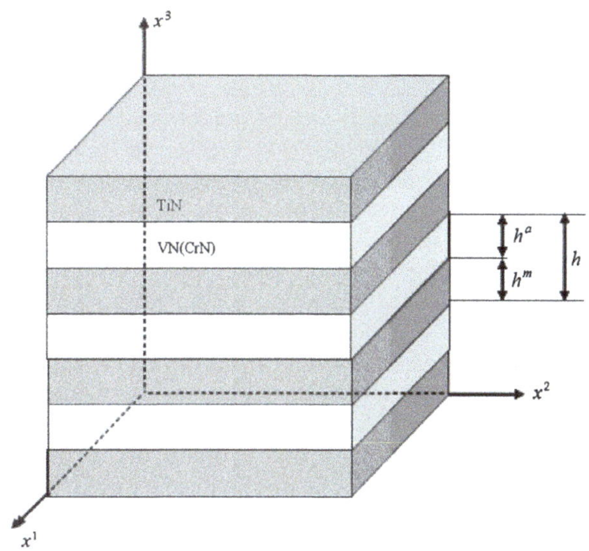

Figure 3. Schematic illustration of an RVE of a multilayered coating.

plane of the coordinate frame is parallel to the layers, and the x_3 axis is perpendicular to the layers. If the thickness of each layer is assumed to be sufficiently small, compared to its in-plane size, the variation of both stress and strain within a phase can be neglected. The thickness h is the modulation period of multilayered coatings. Consequently, the numerical methods applied in these models include mostly, Finite-Element Method (FEM) [59], Finite Difference Method (FDM) and the Boundary-Element Method (BEM) [60, 61]. Typical examples of a multilayered coating, simulated mostly with FEM, are presented in this section, containing various elastic and elastic-plastic material properties, normal and tangential loading conditions, a hard coating/functionally graded substrate system, etc. Finally, a FEM model with cohesive zone modeling, used for the prediction of the formation and propagation of micro-cracks under indentation, is taken as an example of a general procedure for the FE simulation of multilayered coatings.

FEM can be used to simulate the behavior of multilayered coatings under either (nano-) indentation or subjected to a scratch test, mainly concerning:

1. *Stress distribution*. For instance, Djabella and Arnell [62] used a FEM to investigate the contact stresses during the process of elastic compression on coating/substrate systems consisting of a high modulus surface coating on a relatively low modulus substrate wherein two pressure distributions were imposed to consider those arising during the Herzian indentation of a homogeneous elastic half-space by spherical and cylindrical indenters. They found that the stresses of both the outer surface of the coating and the coating-substrate interface are complicated functions of (a) the ratio of the Young moduli of the coating and substrate and (b) the ratio of the coating thickness to the contact halfwidth; and that the stresses which could be responsible for coating failure cannot simultaneously be minimized by appropriate choice of variable. Further, Stephens et al. [63] analyzed initial yielding behavior due to the indentation and friction process between an elastic cylindrical surface and a hard coating/functionally graded substrate system where a thin hard DLC film deposited on a soft Ti-6Al-4V alloy substrate is considered as a model system with consideration of two functional gradient substrate conditions: (a) a gradient in yield strength and (b) a gradient in elastic modulus. They found that in both cases, appropriate gradients result in significant benefit to the reliability of the coated system, compared to the case of an ungraded substrate. Furthermore, Bouzakis et al. [64] introduced a continuous FEM simulation of the nano-indentation hardness test in order to describe the coating's elastic and plastic deformation and herewith, and thereby to extract, precisely and independently of the indentation load, the coating stress-strain curve. Similarly, Tan and Shen [65] employed FEM to examine the relationship between indentation hardness and overall yield strength of multilayered materials, and found that the indentation-derived strength consistently underestimates the composite yield strength within the scope of the continuum-based approach. Additionally, Zhao et al. [66] simulated nano-indentation-induced deformation in TiSiN-based multilayered coatings by means of FEM, and quantified stress distributions under moderate indentation loading in the structure with particular emphasis on the relationship between stress concentrations and crack initiation. The results showed that structural layering can be used to modify the stress distribution and to lower the overall stress level within the coating. In the case of radial tensile stresses at the coating/substrate interface, a reduction of ~50% was achieved through layering.
2. *Hardness measurement*. FEM has been widely used to simulate elastic and plastic deformations beneath a pointed indenter, hardness measurement with a sharp pyramidal indenter, such as Berkovich or Vickers, is one of the most complex mechanical problems, due to the 3D dimensional phenomenon associated with elastic and plastic deformations with large strain fields. As is well known, FEM is used to extract the true hardness of thin coatings, from the composite hardness measurements, based on the assumption that the film and substrate behavior can be modeled with a bilinear law. However, Pelletier et al. [67] showed that the bilinear model does not define correctly the strain hardening, resulting in an overestimation of the yield stress or one need to adjust the target modulus at each indentation depth.
3. *Formation and propagation of cracks*. For examples, Aoki et al. [68] built a FE model for the calculation of the energy release rate during the coating cracking, aiming at showing that SiC/C multilayered coatings successfully suppress through-thickness coating cracks. Furthermore, Chen [69] used a cohesive zone model embedded into a FE model to assess nucleation and propagation of delamination triggered by an indentation in a multilayer stack with relatively weak metal/ceramics interfaces. Holmberg et al. [70] developed a three-dimensional FEM model for calculating the first principal stress distribution in the scratch tester contact of a diamond spherical tip with 200 μm radius sliding with increasing load on a 2 μm thick titanium-nitride coated steel surface, with consideration of the elastic, plastic and fracture behavior of the surface. The results showed that the first crack is initiated at the top of the coating from bending and pulling actions and grows down through the coating also. Other mechanical behavior such as milling, cutting, and sliding have also been investigated with FEM. For instances, Özel et al. [71, 72] applied FEM for predicting the forces, temperatures, and wear rate during the micro-milling of Ti-6Al-4V alloy with fine-grain uncoated and cBN coated micro-end mills, the results revealing advantages of cBN coatings. Grzesik et al. [73] studied the temperature distribution in the cutting zone in the turning processes with differently coated tools and showed the tool-chip interfacial influence on the temperature distributions, as a consequence of using coated tools. Moreover, Coelho et al. [74] combined experiment with FE simulation to investigate the behavior of TiAlN,

TiAlN-nano-coating and AlCrN coated tools in finish-cutting conditions, and showed that the temperature at the tool-chip can be stable and as high as 800 °C when using TiAlN-based coated tool. Additionally, the presence of the coating layer creates an additional thermal resistance between workpiece and the tool, slightly lowering temperature at the substrate. Holmberg et al. [75] carried out a micro-scale finite-element method modelling and stress simulation of a 2 μm TiN-coated steel surface and showed a reduction of the generated tensile buckling stresses in front of the sliding tip, when compressive residual stress of 1 GPa was considered in the model.

4. *Residual Stress*. FEM can also be used to study the residual stress caused by the different thermal expansion coefficients between two phases of multilayered coatings during thermal cyclic loading or a manufacturing process. For instance, Ng and Gan [76] presented finite-element coupled heat transfer and elastic-plastic thermal stress analysis using a general purpose of commercial FEM to simulate the in-situ residual stress generation during plasma spraying of duplex thermal barrier coatings. They also proposed a simple method by post-deposition treatment to effectively reduce the residual stress level. Sarikaya and Celik [77] studied the thermal evolution of MgO-ZrO_2/NiCrAlY coatings on Ni metal and AlSi alloy substrates, using the finite-element method, and showed that stress concentration was at higher levels in the coatings with AlSi substrate than that when with Ni, and that the larger residual stress was obtained with an increase of the coating thickness. Similarly, Teixeira [52, 78] presented a FEM model of the residual stress distribution with a layered metal-ceramic composite coating, based on the difference between the coefficients of thermal expansion of the neighboring layers, obtaining an analytical expression of the residual stress. They also discussed the failure modes of a thin coating under residual stress as: (a) delamination (coating under tensile stress and "weak" interface, i.e. low adherence); (b) perpendicular micro-cracking (coating under tensile stress and a "strong" interface); and (c) buckling and spalling (coating under compressive stress). Moreover, Hsueh and Lee [79] also calculated the elastic thermal residual stress in thermal-barrier coating, which consisted of a substrate, a graded bond coat, and a top coat, with the origin of different coefficients of thermal expansion [80]. Liu et al. [81] investigated the effect of a system composed of Al_2O_3 and SiC on *316L* stainless on the residual stress developed in the composite coatings, and showed that the presence of the graded properties and the compositions within the coating did lead to the reduction of the stress discontinuity at the interfaces between the coating and the substrate whereas the magnitudes of the residual stresses on the coating surface and at the coating/substrate interface were dependent on the Al_2O_3 and SiC coating thickness. Using FEM, Sayman et al. [82] carried out a transient thermal analysis in WC-Co/Cr-Ni multilayered coatings deposited on a *316L* steel substrate during the cooling process from 1 273 K to room temperature. Similarly, Toparli et al. [83] determined the thermal residual stresses in WC-Co/Ni-Al coating layers deposited on 316L stainless steel substrates, developed during and after thermal cycling, showing that the calculated tensile stresses were higher than the compressive stresses. Recently, Far et al. [84] developed a FEM model to evaluate the stress induced by the thermal cycle in a typical plasma-sprayed thermal-barrier coating system, taking into account the effect of thermal and mechanical properties, morphology of the top-coat/bond-coat interface, and oxidation on the local stresses that are responsible for the micro-crack nucleation during cooling, especially near to the metal/ceramic interface.

5. *Other applications*. Apart from the applications described above, FEM has also been used for the following particular applications. For examples, Pan et al. [85] compared the sliding contacts of gradient layer, sandwich layer and ordinary multilayered coatings, subjected to a similar surface load and under elastic-plastic deformations, and found that the positive gradient layer coating can improve the stress and strain fields near the layer-substrate interface, and in the substrate, meanwhile, the sandwich layer coating can greatly reduce the maximum shear stress on the interface of the layer-substrate and the stress at the crack tip. Conversely, the stress and strain distributions of the ordinary layer have not evidently been improved, compared to those of the single layer. Their work provided a useful reference for the selection and design of coatings. Furthermore, Bemporad et al. [86] designed the thickness and Ti distribution in multilayered Ti/TiN coatings with FEM-based optimization leading to a significant increase (45%) in adhesion between the layers and the WC-Co interlayer (Ti-6Al-4V substrate) and load bearing capacity of the coated system, compared to monolayered TiN, without reduction in superficial hardness and in the load bearing capacity. Lakkaraju et al. [87] modeled multilayer Cr/CrN thin films and optimized them to have effective "load support" by the films on stiff A2 steel and compliant 2024-Al substrates. The efficiency of several optimization algorithms such as genetic algorithms and gradient based routines was discussed and the preliminary results were compared to the results of pin-on-disk wear test results of empirically designed coatings. Similarly, Gorishnyy et al. [88] built two-dimensional models, utilizing FEM, for single layer CrN and Cr_2N films and for a number of multilayer combinations of Cr, CrN and Cr_2N layer, aiming at developing an approach for the design of multilayer coatings with enhanced toughness to fracture and improved adhesion for wear-resistant applications.

For developing a generic procedure of FE simulation of multilayered coatings, a parameterized has been developed recently [89]. The method which considered key geometrical, material, interfacial and loading variables as model variables in the FE-based parameterized modelling, was developed with a view to improving the efficiency and accuracy of the analysis and design of multi-layered coating-systems. The method

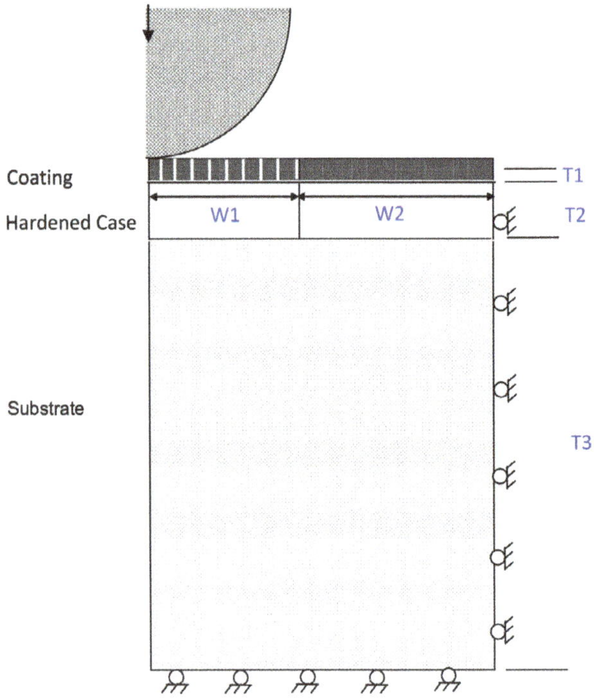

Figure 4. Geometric model of a multi-layered system including the coating, the hardened case and the substrate.

allows for the parameters to be changed easily during a series-analysis. Combined with the capability of the prediction of cracking of the coatings, the developed method/model provides an efficient way for investigating the effects of these parameters on the behavior of multi-layered systems. It was demonstrated with analysis of the coated tool-steel (H11) for: (a) a substrate without being pre-heat-treated and (b) two substrates with a shallow and a deep hardened-case respectively (both are treated by plasma-nitriding). The results showed that the case-hardening of a substrate has a significant influence on the performance of the surface system with coating, especially on its load-bearing capacity and on its the resistance to the development of cracks in the coating.

It is worthy of noting that in reference [89], a bilinear cohesive-zone model was employed to simulate the development of cracks in the coating of a tool subjected to indentation, the general procedure being as follows:

(a) Definition of the geometry model and meshing scheme. In this step, model dimensions and meshing scheme are established. Figure 4 shows the structure of a multi-layered system indented by a spherical indenter and the corresponding geometry. Based on the assumption of an axi-symmetric problem, only one half of this model is shown. The axi-symmetric model is made up of five blocks including: a central and an outer coating layer; a central and an outer hardened case; and the substrate, where: W1 and W2 indicate the width of the central and the outer parts respectively; and T1, T2 and T3 indicate the thickness of the coating, the hardened case and the substrate, respectively.

(b) Materials definition. Any number of materials can be defined for the simulation. Each material is assigned a name. A region in the model can be associated with a particular material through the assignment of its section properties that refer to the material's name.

(c) Definition of the contact and boundary conditions. In this step, the surfaces and interactions between surfaces are created. Boundary conditions, which can be recognized as the interactions between the system and environment, also need to be defined.

(d) Definition of cohesive elements. The cohesive elements, combined with the FE, using a traction-separation relation in order to describe the interface as a continuum of coatings and the host, are defined in order to simulate the formation and propagation of cracks under indentation. Several traction laws have, to-date, been applied for this kind of analysis, either in a form of exponential relationship [90, 91] or in a bilinear relationship [92–94] between the stress and the displacement.

(e) Validation of the FE model. The FE model is validated through a series of indentation tests. A comparison of the experimental results with finite-element results is shown in Figure 5. It should be mentioned that the adhesion energy values used in Cohesive-zone modelling is critical and that these can be obtained either from calculations based on experimental data or by means of computations based on Molecular Dynamics or First principles calculation based on high-precision density functional theory (DFT) [95].

(f) FE analysis. After validation of the model has been done, the model can be used to analyse cases of interest. The analysis can be re-activated for a different set of input-variables, simply by re-setting values of these variables in an input file, without need to modify the model through using a CAD/CAE software.

The project "Multi-scale modelling for the design and analysis of multi-layered engineering surface systems (M3-2S)" was undertaken jointly by 11 international partners, funded through the 7th European Framework Programme, wherein the overall objective of the project was to establish integrated, generic, robust multi-scale modelling techniques, based on atomic FE(nano), crystal plasticity FE (micro) and continuum mechanics FE (macro) modelling. For examples, Leopold et al. [96] developed an advanced adaptive FEM software (AAFEM) for the coating-substrate simulation of superlattice TiN/CrN-coatings; Zhuang et al. [97] developed an user-subroutine to implement a multi-axial, continuum damage mechanics (CDM)-based constitutive model into a commercial FE code, and carried out a series of investigations into the TiN/Cu system; Perucca et al. [98] demonstrated performance of the TiN and TiN/CrN nanoscale multilayered coatings on WC cutting inserts for machining GJL250 cast iron. It was shown that the nano-structured TiN coatings, followed by multilayered coatings, have much better performance, and the FE simulation showed a good agreement between the measured cutting-force and the simulated one; and Karimpour et al. [99] presented an inverse method for

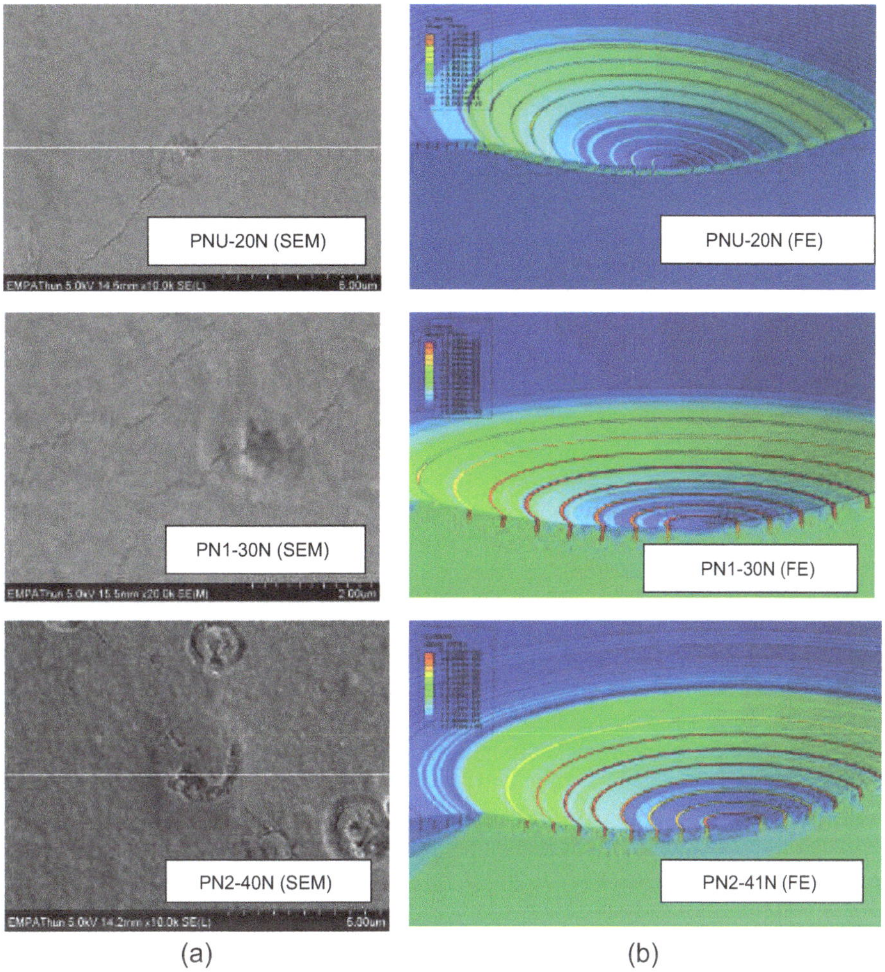

Figure 5. A comparison between (a) Experimental results and (b) Finite Element ones.

identifying properties of the material constituents in their multilayered coatings, combining FE simulations and indentation data, which is extensible to the analysis of multilayered coatings containing any number of constituents.

3. Molecular dynamics

It is well known that the mechanical behavior of a material with a very small volume differs from that which is typically observed for that with a large volume. Analysis of such a phenomenon at the micro-scale has been largely with Molecular Dynamics (MD) Simulation. It is worth noting that there are two basic assumptions made in standard MD simulations [100–102]: (1) Molecules of atoms are described as a system of interacting material points whose motion is decided dynamically by the potential, and (2) No mass changes in the system, i.e., the number of atoms in the system remain the same during the process. Similarly, in continuum models, MD has also been used for the analysis of indentation. For instance, Saraev and Miller [103] used MD simulations to elucidate details of plastic deformation and the underlying deformation mechanisms during the nano-indentation of thin copper films with epitaxial nickel coatings, and observed several deformation mechanisms, such as dislocation pile-up at the interface, dislocation cross-slip and movement of misfit dislocations. Similarly, Cao et al. [104] performed MD simulations of nano-indentation, following a nano-scratching process, to investigate the mechanical and tribological properties of Ni/Al multilayer with a semi-coherent interface. The results showed that the indentation hardness of Ni/Al multilayers is larger than that of a pure Ni thin film, and the remarkable strength of Ni/Al multilayered coatings is due to the semi-coherent interface which acts as a barrier to glide of dislocations during nanoindentation process. Furthermore, Fang and Wu [105] carried out a MD simulation to study effects of indention deformation, contact and adhesion on Al, Ni, and Al/Ni multilayered films. The results showed that when the indention depth of the sample increased, the maximum load, plastic energy, and adhesion increased. Shan et al. [106] studied the effect of a single twin boundary parallel to the indented surface on nanoindentation of Ag(1 1 1) films, and showed that the twin boundary has little influence on the elastic modulus of films. The load for the initial yield is observably reduced when the twin boundary is very near the indented surface due to the nucleation of the glissile dislocations on the slip plane parallel to the surface, rather than the formation of the tetrahedral sessile lock when nano-indentation on

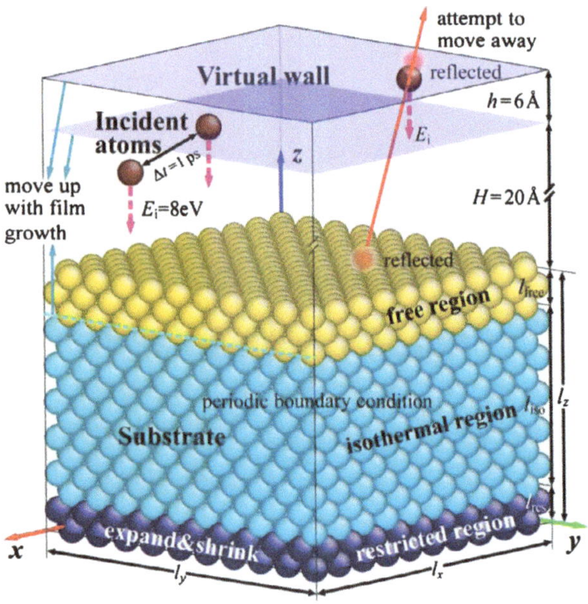

Figure 6. Schematic diagram of the MD model of atomic deposition.

the perfect film. Moreover, they also showed other two commonly-use functions of MD simulation in multilayered coatings:

(a) The tension of multilayered coatings [107]. For instance, mechanical behavior, dislocation nucleation and development, and evolution of an interface in Cu/Ag bilayer films during equal biaxial tension were studied using MD simulations. The results showed that dislocations are prone to nucleate at the interface of Cu/Ag bilayer film, and then propagate towards the free surface. The deformational defects consist of partial dislocations and intrinsic stacking faults, accompanying some extrinsic stacking faults and stacking fault tetrahedral which just occur at the interface and propagate in the Cu layer.
(b) Interface-structure formation [108]. Assisted with MD simulations with EAM potential, the formation of the interface of a Cu/Ag bilayer during sputtering was studied. The method provides a notable way to elucidate the mechanism of interfacial formation, which plays a dominant role in controlling the mechanical properties of multilayered coatings.

A MD model of atomic deposition is shown in Figure 6. Four growth cases, including Ag/Cu(1 1 1), Cu/Ag(1 1 1), Ag/Cu(0 0 1) and Cu/Ag(0 0 1), were simulated under the deposition of E_i = 8 eV (the kinetic energy for incident atoms) and T = 300 K. The results showed that the first- deposited Ag layer grows on the Cu substrate by the layer mode, while the firs deposited Cu layer grows on the Ag substrate by the island mode, determined by the surface energies of the film and substrate.

4. Nano-scale modelling

At the atomistic scale, the first-principles calculation based on the high-precision density functional theory (DFT) [92], is a sophisticated way to provide insight into the bonding configuration, interfacial structure, and electronic properties, which are difficult to obtain by experiments alone at such a scale, as well as the behavior during the ideal tensile/shear process.

Four steps are usually used for building a reasonable interfacial model for multilayered coatings at the atomic scale [109]:

(a) Firstly, the preferred textures of each individual layer for the composited system are made sure, e.g. from experimental observations [110];
(b) Secondly, perform convergence calculations on the surface energy with respect to slab thickness are performed to ensure that individual layers of the multilayered coatings model are thick enough to exhibit *bulk-like* interiors. Taking the TiN/VN interface for examples, the surface energy, (σ_s^0) under the condition that the chemical potential of respective element equals its bulk total energy can be expressed as follows [111]:

$$\sigma_s^0 = \frac{1}{2A}\left[E_{\mathrm{slab}}^{\mathrm{total}} - M_N E_{XN}^{\mathrm{bulk}} - E_X^{\mathrm{bulk}}(M_X - M_N)\right], \quad (1)$$

where $E_{\mathrm{slab}}^{\mathrm{total}}$, E_{XN}^{bulk} and E_X^{bulk} represent the total energies of a formula unit of a slab (TiN or VN), XN bulk, and X bulk (X = Ti or V), respectively, and M_X and M_N denote the number of X and N atoms and A is the surface area. As illustrated in Figures 7b–7g, there are, in principle, six possible terminations for either TiN or VN.
(c) Thirdly, considering the interfacial stacking sequences and various terminations of individual surface, a series of possible candidate models could be constructed. As shown in Figure 8, there are three stacking sequences for TiN/VN multilayered coatings.
(d) Finally, the *adhesion energy* (W_{ad}), a key quantity in predicting the adhesive properties of an interface can be calculated to determine the most energetically stale surface, from [112, 113]:

$$W_{\mathrm{ad}} \equiv \frac{(E_{\mathrm{TIN}} + E_{\mathrm{VN}} - E_{\mathrm{IF}})}{A}, \quad (2)$$

where E_{TIN}, E_{VN}, and E_{IF} are the total energy of isolated TiN, VN slab, and their interface, respectively, and A is the interface area. Two steps were taken to estimate the W_{ad} values. Total energies were first calculated for a series of separations as the two rigid slabs were approaching increasingly closer from a large initial separation. Consequently, the calculated energies behaved like a parabola, passing through a minimum at the equilibrium separation. The *unrelaxed* W_{ad} was obtained by computing the energy difference between the interface at the equilibrium state and the unrelaxed isolated slabs. Then, each isolated slab as well as the interfacial slab was allowed to optimize fully, yielding an estimation of the *relaxed* value of W_{ad}.

More importantly, first-principles calculations can be employed to investigate original mechanisms from atomic

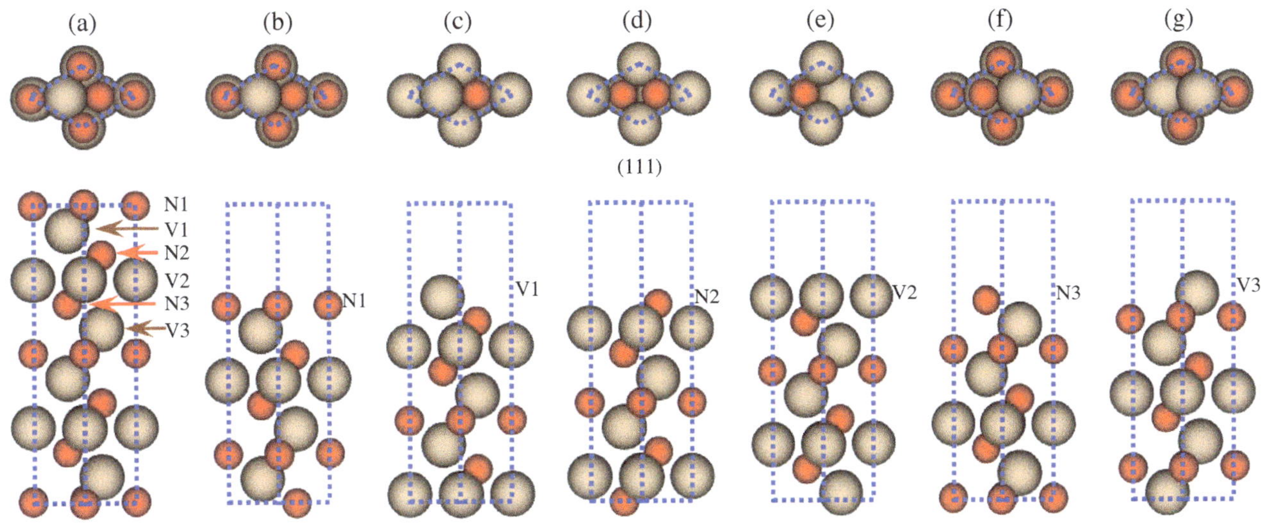

Figure 7. Plots of VN: (a) bulk, (b) N1, (c) V1, (d) N2, (e) V2, (f) N3, and (g) V3 terminations. Upper part shows top view and lower part side view. Only the top nine out of thirteen symmetric (1 1 1) layers are presented for each termination.

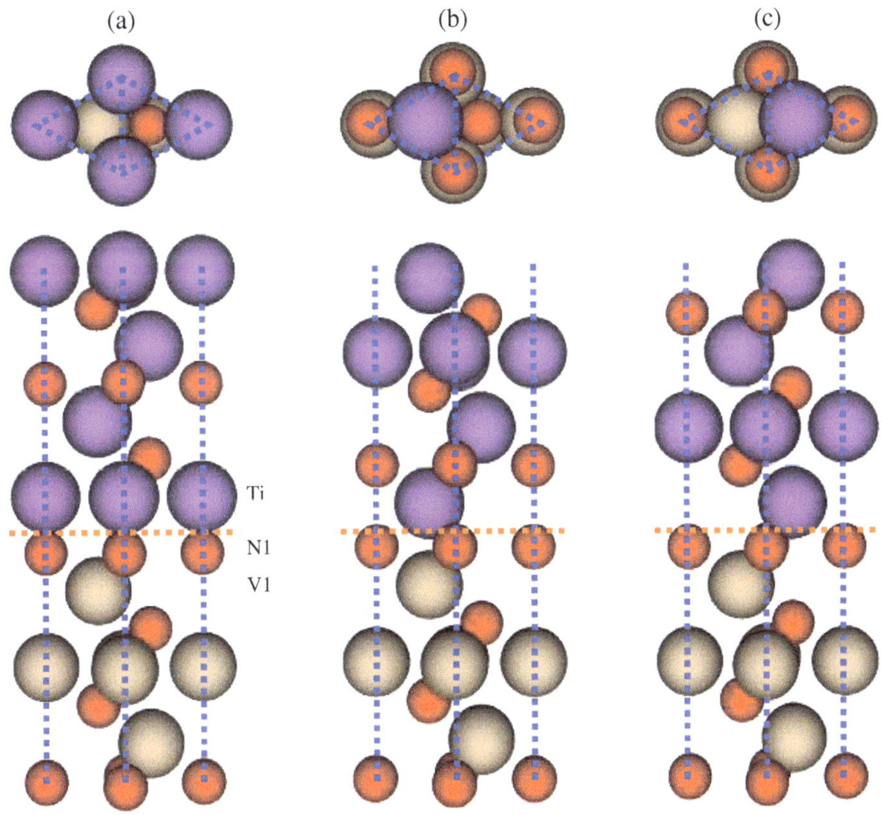

Figure 8. Schematic plots of (a) OT, (b) SL, and (c) TL stacking sequences. Upper part shows top view and lower part side view. The interfaces between N1-terminated VN and Ti-terminated TiN are given only. For top views, only the Ti layer nearest to the interface is shown for clarity. The view direction for bottom panels is along [1 $\bar{1}$ 0] and the location of interface is marked with horizontal dotted lines. The top and bottom portions of the interface have been omitted.

stacking sequences, bonding properties/interactions between neighboring atoms, etc., at the atomistic scale, due to the fact that the atom is the minimum element for the composited systems. For instance, the two mostly addressed topics for multilayered coatings are Template effect and Superhard mechanism. For "Template effect", the metastable phase can form in multilayered coatings due to the template effect, which means that the crystal structure of the newly deposited layer is under a strong influent of the underlying multilayer system. In this way, the newly deposited layer may crystalline in ametastable structure by forming a coherent interface with the rest of the multilayer structure. Such a mechanism often leads to strengthening the coatings and thus is desirable. Taking the representative multilayered coatings TiN/AlN for example, Chawla et al. [114] combined ab initio methods and the finite element (ABAQUS) to study and predict the equilibrium critical thickness up to which the metastable cubic (c) AlN phase is energetically favoured to the stable wurzite(w) variant in TiN/AlN and CrN/AlN bi-layer system. And Zhang and Veprek [115] selected deliberately a couple of representative transition metal (AlN included) as interfacial layers in TiN-based heterostructures, and investigated the thermal stability using first-principles molecular dynamics. The results showed that one-monolayer-thick pseudomorphically stabilized interfacial layer of B1-type AlN is stable within the units within the whole temperature range considered. Moreover, Zhang, Veprek, and Sheng [115–117] have studied metastable phases and spinodal decomposition in $Ti_{1-x}Al_xN$ system by ab initio and thermodynamic modelling compared with the $TiN\text{-}Si_3N_4$ system and found that metastable *fcc*-$Ti_{1-x}Al_xN$ coatings can easily undergo spinodal decomposition into coherent *fcc*-TiN and *fcc*-AlN, but there is a relatively large barrier for the formation of the stable *hcp*-AlN.

Furthermore, a comparison with the $TiN\text{-}Si_3N_4$ system showed that, due to the much higher de-mixing energy of this system as compared to the TiN-AlN one, spinodal decomposition may occur in that system also for semicoherent TiN and Si_3N_4 phases. Holec et al. [118] selected deliberately a couple of representative transition metal aluminium nitride (TM-Al-N) thin films (Ti-Al-N included) and addressed the structure and phase stability. Especially for $Ti_{1-x}Al_xN$, the predicted stability regions of the rock salt cubic structures are $x \leq 0.7$. Similarly, Stampfl and Freeman [119] also investigated the formation, atomic and electronic structure, and stability of metal-nitride interface systems, (1 0 0) AlN/TiN, as well as AlN/VN and VN/TiN in the rocksalt structure, and found that the layer-dependent interaction energy between the adlayer surface and the interface, while typically not taken into account, plays an important role in terms of the formation energy for the initial stages of film growth.

The aforementioned references for elucidating the mechanism of the "Template Effect" are also applied for the "Superhard mechanism" due to the formation of coherent interface, which remarkably decreases the defects, such as initial cracks, voids, etc., as described by Zhang et al. [120]. These authors have studied the intrinsic mechanical properties of two AlN polymorphs (*fcc*- and *hcp*-) and found that, in spite of its higher elastic modulus, the *fcc*-AlN possesses slightly lower shear strength as compared to the stable *hcp*-AlN. Moreover, the calculated electronic density of states indicates a similar covalent bond property for both polymorphs. Consequently, the higher hardness-enhancement of *fcc*-TiN/*fcc*-AlN heterostructures as compared to *fcc*-TiN/*hcp*-AlN nanolaminates is not related to the difference of the intrinsic strength and bonding nature between *fcc*- and *hcp*-AlN, but most likely to the formation of semi-coherent interface.

5. Multiscale models

As the result of the confluence of parallel computing power, experimental capabilities to characterize structure-property relations down to the atomic level, and theories that admit multiple length scales, there has been rapid growth of activities in multiscale modelling. In order to make the computations tractable, multiscale models generally make use of a coarse-fine decomposition. An atomistic simulation method, such as MD, is used in a small subregion of the domain in which it is crucial to capture the individual atomistic dynamics accurately. A continuum simulation is used in all other regions of the domain in which the deformation is considered to be homogeneous and smooth. Since the continuum region is usually chosen to be much larger than the atomistic region, the overall domain of interest can be considerably large. The aforementioned QC methods, based on a combined MD-FE technique, can be applied to a multiscale modelling simulation, where MD is only used in localized regions in which the atomic-scale dynamics are important, and a continuum simulation (FE) used everywhere else. For instance, Sun et al. [121, 122] have applied QC methods to simulate the nanometric cutting of crystal copper. Their study demonstrated that multi-scale simulation is feasible, not withstanding that there is still more work needed to be done to make multi-scale simulation more practical. The QC method is a directly coupled model between two scales (MD and FE). The unique characteristics of such a directly coupled method is that there is a hand shaking (HS) region to connect various scales wherein the message exchange, such as the velocity of particles (atoms or molecules), temperature, etc., is always a big challenge. At the same time, there is another way to utilize the advantages of various single scales, that is to say, the usage of one single scale could provide parameters/evidence for further investigation in another single scale. For instance, both FEM and MD simulation for nano-indentation showed that the calculated mechanical properties intrinsically depend on the interface contact conditions of the film/substrate [123]. Especially, investigations into the "template effect" from first-principles calculations have ascribed the super-hardness of multilayered coatings as the formation of coherent interfaces between the neighboring individual layers. Aiming at evaluating the residual stress caused by lattice mismatch between neighboring layers, a approach based on the representative volume element (RVE), as shown in Figure 3, incorporating with first-principles calculations, is proposed from continuum mechanics principles [124]. At the atomic scale, the evidence for the good bonding between the multilayered coatings has been presented by first-principles calculation, as shown in Figure 9. It can be seen in these figures that, at first, the charge distributions of TiN, CrN show a spherical symmetry. Similar charge distributions can also be observed between the interfacial Ti-N bonds and the Ti-N bonds deeper to TiN: the majority of the charge being located on both Ti and N atoms, with slight but visible distortions directed towards each other. Not surprisingly, for TiN/CrN multilayered coatings, charges that are accumulated on the interfacial *N* come predominantly from adjacent Ti and Cr atoms, as shown in the right panel of Figure 9. It was concluded that the interfacial bonding Ti-N (or Cr-N) is mainly ionic, yet maintains a small degree of covalency, which indicates the well bonded interfaces of the TiN/CrN multilayered coatings. After these, material parameters, such as Young's Modulus (E), Poisson's ratio (v), lattice constant (a), have also been calculated.

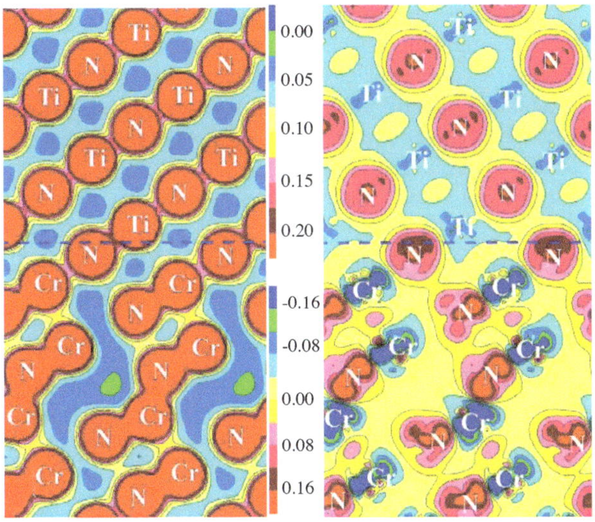

Figure 9. Contour maps of charge density (left) and its difference (right) for TiN/CrN multilayered coatings, taken along the $(1\,\bar{1}\,0)$ plane. The difference in charge density represents redistribution of charge in the interface relative to the isolated system. The interface is indicated by a horizontal line and atoms that intersect the contour plane are labeled. The upper scale denotes the magnitude of charge in the left panel and lower scale that in the right panel scale.

In the following, the constitutive relationship of each phase can be expressed as:

$$\Delta\sigma^m = L^m : \Delta\varepsilon^m \quad (\text{for the } m\text{-phase (TiN)}), \tag{3}$$

$$\Delta\sigma^a = L^a : \Delta\varepsilon^a = L^a : (\Delta\varepsilon^m + \Delta\varepsilon^*) = L^a : \Delta\varepsilon^m + L^a : \Delta\varepsilon^* \quad (\text{for the } a\text{-phase (CrN)}), \tag{4}$$

where $\Delta\varepsilon^*$ is mismatch strain induced by the lattice mismatch between neighbouring layers. Based on the assumption that the in-plane strain and the out-of-plane stress components in both TiN and CrN phases should be identical to the corresponding values in the RVE to meet the compatibility and equilibrium conditions [125], the analytical solution of residual stress can be obtained and the relationship between various parameters can also be described. The whole process presents another way to take advantage of modelling at various length-scales either in a top-down approach or in a bottom-up one.

6. Discussion

Multi-scale modelling can synthesizes atomistic and continuum methods to provide a more natural and coherent description of material phenomenon than that obtained from the fragmented perspective for a limited individual scale. It has experienced some success with metals, but the usage for transition nitride multilayered coatings has been lacking. Especially, there is essentially a request for research related to multiscale modelling to deal with defect/boundary/cracks. The challenges on the future development of numerical modelling for multilayered coatings may be from the following aspects:

1. *Continuum models.* Over the past few decades, continuum models have dominated materials modelling research. This approach of predicting material deformations and failures by implicitly averaging atom-scale dynamics and defect evolutions over time and space, however, is valid only for large enough systems that include a substantial number of defects. As a result, numerous experimental observations of multilayered coatings cannot be readily explained within the continuum mechanics framework: dislocation patterns in fatigue and creep, surface roughening and crack nucleation in fatigue, inherent inhomogeneity of plastic deformation, the statistical nature of brittle failure, plastic flow localization in shear bands, and the effects of size, geometry and stress state on yield properties [126], etc. Moreover, the accuracy of the models needs to be validated for each case dealt with.
2. *Molecular dynamics models.* A MD model could only explore the phenomenon occurring at nanometric scale. Its applications to the analysis of multilayered coatings are limited by two factors: (a) Availability of the values of potentials for the atomic interactions to be examined. As well known, a key factor that determines the accuracy of a MD simulation significantly is the potential used to describe the interaction between atoms. A great effort has been made to develop various interatomic potentials fitted to different first principles, empirical and semi-empirical models, but only a few potentials for relatively simple material systems have been developed until now. Typically, multilayered coatings consist of at least two elements, and fewer potentials are available due to the increased complexity of interatomic action. On the other hand, simulations will often give acceptable results only if the potential was fitted for the problem under consideration, or one that is very similar. In other words, interfaces of multilayers play an important role in their properties. Therefore, more attention should be paid to choice and development of potentials with proper descriptions of interfacial properties before these multilayered systems could be simulated with MD accurately. (b) A typical MD simulation is still restricted to a very small system consisting of several million atoms or less and timescales in the order of picoseconds. A general acceleration methodology would have revolutionary implications that would relax current limitations on the use of MD.
3. *First principles calculations.* Many of the first principles calculations are confined to zero temperature, wherein the basic quantity is a Hamiltonian for the system expressed, in terms of the appropriate degrees of freedom. In principle, Hamiltonian methods can be extended to equilibrium at finite temperatures by using the free energy. Inherently non-equilibrium situations are, however, fundamentally different and a general approach is not yet within grasp. At the same time, the limitation to the computational scale, i.e. the maximum number of atoms which can be included in a numerical model for first principles, is limited. Generally, the maximum number of atoms used in a model does not exceed 100.

The concept of multi-scale modelling embodies a comprehensive description of a material, which requires understanding over both multiple time and length scales. Although the modelling for crystalline materials has coupled the nano- and intermediate-scale in a highly effective manner, the challenges remaining are the for longer time and length scales, largely due to the high requirement on computational effort associated with high-resolution calculations over such scales. In addition, multi-scale models also require that descriptions at all levels be consistent with each other. The requirement for consistency often demands the models for different process or pieces of a system to be coupled and solved concurrently. Consequently, multiscale modelling, on one hand, would still need more effective coupling methods as an enabler. On the other hand, it does not satisfactorily address issues such as disparate time scales in the two regions, and it provides only a rather simplified treatment of the interfaces between the atomistic and continuum regimes.

For the aforementioned unique properties of multilayered coatings: the thickness of the individual layers of a coating would be at the nano-scale while the entire thickness of the coating is at the macro- or meso-scale. Single-scale methods such as first-principles calculations, Molecular Dynamics and the Finite Element Method will have difficulty in analyzing such multilayered coatings alone, due to the limitations to the time and length scales that each method is confined to. With further development of accurate inter-atomic potentials for the material-components in multilayered coatings and further development of time-integration algorithms used in the simulation, the MD and Multiscale modelling (MD coupled) will become a prominent tool for elucidating complex physical phenomenon. Although multi-scale modelling is important for the design and analysis of multilayered coatings, effort for the development of the modelling at individual length-scales could not be neglected, that is to say, efficiency and effectiveness of multiscale modelling not only concern how the models for individual length-scales are coupled/linked but also depend on how individual modelling methods and techniques for a particular length-scale could advance.

7. Conclusions and remarks

Multilayered coatings are currently a subject of intensive research because of their potential for a wide range of technological applications and also due to their scientific importance. In this paper, development of the following numerical methods and models have been reviewed: (i) Continuum models; (ii) Molecular Dynamics simulations; and (iii) Nano-scale (first-principles calculations) modelling, for which examples of typical applications have been given.

In continuum models, FEM has been applied mainly for investigating the stress distribution, formation and propagation of crack under the (nano-) indentation or scratch test, formation of residual stresses, etc. Similarly, Molecular Dynamics (MD) simulations have also been applied mainly to the analysis of nano-indentations. Applications of first principles calculations are mainly presented on two kinds of mechanisms: "template effect" and "super-hardness". Multiscale modelling, such as that with QC methods, has also been discussed.

In summary, multi-scale modelling is unambiguously a particularly vibrant field of research related to engineering and sciences. Much of the excitement is being fuelled by increasingly sophisticated experiments that are capable of probing mater at the level of individual atoms or molecules. Interpretation of their experimental results require molecular and multiscale models to be connected with experimental observations. It is felt that more effort on the numerical modelling of multilayered coatings is needed. For example, when 400 nano-layers coated onto a substrate material needs to be modelled, there has to be a more effective method and more efficient way to achieve it. Developing accurate descriptions of inter-atomic potentials for multi-material elements, improving time-integration algorithms used in numerical simulations and developing more effective and efficient coupling methods for multiscale modelling are urgent tasks to be undertaken in order to develop a powerful numerical tool for high-efficiency performance prediction and process design for multi-layered coatings on engineering components, machining and forming tools.

Acknowledgements. The authors would like to acknowledge the European Commission for financial support to the research reported through the EU FP7 M3-2S project and acknowledge the permission for using the CASTEP code at the University of Strathclyde as a registered academic license of CASTEP under the STFC/York/Accelrys agreement (UK). The authors would also like to thank all partners of the EU FP7 M3-2S project for their collaboration and encouragement for conducting this review work.

References

1. P. Hovsepian, D. Lewis, W. Münz, Surf. Coat. Technol. 133–134 (2000) 166.
2. Z. Zhou, W. Rainforth, Q. Luo, P. Hovsepian, J. Ojeda, M. Romero-Gonzalez, Acta Materialia 58 (2010) 2912.
3. C. Ma, J. Huang, H. Chen, Thin Solid Films 446 (2004) 184.
4. V. Flores, M. Roldán, C. Real, A. Páez, G. Castro, J. Appl. Phys. 104 (2008) 023519.
5. P. Yashar, W. Sproul, Vacuum 55 (1999) 179.
6. C. Mendibide, J. Fontaine, P. Steyer, C. Esnouf, Tribology Lett. 17 (2004) 779.
7. K. Holmberg, A. Mattewst, H. Ronkainen, Tribol. Int. 31 (1998) 107.
8. J. Chen, R. Ji, R. Khan, X. Li, B. Beake, H. Dong, Surf. Coat. Technol. 206 (2011) 522.
9. S. Yang, X. Li, K. Cooke, D. Teer, Appl. Surf. Sci. 258 (2012) 2062.
10. C. Ma, J. Huang, H. Chen, Surf. Coat. Technol. 133–134 (2000) 289.
11. Q. Sun, Z. Fu, Electrochim. Acta 54 (2008) 403.
12. U. Helmersson, S. Todorova, S. Barnett, J. Sundgren, L. Markert, J. Greene, J. Appl. Phys. 62 (1987) 481.
13. X. Chu, M. Wong, W. Sproul, S. Rohde, S. Barnett, J. Vac. Sci. Technol. A 10 (1992) 1604.
14. X. Chu, S. Barnett, Surf. Coat. Technol. 57 (1993) 13.
15. W. Sproul, Vacuum 51 (1998) 641.

16. Z. Zhou, W. Rainforth, D. Lewis, S. Creasy, J. Forsyth, F. Clegg, A. Ehiasarian, P. Hovespian, W. Münz, Surf. Coat. Technol. 177–178 (2004) 198.
17. Q. Luo, Z. Zhou, W. Rainforth, P. Hovsepian, Tribology Lett. 24 (2006) 171.
18. Z. Zhou, C. Calvert, W. Rainforth, Q. Luo, L. Chen, P. Hovsepian, J. Phys.: Conf. Ser. 26 (2006) 95.
19. P. Hovsepian, Q. Luo, G. Robinson, M. Pittman, M. Howarth, D. Doerwald, R. Tietema, W. Sim, A. Deeming, T. Zeus, Surf. Coat. Technol. 201 (2006) 265.
20. Y. Chang, S. Yang, D. Wang, Thin Solid Films 515 (2007) 4722.
21. D. Lee, M. Kim, Y. Lee, S. Kwon, Surf. Coat. Technol. 141 (2001) 232.
22. D. Lee, M. Kim, S. Kwon, Met. Mater. Int. 7 (2001) 375.
23. Y. Otani, S. Hofmann, Thin Solid Films 287 (1996) 188.
24. S. Aouadi, K. Wong, K. Mitchell, F. Namavar, E. Tobin, D. Mihut, S. Rohde, Appl. Surf. Sci. 229 (2004) 387.
25. V.M. Vishnyakov, V. Bachurin, K. Minnebaev, R. Valizadeh, D. Teer, J. Colligon, V.V. Vishnyakov, V. Yurasova, Thin Solid Films 497 (2006) 189.
26. S. PalDey, S. Deevi, Mater. Sci. Eng. A 342 (2003) 58.
27. D. Lee, Surf. Coat. Technol. 173 (2003) 81.
28. C. Mendibide, P. Steyer, C. Esnouf, P. Goudeau, D. Thiaudière, M. Gailhanou, J. Fontaine, Surf. Coat. Technol. 200 (2005) 165.
29. C. Mendibide, P. Steyer, J. Fontaine, P. Goudeau, Surf. Coat. Technol. 201 (2006) 4119.
30. C. Mendibide, P. Steyer, J. Millet, Surf. Coat. Technol. 200 (2005) 109.
31. M. Ramzan, A. Rana, M. Hafeez, E. Ahmed, A. Bhatti, M. Wasiq, M. Nadeem, Acta Chim. Slov. 61 (2014) 80.
32. X. Wang, H. Masumoto, Y. Someno, T. Hirai, Appl. Phys. Lett. 72 (1998) 3264.
33. D. Hinczewski, M. Hinczewski, F. Tepehan, G. Tepehan, Solar Energy Mater. Solar Cells 87 (2005) 181.
34. J. Selj, T. Mongstad, R. Søndenå, E. Marstein, Solar Energy Mater. Solar Cells 95 (2011) 2576.
35. S. Chhajed, D. Poxson, X. Yan, J. Cho, E. Schubert, R. Welser, A. Sood, J. Kim, Appl. Phys. Expr. 4 (2011) 052503.
36. J. Zhang, Y. Xie, X. Cheng, H. Jiao, Z. Wang, Appl. Opt. 52 (2013) 5788.
37. J. Hiller, J. Mendelsohn, M. Rubner, Nat. Mater. 1 (2002) 59.
38. V. Medvedev, A. Boogaard, R. Meer, A. Yakshin, E. Louis, V. Krivtsun, F. Bijkerk, Optics Express 21 (2013) 16964.
39. H. Raut, V. Ganesh, A. Nair, S. Ramakrishna, Energy Environ. Sci. 4 (2011) 3779.
40. E. Louis, A. Yakshin, T. Tsarfati, F. Bijkerk, Prog. Surf. Sci. 86 (2011) 255.
41. G. Salazar-Alvarez, H. Lidbaum, A. López-Ortega, M. Estrader, K. Leifer, J. Sort, S. Suriñach, M. Baró, J. Nogués, J. Am. Chem. Soc. 133 (2011) 16738.
42. M. Yallapu, S. Othman, E. Curtis, B. Gupta, M. Jaggi, S. Chauhan, Biomaterials 32 (2011) 1890.
43. J. Li, H. Gao, Z. Chen, X. Wei, C. Yang, Anal. Chim. Acta 665 (2010) 98.
44. P. Pilla, V. Malachovská, A. Borriello, A. Buosciolo, M. Giordano, L. Ambrosio, A. Cutolo, A. Cusno, Optics Express 19 (2011) 515.
45. S. Nair, G. Pookat, V. Saravanan, M. Anantharaman, J. Appl. Phys. 114 (2013) 064309.
46. J. Chen, G. Bell, B. Beake, H. Dong, Tribology Lett. 43 (2011) 351.
47. L. Fan, F. Zhou, C. Wang, H. Gao, S. Zhang, H. Zhang, X. Wang, IEEE Transactions on Ultrasonics, Ferroelectrincs, and Frequency Control 58 (2011) 451.
48. S. Zhang, D. Sun, Y. Fu, H. Du, Surf. Coat. Technol. 198 (2005) 74.
49. M. Nordin, M. Herranen, S. Hogmark, Thin Solid Films 348 (1999) 202.
50. M. Nodin, M. Larsson, S. Hogmark, Wear 232 (1999) 221.
51. M. Nodin, M. Larsson, S. Hogmark, Surf. Coat. Technol. 106 (1998) 234.
52. V. Teixeira, Thin Solid Films 392 (2001) 276.
53. P. Steyer, A. Mege, D. Pech, C. Mendibide, J. Fontaine, J. Pierson, C. Esnouf, P. Goudeau, Surf. Coat. Technol. 202 (2008) 2268.
54. M. Stueber, H. Holleck, H. Leiste, K. Seemann, S. Ulrich, C. Ziebert, J. Alloys Compd. 483 (2009) 321.
55. C. Lu, Y. Mai, Y. Shen, J. Mater. Sci. 41 (2006) 937.
56. S. Zhang, D. Sun, Y. Fu, H. Du, Surf. Coat. Technol. 198 (2005) 2.
57. K. Holmberg, H. Ronkainen, A. Laukkanen, K. Wallin, Surf. Coat. Technol. 202 (2007) 1034.
58. Z. Liu, C. Zhang, Y. Shen, Y. Mai, J. Appl. Phys. 95 (2004) 758.
59. J. Mackerle, Model. Simul. Mater. Sci. Eng. 13 (2005) 935.
60. B. Bhushan, W. Peng, Appl. Mech. Rev. 55 (2002) 435.
61. J. Luo, Y. Liu, E. Berger, Computational Mechanics 24 (2000) 448.
62. H. Djabella, R. Arnell, Thin Solid Films 213 (1992) 205.
63. L. Stephens, Y. Liu, E. Meletis, J. Tribol. 122 (2000) 381.
64. K. Bouzakis, N. Michailidis, G. Erkens, Surf. Coat. Technol. 142–144 (2001) 102.
65. X. Tan, Y. Shen, Compos. Sci. Technol. 65 (2005) 1639.
66. X. Zhao, Z. Xie, P. Munroe, Mater. Sci. Eng. A 528 (2011) 1111.
67. H. Pelletier, J. Krier, A. Cornet, P. Mille, Thin Solid Films 379 (2000) 147.
68. T. Aoki, H. Hatta, T. Hitomi, H. Fukuda, I. Shiota, Carbon 9 (2001) 1477.
69. J. Chen, S. Bull, Thin Solid Films 517 (2009) 3704.
70. K. Holmberg, A. Laukkanen, H. Ronkainen, K. Wallin, Tribol. Int. 38 (2005) 1035.
71. T. Özel, T. Thepsonthi, D. Ulutan, B. Kaftanoglu, CIRP Ann. – Manuf. Technol. 60 (2011) 85.
72. T. Özel, M. Sima, A.K. Srivastava, B. Kaftanoglu, CIRP Ann. – Manuf. Technol. 59 (2010) 77.
73. W. Grzesik, M. Bartoszuk, P. Nieslony, J. Mater. Process. Technol. 164–165 (2005) 1204.
74. R. Coelho, E. Ng, M. Elbestawi, Int. J. Mach. Tools Manuf. 47 (2007) 263.
75. K. Holmberg, H. Ronkainen, A. Laukkanen, K. Wallin, S. Hogmark, S. Jacobson, U. Wiklund, R. Souza, P. Ståhlc, Wear 267 (2009) 2142.
76. H. Ng, Z. Gan, Finite Elem. Anal. Des. 41 (2005) 1235.
77. O. Sarikaya, E. Celik, Mater. Design 23 (2002) 645.
78. V. Teixeria, Vacuum 64 (2002) 393.
79. C. Hsueh, S. Lee, Composites: Part B 34 (2003) 747.
80. S. Suresh, Progr. Mater. Sci. 42 (1997) 243.

81. H. Liu, J. Tao, Y. Gautreau, P.Z. Zhang, J. Xu, Mater. Design 30 (2009) 2785.
82. O. Sayman, F. Sen, E. Celik, Y. Arman, Mater. Design 30 (2009) 770.
83. M. Toparli, F. Sen, O. Culha, E. Celik, J. Mater. Process. Technol. 190 (2007) 26.
84. M. Far, J. Absi, G. Mariaux, F. Dubois, Mater. Design 31 (2010) 772.
85. X. Pan, L. Yan, J. Xu, Thin Solid Films 354 (1999) 154.
86. E. Bemporad, M. Sebastiani, F. Casadei, F. Carassiti, Surf. Coat. Technol. 201 (2007) 7652.
87. R. Lakkaraju, F. Bobaru, S. Rohde, J. Vac. Sci. Technol. A 24 (2006) 146.
88. T. Gorishnyy, L. Olson, M. Oden, S. Aouadi, S. Rohde, J. Vac. Sci. Technol. A 21 (2003) 332.
89. J. Feng, Y. Qin, Q. Zeng, E. Almandoz, G. Fuente, H. Dong, R. Koodakal, J. Michler, J. Multiscale Modelling 03 (2011) 1.
90. X. Xu, A. Needleman, Model. Simul. Mater. Sci. Eng. 1 (1993) 111.
91. X. Xu, A. Needleman, J. Mech. Phys. Solids 42 (1994) 1397.
92. P. Geubelle, J. Baylor, Composites: Part B 29B (1998) 589.
93. N. Chandra, H. Li, C. Shet, H. Ghonem, Int. J. Solids Struct. 39 (2002) 2827.
94. S. Nekkanty, M. Walter, R. Shivpuri, J. Mechanics Mater. Struct. 2 (2007) 1231.
95. M. Payne, M. Teter, D. Allan, T. Arias, J. Joannopoulos, Rev. Mod. Phys. 64 (1992) 1045.
96. J. Leopold, K. Heller, A. Meyer, R. Wohlgemuth, J. Multiscale Modelling 3 (2011) 91.
97. Y. Chen, W. Zhuang, S. Wang, J. Lin, D. Balint, D. Shan, Chinese J. Mech. Eng. 25 (2012) 860.
98. M. Perucca, S. Durante, U. Semmler, C. Rüger, G. Fuentes, E. Almandoz, IOP Conf. Series, Mater. Sci. Eng. 40 (2012) 012004.
99. M. Karimpour, D. Balint, K. Rzepiejewska-Malyska, A. Szerling, J. Michler, J. Lin, Comput. Mater. Sci. 68 (2013) 384.
100. J. Haile, Molecular dynamics simulation elementary methods, 1st edn., John Wiley & Sons, Inc., New York, NY, USA, 1992.
101. M. Allen, D. Tildesley, Computer simulation of liquids, 1st edn., Oxford University Press, New York, 1987
102. D. Rapaport. The art of Molecular Dynamics Simulation, in: Dennis Rapaport (Ed.), 2nd edn., Cambridge University Press, New York, 2004
103. D. Saraev, R. Miller, Acta Mater. 54 (2006) 33.
104. Y. Cao, J. Zhang, Y. Liang, F. Yu, T. Sun, Appl. Surf. Sci. 257 (2010) 847.
105. T. Fang, J. Wu, Comput. Mater. Sci. 43 (2008) 785.
106. L. Yuan, Z. Xu, D. Shan, B. Guo, Appl. Surf. Sci. 258 (2012) 6111.
107. L. Yuan, Z. Xu, D. Shan, B. Guo, Appl. Surf. Sci. 282 (2013) 450.
108. Z. Xu, L. Yuan, D. Shan, B. Guo, H. Dong, X. Li, J. Multiscale Modelling 3 (2011) 23.
109. D. Yin, X. Peng, Y. Qin, Z. Wang, J. Appl. Phys. 108 (2010) 033714.
110. X. Li, W. Wu, H. Dong, Surf. Coat. Technol. 205 (2011) 3251.
111. Z. Wang, S. Tsukimoto, M. Saito, Y. Ikuhara, Phys. Rev. B 79 (2009) 045318.
112. M. Finnis, J. Phys.: Condens. Matter 8 (1996) 5811.
113. L. Liu, S. Wang, H. Ye, Surf. Inter. Anal. 35 (2003) 835.
114. V. Chawla, D. Holec, P. Mayrhofer, J. Phys. D: Appl. Phys. 46 (2013) 045305.
115. R. Zhang, S. Veprek, Mater. Sci. Eng. A 448 (2007) 111.
116. R. Zhang, S. Veprek, Acta Mater. 55 (2007) 4615.
117. S. Sheng, R. Zhang, S. Veprek, Acta Mater. 56 (2008) 968.
118. D. Holec, R. Rachbauer, L. Chen, L. Wang, D. Luef, P. Mayrhofer, Surf. Coat. Technol. 206 (2011) 1698.
119. C. Stampfl, A. Freeman, Appl. Surf. Sci. 258 (2012) 5638.
120. R. Zhang, S. Sheng, S. Veprek, Appl. Phys. Lett. 91 (2007) 031906.
121. X. Sun, S. Chen, K. Cheng, D. Huo, W. Chu, J. Eng. Manuf. 220 (2006) 1217.
122. X. Sun, K. Cheng, Int. J. Adv. Manuf. Technol. 47 (2010) 891.
123. X. Huanag, A. Pelegri, Compos. Sci. Technol. 67 (2007) 1311.
124. D. Yin, X. Peng, Y. Qin, Z. Wang, J. Multiscale Modelling 3 (2001) 65.
125. X. Peng, J. Fan, J. Zeng, Int. J. Solids Struct. 39 (2002) 435.
126. W. Liu, E. Karpov, S. Zhang, H. Park, Comput. Methods Appl. Mech. Eng. 193 (2004) 1529.

4

Micro-texturing into DLC/diamond coated molds and dies via high density oxygen plasma etching

Ersyzario Edo Yunata[1] and Tatsuhiko Aizawa[2,*]

[1] Department of Regional Environment, Shibaura Institute of Technology, 3-9-14 Shibaura, Minato-City, Tokyo 108-8548, Japan
[2] Department of Engineering and Design, Shibaura Institute of Technology, 3-9-14 Shibaura, Minato-City, Tokyo 108-8548, Japan

Abstract – Diamond-Like Carbon (DLC) and Chemical Vapor Deposition (CVD)-diamond films have been widely utilized not only as a hard protective coating for molds and dies but also as a functional substrate for bio-MEMS/NEMS. Micro-texturing into these hard coated molds and dies provides a productive tool to duplicate the original mother micro-patterns onto various work materials and to construct any tailored micro-textures for sensors and actuators. In the present paper, the high density oxygen plasma etching method is utilized to make micro-line and micro-groove patterns onto the DLC and diamond coatings. Our developing oxygen plasma etching system is introduced together with characterization on the plasma state during etching. In this quantitative plasma diagnosis, both the population of activated species and the electron and ion densities are identified through the emissive light spectroscopy and the Langmuir probe method. In addition, the on-line monitoring of the plasmas helps to describe the etching process. DLC coated WC (Co) specimen is first employed to describe the etching mechanism by the present method. Chemical Vapor Deposition (CVD) diamond coated WC (Co) is also employed to demonstrate the reliable capacity of the present high density oxygen plasma etching. This oxygen plasma etching performance is discussed by comparison of the etching rates.

Key words: Micro-texturing, Diamond like carbon, CVD-diamond, Oxygen plasma, Etching, Mold-dies

1. Introduction

Diamond-Like Carbon (DLC) and Chemical Vapor Deposition (CVD)-diamond films have been widely utilized as a protective coating of mechanical, functional and fashion-oriented parts besides tools and dies [1]. For an example, the boron-doped diamond or nano-crystalline diamond coatings are favored for application to electrically conductive substrates and bio-actuators [2, 3]. In addition, micro-textures or micro-patterns on the metallic or polymer products and tools, work to reduce the friction and wear of their surfaces under the presence of minimum quantity lubrication (MQL) [4, 5]. Hence, the micro-textured DLC- and diamond-coatings are expected not only to further reduce the friction and wear under MQL but also to work as a mold-die to duplicate their micro-textures onto the metallic and polymer sheets even without lubricating oils [6–8].

In order to put the above manufacturing policy into practice, reliable machining methods have to be cultivated as a productive means to make micro-texturing onto the DLC- and diamond-coatings. Micro-EDM (electrical discharge machining) is effective to make micro-patterns onto the electrically conductive DLC- and diamond-films [9]. Its geometric accuracy in cutting the films is determined by the diameter of wires in EDM; when the size and dimension of micro-pattern unit becomes less than 100 μm, EDM is difficult to be applied in practice. In addition, large amount of leading time is needed to make machining a thousand or a million of micro-textures onto these films.

Chemical etching by $CF_4 + O_2$ mixture gas or lithography is also available to make fine micro-textures with aid of masking [10]. Its slow etching rate by 1.8 μm/h is unsuitable to practical micro-texturing with high aspect ratio. Since CF_4 is a typical hazard gas to be post-treated in the closed system to avoid from direct emission in air. The authors have been developing the high density oxygen plasma etching and ashing method for DLC- and diamond-coated tools and dies [11–15]. Since only a pure oxygen gas is used in practice, every etching process can be done in open to air. To be discussed in later, its fast etching rate is also suitable to industrial application. Furthermore, the oxygen plasma state is controllable by using the devices in suitable to each application.

In the present paper, this plasma etching process is applied to make fast-rate micro-texturing of the DLC- and

*email: taizawa@sic.shibaura-it.ac.jp

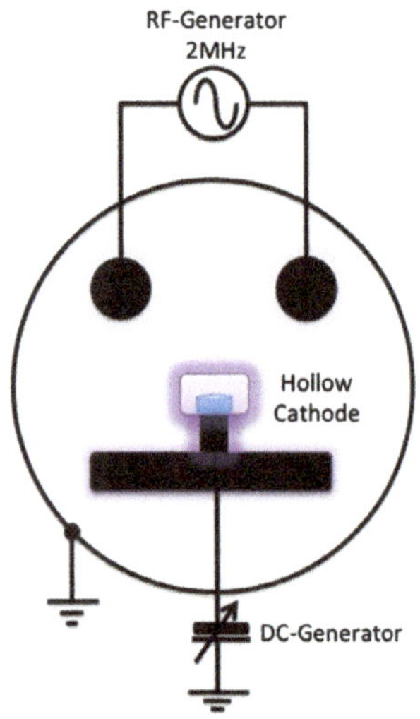

Figure 1. High density oxygen plasma etching system with a hollow cathode.

diamond-coatings. First, a high density oxygen plasma generation is quantitatively discussed by using the plasma diagnosis instrument. Both the emissive-light optical spectroscopy and the Langmuir probe method are utilized for this diagnosis. As a device to control the oxygen ion and electron densities in plasmas, the hollow cathode device is used to investigate the effect of oxygen ion density on the etching behavior.

Two kinds of DLC-coated silicon specimens are first employed to describe the fast-rate and homogeneous etching behavior: e.g. DLC coating with hydrogen content and hydrogen-free DLC. The effect of the hydrogen content in DLC on the etching rate is investigated to describe the etching performance of carbon based coatings. CVD-diamond coated WC (Co) specimens are also used to demonstrate how fine and fast the present oxygen plasma etching makes micro-textures onto the diamond coating.

2. Experimental procedure

2.1. Oxygen plasma etching system

The high density oxygen plasma etching system consists of the vacuum chamber, the plasma generator, the control unit, the carrier gas supply, and the plasma diagnosis instrument. The chamber is neutral in electricity; RF dipole electrodes and DC-bias work independently to generate RF and DC plasmas, respectively. Hence, the ionized species and activated radicals in the RF plasma are attracted to this DC biased plate with kinetic energy, as illustrated in Figure 1. Either RF-plasma or DC plasma or, both are ignited by switching on either or both on the control panel. In addition, there is no mechanical matching box for RF plasma generation in this system. Both the input and output powers are automatically matched by frequency adjustment around 2 MHz. A standard experimental set up is summarized as follows. The base pressure is less than 5×10^{-3} Pa, and pure oxygen gas (purity; 99.99%) is only used as a carrier gas. RF voltage, DC bias, and pressure are varied in a range from 100 V to 250 V, from −400 V to −600 V, and from 25 Pa to 100 Pa, respectively.

A hollow-cathode device is also employed in this system to control the external plasma condition to etching. As shown in Figure 1, the generated RF-plasma is confined in this hollow so that higher ion and electron densities are attained for etching the DLC- and diamond films. In the following plasma diagnosis, both the emissive-light optical spectroscopic (EOS) analyzer and the Langmuir probe are utilized to quantitatively describe the plasma state for etching. Figure 2a depicts the EOS system to detect the population of activated species in the plasma. With comparison to the reference data, each activated species is identified by using the peak position and intensity, respectively. On the other hand, the electron and oxygen ion densities as well as the electron temperature are measured by using the Langmuir probe in Figure 2b. The generated plasma state is characterized by the variation of electron and ion densities with the pressure in plasmas. In particular, the measured densities by the Langmuir probe method are compared between the oxygen plasmas with and without the hollow cathode.

2.2. Specimens

Two types of DLC-coated silicon specimens were respectively prepared by RF-sputtering and un-balanced magnetron sputtering to have 5 μm in thickness: the hydrogen-terminated DLC film and the hydrogen free one. Both are symbolized by a-C:H and a-C, respectively in the following. The diamond-coated WC (Co) specimens are also prepared by CVD processing to have 20 μm thick diamond coating.

In the following micro-texturing onto DLC-coatings, two-types of resin-type masking patterns were prepared: the micro-lattice pattern with the line-space of 3 μm and the line-and space masking pattern, as shown in Figure 3. The former was utilized to describe the difference in micro-texturing process between a-C:H and a-C films with and without use of the hollow cathode. The latter is used to describe the fine micro-texturing into a-C films.

Next, a stainless sheet mask with the thickness of 50 μm is first used for masking on the diamond-coated specimens to make micro-line patterning with the width of 100 μm.

2.3. Observation and measurement

Optical microscope and scanning electron microscope (SEM) were used to observe the etched surface. The surface profilometer (Keyence, Co. Ltd.) and the laser-microscope (Nicon, Co. Ltd.) were employed to measure the surface profile of etched micro-texture. Raman spectroscopy (Renishaw, Co. Ltd.) was also utilized to prove the perfect removal of DLC and diamond coatings from mold substrates.

3. Experimental results

3.1. Plasma diagnosis

Both the emissive light spectroscopy and the Langmuir probe method were utilized to analyze the generated oxygen

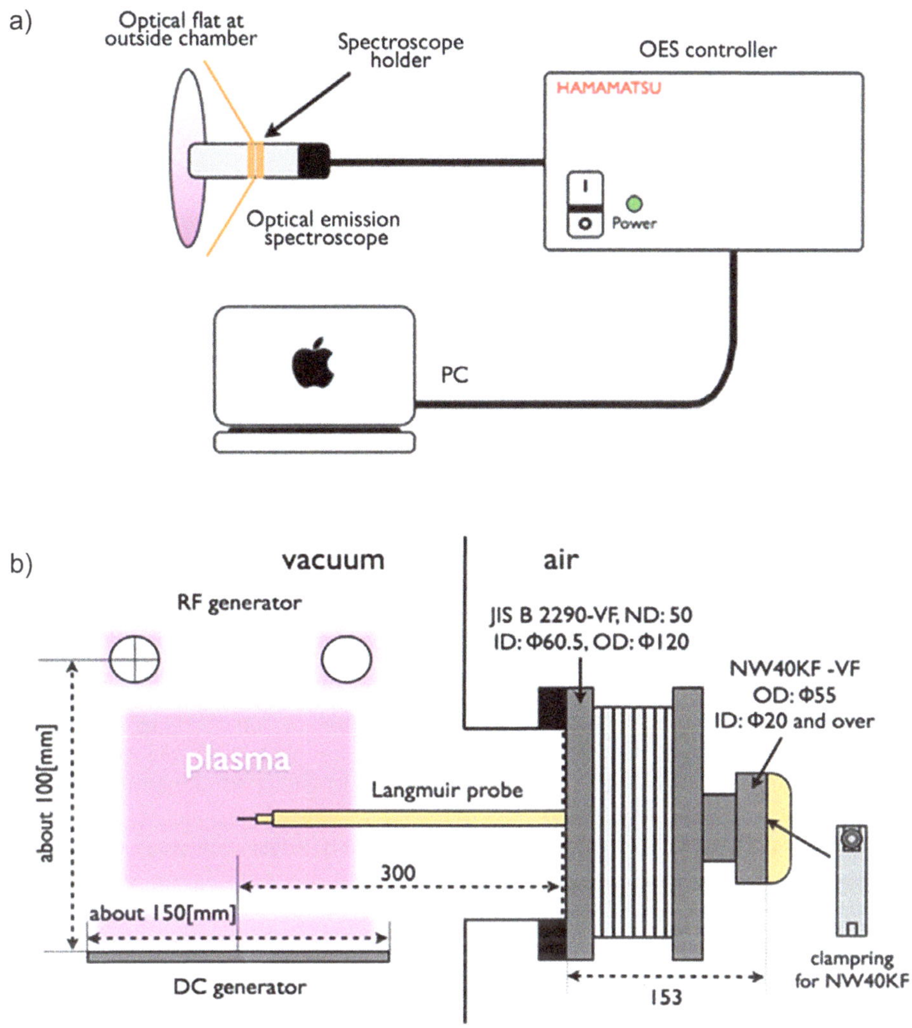

Figure 2. Quantitative plasma diagnosis instrumentation equipped to the high density plasma etching system. (a) Langmuir probe method and (b) emissive-light optical spectroscopy.

plasma state with and without the hollow cathode. Figure 4 compares the emissive light spectra from plasmas with and without the hollow cathode. The oxygen ions $\{O^+, O^{2+}, O^{3+}\}$ or {OII, OIII, OIV} and activated oxygen molecules O_2^+ are seen in the spectrum without the hollow cathode, besides the activated oxygen atom O* or OI. When using the hollow cathode, two strong peaks are seen in the spectrum: OI peaks are mainly observed at $\lambda = 776.34$ nm and 843.78 nm, respectively. This reveals that much amount of oxygen atoms are available for etching by using the hollow cathode oxygen plasma.

These two peaks in Figure 4 correspond to the oxygen atom transition O ($3p^5P \rightarrow 3s^5S$) and O ($3p^3P \rightarrow 3s^3S$), respectively. The peak intensity of activated oxygen atoms at 776.34 nm without the hollow cathode is only 2.5×10^3 counts; while it reaches to around 1.5×10^4 counts with use of the hollow cathode. The high intensity indicates the activated oxygen atom is a main species in the generated oxygen plasma and that its population is enhanced by using the hollow cathode. That is, the oxygen atom flux to the specimen on the cathode becomes lower without the hollow cathode, since the ionization process takes place in all area of the chamber. On the other hand, the ionization advances only in the hollow tube; this confinement of plasmas results in higher oxygen atom flux to the specimen. Next, this difference of plasma state is described by using the Langmuir probe.

Figure 5 compares the variation of electron and ion densities or Ne and Ni with the pressure, p, in the oxygen plasmas with and without the hollow cathode. In cases without the hollow cathode, Ne increases and Ni decreases significantly with increasing the pressure. For an example, Ne = 1.35×10^{15} m^{-3} and Ni = 9.60×10^{16} m^{-3} at $p = 65$ Pa while Ne = 3.30×10^{16} m^{-3} and Ni = 3.00×10^{16} m^{-3} at $p = 105$ Pa. This pressure dependency of Ni and Ne is common to the high density plasmas, irrespective of the cathode configuration. On the other hand, using the hollow cathode, both Ni and Ne become higher than those without the hollow cathode. For an example, Ne = 1.07×10^{16} m^{-3} and Ni = 6.99×10^{17} m^{-3} at $p = 65$ Pa, and, Ne = 7.51×10^{16} m^{-3} and Ni = 1.41×10^{17} m^{-3} at $p = 105$ Pa.

The hollow cathode effect on the increase of electron and ion densities is caused by confinement of plasmas in the hollow

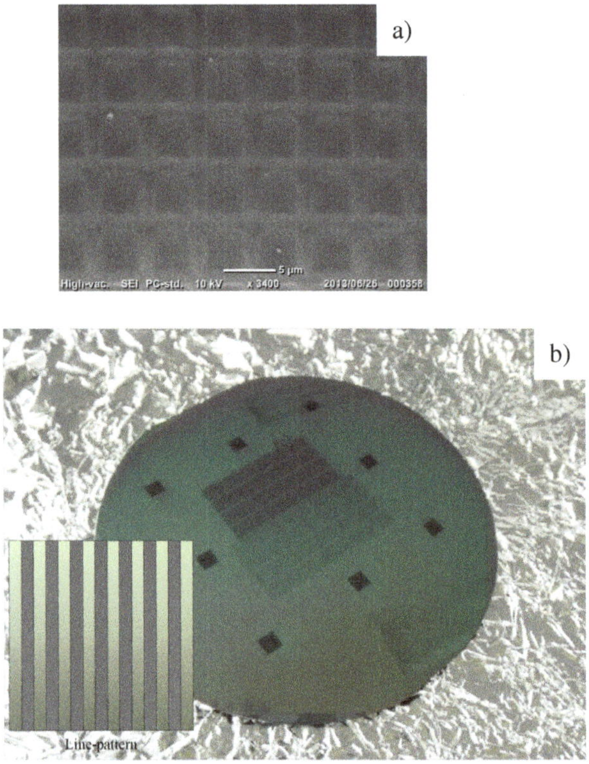

Line-pattern

Figure 3. Two types of resin-type masking patterns for etching the DLC-coated silicon substrates. (a) SEM image of micro lattice patterns with the width of 3 μm and (b) micro-line and space patterns for various line widths.

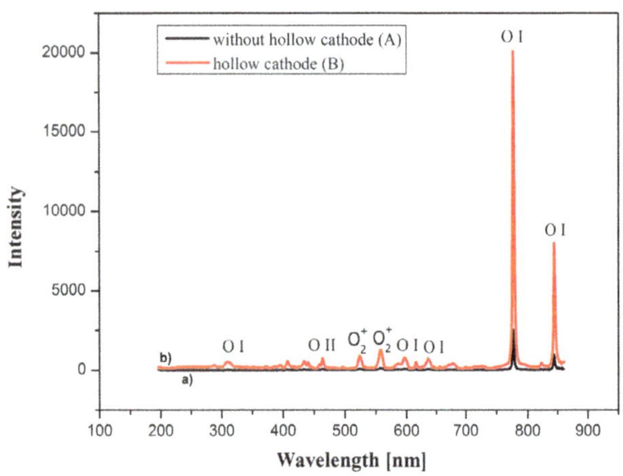

Figure 4. Comparison of the emissive light spectra from plasmas without and with the hollow cathode.

cathode. Electrons in the plasmas have more reaction cross-section with the oxygen molecules to generate more population of oxygen ions. This results in densification both in the electron and oxygen ion state. In the following experiments, this effect of oxygen ion density in plasmas on the etching behavior is experimentally investigated to demonstrate that densification in plasmas is needed for fast-rate etching of DLC- and CVD-diamond coatings.

3.2. Micro-texturing into DLC-coatings

Two DLC-coated silicon specimens were employed to describe the difference in etching behavior at the same processing conditions: a-C:H film with the hydrogen content of 10–15 at%, and a-C, hydrogen free DLC-film. The resin-masking was commonly used to make lattice pattern. The film thickness is constant, 5 μm. Figure 6 compares the etched a-C:H films after etching with and without use of the hollow cathodes after 500 s.

The average etching rate becomes 1.9 μm/h without the hollow cathode and 2.1 μm/h with the hollow cathode, respectively. Since the resin mask has insufficient toughness against the oxygen atom flux, the corners of square patterns are severely over-etched to be round. Next, a-C film is also etched by the hollow cathode method under the same condition as in the above. The average etching rate reaches to 9.2 μm/h.

Finer etched surfaces are attained when hollow-cathode etching of a-C films. Then, the micro-line masking pattern with the width of 7 μm was used to investigate the etching profile in depth of this a-C films. Figure 7 depicts the SEM image of micro-grooves, etched into the depth of a-C film for 3 ks, with the skewed angle of 30°. Since the un-masked a-C film is completely etched away, fine rectangular micro-grooves are formed to have sharp edges with some over-etching. Both the etched bottom and top surfaces are shaped flat with the depth of 4.8 μm, which corresponds to the a-C film thickness. This proves that the present oxygen plasma etching advances anisotropically into the depth of a-C films.

3.3. Micro-texturing into CVD-diamond coatings

CVD-diamond coated WC (Co) specimen was used for oxygen plasma etching with use of the hollow cathode. In this case, the stainless steel sheet with the micro-line patterns was employed as a mask; e.g. the line width was 100 μm and the mask width, 20 μm. The CVD-diamond film is 20 μm thick. The experimental conditions are listed in the following: RF (250 V), DC(−550 V) and P (30 Pa).

The etched diamond-film after 7.2 ks is shown in Figure 8. The unmasked CVD-diamond film is selectively etched away. Different from the etching process of DLC films with use of the resin masking, no over-etching is seen but a sharp edge is formed at the border between the masked and unmasked regions of diamond film. The surface profile in Figure 8b reveals that the gradient of sharp edge by the present etching could be estimated by $\tan(\theta) = 20$ μm/4 μm; $\theta = 80°$. Considering that the metal-mask thickness of 50 μm and the clearance between the mask and the diamond film, this edge sharpness is favored for practical micro-patterning into the diamond coatings.

In this rectangular trench with 100 × 20 μm^2 in Figure 8, its bottom surface turns to be bare WC (Co) substrate surface since almost of all the diamond films are removed away to form this trench or micro-groove. In fact, the measured

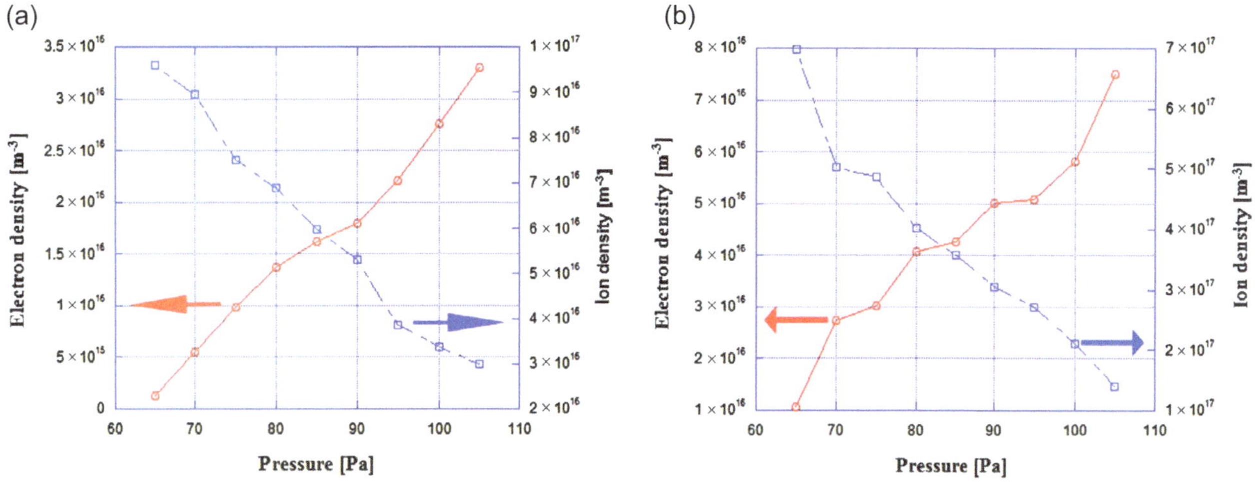

Figure 5. Comparison on the variation of electron and ion densities with the pressure in the plasmas without and with the hollow cathode. (a) without the hollow cathode and (b) with the hollow cathode.

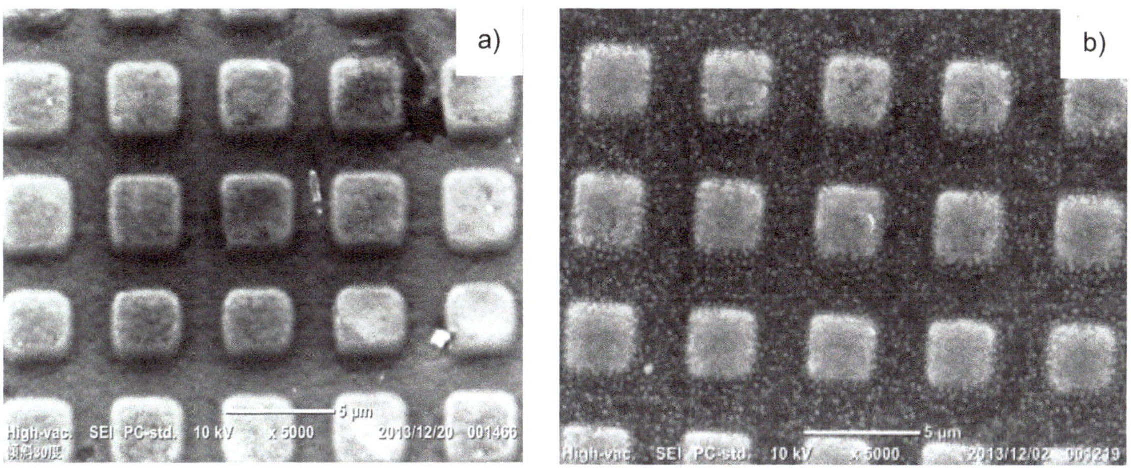

Figure 6. Comparison of SEM images foe etched a-C:H films with and without the hollow cathode. (a) Without the hollow cathode and (b) With the hollow cathode.

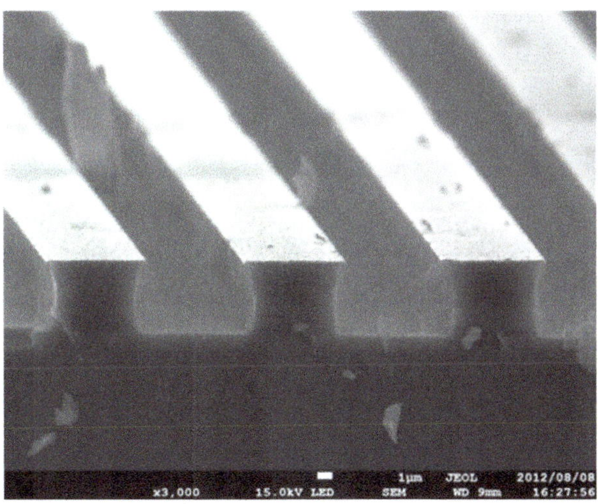

Figure 7. SEM image of etched micro-grooves into the a-C coated silicon substrate with use of micro-line pattern with the line width of 7 μm.

hardness on this bottom surface is nearly constant by 2000 Hv, which is equivalent to the original hardness of WC (Co). On the other hand, the masked top surfaces are still diamond coating. This implies that diamond lines are formed by complete ashing of un-masked regions.

4. Discussion

Different from the plasma enhanced chemical etching by $CF_4 + O_2$, the etching process only by O_2 is driven by chemical reaction and physical bombardment. In the former, direct oxidation of carbon based films by C (in DLC and diamond films) + O (in plasmas) $\rightarrow$ CO (carbon mono-oxide) is responsible for etching. Hence, if this direct oxidation process were retarded by another chemical reaction, the etching rate could be lowered. In the latter, the physical bombardment effect in etching is enhanced by increasing the DC-bias to accelerate the flux of oxygen ions onto the specimen surface. In particular, the confined plasmas in the hollow-cathode

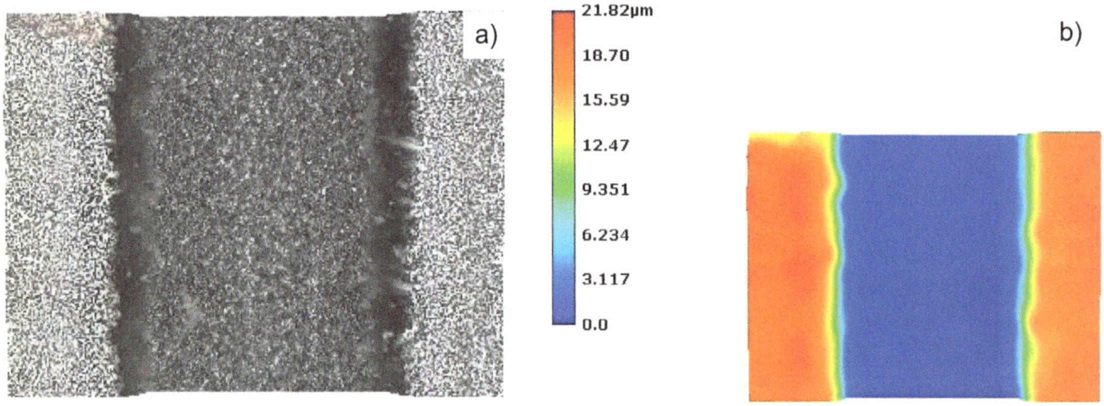

Figure 8. Microgroove formation into the CVD diamond coating by the hollow cathode oxygen plasma etching for 7.2 ks. (a) Optical microscopic image of etched microgroove and (b) surface profile across the microgroove.

device has a capacity to drive the above two etching processes; especially, this device is preferable for anisotropic etching of DLC and diamond coatings. In fact, the etched micro-lattice and micro-groove patterns has the same size and dimension as the original mask patterns. Furthermore, the depth profile of etched micro-patterns is uniform; e.g. the rectangular cross-section of micro-groove is formed by the present etching process.

The hydrogen free DLC with the type of a-C and the CVD-diamond films are etched into micro-textures in fast-rate as predicted by the previous studies in reference [12]. In fact, the etching rates were 9.2 μm/h and 10 μm/h for a-C and diamond coatings, respectively. This high etching performance is driven by the direct oxidation between the sputtered carbon from a-C and diamond and the activated oxygen flux. In application of high DC-bias, the sputtered carbon by the oxygen ions reacts with oxygen atoms at the etching front end; the resultant carbon monoxide is ejected from the formed micro-textures. This efficient route of reactions results in higher etching rate for a-C and diamond coatings.

However, in etching the hydrogenated amorphous carbon films, the hollow cathode discharging approach has little influence on the acceleration of etching rate. Using the same resin masks in etching of a-C and a-C:H films with the same thickness, the etching rates were 2.1 μm/h for a-C:H and 9.2 μm/h for a-C. Although the etching rate against a-C:H films is retarded by the resin masking procedure, a relatively high content of hydrogen in a-C:H has possibility to prevent the oxygen atom flux from direct reaction with carbon in a-C:H just like the metal dopes (M) in a-C:H:M.

The hydrogen free a-C type DLC's and CVD-diamond films can be etched in fast-rate as predicted by the previous studies [12]. In particular, higher etching rate of diamond coatings than the conventional etching methods with use of CF_4 is preferable for practical applications. As had reported in reference [15], no residuals of diamond films are analyzed to be present by the Raman spectroscopy. Sharp edge formation as well as no residual formation of diamonds by the present etching must be key-technological items in fabrication of diamond bio-MEMS and NEMS.

5. Conclusion

The hollow cathode oxygen plasma etching method is proposed as an effective means for etching of DLC- and CVD-diamond coatings. Although the etching process might be retarded by the presence of high content hydrogen in DLC films, fast-rate etching is attained by this method both in the hydrogen-free DLC's and CVD-diamond coating. In particular, micro-textures are machined into the CVD-diamond coatings with sufficiently sharp edges; i.e. fine rectangular micro-groove is cut into the diamond films with the depth of film thickness. In this approach, the resolution of micro-textures is mainly determined by the geometric accuracy in masking. Various kinds of fine masking are in trial to aim at higher and more precise resolution in this micro-texturing.

Implications and influences

The present paper has contribution to provide a means for micro-texturing into the CVD diamond coatings via the high density oxygen plasma etching. This method is useful not only for fabrication of dies and molds to duplicate the original micro-textures onto the metallic and polymer sheets but also for construction of sensors and actuators in bio-MEMS and bio-NEMS.

Acknowledgements. The authors would like to express their gratitude to Ms. R. Tsuda (SIT), Mrs. S. Ueda (Fujikoshi, Co. Ltd.), and, Mrs. M. Misawa and Ms. K. Ando (Renishow, Co. Ltd.) for their help in the etching experiments, in CVD-coating and in Raman spectroscopy, respectively. This study is financially supported in part by the MEXT-project with the contract of # 411419.

References

1. A. Erdemir, C. Donnet, Journal of Physics D Applied Physics 39 (2006) 311–327.
2. E. Kohn, P. Gluche, M. Adamschik, Diamond and Related Materials 8 (1999) 934–940.

3. M.W. Varney, D.M. Aslam, A. Janoudi, H.-Y. Chan, D.H. Wang, Biosensors 1 (2011) 118–133.
4. N. Kawasegi, H. Sugimori, H. Morimoto, N. Morita, I. Hori, Precise Engineering 33 (2009) 248–254.
5. T. Enomoto, T. Sugihara, Manufactruring Technology 59 (2010) 597–600.
6. T. Aizawa, Applied Mechanics and Materials 289 (2013) 23–37.
7. T. Aizawa, T. Fukuda, Research Report of Shibaura Institute of Technology 57 (2013) 1–10.
8. T. Aizawa, T. Fukuda, Surface and Coating Technology 215 (2013) 364–368.
9. P. Agrawal, A. Manoria, R. Thakur, H. Jain, International Journal of Mechanical Engineering and Robotics Research 1 (2012) 121–128.
10. Z. Cao, D.M. Aslam, Proceedings of IEEE on nano-electro mechanical system, Xiamen, China, 2010.
11. T. Aizawa, Proceedings of the 5th Southern East Asia Technology University Consortium Conference, Hanoi, Vietnam, 2011, CD-ROM.
12. T. Aizawa, K. Mizushima, T.N. Redationo, M. Yang, Proceedings of the 8th International Conference on Mciro-Manufacturing 8 (2013) 459–466.
13. T. Aizawa, Y. Sugita, Research Report of Shibaura Institute of Technology 55 (2011) 13–22.
14. T. Aizawa, T. Fukuda, Research Report of Shibaura Institute of Technology 56 (2012) 47–56.
15. E.E. Yunata, T. Aizawa, Manufacturing Letters (2015) (to be published).

5

Hot semi-punching and cold scrap removing processes for hot stamping of ultra-high strength steel parts

Ken-ichiro Mori*, Tomoyoshi Maeno, and Takuya Suganami

Department of Mechanical Engineering, Toyohashi University of Technology, Toyohashi, Aichi 441-8580, Japan

Abstract – Hot semi-punching and cold scrap removing processes for hot stamping of ultra-high strength steel parts were developed to eliminate laser cutting having low productivity after hot stamping. A quenchable steel sheet is semi-punched without separation of punching scraps during hot stamping, and subsequently, the scraps are removed from the hot-stamped part at room temperature. For hot semi-punching, additional channels for taking the punching scraps out of dies are not required. Minimum remainder without separation of punching scraps and no clearance between the die and punch were optimal for the hot semi-punching process. The hot punching and cold removing loads of the quenched sheet were considerably smaller than the cold punching load, the quality of the hole edge was high and the delayed fracture around the sheared edge was prevented. Hot semi-punching and cold scrap removal were applied to a hot hat-shaped bending process.

Key words: Hot stamping, Ultra-high strength steel part, Hot semi-punching, Cold scrap removing

1. Introduction

To improve the passenger protection for collision of automobiles, the use of high strength steel parts in the body-in-white increases, and hot stamping of quenchable steel sheets is appropriate for producing these parts. Eriksson et al. [1] measured material data in hot stamping of boron steel sheets and simulated a hot bending process of a tubular beam. Hein [2] explained the specific requirements for simulation of hot stamping processes. Geiger et al. [3] measured basic characteristics in hot stamping of the boron steel sheet. Mori et al. [4] prevented the springback in bending by heating ultra-high strength steel sheets. Bariani et al. [5] examined the improvement of formability by hot stamping. Merklein et al. [6] observed deformation behaviour of a quenchable steel sheet in hot stamping.

Although stamped automobile body panels are punched to make many holes for joining, paint removing, attachment, etc., it is not easy to punch die-quenched parts having high strength. The tool life is remarkably reduced by large punching load, and worn tools bring about the deteriorations in dimensional accuracy of the punched hole and in quality of the sheared edge, as described by So et al. [7]. Valls et al. [8] evaluated wear resistance in cold shearing for various tool materials and coatings, and Nothhaft et al. [9] examined the influence of the punch chamfer. In addition, punching of the die-quenched parts has a risk of delayed fracture induced by tensile residual stress and rough fracture surface of the sheared edge. Although laser cutting is generally employed for the die-quenched parts, the productivity is low and the production cost is high, particularly for many small holes in automobile body panels. Fritz [10] summarised recent improvements of productivity and production cost for laser cutting of the die-quenched parts. The laser cutting length of the hot-stamped parts is minimised by predicting the size and shape of the blank from those of the product in order to reduce the production cost.

If hardening of punching portions in hot stamping is locally prevented, the problems for cold shearing of hot-stamped parts can be solved, as described by Picas et al. [11]. Mori and Okuda [12] developed a tailored tempering process with grooved tools for forming a product having a strength distribution. Mori et al. [13] developed a tailored die quenching process of steel parts having strength distribution using bypass resistance heating in hot stamping. Maeno et al. [14] prevented hardening of a trimmed flange by slow cooling using spacers thicker than the sheet. On the other hand, Mori et al. [15] developed a punching process of ultra-high strength steel products using local resistance heating, and Mori et al. [16] extended local resistance to punching of small holes of a die-quenched sheet by controlling the temperature distribution.

Since heated sheets are soft, So et al. [17] punched the sheets during hot stamping. In punching during hot stamping, the structure of tools becomes complicated, because hot stamping including die quenching is generally a one-shot process.

*e-mail: mori@plast.me.tut.ac.jp

A large cooling speed in die quenching is required to attain high strength of stamped products. In addition, it is not easy to design additional channels for taking out punching scraps from dies, because dies used for hot stamping have cooling channels for die quenching. Choi et al. [18] half-trimmed a sheet during hot stamping to improve tool performance, and subsequently performed complete trimming at room temperature.

In the present study, hot semi-punching and cold scrap removing processes for hot stamping of ultra-high strength steel parts are developed to eliminate laser cutting having low productivity after hot stamping. The deformation behaviour of a quenchable steel sheet in hot semi-punching and cold scrap removing was examined.

2. Hot semi-punching and cold scrap removing processes

To eliminate laser cutting generally used in the hot stamping operation, a quenchable steel sheet is semi-punched during hot stamping, whereas the punching scraps are not removed from the hot-stamped part as shown in Figure 1. Subsequently, the scraps are removed from the hot-stamped part at room temperature. The semi-punching scraps slightly remain in the hot-stamped part to transfer to the cold scrap removing process. Since the scraps are not removed during hot stamping, additional channels for taking out punching scraps are not necessary for the hot-stamping dies. The cold scrap removing load is small due to the slight remainder. Naturally the hot semi-punching load is also small due to small flow stress of the heated sheet.

Firstly, a hot semi-punching operation without forming was performed to examine the deformation behaviour in semi-punching and cold scrap removing. The experimental setups of hot semi-punching and cold scrap removing are shown in Figure 2. A non coated quenchable steel sheet 22MnB5 (C: 0.21, Mn: 1.2, Si: 0.25, B: 0.0014 mass%) having 1.6 mm in thickness was employed for the experiment, and the tensile strength and elongation of this sheet measured from the tensile test are shown in Figure 3. The 70 mm square sheet was heated at 1100 °C for 240 s with an electrical furnace and was punched at 850 °C after manual transference to the dies. Since the square sheet is too small, the temperature of the sheet rapidly decreased for 5 s during the manual transference. The sheet was fully austenitised because the Vickers hardness of the die-quenched sheet was about 470 HV0.3. A 1500 kN CNC servo press was used for stamping. The amount of remainder in semi-punching was adjusted by inserting a shim spacer between the punch and the press slide, and the remainder rate r to the sheet thickness was changed. For example, $r = 10\%$ means $1.6 \times (1 - 0.1) = 1.44$ mm in punch stroke. The diameters of the punches used in the hot semi-punching and cold scrap removing operations were equal, and the diameter of the die hole for cold scrap removing was larger than that for hot semi-punching due to easy positioning. It is easy to remove the scrap due to a small amount of remainder.

The conditions used for hot semi-punching are shown in Table 1. Since the sheet is not completely punched in semi-punching, no clearance between the punch and die is available without collision of the punch with the die. The diameter of the die shown in Figure 2a is determined from the clearance of hot semi-punching.

3. Results of hot semi-punching and cold scrap removing

3.1. Hot semi-punching

The hot semi-punched sheet for $c = 0\%$ and $r = 10\%$ is shown in Figure 4. The punching scrap remains in the sheet. A remarkable oxidation scale appeared on the surfaces because of the non-coated steel sheet.

The range of separation of the scrap for the different remainder rates and clearances is shown in Figure 5. For $r = 10\%$, the scrap separated from the sheet in $c = 10\%$, complete punching, whereas no separation occurred in $c = 0\%$.

The diameters of the semi-punched hole and scrap are given in Table 2. The punched holes expanded from the diameter of the punch of 5.95 mm, because the periphery of the sheet was early cooled by sandwiching between the upper and lower dies, and then the region around the hole was cooled. On the other hand, the scrap became smaller due to the shrinkage.

The deformation behaviour during hot semi-punching was calculated by the commercial FEM software LS-Dyna. The distribution of normal stress in the radial direction is illustrated in Figure 6. The normal stress in the radial direction the shearing region for $c = 0\%$ is highly compressive, and the compressive stress is larger than that for $c = 10\%$. This leads to delay of the separation.

The relationships between the hot semi-punching load and the remainder rate and between the cold scrap removing load and the remainder rate for $c = 0\%$ are shown in Figure 7, where the cold punching load of the die-quenched sheet is shown as a comparison. The hot semi-punching load is considerably lower than the cold punching load. In addition, the cold scrap removing load is also small, less than 1 kN, and thus pneumatic and hydraulic cylinders are enough for cold scrap removing.

3.2. Cold scrap removing

The cross-sections and surfaces of the holes after cold scrap removing are shown in Figure 8. For $c = 10\%$ and $r = 10\%$, the scrap separated from the sheet as shown in Figure 5, complete and single punching. As the remainder rate decreases, the quality of the hole improves, because the surfaces of the hot-punched and cold-removed edges are burnished and fracture ones, respectively, i.e. the quality of edges by cold punching of die-quenched sheets is considerably low. For $c = 0\%$ and $r = 10\%$, the surface is almost burnished. The semi-punched hole was hardly deformed by cold scrap removing due to the small remainder. No burr appeared, whereas the rollover became large due to hot semi-punching. Since the remainder is necessary only for the transference to the cold scrap removing process, the minimum remainder rate

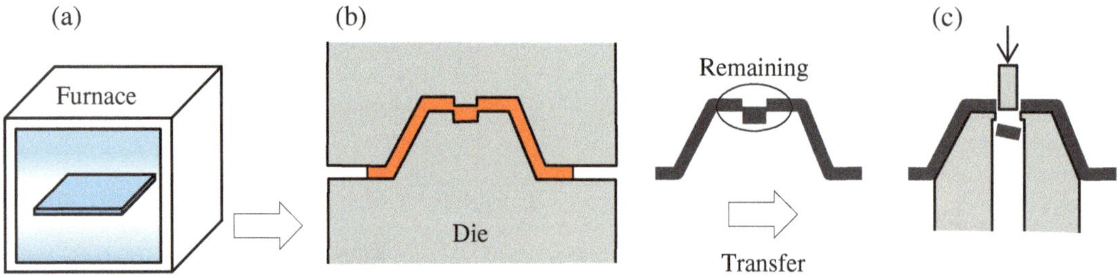

Figure 1. Hot semi-punching and cold scrap removing in hot stamping of ultra-high strength steel part. (a) Heating, (b) hot semi-punching in hot stamping and (c) cold scrap removing.

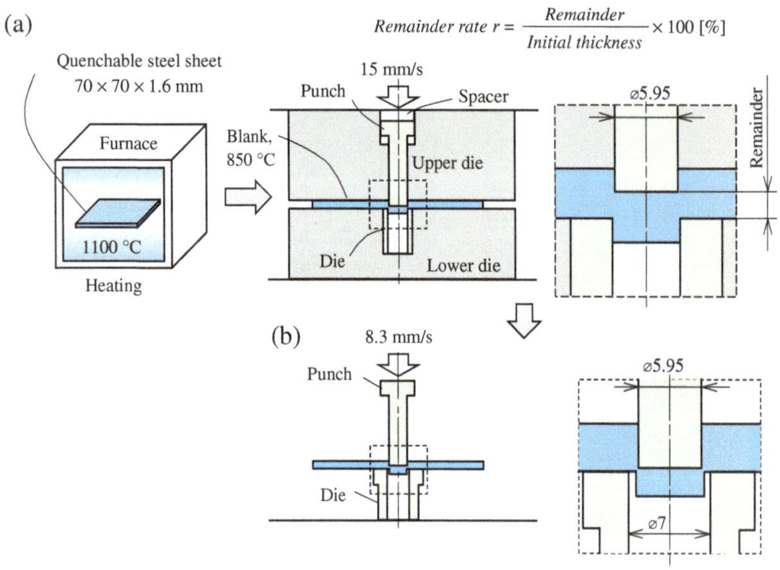

Figure 2. Experimental setup of (a) hot semi-punching and (b) cold scrap removing.

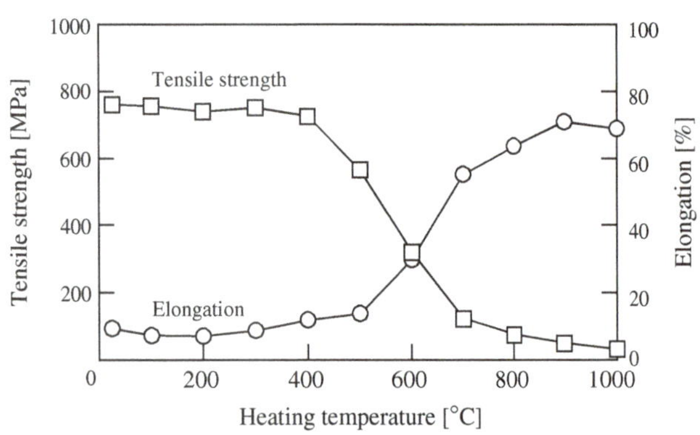

Figure 3. Variations of tensile strength and elongation with heating temperature for quenchable steel sheet 22MnB5.

Table 1. Conditions used for hot semi-punching.

Heating temperature	1100 °C
Forming temperature	850 °C
Punch speed in hot semi-punching	15 mm/s
Holding time at bottom dead centre	5 s
Remainder rate	r = 5–50%
Clearance	c = 0, 10%

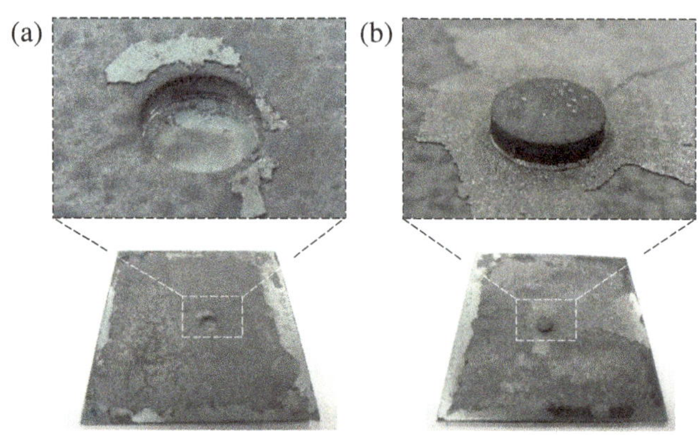

Figure 4. Hot semi-punched sheet for c = 0% and r = 10%. (a) Punch and (b) die sides.

is optimal, i.e. c = 0% and r = 10%. On the other hand, the burr for single punching of c = 10% and r = 10% is larger than that for c = 0% and r = 10%.

The surfaces of the holes after cold scrap removing and laser cutting are shown in Figure 9. In laser cutting, the power was 280 W and the speed was 800 mm/min. Although the rollover was prevented for laser cutting, a mark was

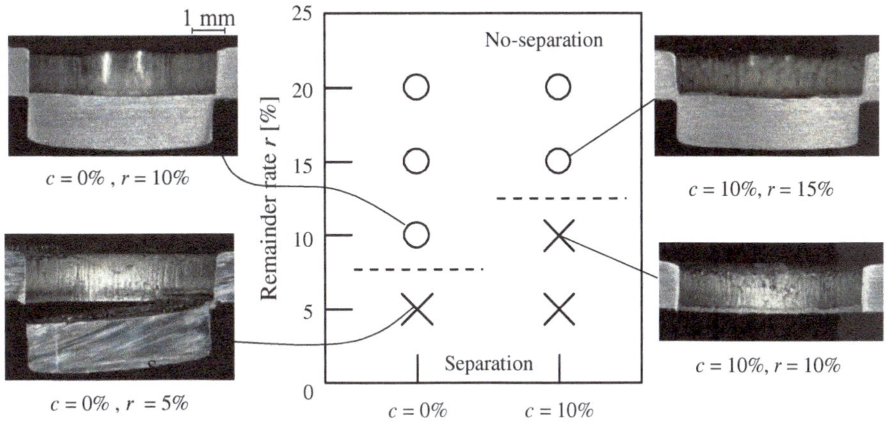

Figure 5. Range of occurrence of separation for different remainder rates and clearances.

Table 2. Diameters of semi-punched hole and scrap.

c	0%				10%		
r	10%	15%	20%	50%	15%	20%	50%
D_h [mm]	6.1	6.15	6.1	6.15	6.15	6.1	6.15
D_s [mm]	5.9	5.85	5.85	5.9	6.25	6.25	6.2

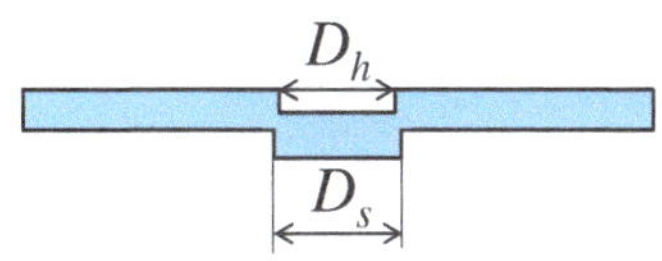

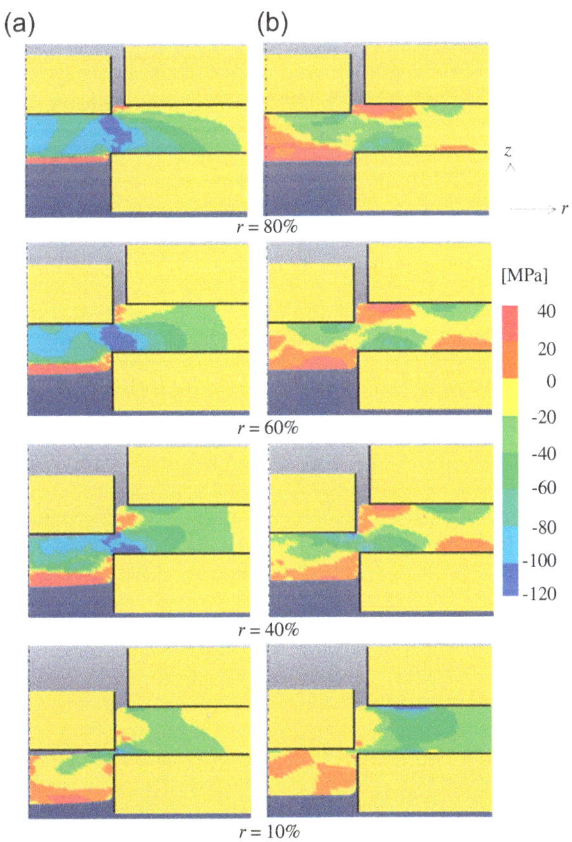

Figure 6. Distribution of normal stress in radial direction calculated by FEM software. (a) c = 0% and (b) c = 10%.

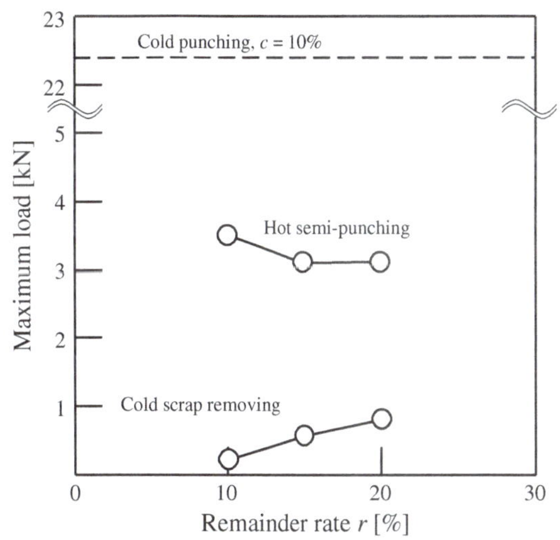

Figure 7. Relationships between hot semi-punching maximum load and remainder rate and between cold scrap removing maximum load and remainder rate for c = 0%.

caused at the end of cutting by scanning in the hoop direction.

The distribution of Vickers hardness in the thickness direction around the hole edge is shown in Figure 10, where the hardness of the die-quenched sheet without punching, fully far from the hole, is indicated as a comparison. Although the hardness was decreased from that of die quenching by laser cutting, the drop in hardness was not observed for hot semi-punching and cold scrap removing.

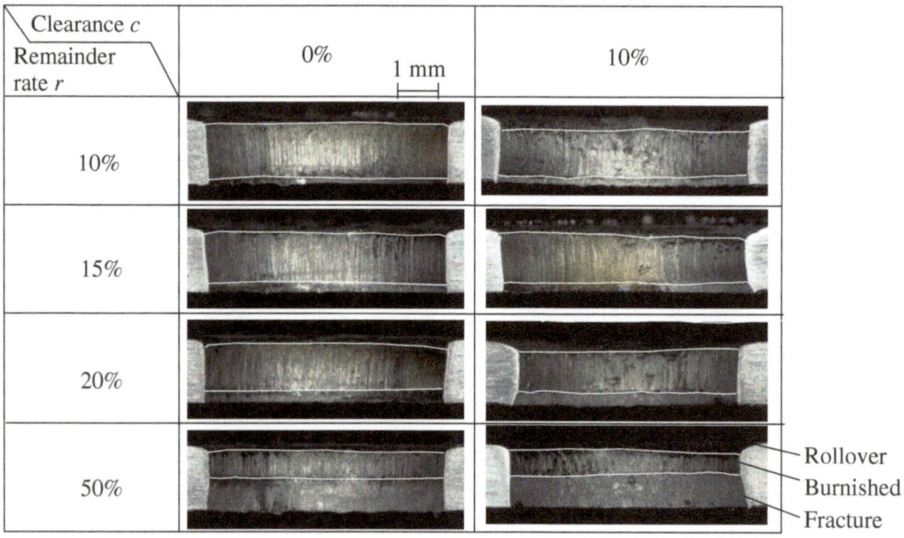

Figure 8. Cross-sections and surfaces of holes after cold scrap removing.

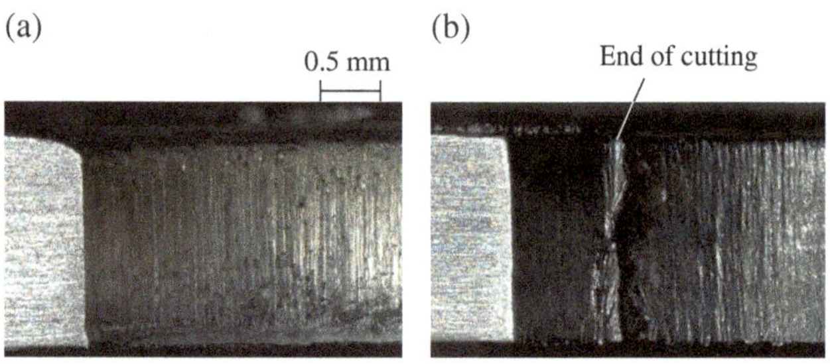

Figure 9. Surfaces of holes after (a) cold scrap removing for c = 0% and r = 10% and (b) laser cutting.

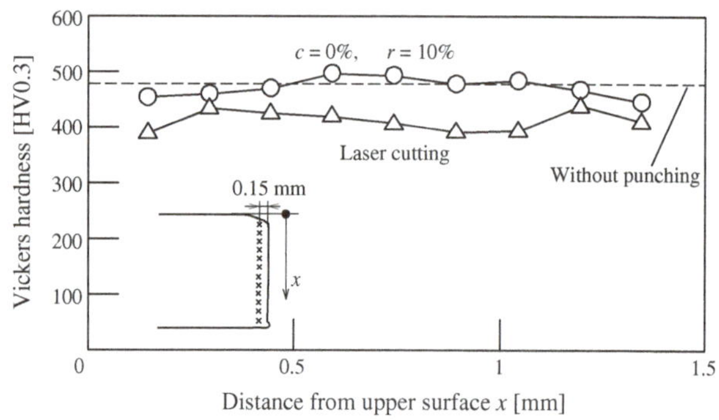

Figure 10. Distribution of Vickers hardness in thickness direction around hole edge.

The delayed fracture times around the hole edge for cold punching and hot semi-punching and cold scrap removing are shown in Figure 11, where the delayed fracture time is the time from the soak of the sheet in the 35% hydrochloric acid to the visual observation of cracks. Although cracks were caused for cold punching, the occurrence of cracks was prevented by heating for hot semi-punching.

4. Hot hat-shaped bending including semi-punching

Hot semi-punching and cold scrap removing were applied to the hot hat-shaped bending process shown in Figure 12, and the flat top and inclined flange were semi-punched. The non-coated quenchable steel sheet 22MnB5 having 1.6 mm in thickness used in Chapters 2 and 3 was employed for the experiment. The sheet was resistance-heated for hot hat-shaped bending including semi-punching. The heated blank was transferred to the bending dies with a pneumatic cylinder. The conditions used for the hot hat-shaped bending process including semi-punching are given in Table 3.

To prevent the collision of the upper die with the lower die, the remainder rate for c = 0% was increased to 15%. For conventional sheets 22MnB5 sheets having a thickness between 1.0 and 2.6 mm for hot stamping, c = 0% and r = 15% are appropriate. The punch head for the inclined flange was inclined to 10° parallel to the inclined flange.

The hat-shaped sheet having the semi-punched top and flange is shown in Figure 13. Since the heating time was considerably shortened by resistance heating, 3.3 s, the oxidation of the hat-shaped sheet was comparatively small. Not only the flat top but also the inclined flange was semi-punched with the inclined head punch without separation from the sheet. Cold-rolled sheets without annealing are austenitised even for the short heating time of 3.3 mm, because the grains are fine.

The cross-sections and surfaces of the top and flange holes after cold scrap removing are shown in Figure 14. The top hole is similar to that without forming shown in Figure 8, whereas a non-uniform burr in the hoop direction was caused for the flange hole. Since the diameter of the semi-punched hole increases as shown in Table 2, the burr in the flange hole is prevented by increasing the diameter of the punch for that of the hole. It is desirable to optimise the dimensions of the tools used for cold scrap removing.

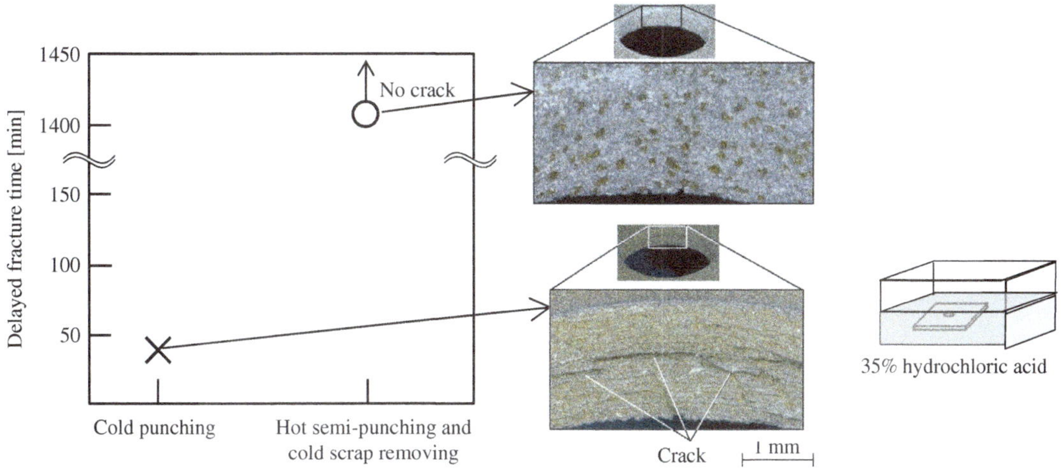

Figure 11. Delayed fracture times around hole edge for cold punching and hot semi-punching and cold scrap removing.

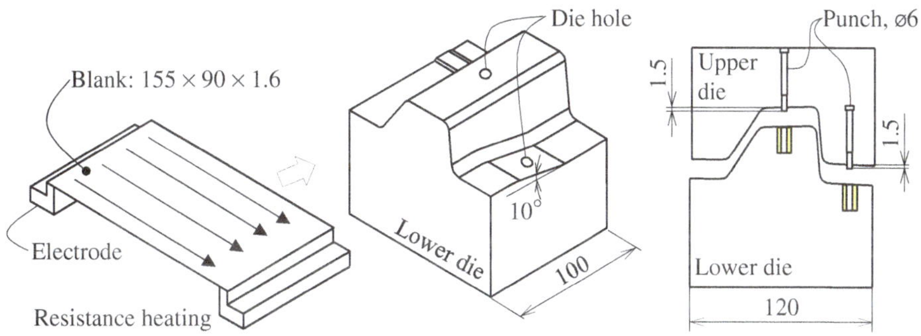

Figure 12. Hot hat-shaped bending including semi-punching.

Table 3. Conditions used for hot hat-shaped bending including semi-punching.

Heating temperature and time	930 °C and 3.3 s
Current for heating	7.7 kA
Transferring time	2.0 s
Bending temperature	850 °C
Holding time at bottom dead centre	5 s
Bending speed	70 mm/s
Clearance	$c = 0\%$
Remainder rate	$r = 15\%$

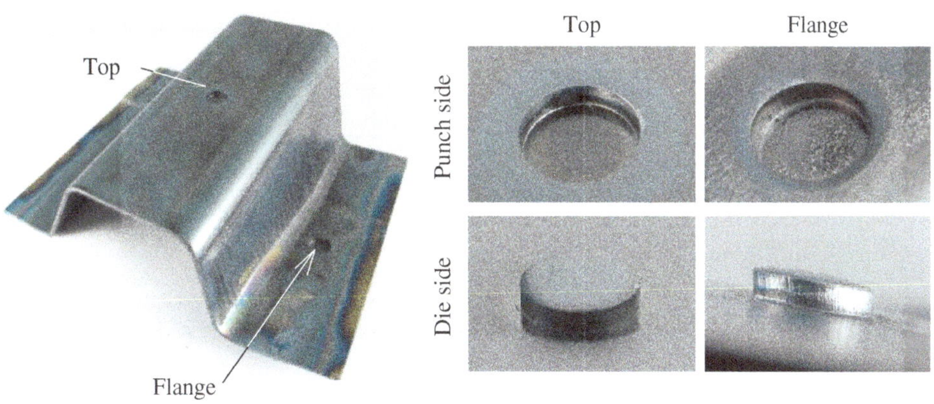

Figure 13. Hat-shaped sheet having semi-punched top and flange.

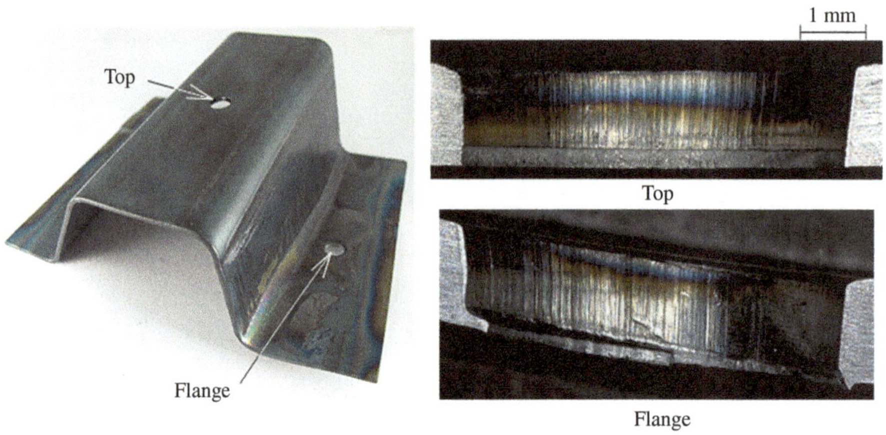

Figure 14. Cross-sections and surfaces of bottom and flange holes after cold scrap removing.

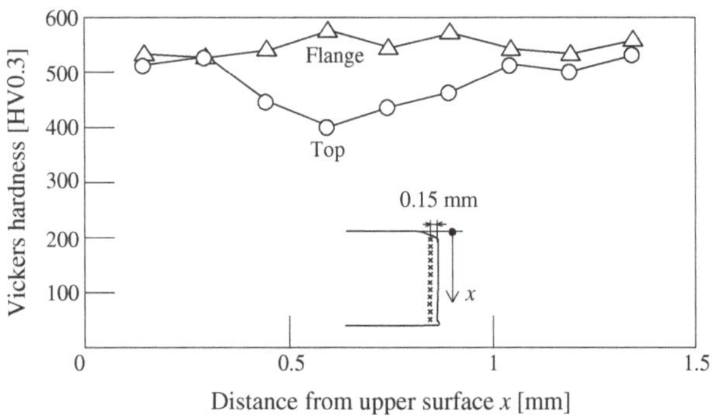

Figure 15. Distributions of Vickers hardness in thickness direction around edges of top and flange holes.

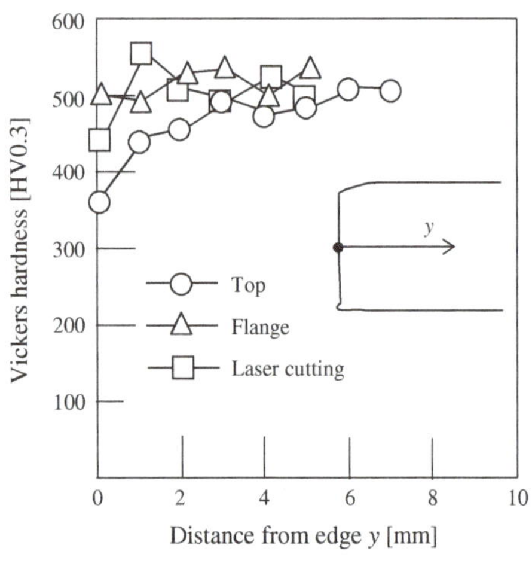

Figure 16. Distributions of Vickers hardness at middle of thickness from edges of top and flange holes.

The distributions of Vickers hardness in the thickness direction around the edges of the top and flange holes are shown in Figure 15. The hardness for the top hole decreased in the middle of the thickness, because the hole edge was detached from the punch by the thermal shrinkage during die quenching and the cooling speed was low.

The distributions of Vickers hardness at the middle of the thickness from the edges of the top and flange holes are shown in Figure 16. Although the hardness around 500 HV0.3 for laser cutting is close to the hole edge, that for the top hole is above 3 mm from the edge. In semi-punching, the hardness around the hole edge is influenced by the contact with the punch.

5. Conclusions

Although the use of hot stamping processes of ultra-high strength steel parts considerably increases, punching and trimming of stamped parts are still a bottleneck due to low productivity of laser cutting. Hot punching and trimming included in hot stamping processes are attractive for reducing the production cost, whereas the design of hot stamping processes becomes complex due the addition of shearing operations in the one-shot process. The hot punching process is more complicated than the hot trimming process, because tools are positioned inside and many holes are made. The hot semi-punching process is simpler than the hot full punching one because of no treatment of punching scraps.

Hot punching and trimming have some problems for the application to actual hot stamping operations. Tool materials and coating effective in hot punching and trimming are necessary because tools undergo large shearing force at high temperatures. The design of the stamping process for including punching and trimming tools is significantly important. The timing of punching and trimming operations is also adjusted, e.g. the punches are not taken from holes by thermal shrinkage after die quenching. The productivity and production cost of hot stamping operations are improved by hot punching and trimming, and the hot semi-punching and cold scrap removing processes are a realistic approach.

Acknowledgements. This work was supported in part by the Japanese Ministry of Education, Culture, Sports, Science and Technology with Grant-in-Aid for Scientific Research (B) (No. 21360360) and Osawa Scientific Studies Grants Foundation. The authors wish to thank Mr. M. Sakagami for his support.

References

1. M. Eriksson, M. Oldenburg, M.C. Somani, L.P. Karjalainen, Testing and evaluation of material data for analysis of forming and hardening of boron steel components, Modelling and Simulation in Materials Science and Engineering 10 (2002) 277–294.
2. P. Hein, A global approach of the finite element simulation of hot stamping, Advanced Materials Research 6 (2005) 763–770.
3. M. Geiger, M. Merklein, C. Hoff, Basic investigations on the hot stamping steel 22MnB5, Advanced Materials Research 6 (2005) 795–802.
4. K. Mori, S. Maki, Y. Tanaka, Warm and hot stamping of ultra high tensile strength steel sheets using resistance heating, CIRP Annals – Manufacturing Technology 54 (2005) 209–212.
5. P.F. Bariani, S. Bruschi, A. Ghiotti, A. Turetta, Testing formability in the hot stamping of HSS, CIRP Annals – Manufacturing Technology 57 (2008) 265–268.
6. M. Merklein, J. Lechler, M. Geiger, Investigation of the thermo-mechanical properties of hot stamping steels, Journal of Materials Processing Technology 174 (2006) 452–455.
7. H. So, H. Hoffmann, R. Golle, Blanking of press hardened ultra high strength steel, Proceedings of 2nd International Conference on Hot Sheet Metal Forming of High-Performance Steel, Luleå, 2009, pp. 137–146.
8. I. Valls, B. Casas, N. Rodríguez, Improving die durability in hot stamping and hard cutting, Proceedings of 2nd International Conference on Hot Sheet Metal Forming of High-Performance Steel, Luleå, 2009, pp. 331–338.
9. K. Nothhaft, J. Suh, M. Golle, I. Picas, D. Casellas, W. Volk, Shear cutting of press hardened steel: influence of punch chamfer on process forces, tool stresses and sheared edge qualities, Production Engineering 6 (2012) 413–420.
10. M. Fritz, Process optimization of laser cutting and in the heating process, Proceedings of 3rd International Conference on Hot Sheet Metal Forming of High-Performance Steel, Kassel, 2011, pp. 239–245.
11. I. Picas, R. Hernández, D. Casellas, I. Valls, Cold cutting of microstructurally tailored hot formed components, Proceedings of 2nd International Conference on Hot Sheet Metal Forming of High-Performance Steel, Luleå, 2009, pp. 115–125.
12. K. Mori, K. Okuda, Tailor die quenching in hot stamping for producing ultra-high strength steel formed parts having strength distribution, CIRP Annals – Manufacturing Technology 59 (2010) 291–294.
13. K. Mori, T. Maeno, K. Mongkolkaji, Tailored die quenching of steel parts having strength distribution using bypass resistance heating in hot stamping, Journal of Materials Processing Technology 213 (2013) 508–514.
14. T. Maeno, K. Mori, T. Nagai, Improvement in formability by control of temperature in hot stamping of ultra-high strength steel parts, CIRP Annals – Manufacturing Technology 63 (2014) 301–304.
15. K. Mori, T. Maeno, S. Fujisaka, Punching of ultra-high strength steel sheets using local resistance heating of shearing zone, Journal of Materials Processing Technology 212 (2012) 534–540.
16. K. Mori, T. Maeno, Y. Maruo, Punching of small hole of die-quenched steel sheets using local resistance heating, CIRP Annals – Manufacturing Technology 61 (2012) 255–258.
17. H. So, D. Faßmann, H. Hoffmann, R. Golle, M. Schaper, An investigation of the blanking process of the quenchable boron alloyed steel 22MnB5 before and after hot stamping process, Journal of Materials Processing Technology 212 (2012) 437–449.
18. H.S. Choi, B.M. Kim, D.H. Kim, D.-C. Ko, Application of mechanical trimming to hot stamped 22MnB5 parts for energy saving, International Journal of Precision Engineering and Manufacturing 15 (2014) 1087–1093.

6

Punching process including thickening of hole edge for improvement of fatigue strength of ultra-high strength steel sheet

Purwo Kadarno*, Ken-Ichiro Mori, Yohei Abe, and Tatsuro Abe

Department of Mechanical Engineering, Toyohashi University of Technology, Toyohashi, Aichi 441-8580, Japan

Abstract – A punching process including thickening of a hole edge of an ultra-high strength steel sheet with a taper punch and step die was developed to improve fatigue strength of the punched sheet. In this process, the sheet was punched by the bottom of the punch, and subsequently the hole edge was thickened by the taper of the punch and the corner step of the die. The taper angle of the punch and the step height of the die were optimised to increase the amount of thickening. The quality of the sheared edge for the thickened punched sheet was improved by ironing with the taper of the punch, and compressive residual stress was generated around the hole edge. Seizure on the surface of the punch was prevented by VC-coating for repeated punching and realistic punching speed. It was found that the present punching process including thickening is effective in improving the fatigue strength of the punched ultra-high strength steel sheet and in preventing occurrence of delayed fracture.

Key words: Punching, Thickening, Hole edge, Ultra-high strength steel sheet, Fatigue strength

1 Introduction

To improve the fuel efficiency of automobiles, the reduction in weight of body panels has become a high priority for a vehicle development. Kleiner et al. [1] have reviewed technologies of metal forming processes of lightweight components for the reduction. Among lightweight materials, high strength steel sheets are the most attractive material for body-in-white parts due to the high specific strength and cost competitiveness. In stamping operations of high strength steel sheets, large springback, small formability, short tool life, etc. are problematic, particularly for ultra-high strength steel sheets having a tensile strength above 1 GPa. Mori et al. [2] have decreased springback in bending of ultra-high strength steel sheets by a small additional reduction in thickness at the bottom dead weight centre.

Body-in-white parts are generally punched to make many holes for joining, paint removing, attachment, reduction in weight, etc. In punching of high strength steel sheets, tools tend to wear and chip due to large punching load, and thus the tool life is short. Eriksson and Olsson [3] have improved the tool life in stamping of high strength steel sheets by the PVD coatings. Inoue et al. [4] have observed damage of VC and TiC coated tools in stamping of ultra-high strength steel sheets. In addition, the quality of the sheared edge deteriorates due to early onset of cracks for the small ductility, the increase in rough fracture surface.

In high strength steel sheets, the static strength of formed products is almost proportional with the strength of the sheet, whereas Murakami et al. [5] have reported that the increase in the fatigue strength becomes gradually small. Thomas et al. [6] have observed that onset and progress of fatigue cracks of punched high strength steel sheets are accelerated around holes due to concentration of stress, particularly for rough fracture surface and sharp burr of the sheared edge. Therefore, the improvement of the quality of the sheared edge of punched sheets is useful for increasing the fatigue life. Thipprakmas et al. [7] have improved the quality of the punched edge of aluminium sheets by shaving with a taper punch. Kim et al. [8] have developed a burr-free punching process consisting of mechanical half piercing and hydro counter punching. Mori et al. [9] have smoothed the fracture surface of the sheared edge using the conical punch to improve the hole expansion ratio of the punched ultra-high strength steel sheet. Mori et al. [10] have improved a sheared edge of ultra-high strength steel sheets in slight clearance punching with a punch having a small round edge. Abe et al. [11] have developed a gradually contacting punch for improving stretch flangeability of ultra-high strength steel sheets. Mori et al. [12] have developed a warm and hot

*e-mail: kadarno@plast.me.tut.ac.jp; pkadarno@gmail.com

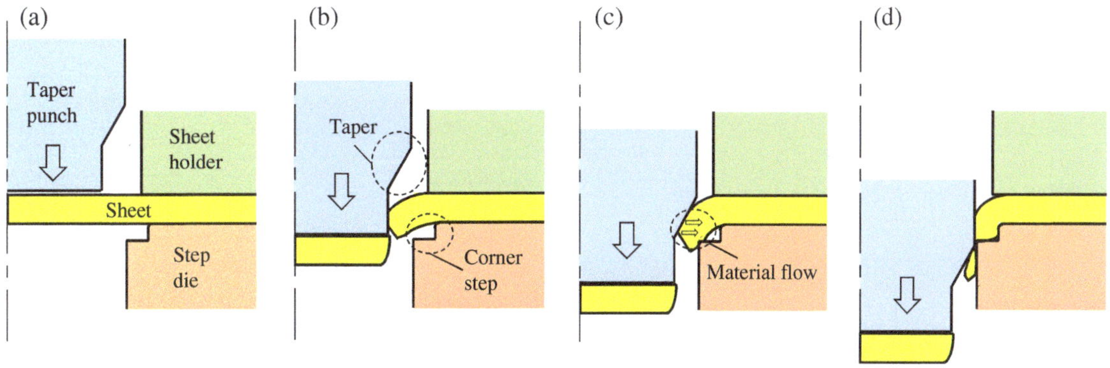

Figure 1. Punching process including thickening of hole edge for increasing fatigue life of ultra-high strength steel sheet. (a) Initial, (b) punching, (c) thickening and (d) cropping stage.

punching process using resistance heating to improve the quality of sheared edge of the ultra-high strength steel sheet. Thomas [13] has examined the effect of the clearance ratio on the quality of the sheared edge, the formability and the fatigue life of the punched sheets in punching of high strength steel sheets. Matsuno et al. [14] have employed a punch having a round corner and an inclined bottom for punching of thick high strength steel sheets and examined the fatigue properties of the punched sheets. Although the fatigue strength is improved by the quality of the sheared edge, the improvement is not enough to extend the applicable range of ultra-high strength steel sheets.

The concentration of stress around the punched hole is relieved by increasing the thickness around the hole edge, and thus the fatigue strength may be improved. Nakano [15] has reviewed plate forging processes including local thickening. Merklein et al. [16] have reviewed sheet-bulk metal forming processes for controlling wall thickness of formed products. Although incremental forming processes such as flow forming, shear spinning and orbital forming have high ability in controlling metal flow due to repeated local deformation, the productivity is not high. It is desirable in industry to develop a stamping process having high productivity. Kadarno et al. [17] have examined the effect of the thickening of the hole edge by flanging on the fatigue strength of punched high strength steel sheets. Thipprakmas et al. [18] have examined the effect of the quality of the sheared edge of the initial hole on the flanged shape in hole flanging. Since the hole flanging stage is added to the punching process, the number of stages increases by one.

In the present study, a punching process including thickening of a hole edge of ultra-high strength steel sheets was developed to improve fatigue strength of the punched sheets. Shapes of a punch and die were design for succession of punching and thickening stages. The effect of the thickening on the strength of the punched sheets was evaluated from the experiment.

2 Punching process including thickening of hole edge

To improve the fatigue strength of punched ultra-high strength steel sheets, an edge of the punched hole was thickened. A hole flanging operation is conventionally performed

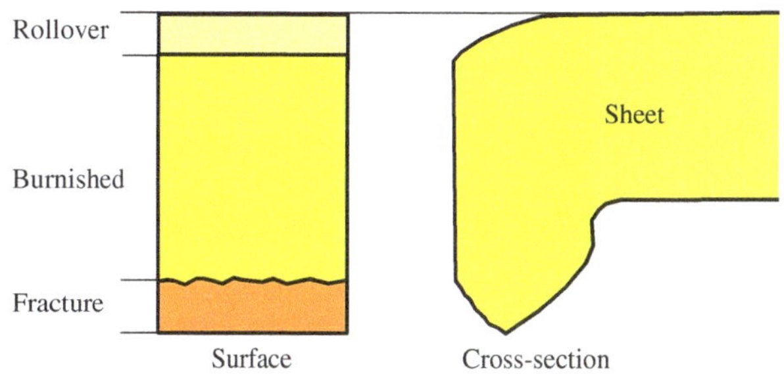

Figure 2. Surface and cross-section of thickened hole edge without compression of sharp bottom for excessive height of corner step.

subsequently to a punching operation, and thus the number of stages increases by one. In the present study, a pair of taper punch and step die was designed to include thickening of the hole edge in the punching process through one shot as shown in Figure 1. The sheet is first punched by the tension between the taper punch and the step die. Subsequently, the hole edge is bent by the taper of the punch into the corner step of the die and the surface of the sheared edge is ironed. After sufficient filling into the corner step of the die, the excessive material is then sheared. The corner step of the die has the function of not only forming the thickened edge but also compressing the edge bottom. The function is effective for ultra-high strength steels sheets having low ductility. Although the firstly punched hole is smaller than the desired hole, the hole is formed into the desired one by the subsequent thickening stage. A part of the punching scrap is effectively utilised as thickening material in the present process. This is an ecological process for improving material efficiency. In addition, it is easy to install the present punching process in the stamping sequence.

The amount of thickening of the hole edge in the present punching process is influenced by the shapes of the punch and die. As the diameter of the punch bottom decreases, the material for forming the thickened edge increases due the decrease in diameter of the punched hole, and thus the amount of thickening becomes large. Although this material is filled into the corner step of the die, the sharp edge bottom having remaining fracture surface is not compressed for excessive height and width of the corner step as shown in Figure 2. This leads to onset of fatigue cracks.

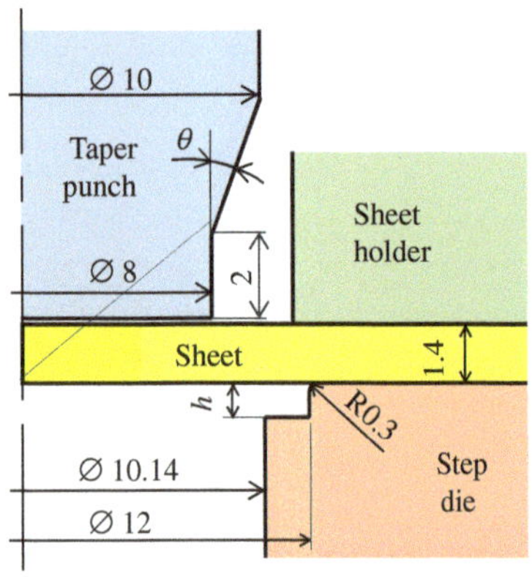

Figure 3. Dimensions of tools used in punching process including thickening of hole edge of ultra-high strength steel sheet.

Table 1. Conditions of punching process of ultra- high strength steel sheet.

Taper angle of punch θ	10°, 13° and 20°
Step height of die h	0.5, 0.7, 0.9 and 1.1 mm
Punch	SKD11, VC coating
Die	SKD11, no coating
Punching speed	0.08 mm/s
Lubrication	Rust prevention oil

In the present study, the diameter of the punched hole was set to be 10 mm typical of body-in-white parts. The dimensions of the tools are shown in Figure 3. The diameter of the punch bottom was smaller than that of the desired hole to thicken the hole edge. For a diameter of the punch bottom of 8 mm, the theoretical amount of thickening for complete filling without cropping loss is 1.1 mm. The amount of thickening was adjusted by the step height of the die h and the taper angle of the punch θ under fixing the step diameter at 12 mm.

The conditions of the punching process including thickening of the hole edge of an ultra-high strength steel sheet are given in Table 1. The sheets were punched by a 50 kN screw driven type universal testing instrument. A rust prevention oil was applied to the sheets and an additional lubricant was not employed. Each punching test was performed at least two times.

The ultra-high strength steel sheet JSC980 having a nominal tensile strength of 980 MPa and was used in the experiment, where the high strength steel sheet JSC590 was punched as a comparison. The 980 MPa sheets having 1.4 mm in thickness are the commonest ultra-high strength steel sheets and application to automobile body panels is recently increasing. The mechanical properties and the flow stress curves of these sheets measured from the tensile test are given in Table 2 and Figure 4, respectively. The specimens were cut in the 0°, 45° and 90° directions with respect to the rolling direction of the sheets, and the averages of the measured values are shown.

Finite element simulation using the commercial software ABAQUS was performed under the assumption of axi-symmetric deformation to examine deformation behaviour during the punching operation. The sheet was modelled to be

Table 2. Mechanical properties of ultra-high strength steel sheet.

	JSC980	JSC590
Thickness	1.4 mm	
Flow stress curves	$\sigma = 1574\varepsilon^{0.14}$ MPa	$\sigma = 979\varepsilon^{0.16}$ MPa
Tensile strength	1071 MPa	632 MPa
Elongation	16%	24%
Reduction of area	41%	55%

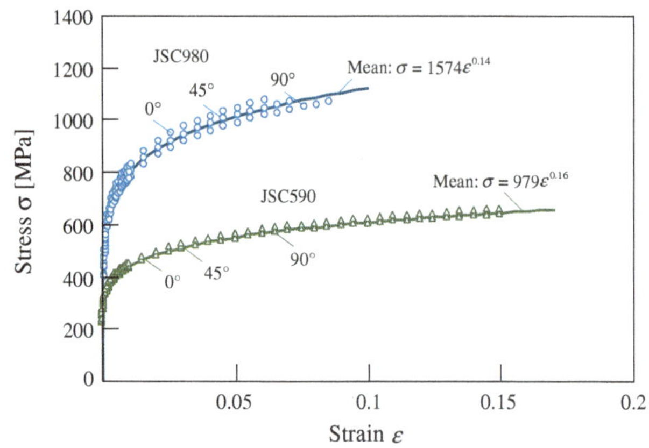

Figure 4. Flow stress curves for JSC980 and JSC590 measured from tensile test.

Table 3. Material constants for JSC980 used in calculation.

Young's modulus	210 GPa
Poisson's ratio	0.3
Initial relative density	0.995 (f_0 = 0.005)
Void nucleation parameters	$\varepsilon_N = 0.3$, $s_N = 0.1$ and $f_N = 0.04$
Porous failure criteria	$f_F = 0.25$ and $f_c = 0.15$

elastic-plastic and isotropic, whereas the die, sheet holder and punch were assumed to be rigid. The sheet was divided into 4-node quadrilateral ring elements. The coefficient of friction at the interfaces between the tools and sheet was assumed to be 0.1. ALE adaptive remeshing was employed to simulate severe plastic deformation in the punching process, and the Gurson damage model developed by Tvergaard [19] was included to simulate fracture in punching.

Gurson's yield criterion is defined as the following function of a void volume fraction f:

$$\Phi = \left(\frac{\bar{\sigma}}{\sigma_y}\right)^2 + 2q_1 f \cosh\left(q_2 \frac{3\sigma_m}{2\sigma_y}\right) - \left(1 - q_3 f^2\right) = 0,$$

where $\bar{\sigma}$ is the equivalent stress, σ_y is the tensile yield stress and σ_m is the hydrostatic stress. Tvergaard (1991) determined the values of the material parameters, $q_1 = 1.5$, $q_2 = 1$ and $q_3 = {q_1}^2 = 2.25$.

The total change in the void volume fraction is given as:

$$\dot{f} = \dot{f}_{\text{gr}} + \dot{f}_{\text{nucl}},$$

where $\dot{f}_{\text{gr}}$ is the change due to growth of existing voids and $\dot{f}_{\text{nucl}}$ is the change due to nucleation of new voids given by the strain-controlled relationship:

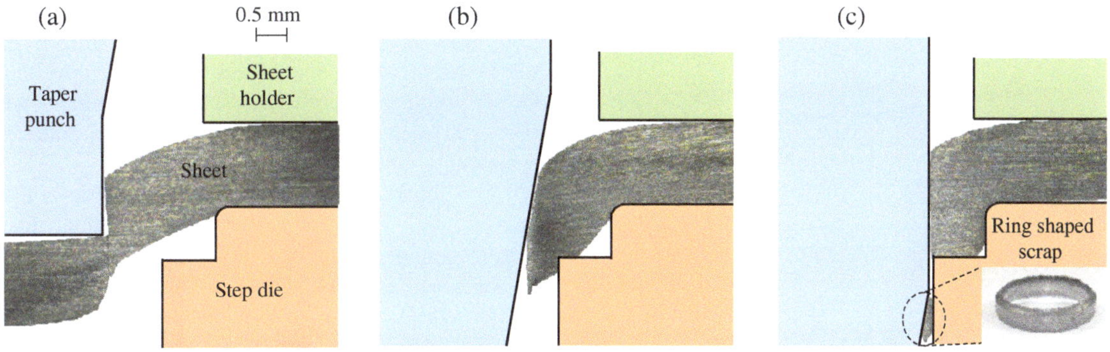

Figure 5. Deformation behaviour of sheet during punching obtained from experiment for $\theta = 10°$, $h = 0.9$ mm, JSC980 and (a) $s = 2$ mm, (b) $s = 7$ mm and (c) $s = 10$ mm.

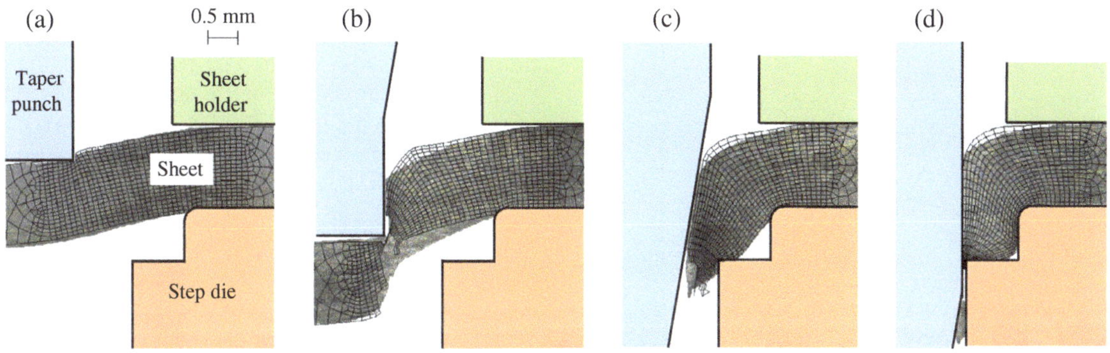

Figure 6. Comparison between deformation behaviours obtained from calculation and experiment for $\theta = 10°$, $h = 0.9$ mm, JSC980 and (a) $s = 0.55$ mm, (b) $s = 2$ mm, (c) $s = 7$ mm and (d) $s = 10$ mm.

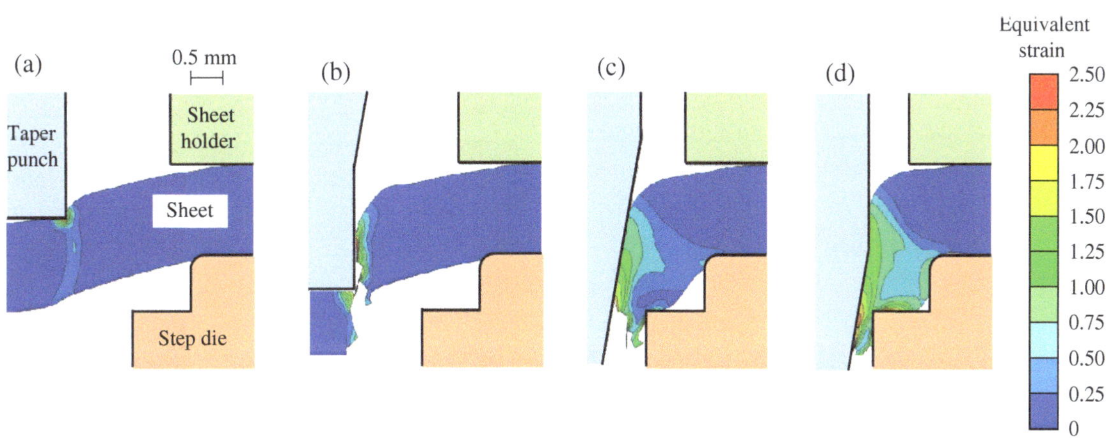

Figure 7. Distribution of plastic equivalent strain in sheet calculated by finite element simulation for $\theta = 10°$, $h = 0.9$ mm, JSC980 and (a) $s = 0.85$ mm, (b) $s = 2$ mm, (c) $s = 7$ mm and (d) $s = 9$ mm.

$$\dot{f}_{\text{nucl}} = A\dot{\bar{\varepsilon}}_m^{pl},$$

where

$$A = \frac{f_N}{s_N\sqrt{2\pi}} \exp\left[-\frac{1}{2}\left(\frac{\dot{\bar{\varepsilon}}_m^{pl} - \varepsilon_N}{s_N}\right)^2\right],$$

where ε_N is the mean value for void nucleation, s_N is the standard deviation and f_N is the volume fraction of the nucleated voids.

The material constants for JSC980 used in the calculation are shown in Table 3. The input parameters for the damage model were determined by equalising the calculated shape of

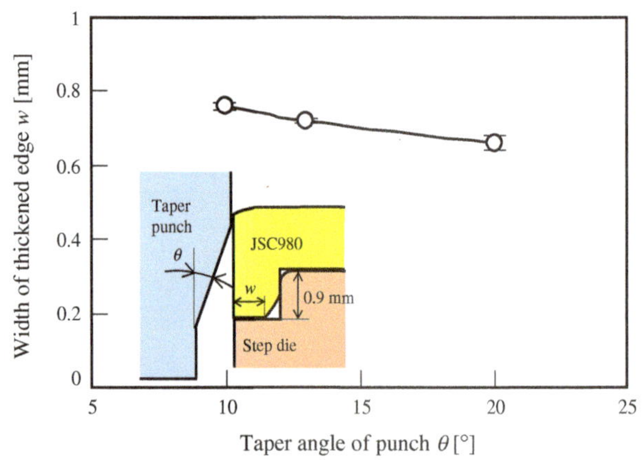

Figure 8. Relationship between width of thickened edge and taper angle of punch obtained from experiment for h = 0.9 mm and JSC980.

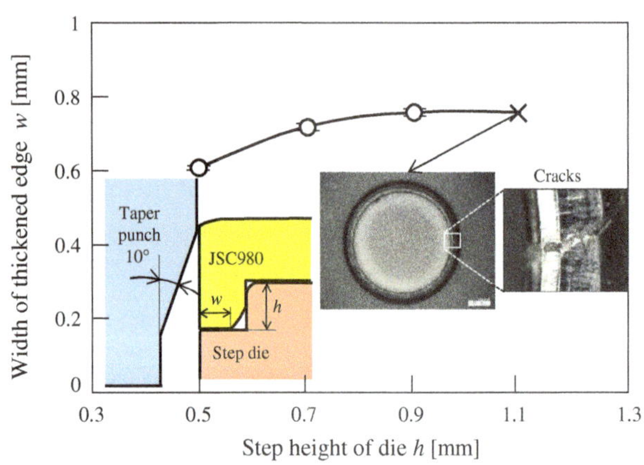

Figure 9. Relationship between width of thickened edge and step height of die obtained from experiment for θ = 10° and JSC980.

the sheared surface with that for the experimental result in the simulation of the punching process using a flat punch.

3 Results of punching including thickening of hole edge

3.1 Deformed shape

The deformation behaviour of the sheet during punching obtained from the experiment for the taper angle of the punch θ = 10°, the step height of the die h = 0.9 mm and JSC980 is shown in Figure 5, where s is the punch stroke. The sheet is bent and punched in the former stage by the bottom of the punch, and then the hole edge is thickened by the taper of the punch and the corner step of the die. Since the material is sufficiently filled into the corner step, a ring shaped scrap is caused.

The deformation behaviour during punching was calculated from the finite element simulation in detail. A comparison between the deformation behaviours obtained from the calculation and experiment for θ = 10°, h = 0.9 mm and JSC980 is shown in Figure 6, where the calculation results is shown as a top layer on the cross section of the experiment results. The deformation behaviour of the sheet during punching obtained by the calculation is in almost good agreement with that by the experiment, whereas the experimental rollover is more remarkable.

The distribution of plastic equivalent strain in the sheet calculated by the finite element simulation for θ = 10°, h = 0.9 mm and JSC980 is shown in Figure 7. As the punch stroke increases, the equivalent strain concentrates around the bottom corner of the punch, and the sheet is punched. The sheared edge undergoes severe deformation with the taper and upper corner of the punch.

The relationship between the width of the thickened edge and the taper angle of the punch obtained from the experiment for h = 0.9 mm and JSC980 is shown in Figure 8. As the taper angle of the punch decreases, the width of the thickened edge increases, whereas the taper angle was limited to θ = 10° due the increase in stroke required for punching. For the small taper angle, the material flow in the circumferential direction is accelerated, and thus the material filling into the corner step of the die increases. The width of the thickened edge for θ = 10° is the largest, and thus θ = 10° was fixed in the following experiments.

The relationship between the width of the thickened edge and the step height of the die obtained from the experiment for θ = 10° and JSC980 is shown in Figure 9. As the step height of the die increases, the width of the thickened edge increases. When the step height of the die is small, the sheared edge is not sufficiently flanged because of early contact with the step of the die. For h = 1.1 mm, cracks occurred at the bottom of the thickened edge because of late contact with the die step.

The distributions of void volume fraction in the sheet calculated by the finite element simulation for θ = 10°, s = 9 mm, JSC980 and h = 0.5, 0.9 and 1.1 mm are given in Figure 10. When the void fraction exceeds a critical value, the fracture occurs in the simulation of the punching stage, and thus the void fraction is extremely high around the sheared edge. The void fraction in the bottom of the thicken edge for h = 1.1 mm becomes high, whereas that in the top surface is not very high. This process can be applied to a less ductile material by decreasing the step height of the die, while the amount of thickening decreases.

The forming load obtained from the experiment for θ = 10°, h = 0.9 mm and JSC980 is compared with that without thickening using a flat punch without a taper in Figure 11, where the ratio of clearance between the punch and die to the sheet thickness for the flat punch was 20%. The result for the non-coated punch was added as a comparison. The forming load with thickening has two peaks. The first peak is caused by punching the sheet and is smaller than that without thickening and the second peak. The second peak is mainly generated by ironing the sheared edge and by shearing excessive material after sufficient filling into the corner step of the die. Since the forming loads with and without coating are almost similar, the frictional effect is not large. The punching stroke with thickening is much larger than that without thickening.

The maximum forming load and forming energy with and without thickening obtained from the experiment for θ = 10°

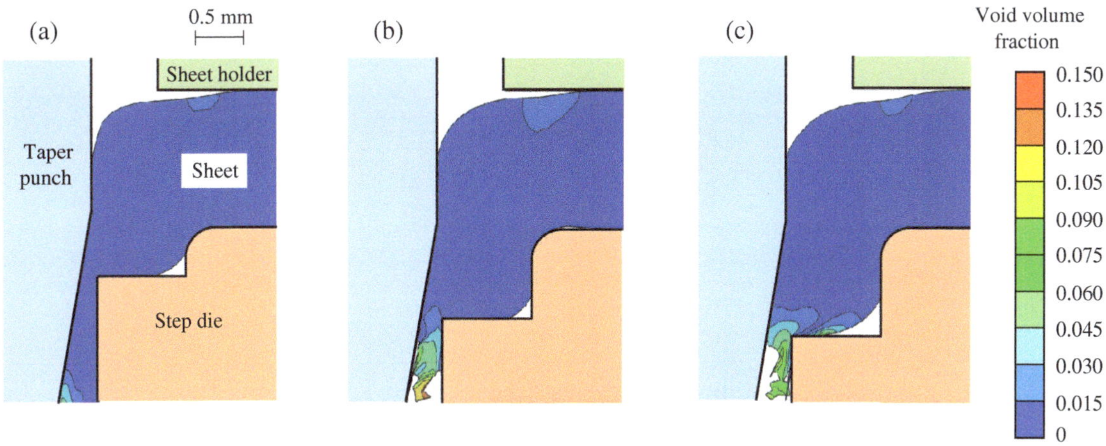

Figure 10. Distributions of void volume fraction in sheet calculated by finite element simulation for $\theta = 10°$, $s = 9$ mm, JSC980 and (a) $h = 0.5$ mm (b) $h = 0.9$ mm and (c) $h = 1.1$ mm.

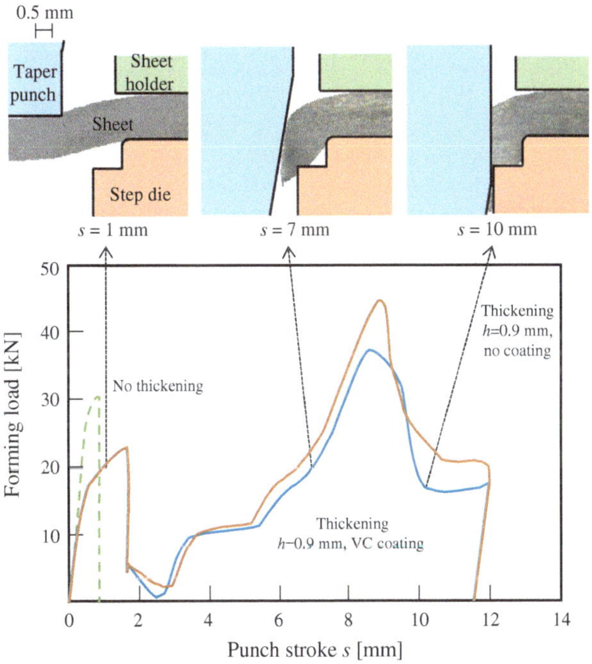

Figure 11. Forming load-punch stroke curves with thickening for $\theta = 10°$, $h = 0.9$ mm and JSC980 and without thickening obtained from experiment.

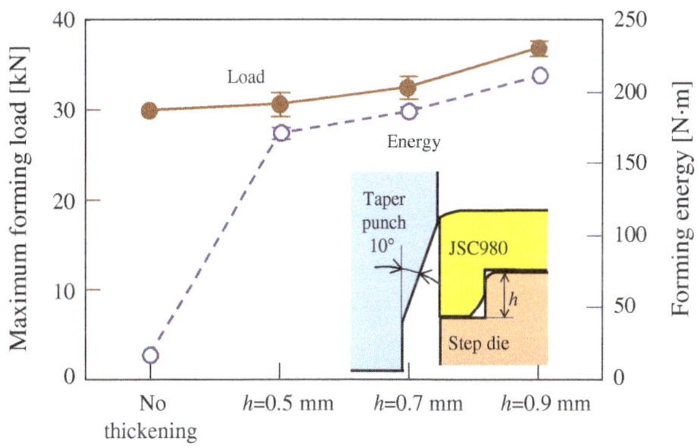

Figure 12. Maximum forming load and forming energy with and without thickening obtained from experiment for $\theta = 10°$ and JSC980.

and JSC980 are illustrated in Figure 12. The forming energy was calculated by integrating the forming load-punch stroke curve shown in Figure 11. As the step height of the die increases, the forming load and energy increase. The forming energy with thickening becomes considerably large due to the enlargement of the punch stroke.

3.2 Quality of sheared edge

The surfaces and cross-sections of the sheared edge with and without thickening obtained from the experiment for $\theta = 10°$ and JSC980 are shown in Figure 13. As the step height of the die increases, the amount of thickening increases. The area of the rough fracture surface without thickening is considerably improved by thickening. Although the depth of the rollover for thickening is larger than that without thickening due to bending during punching by the bottom of the punch, the depth is almost constant for the step height of the die. The rollover can be reduced by separating the step portion from the die and by pressing back the sheared edge portion with the separated step portion after thickening. Since this separation complicates structure of tools, the effect of thickening on the fatigue strength was evaluated without improvement of the rollover.

The distributions of the arithmetic mean value of the surface roughness of the sheared edge in the thickness direction with and without thickening obtained from the experiment for $\theta = 10°$, $h = 0.9$ mm and JSC980 are shown in Figure 14. The surface roughness was measured at intervals of 0.1 mm in the thickness direction. The surface roughness of the thickened sheared edge is considerably smaller than that without thickening due to the elimination of fracture surface. Although the surface roughness without thickening scatters, the scatter with thickening is very small.

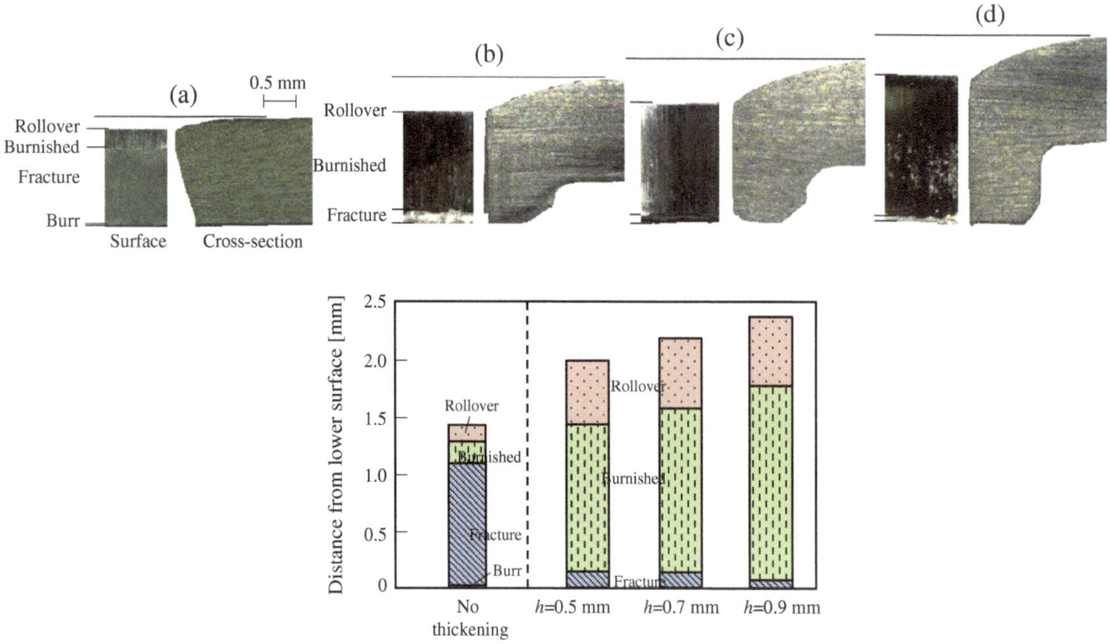

Figure 13. Surface and cross-section of sheared edge obtained from experiment for $\theta = 10°$, JSC980, (a) no thickening, (b) $h = 0.5$ mm, (c) $h = 0.7$ mm and (d) $h = 0.9$ mm.

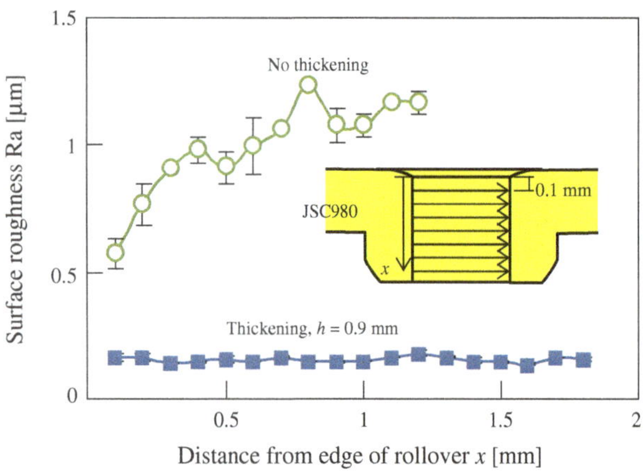

Figure 14. Distributions of arithmetic mean value of surface roughness of sheared edge in thickness direction with and without thickening obtained from experiment for $\theta = 10°$, $h = 0.9$ mm and JSC980.

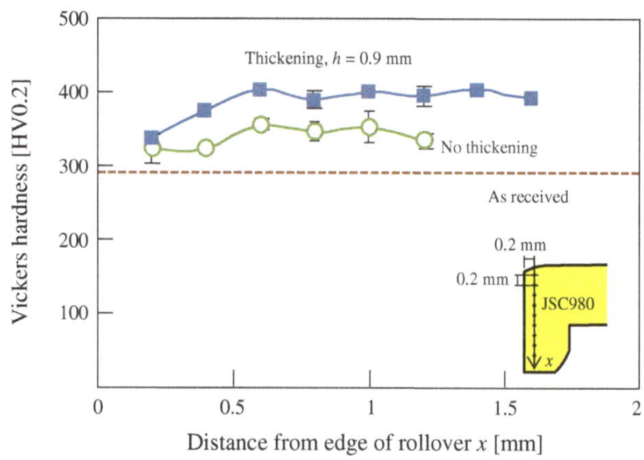

Figure 15. Distributions of Vickers hardness in thickness direction around sheared edge with and without thickening obtained from experiment for $\theta = 10°$, $h = 0.9$ mm and JSC980.

The distributions of Vickers hardness in the thickness direction around the sheared edge with and without thickening obtained from the experiment for $\theta = 10°$, $h = 0.9$ mm and JSC980 are shown in Figure 15. The hardness was measured in the cross-section at 0.2 mm from the sheared edge with intervals of 0.2 mm in the thickness direction. The hardness with thickening is higher than that without thickening due to ironing with the taper of the punch.

To examine the work-hardening behaviour around the sheared edge due to the punching and thickening, the distributions of Vickers hardness was measured for different distances from the sheared edge y, $\theta = 10°$, $h = 0.9$ mm and JSC980 as shown in Figure 16. The hardness was measured in the cross-section at 0.2, 0.4 and 0.6 mm from the sheared edge at intervals of 0.2 mm in the thickness direction. As the distance from the sheared edge increases, the hardness decreases. The hardness near the sheared edge was the largest due to the influence of ironing with the taper of the punch.

The residual stress in the sheared edge of the punched sheet was measured by the X-ray diffraction. A diffraction angle 2θ of the X-ray was set between 153.0° and 157.8° at intervals of 0.02°. The residual stress on the surface of the sheared edge for $\psi = 90°$ was obtained by extrapolating the stress obtained

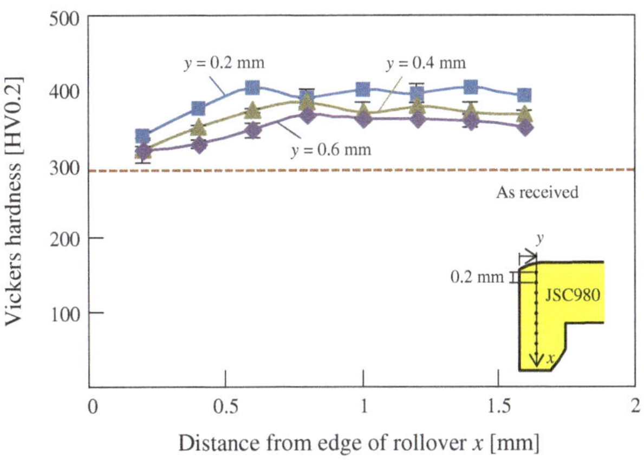

Figure 16. Distributions of Vickers hardness in thickness direction around sheared edge obtained from experiment for different distance from sheared edge y, $\theta = 10°$, $h = 0.9$ mm and JSC980.

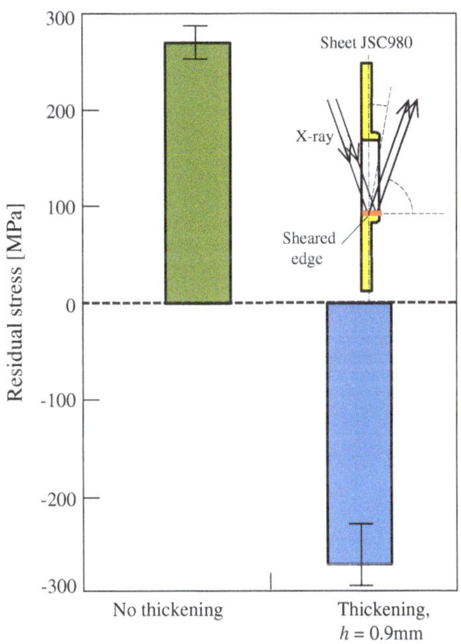

Figure 17. Residual stresses in thickness direction in sheared edge of punched sheets with and without thickening measured by X-ray diffraction for $\theta = 10°$, $h = 0.9$ mm and JSC980.

for inclination angles ψ of 0°, 18°, 27°, 33° and 45°. The residual stresses in the thickness direction in the sheared edge of the punched sheets with and without thickening for $\theta = 10°$, $h = 0.9$ mm and JSC980 are shown in Figure 17. Although the residual stress without thickening is tensile, the stress with thickening becomes compressive due to the ironing during the thickening stage.

4 Strength of punched sheet

4.1 Fatigue strength

The bending and tensile fatigue tests of the punched sheets with the thickened hole edge were performed. The procedure and dimensions of specimens for the two fatigue tests are shown in Figure 18. The fatigue test was performed each condition at least two times and the averages of the measured values are shown. The frequencies for the alternative bending and repeated tensile fatigue tests were set at 25 and 50 Hz, respectively. The fatigue test was ended when the sheet was ruptured. The maximum number of cycles for the fatigue tests was 10^7. Since the stress is complexly distributed in the punched sheet during the fatigue tests due to stress concentration around the punched holes, the nominal stress calculated for a sheet without a hole and thickening is indicated in the following results. For bending, the maximum bending stress calculated on the surface of the sheet is indicated.

The plots of the nominal stress amplitude versus the number of cycles to failure for $\theta = 10°$, the bending and tensile fatigue tests of the punched sheets with thickening are compared with those without thickening in Figure 19, where no rupture occurred for the numbers of cycles of 10^7. As the step height of the die increases, the fatigue strength increases. The fatigue strength of the punched sheets with thickening is considerably higher than that without thickening. The increase in fatigue strength with thickening for JSC980 was larger than that for JSC590. The improvement of the fatigue strength for thickening is due to the decrease in stress by the thickening, the compressive stress, the small surface roughness and the large hardness around the sheared edge. It was found that thickening of the hole edge is useful for improving the fatigue strength of punched ultra-high strength steel sheet.

4.2 Delayed fracture

Mori et al. [10] reported that punched ultra-high strength steel sheet have the risk of delayed fracture. The occurrence of delayed fracture on the sheared edge was visually observed by keeping the punched sheets in 35% concentration hydrochloric acid at room temperature. The occurrence of delayed fracture was accelerated with the high concentration of the acid.

The effect of thickening of the hole edge on the occurrence of the delayed fracture time obtained from the experiment for $\theta = 10°$, $h = 0.9$ mm and JSC980 is given in Figure 20, where the delayed fracture time is the time from the soak of the sheet in the acid. Although cracks were observed after 2.5 h for the punched sheet without thickening, no cracks were observed up to 48 h for the thickened sheet. For no thickening, the rough fracture surface in the sheared edge containing microcracks becomes large, and thus the hydrogen tends to diffuse. Moreover, the tensile residual stress in the sheared edge accelerates the occurrence of the delayed fracture. The delayed fracture for the thickening was prevented due to the large compressive stress and the small surface roughness.

4.3 Static strength

To investigate the effect of the thickening on the static strength of the punched sheets, the static tensile test was carried out. The procedure and dimensions of specimen for the static test were the same with those for the fatigue tensile test shown in Figure 18b and 18c. The static test was performed each

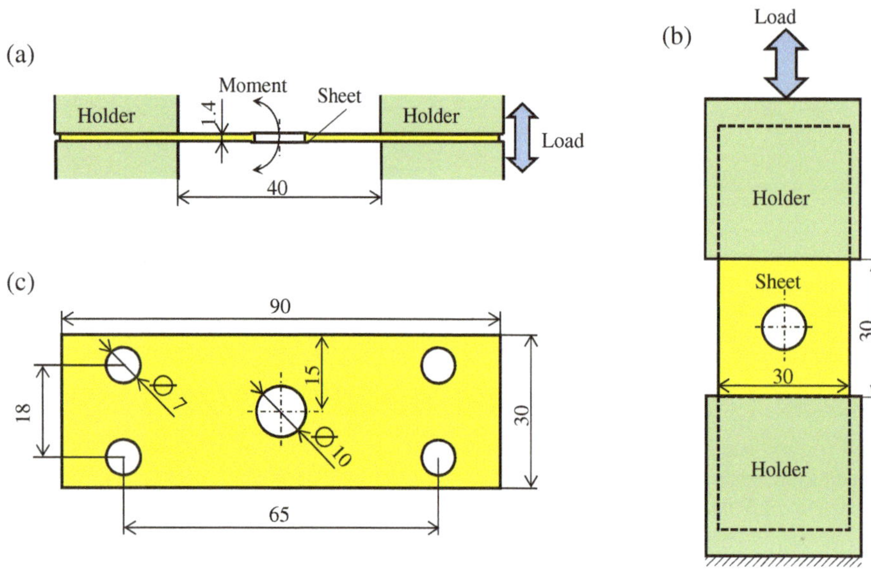

Figure 18. Procedure of (a) bending and (b) tensile fatigue tests and (c) dimensions of specimens.

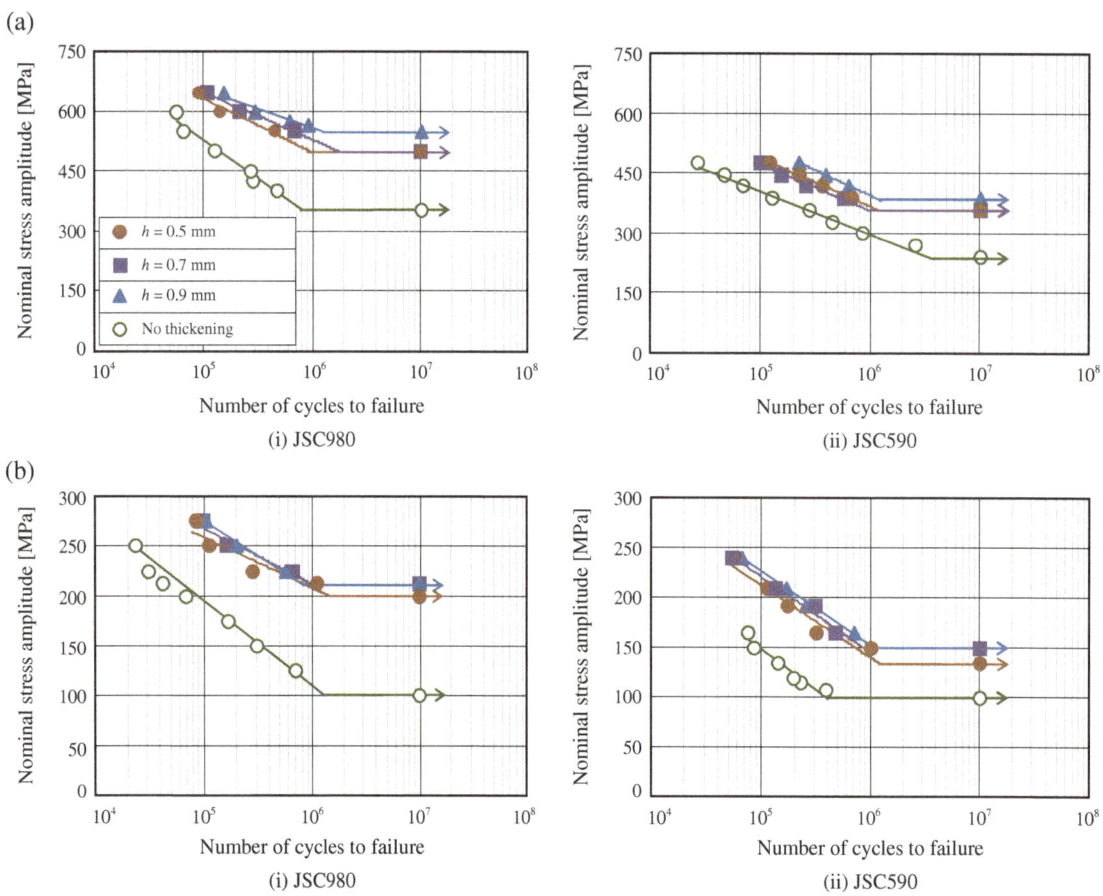

Figure 19. Plots of nominal stress amplitude versus number of cycles to failure of punched sheets with and without thickening for $\theta = 10°$, (a) bending and (b) tensile fatigue tests.

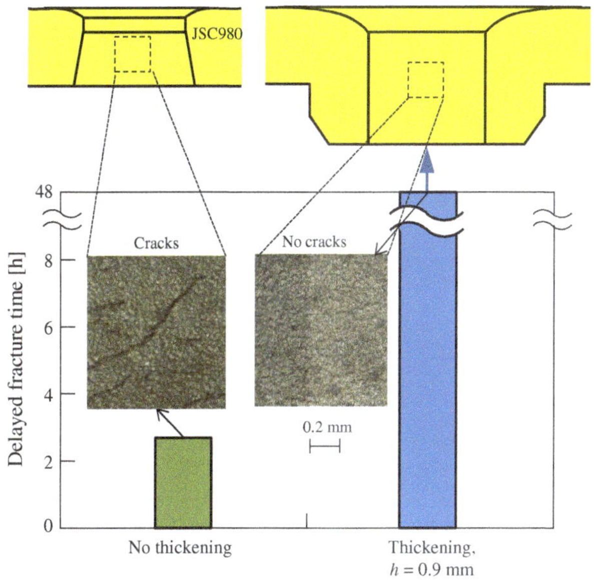

Figure 20. Effect of thickening of hole edge on occurrence of delayed fracture time obtained from experiment for $\theta = 10°$, h = 0.9 mm and JSC980.

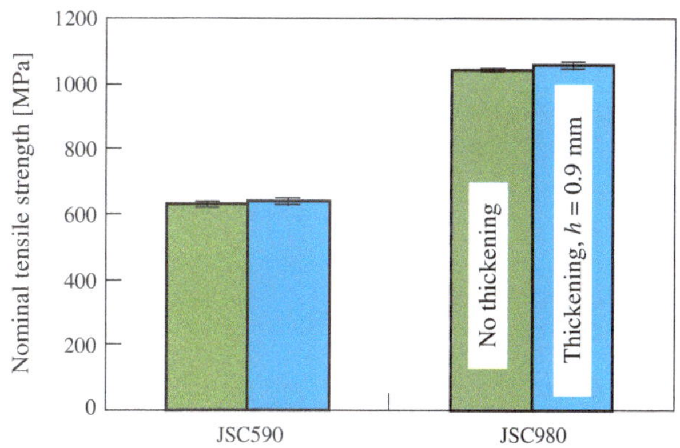

Figure 21. Nominal tensile strength of punched sheets with and without thickening obtained from experiment for $\theta = 10°$, h = 0.9 mm.

condition at least two times and the averages of the measured values are shown.

The nominal tensile strength of the punched sheets with and without thickening obtained from the experiment for $\theta = 10°$, h = 0.9 mm is shown in Figure 21. The increase in the static strength for the thickening for both JSC590 and JSC980 was comparatively small due to the small increase in area, only 5% increase.

5 Repeated punching including thickening of hole edge

Since the above mentioned results were obtained for a slow punching speed of 0.08 mm/s, repeated punching was carried out under a realistic punching speed of 100 mm/s to check the applicability of the present punching process to actual stamping operations. The ultra-high strength steel sheet JSC980 was punched for $\theta = 10°$ and h = 0.9 mm with a CNC servo press having a capacity of 800 kN. The punch was VC-coated by Toyota diffusion coating process and the hardness was 3500 HV. The wear and seizure resistances of this coating are very high and thus the required amount of lubricant could be reduced.

The surface and cross-section of the sheared edge in repeated punching for $\theta = 10°$, h = 0.9 mm and JSC980 are shown in Figure 22, where n is the number of strikes. The surface quality of the sheared edge is high and the geometry of the cross section is hardly changed even for n = 100.

The surface of the punch in repeated punching for $\theta = 10°$, h = 0.9 mm and JSC980 is given in Figure 23. The surface of the VC-coated punch hardly changed even for n = 100,

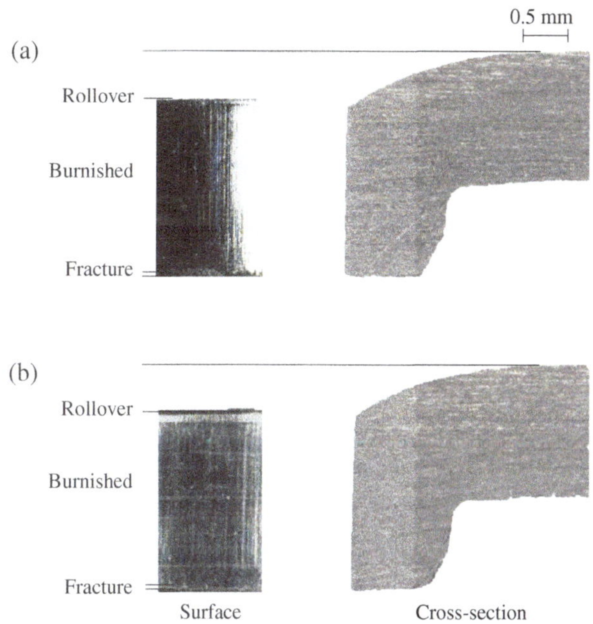

Figure 22. Surface and cross-section of sheared edge in repeated punching for $\theta = 10°$, h = 0.9 mm, JSC980, (a) n = 1 and (b) n = 100.

whereas seizure occurred due to severe shear deformation for the non-coated punch and n = 1.

6 Conclusions

A punching process including thickening of a hole edge of ultra-high strength steel sheet was developed to improve fatigue strength of the punched sheet. A pair of taper punch and step die was designed to include thickening of the hole edge in the punching process and easily installed in conventional die sets. A part of the punching scrap was effectively utilised as

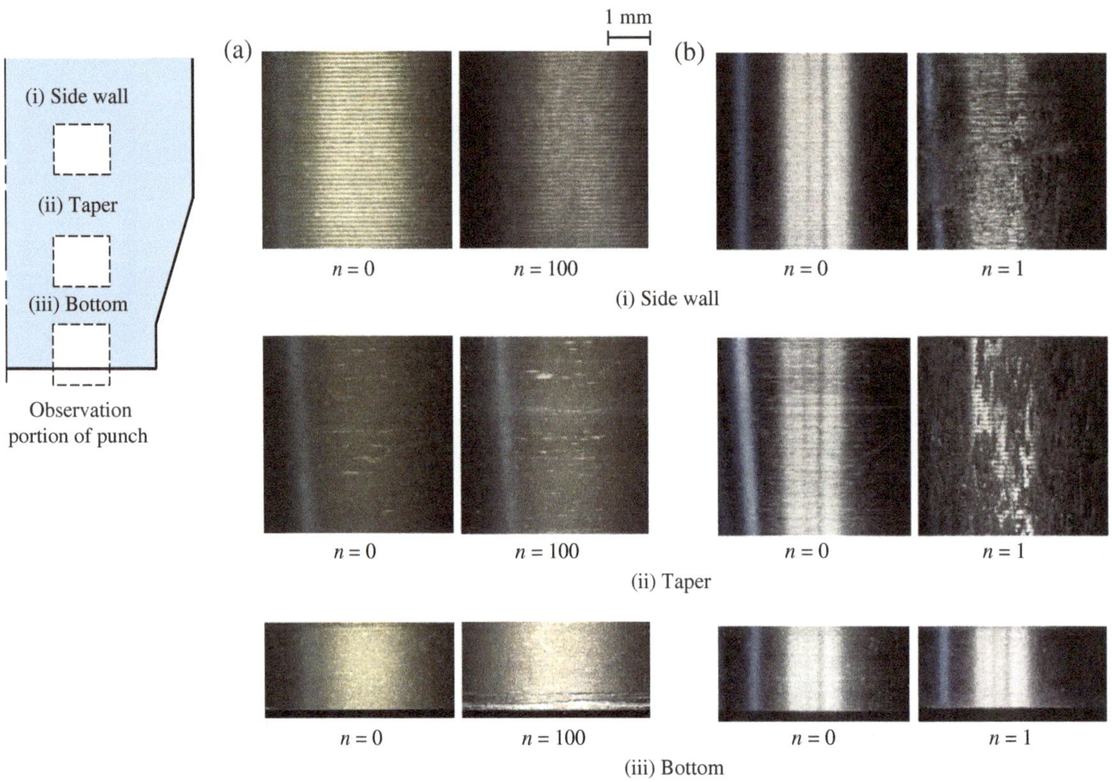

Figure 23. Surface of punch in repeated punching for $\theta = 10°$, $h = 0.9$ mm, JSC980, punches (a) with and (b) without VC coating.

thickening materials in the present process. This is an ecological process for improving material efficiency. The results are summarised as follows:

1. The sheet was punched in the former stage by the bottom of the punch, and then the hole edge was thickened by the taper of the punch and the corner step of the die.
2. The surface quality of the sheared edge with thickening was improved due to ironing with the taper of the punch during the thickening stage.
3. The fatigue strength of the punched sheets with thickening was considerably higher than that without thickening due to the decrease in stress by the thickening, the compressive stress, the small surface roughness and the large hardness around the sheared edge.
4. The delayed fracture was prevented by thickening due to the large compressive residual stress and the small surface roughness.

Implications and influences

The use of the ultra-high strength steel sheets for automobile parts is increasing, whereas the applicable range is still limited to the fatigue strength. For example, the ultra-high strength steel sheets are not applied to automobile wheels subjected to cyclic loading due to insufficient fatigue strength. The fatigue strength of the ultra-high strength steel sheets is not increased as much as the static strength. Although microstructure in the ultra-high strength steel sheets is controlled in steel making industry to heighten the fatigue strength, forming of products having high fatigue strength is an alternative approach. The fatigue strength can be increased by relieving concentration of tensile stress during loading, particularly relieving around sheared edges such as the present process is considerably effective. Since the thicken hole edges becomes projections, this side is required to be designed inside parts.

References

1. M. Kleiner, M. Geiger, A. Klaus, Bulk sheet metal forming, CIRP Annals – Manufacturing Technology 52 (2003) 725–745.
2. K. Mori, K. Akita, Y. Abe, Springback behaviour in bending of ultra-high-strength steel sheets using CNC servo press, International Journal of Machine Tools and Manufacture 47 (2007) 321–325.
3. J. Eriksson, M. Olssson, Tribological testing of commercial CrN, (Ti,Al)N and CrC/C PVD coatings – Evaluation of galling and wear characteristics against different high strength steels, Surface & Coating Technology 205 (2011) 4045–4051.
4. K. Inoue, M. Suzuki, S. Nishino, K. Ohya, Y. Tomota, Effect of coating microstructure of press-working dies on sliding damage, Steel Research International 81, Supplement Metal Forming 2010 (2010) 849–852.
5. Y. Murakami, S. Kodama, S. Konuma, Quantitative evaluation of effects of non-metallic inclusions on fatigue strength of high strength steels. I: Basic fatigue mechanism and evaluation of correlation between the fatigue fracture stress and the size and location of non-metallic inclusions, International Journal of Fatigue 11 (1989) 291–298.

6. D.J. Thomas, M.T. Whittaker, G. Bright, Y. Gao, The influence of mechanical and CO_2 laser cut-edge characteristics on the fatigue life performance of high strength automotive steels, Journal of Materials Processing Technology 211 (2011) 263–274.
7. S. Thipprakmas, S. Rojananan, P. Paramaputi, An investigation of step taper-shaped punch in piercing process using finite element method, Journal of Materials Processing Technology 197 (2008) 132–139.
8. S.S. Kim, C.S. Han, Y.S. Lee, Development of a new burr-free hydro mechanical punching, Journal of Materials Processing Technology 162–163 (2005) 524–529.
9. K. Mori, Y. Abe, Y. Suzui, Improvement of stretch flangeability of ultra-high strength steel sheet by smoothing of sheared edge, Journal of Materials Processing Technology 210 (2010) 653–659.
10. K. Mori, Y. Abe, Y. Kidoma, P. Kadarno, Slight clearance punching of ultra-high strength steel sheets using punch having small round edge, International Journal of Machine Tools and Manufacture 65 (2013) 41–46.
11. Y. Abe, K. Mori, K. Norita, Gradually contacting punch for improving stretch flangeability of ultra-high strength steel sheets, CIRP Annals – Manufacturing Technology 62 (2013) 263–266.
12. K. Mori, S. Saito, S. Maki, Warm and hot punching of ultra-high strength steel sheet, CIRP Annals – Manufacturing Technology 57 (2008) 321–324.
13. D.J. Thomas, Effect of mechanical cut-edges on the fatigue and formability performance of advanced high-strength steels, Journal of Failure Analysis and Prevention 12 (2012) 518–531.
14. T. Matsuno, Y. Kuriyama, H. Murakami, S. Yonezawa, H. Kanamaru, Effects of punch shape and clearance on hole expansion ratio and fatigue properties in punching of high strength steel sheets, Steel Research International 81, Supplement Metal Forming 2010 (2010) 853–856.
15. T. Nakano, Introduction of Flow Control Forming (FCF) for sheet forging and new presses, in Proceedings of the International Seminar on Precision Forging, Kyoto, 1973, p. 35–40.
16. M. Merklein, J.M. Allwood, B.-A. Behrens, A. Brosius, H. Hagenah, K. Kuzman, K. Mori, A.E. Tekkaya, A. Weckenmann, Bulk sheet metal forming, CIRP Annals – Manufacturing Technology 61 (2012) 725–745.
17. P. Kadarno, K. Mori, Y. Abe, Two-stage plate forging for thickening hole edge of punched high strength steel sheet, in Proceedings of 14th International Conference on Metal Forming, Krakow, 2012, p. 899–902.
18. S. Thipprakmas, M. Jin, M. Murakawa, Study on flanged shapes in fineblanked-hole flanging process (FB-hole flanging process) using finite element method (FEM), Journal of Materials Processing Technology 192–193 (2007) 128–133.
19. V. Tvergaard, Mechanical modelling of ductile fracture, Meccanica 26 (1991) 11–16.

7

Innovative cold joining technologies based on tube forming

Luis M. Alves, Carlos M.A. Silva, and Paulo A.F. Martins*

IDMEC, Instituto Superior Técnico, Universidade de Lisboa, Av. Rovisco Pais, 1049-001 Lisboa, Portugal

Abstract – This paper is focused on innovative cold joining technologies for connecting tubes and fixing tubes to sheets. The proposed technologies are based on the utilization of plastic instability waves in thin-walled tubes subjected to axial compression and may be seen as an alternative to conventional joining technologies based on mechanical fixing with fasteners, welding and structural adhesive bonding. Besides allowing connecting dissimilar materials and being successfully employed in fixture conditions that are difficult and costly to achieve by means of conventional joining the new proposed technologies also cope with the growing concerns on the demand, lifecycle and recycling of materials.

Key words: Joining, Cold forming, Tubes, Experimentation, Finite element modelling

1. Introduction

Modern lightweight structures employed in architecture, engineering and building construction make use of tubular trusses in which tubes are connected to each other or fixed to sheets. Tubular joints are also essential elements in air-conditioning, refrigeration, heat-exchangers, supply lines and pipelines to convey fluids from one location to another.

In case of connecting tubes, there are several well-known types of joints based on commercially available tee fittings, saddle adapters and weld-o-lets for standard geometries and materials, such as carbon steel, stainless steel, copper and polyethylene, among other thermoplastics (Figures 1a–1c).

The main advantage of connecting tubes by means of commercially available tee joints is that it is a cost effective solution whenever the quantity of connections to be performed is significant. However, the utilization of these types of joints is limited by industry standards and always requires cutting and preparation of tube ends, welding or brazing and, in most cases, quality inspection of the welds.

Custom based solutions are generally based on the utilization of on nozzle-weld and spin-formed joints (Figures 1d and 1e). Nozzle-weld joints (Figure 1d) require cutting a hole in the main tube, shaping a contoured end in the branch tube to match the diameter of the main tube and welding along the contour. Spin-formed joints (Figure 1e) also requires cutting a hole in the main tube but the difference is that material around that hole is subsequently shaped into a tee fitting where the branch tube will be brazed or welded.

In case of fixing tubes to sheets, the most widespread technologies are based on the utilization of fasteners (nuts and bolts or rivets), welding and structural adhesive bonding (Figures 1g–1i). The majority of these joints are simple to design and easy to assemble and disassemble but their range of application is often limited by various technical constraints.

For example, mechanical couplings based on fasteners are limited by the maximum loading capacity that nuts, bolts and rivets can support safely.

Welding joints suffer from dimensional inaccuracies and heat-affected zones resulting from the heat-cooling cycles, from difficulties of welding dissimilar materials (e.g., joining steel or aluminium tubes to aluminium or copper sheets), from the production of undesirable fumes and smokes in fabrication, and from expensive and time consuming routes related to the above mentioned quality inspection procedures.

Structural adhesive bonding requires careful surface preparation with tight tolerances and is accomplished by placing a thin film of liquid or semisolid adhesives between the counterfacing surfaces of the tubes and sheets to be joined. The procedure needs time for the adhesive to cure and the resulting joints may experience decrease in performance over time under adverse environmental conditions.

From what was mentioned before, it can be concluded that no matter the application and the differences between the technologies that are currently available for connecting tubes (Figures 1a–1e) and fixing tubes to sheets (Figures 1g–1i), their universe of application is always limited by aesthetic, physical, chemical, and mechanical requirements.

*e-mail: pmartins@ist.utl.pt

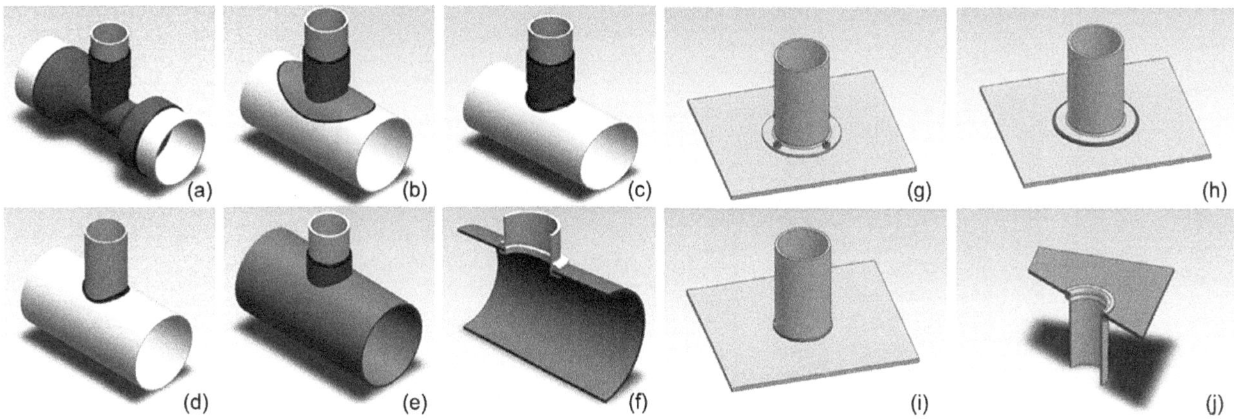

Figure 1. Conventional and new proposed technologies for connecting tubes and fixing tubes to sheets.

Table 1. Summary of the main features of joining by forming and by welding.

	Joining by forming	Joining by welding
Mechanism	Plastic deformation	Melting with addition of filler materials
Shape of the connections	Arbitrary geometries	Limited to butt, lap, corner and edge joints
Operating temperature	Ambient	Melting point
Heat-affected zones	No	Yes
Shielding gases	No	Yes
Materials	Metals and polymers	Metals (similar)
Coated materials	Possible	Very difficult or impossible
Energy consumption	Less	More
Productivity	More	Less
Cost	Less	More
Environmental friendliness	More	Less

Recent developments in joining by forming that are comprehensively systematized in the state-of-the-art reviews by Mori et al. [1] and Groche et al. [2] allow concluding that plastic deformation offers great potential to connect tubes and to fix tubes to sheets. Moreover, joining by forming combines the growing demands for high productivity, low fabrication costs and environmental friendliness with high performance and material versatility. Table 1 summarizes the main differences between joining by forming and by welding.

This paper is built upon recent developments in joining by forming and presents innovative technologies that make use of axisymmetric or asymmetric plastic instability waves in thin-walled tubes subjected to axial compression for connecting tubes and fixing tubes to sheet panels at room temperature in situations where the axis of the branch tube or sheet is perpendicular or inclined to the axis of the main body tube. In case of connecting tubes by cold forming, the paper will also be focus on recent developments in the joining of tubes by their ends.

This paper aggregates several new technologies of joining by forming that were recently developed and published by the authors [3, 5]. These technologies are not only capable of connecting tubes and fixing tubes to sheets made from dissimilar materials (e.g. metals and polymers) as they can be successfully employed in fixture conditions that are difficult and costly to achieve by means of conventional joining technologies.

Figures 1f and 1j present schematic examples of the application of the new proposed technologies for cases in which the axis of the branch tube or sheet is perpendicular to the axis of the main body tube that will be comprehensively described in the presentation.

2. New joining technologies

2.1. Inclined connections

The new proposed technologies for connecting tubes [3] and fixing tubes to sheets [4] are schematically shown in Figure 2. The idea behind the technologies is based on controlling the development and propagation of plastic instability waves in thin-walled tubes subjected to axial compression beyond the bifurcation point of the load-displacement curve.

In case of tube joining (Figure 2a) the left and middle schematic drawings show the upper and lower dies that are needed to trigger and propagate inclined, out-of-plane, instability waves between contoured dies at the open and closed positions. The rightmost schematic drawing shows an application of this

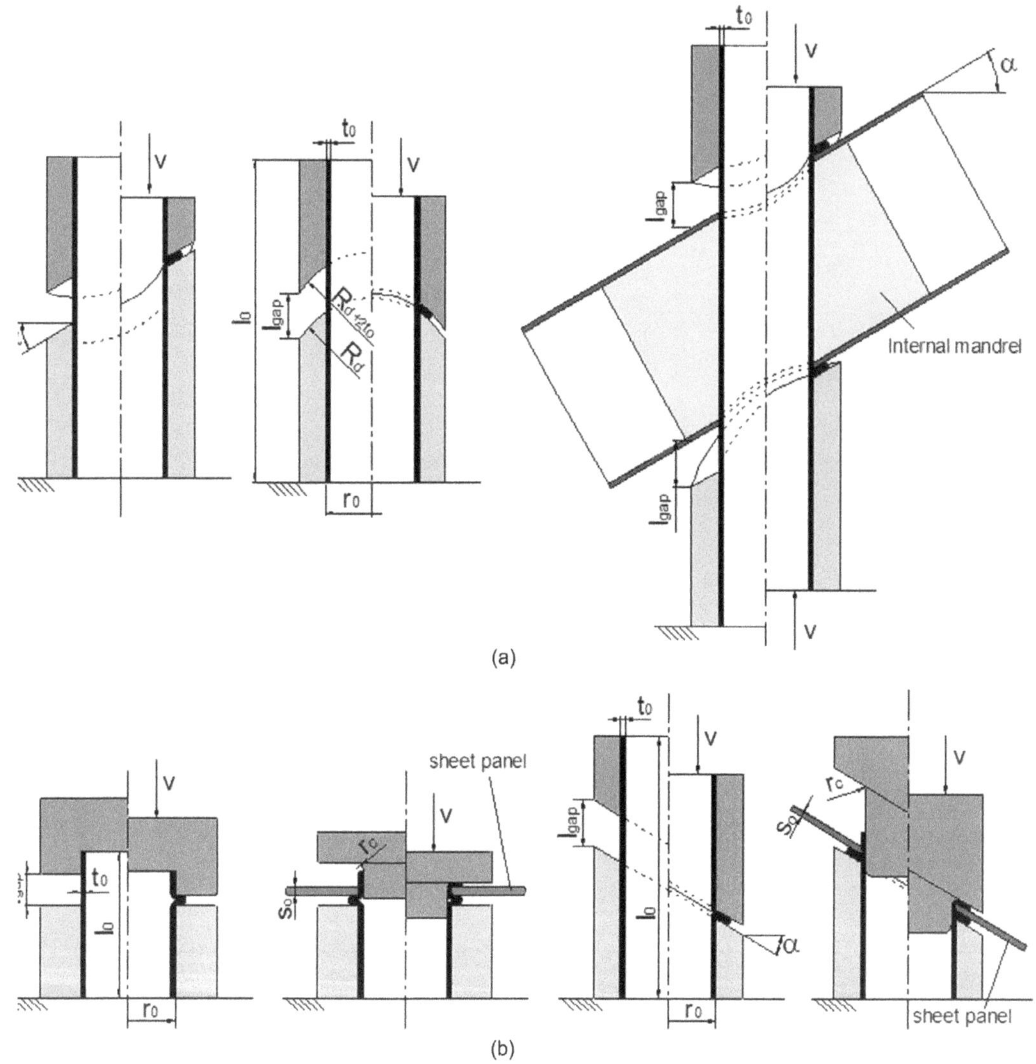

Figure 2. (a) Sectional schematic views of the tool system to develop and propagate inclined, out-of-plane, instability waves in tubes and to produce inclined tube joints. (b) Sectional schematic views of the tool system to develop and propagate inclined, in-plane, instability waves to fix tubes to sheets.

concept for producing inclined tube joints. The sectional views show the active tool components consisting of the upper and lower contoured dies and the internal mandrels (if present). The internal diameter of the dies is dedicated to a specific reference radius r_0 of the main body tube. The radius R_d of the parting out-of-plane surface of the dies together with its inclination α to the axis of the main body tube are dedicated to a specific instability wave or, in case of tube joining, to a specific radius of the branch tube. The difference between the radius R_{d+2t_0} and R_d of the upper and lower parting surfaces is crucial to accommodate the plastic compression bead at the end of stroke. The initial gap opening l_{gap} between the upper and the lower contoured dies controls triggering and propagation of the plastic instability waves namely, the number, width and relative position of the compression beads along the axis of the main tube.

Figure 2b shows two different tool systems utilized for fixing tubes to sheets that also make use of axisymmetric (leftmost setup) or asymmetric (rightmost setup), in-plane, plastic instability waves. The joints are accomplished by compression beading and axisymmetric or asymmetric tube end flaring. The plastic instability waves are produced in appropriate flat or contoured dies whereas flaring is accomplished by compressing the upper tube end with and appropriate radiused punch in order to expand material outwards and to form a single-lap inclined flange.

In connection to what was said above, the main operating parameters that are considered relevant to the new technologies for connecting tubes and fixing tubes to sheets are: (i) the slenderness ratio l_{gap}/r_0 between the initial gap opening and the outer radius of the tube, (ii) the ratio t_0/r_0 between the wall thickness and the outer radius of the tube, (iii) the mandrel (if present) and (iv) the tribological conditions.

Other relevant operating parameters that are process dependent are: (i) the inclination angle α of the contoured dies, (ii) the radius r_c of the flaring punch and (iii) the angle β of the tube end chamfers.

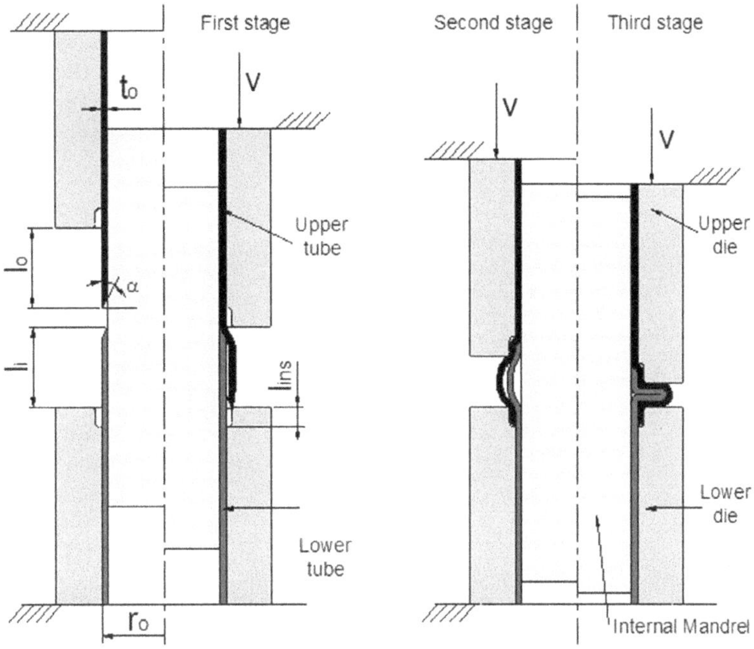

Figure 3. Sectional schematic views of the tool system to connect two tubes by their ends.

2.2. End-to-end connections

The new proposed technology for joining two tubes by their ends [5] is schematically shown in Figure 3. As seen from the open, intermediate and closed positions of the tool system, joining is accomplished by a sequence of three different cold forming stages that are carried out sequentially in a single stroke: (i) expansion, (ii) local buckling and (iii) clamping by mechanical locking.

Expansion is performed by forcing the upper tube against the chamfered end of the lower tube in order to enlarge the unsupported height l_0 of the upper tube radially and to create adjacent counterfacing surfaces between the two tubes to be joined. During this stage the chamfered end of the lower tube acts like a tapered punch (refer to "first stage" in Figure 3).

Once the unsupported height l_0 of the upper tube reaches the end of the depth of insertion l_{ins} resulting from the radial clearance between the tube and the upper end of the lower die, there is a sudden change in material flow and plastic instability waves are triggered as a result of local bucking under axial compression loading (refer to "second stage" in Figure 3). The internal mandrel ensures that the plastic instability waves are formed exclusively outwardly so that design specifications of the inner diameter of the tube joint are met and alterations in the flow of liquids or gases are prevented.

Finally, propagation of the plastic instability waves under continuous axial compression loading clamps the adjacent counterfacing surfaces of the two tubes by mechanical locking (refer to "third stage" in Figure 3). It is worth noting that although the new proposed joining process has been developed for carbon steel tubes it can also be used in tubes made from other metals or thermoplastics such as polyethylene (PE) and polypropylene (PP) [6].

The main operating parameters of this technology are similar to those listed in Section 2.1 but it is worth noticing that $l_{gap} \cong l_0 + l_i + 2l_{sc}$ in case of joining tubes by their ends and that the depth of the side clearance l_{sc} that controls the insertion of the tubes in the lower and upper dies also plays a role in the process.

3. Experimentation

3.1. Mechanical characterization of the materials

The experimental development of the new proposed technologies were performed on commercial S460MC (carbon steel) welded tubes and commercial PVC-U PN10 (unplasticized polyvinylchloride). The S460MC tubes were supplied with an outer radius $r_0 = 16$ mm and a wall thickness $t_0 = 1.5$ mm, and were utilized in the "as-received" condition. The PVC-U PN10 tubes were supplied with an outer radius $r_0 = 16$ mm and a wall thickness $t_0 = 1.6$ mm, and were manufactured in compliance with EN 1452-2 standard for a maximum operating pressure of 10 bar.

The stress-strain curve of the S460MC tubes was determined by combining tensile and stack compression tests carried out at room temperature on a hydraulic testing machine (Instron SATEC 1200 kN). In contrast, the strength differential effect of PVC requires determination of the individual stress-strain curves for tension and compression by means of tensile and stack compression tests (Figure 4a).

The tensile test specimens were machined from the supplied tube stock in accordance to the ASTM D 638 (PVC-U) and ASTM E8/E8M–09 (S460MC) standards. The stack compression test specimens consisted of multi-layer cylinders that were assembled by pilling up 3 circular discs with 12 mm

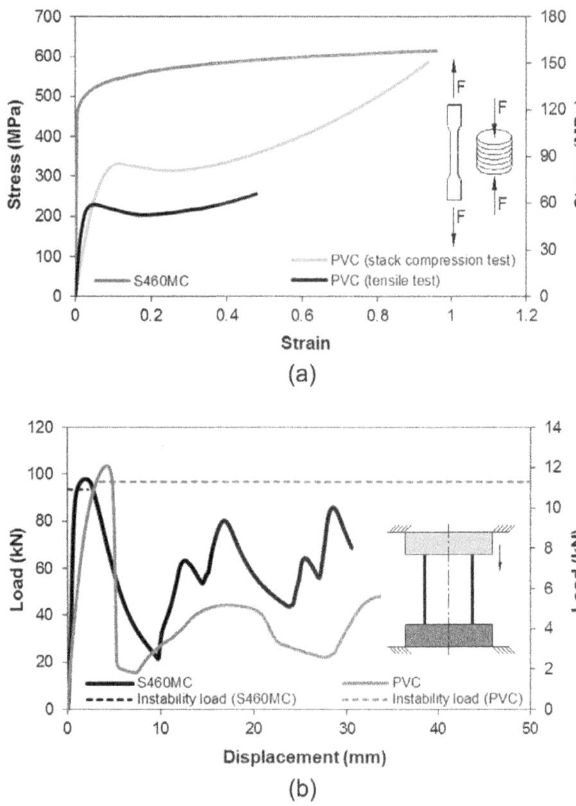

Figure 4. Mechanical characterization of the commercial S460MC and PVC tubes. (a) Stress-strain curves and (b) load-displacement curves and critical instability loads for the axial compression of tubes between flat parallel platens. The differences in strength of both materials justify the existence of two vertical axes in both plots. S460MC carbon steel refers to the left vertical axis while PVC refers to the right vertical axis.

diameter and 1.6 mm (PVC) or 1.5 mm (S460MC) thickness cut out from the supplied tubes by a hole-saw. All the tests were carried out at room temperature on a hydraulic testing machine (Instron SATEC 1200 kN) with a cross-head speed equal to 10 mm/min.

The critical instability load for the occurrence of local buckling was determined by compressing PVC and S460MC steel tubes with 100 mm initial length between flat parallel platens. As seen in Figure 4b, the shape of the experimental load-displacement curves that were obtained for the tubular specimens of each material are similar but the absolute values of their critical instability loads F_{cr} are very different. In fact, the critical instability load of S460MC carbon steel tubes is 93.5 kN while that of PVC is 11.3 kN.

3.2. Testing conditions

The experimental setups associated to the different joining technologies that are schematically shown in Figures 2 and 3 were installed in the hydraulic testing machine (Instron SATEC 1200 kN) where the mechanical characterization of the tubes had previously been performed. The experiments were carried out at room temperature with a cross head speed equal to that utilized in the mechanical characterization of the tube materials with the purpose of identifying the modes of deformation and setting up the process feasibility window of the new joining technologies as a function of a selected number of operating parameters. The influence of some of these parameters will be provided later in Section 5.

4. Numerical modelling

Because the experimental development of the new joining technologies for connecting tubes and fixing tubes to sheets was performed with a quasi-static constant displacement rate of the upper-table of the hydraulic testing machine, no inertial effects on plastic deformation are likely to occur and therefore no dynamic effects in the joining process are needed to be considered. These operating conditions allowed numerical modelling to be performed with the finite element flow formulation and enabled the authors to utilize their in-house computer program I-form that has been extensively validated against experimental measurements of metal forming processes since the end of the 1980's [7].

I-form is based on the finite element flow formulation and includes an extension to pressure-sensitive polymers that allows modelling metal and polymer tubes simultaneously. The extension of the flow formulation to pressure-sensitive polymers is based on the utilization of a non-associated flow rule in which the plastic potential $Q = J_2$ commonly utilized in metallic materials is not retrieved from the yield function $F(\sigma_{ij})$ that explicitly accounts for the strength-differential effect resulting from the differences between tensile σ_T and compressive σ_c flow stresses [8],

$$F(\sigma_{ij}) = \bar{\sigma}^2 - \sigma_C \cdot \sigma_T + (\sigma_C - \sigma_T)\,\sigma_{kk} = 0, \tag{1}$$

where $\bar{\sigma} = \sqrt{\frac{3}{2}\sigma'_{ij}\sigma'_{ij}}$ is the effective stress and $\sigma_{kk} = \delta_{ij}\,\sigma_{ij}$.

The non-associated flow rule where $Q = J_2$ and $\dot{\lambda}$ is a scalar factor of proportionality is given by,

$$\dot{\varepsilon}^p_{ij} = \dot{\lambda}\frac{\partial Q(\sigma_{ij})}{\partial \sigma_{ij}}. \tag{2}$$

The finite element flow formulation giving support to I-form is built upon the following extended variational statement that accounts for contact with friction between rigid and deformable objects:

$$\Pi = \int_V \bar{\sigma}\,\dot{\bar{\varepsilon}}\,\mathrm{d}V + \frac{1}{2}K\int_V \dot{\varepsilon}_v^2\,\mathrm{d}V - \int_{S_T} T_i u_i\,\mathrm{d}S + \int_{S_f}\left(\int_0^{|u_r|}\tau_f\,\mathrm{d}u_r\right)\mathrm{d}S, \tag{3}$$

where, $\dot{\bar{\varepsilon}}$ is the effective strain rate, $\dot{\varepsilon}_v$ is the volumetric strain rate, K is a large positive constant enforcing the incompressibility constraint of both metals and polymers and V is the control volume limited by the surfaces S_U and S_T, where velocity and traction are prescribed. Friction at the contact interfaces S_f is treated as a traction boundary condition and the additional power consumption term is modelled through the utilization of the law of constant friction $\tau_f = mk$.

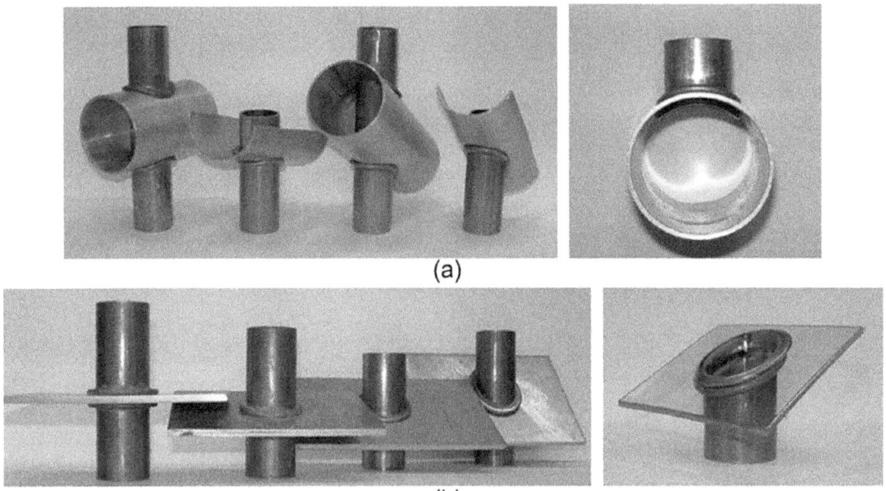

Figure 5. Application of the new proposed technology for (a) connecting full-size and half-sectioned tubes and for (b) fixing tubes to sheets (metallic and polymeric) at different inclination angles.

The contact between deformable bodies (e.g. between the counterfacing surfaces of the tubes or between the tubes and sheets, among others) was modelled by means of a non-linear procedure based on a penalty approach. The procedure is comprehensively described by Nielsen et al. [9] and has the advantage of avoiding additional degrees of freedom as in case of alternative solutions based on the utilization of Lagrange multipliers.

The tubes and sheets were discretized by means of 2D and 3D finite element models as a function of the rotational symmetry conditions of the joining operation. In case of rotational symmetry, the cross section of the tubes and sheets was discretized by means of quadrilateral elements across the thickness. In the absence of rotational symmetry, the discretization is fully three-dimensional and performed by means of hexahedral elements. The dies and mandrels were treated as rigid bodies.

5. Results and discussion

5.1. Inclined connections

Figure 5a shows applications of the proposed technology for connecting tubes (or half-tubes) in situations where the axis of the branch tube is perpendicular or inclined to the axis of the main body tube. Figure 5b shows applications for fixing tubes to sheets in situations where the axis of the sheet is perpendicular or inclined to the axis of the main body tube. The solutions can, for example, be successfully employed in the seat bottom frame of automotives as an alternative to welding in order to reduce costs and eliminate heat distortion problems. All the connections shown in Figure 5 were performed between welded carbon S460MC steel tubes, seamless aluminium AA6062 tubes, aluminium AA5754 sheets and polycarbonate sheets.

Figure 6 shows the results obtained from the numerical modelling of inclined connections. As seen, Figure 6a presents the initial, intermediate and final predicted geometry of a tube attachment where the axis of the branch tube (labelled as "Tube B") is inclined by 30° to the axis of the main body tube (labelled as "Tube A"). Joining is performed by means of two inclined, out-of-plane, plastic instability waves that closely match the intersection of the two tubes. Two different types of mandrels are employed; (i) a conventional internal mandrel placed inside the main body tube ("Mandrel A") and (ii) a special purpose internal sectioned mandrel made of two different parts ("Mandrel B") placed inside the branch tube. The conventional mandrel avoids the development of unacceptable inward plastic flow during triggering and propagation of the plastic instability waves whereas the special purposed sectioned mandrel (allowing for the easy removal of mandrel in practice) prevents the compression beads to plough into the branch tube. The branch tube behaves as a sleeve during the entire joining process and its internal sectioned mandrel contributes decisively to ensure the overall success of the inclined joining process.

Figure 6b, allows understanding the role played by the internal mandrel in preventing the development of plastic instability waves exhibiting both inward and outward plastic flow during the fixing of tubes to sheets in situations where the axis of the sheet is perpendicular or inclined to the axis of the main body tube (refer also to Figure 5b). The compression beads resulting from these plastic instability waves exhibiting both inward and outward plastic flow would lead to non-acceptable joints and, therefore, justify the reason why the utilization of internal mandrels is mandatory in order to ensure the overall quality and tolerances that are required for the tube-sheet inclined connections.

5.2. End-to-end connections

Figure 7a shows a photograph of several tube specimens that were connected by their ends using the new proposed technology. The observation of the real and finite element predicted

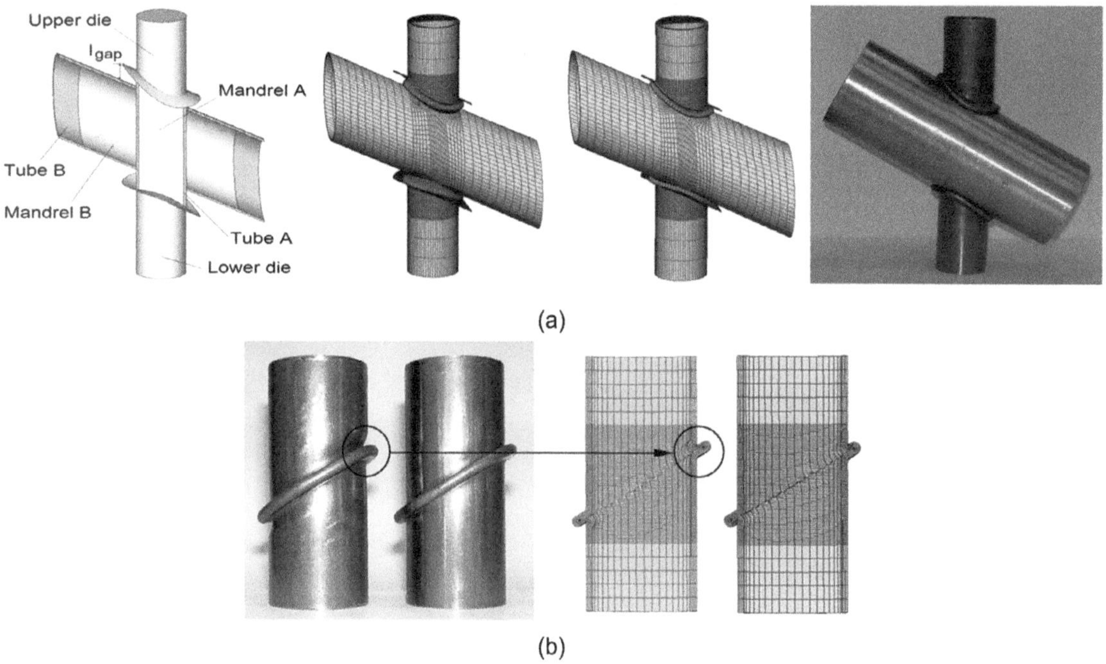

Figure 6. (a) Finite element model and predicted geometries at the middle and end of stroke with photograph of an inclined connection ($\alpha = 30°$) between the two tubes. (b) Photograph and finite element predicted geometries disclosing the influence of the internal mandrel in the development of sound asymmetric, in-plane, plastic instability waves for fixing tubes to sheets.

Figure 7. Modes of deformation associated to end-to-end joining of S460MC tubes by cold forming. (a) Experimental joints and (b) experimental and finite element predicted geometries of the cross sections for specimens with different values of the slenderness ratio l_{gap}/r_0.

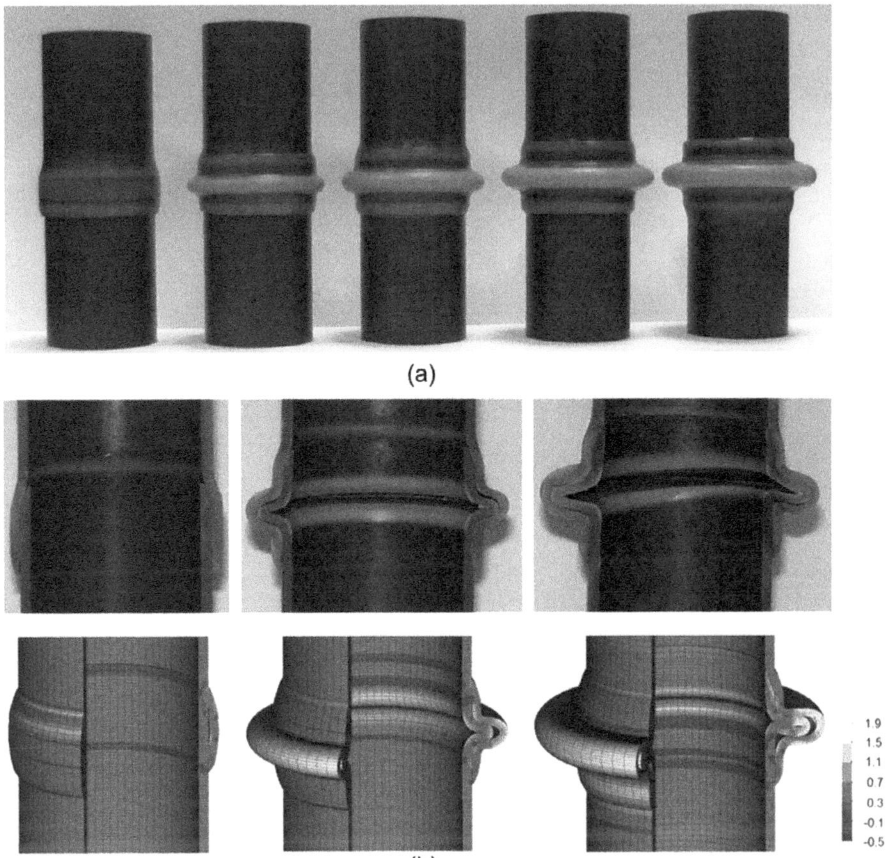

Figure 8. Modes of deformation associated to end-to-end joining of PVC tubes by cold forming. (a) Experimental joints and (b) experimental and finite element predicted geometries of the cross sections for specimens with different values of the slenderness ratio l_{gap}/r_0.

cross sections in Figure 7b allows concluding that the leftmost test sample (corresponding to $l_{gap}/r_0 = 1.9$) does not ensure locking between the two tubes whereas the rightmost test sample (corresponding to $l_{gap}/r_0 = 4.4$) presents a joint with two compressions beads instead of one.

In case of the leftmost test sample, the absence of connection is because the initial unsupported gap height l_{gap} is not big enough to allow compression beads to develop and lock with each other by plastic instability. In case of the rightmost test sample, the formation of two compression beads instead of one is due to the fact that high values of the initial unsupported gap height l_{gap} provide conditions for the development of multiple compression beads that will interfere and be placed on top of each other, as they are formed in-between the upper and lower dies.

Neither the operative conditions corresponding to the leftmost test sample nor those corresponding to the rightmost test sample are acceptable for connecting the two tubes by their ends. The process window is therefore restricted to values of the slenderness ratio l_{gap}/r_0 in the range between the two abovementioned limits. However, it is worth noting that the process window must not be confused with the potential range of applicability because the limits on the slenderness ratio l_{gap}/r_0 only define the range of values of the unsupported heights of the upper and lower tubes that need to be utilized for successfully connecting any two tubes by their ends.

Figure 8 shows the application of the proposed technology to the connection of two PVC tubes by their ends. Similarly to what authors observed in metallic tubes, the end-to-end joining of PVC tubes by cold forming is dependent upon the development and propagation of sound axisymmetric plastic waves at the gap opening l_{gap} between the upper and lower dies.

When the slenderness ratio l_{gap}/r_0 has a value smaller than the minimum threshold it is not possible to connect the PVC tubes by their ends because the initial gap opening l_{gap} is not big enough to allow plastic instability waves to develop and clamp the two tubes by mechanical locking (refer to the leftmost sample of Figure 8a, where $l_{gap}/r_0 = 0.56$. In contrast, when the slenderness ratio l_{gap}/r_0 has a value larger than the maximum threshold it is not possible to ensure a sound and reliable connection of the PVC tubes by their ends (refer to the rightmost sample in Figure 8a where $l_{gap}/r_0 = 3.75$). Values of the slenderness ratio l_{gap}/r_0 in-between the leftmost and rightmost samples in Figure 8a are acceptable for connecting the two PVC tubes by their ends by means of cold forming. The process window is therefore characterized by values of the slenderness ratio l_{gap}/r_0 in the range between 1.8 and 3.2.

The observation of the different modes of deformation in the end-to-end joining of PVC tubes by their ends adds a new physical phenomenon known as "stress whitening" that does not occur in metallic tubes (Figure 8b). Stress whitening consists in the change of color with plastic deformation and is attributed to the occurrence of crazing in the regions of the PVC tubes subjected to high values of tensile stresses. This phenomenon is particularly important in PVC and Figure 8b includes the finite element predicted distribution of crazing in accordance with the normalized version of Bucknall's modification of the stress-bias criterion due to Sternstein and Ongchin [10].

6. Conclusions

The new proposed technologies for connecting tubes and tubes to sheets by triggering and controlling the propagation of plastic instability waves in thin-walled tubes offer several advantages as compared with conventional technologies based on mechanical fixing with fasteners, welding or structural adhesive bonding because they are:

- flexible solutions capable of handling small, medium or large batch sizes with different geometries and high levels of repeatability in production line;
- environmentally friendly solutions that allow savings in raw material and eliminate filler materials and shielding gases;
- energy saving solutions that eliminate heat-cooling cycles as well as heat affected zones in the regions of the tubes and sheet panels that are joined together;
- value added solutions that are capable of connecting tubes and fixing tubes to sheet panels made of dissimilar materials;
- cost-efficient solutions that require low amount of capital investment because they can be designed to operate with existing machine-tools.

Because the new proposed technologies can be successfully employed in fixture conditions that are difficult and costly to achieve by means of conventional technologies they can also be utilized to foster innovative ideas in product development.

Acknowledgements. The authors would like to acknowledge the support provided by Fundação para a Ciência e a Tecnologia of Portugal under LAETA – UID/EMS/50022/2013 and by MCG – Mind for Metal, Carregado, Portugal.

References

1. K.I. Mori, N. Bay, L. Fratini, F. Micari, A.E. Tekkaya, Joining by plastic deformation, CIRP Annals of Manufacturing Technology 62 (2013) 673–694.
2. P. Groche, S. Wohletz, M. Brenneis, C. Pabst, F. Resch, Joining by forming – A review on joint mechanisms, applications and future trends, Journal of Materials Processing Technology 214 (2014) 1972–1994.
3. A. Gonçalves, L.M. Alves, P.A.F. Martins, Tube joining by asymmetric plastic instability, Journal of Materials Processing Technology 214 (2014) 132–140.
4. A. Gonçalves, L.M. Alves, P.A.F. Martins, Inclined tube-sheet plastically deformed joints, Steel Research International 85 (2014) 67–75.
5. C.M.A. Silva, C.V. Nielsen, L.M. Alves, P.A.F. Martins, Environmentally friendly joining of tubes by their ends, Journal of Cleaner Production 87 (2015) 777–786.
6. L.M. Alves, C.M.A. Silva, P.D. Santos, P.A.F. Martins, Mechanical joining of PVC tubes by their ends, Journal of Materials: Design and Applications, in press (2015), DOI: 10.1177/1464420715586609.
7. M.L. Alves, J.M.C. Rodrigues, P.A.F. Martins, Three-dimensional modelling of forging processes by the finite element flow formulation, Journal of Engineering Manufacture 218 (2004) 1695–1707.
8. L.M. Alves, P.A.F. Martins, Nosing of thin-walled PVC tubes into hollow spheres using a die, International Journal of Advanced Manufacturing Technology 44 (2009) 26–37.
9. C.V. Nielsen, W. Zhang, L.M. Alves, N. Bay, P.A.F. Martins, Modelling of thermo-electro-mechanical manufacturing processes with applications in metal forming and resistance welding, Springer-Verlag, London, UK, 2013.
10. C.B. Bucknall, New criterion for craze initiation, Polymer 48 (2007) 1030–1041.

8

The influences of filament temperature on the structure of boron nitride films and its tribological characterization for microforming die application

Yong Jin[1,*], Shigeo Yasuhara[2], Tetsuhide Shimizu[1], and Ming Yang[1]

[1] Graduate School of System Design, Tokyo Metropolitan University, Hino 191-0065, Japan
[2] Japan Advanced Chemicals, 3131-4 Tana, Chuo-ku, Segamihara-shi, Kanagawa 252-0244, Japan

Abstract – Boron nitride film was deposited on ⟨1 0 0⟩-oriented silicon substrate by hot filament assisted chemical vapor deposition. The $B[N(CH_3)_2]_3$ (Tris(dimetylamino)borane, TDMAB) was used as the single source precursor both for boron and nitride, and ammonia gas was used as the extra source to increase the N concentration in the films. Elemental composition of the films deposited under different filament temperatures were measured by energy dispersive X-ray (EDX) analysis, and the structure of the films were measured by X-ray diffraction (XRD) and Fourier transform infrared spectroscopy (FT-IR). The boron nitride films deposited under lower filament temperature was amorphous, while BN films contain hexagonal structure were deposited at higher filament temperature. To verify whether the films can be applied for microforming die, ball on disk test was carried out using pure titanium ball as the counterpart to investigate the interfacial behavior between the films and pure titanium. The results show that the reaction between the films and pure titanium was low as there was no titanium adhesion on the wear track when the film was remained.

Key words: Filament temperature, Boron nitride, Structure, Tribological, Microforming

1. Introduction

Because of the similar crystal structure and properties as diamond, cubic boron nitride (cBN) has gained more and more attention from the last decades. And some properties of cBN are superior to diamond, like high temperature oxidation resistance and chemical inertness to ferrous alloys. Due to these properties, cBN has a wide application field as cutting tools and wear resistant coatings used in metal forming [1].

Various techniques were used to deposit cBN films, such as physical vapor deposition (PVD) and chemical vapor deposition (CVD) methods. These methods like ion-beam assisted deposition [2], RF sputtering [3] and ion-assisted CVD [4] have been succeeded in depositing nano-cBN films. However, these methods involve energetic ion bombardment, which causes the high compressive stress in the films. This internal stress leads the films cracking and delamination when the films thickness exceeds to several hundreds of nanometers, and it is difficult to deposit thick cBN films over 1 μm without delamination.

Compared with the PVD methods or plasma assisted CVD methods with high stress in the films, a simple chemical way for deposition of thick cBN films would be of high interest for industrial applications because of reducing the stress in the film. Weissenbacher and Haubner [5] used hot filament CVD to decompose the Triethylborazine ($C_6H_{18}B_3N_3$) source on various substrates, and obtained the B, C and N containing layers under different deposition conditions. Dumont et al. [6] used the Tris(dimethylamino)borane (TDMAB) as the single source precursor, and deposited the BN films on Si substrate under different deposition temperature. However, some difficulties appear in the deposition of cBN film by CVD without energetic plasma bombardment, and there has no available thermodynamic data as well as the exact transition temperature between hBN and cBN.

In this work, BN films were deposited by hot filament assisted CVD method, and the TDMAB gas was used as the single source precursor which was decomposed and activated by the hot filament. The influence of different filament

*e-mail: jinyonghit@gmail.com

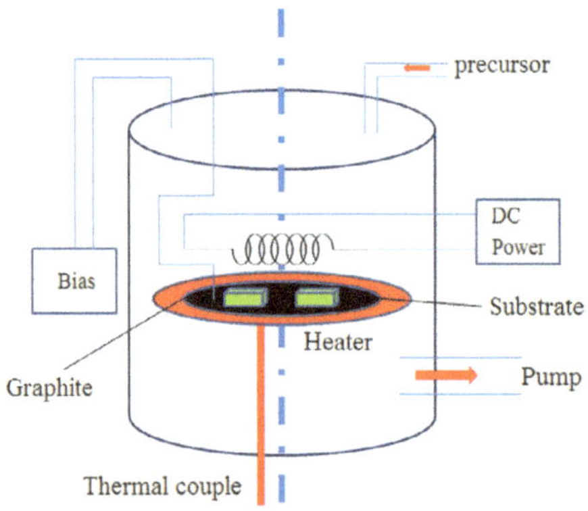

Figure 1. Schematic diagram of the deposition unit.

Table 1. BN films deposition conditions.

Substrate	*p*-type Si (1 0 0)
Working pressure (Pa)	40
Substrate temperature (°C)	700
Carrier gas flow rate (sccm)	30
Ammonia gas flow rate (sccm)	300
Bias voltage (V)	100
Filament temperature (°C)	1500, 1800, 2000
Filament distance (mm)	10
Deposition time (min)	30

Table 2. Ball-on-disk test conditions.

Ball material	Pure titanium
Ball diameter	6 mm
Rotation speed	50 rpm
Rotation radius	3 mm
Loading	1 N
Ending condition	100 laps

temperature on the structure of BN film was studied. On the other hand, in order to investigate the possibility of the films applied to microforming die for forming pure titanium, the tribological properties and interfacial behavior between the films and pure titanium were investigated by using ball on disk test.

2. Experiment

BN films were deposited on ⟨1 0 0⟩-oriented silicon substrate by hot filament assisted CVD. Figure 1 shows the schematic diagram of the experimental set up which was designed by Japan Advanced Chemicals Ltd. The $B[N(CH_3)_2]_3$ (TDMAB) gas was chose as the single source precursor since its nontoxic properties and the fixed ratio of boron and nitride. BN films were deposited under different filament temperatures which were controlled by a direct-current supply power. Since the N content in the films were much lower than the B content, ammonia gas was used as the extra gas to increases the N content in the films. Si substrates were cleaned by acetone and ethanol in an ultrasonic bath before deposition, and put on a graphite substrate which was heated to 700 °C by a SiC ceramic heater. The detail deposition conditions are summarized in Table 1.

All the films obtained were analyzed by energy dispersive X-ray (EDX) analysis to estimate the elemental compositions. X-ray diffraction (XRD) and Fourier transform infrared spectroscopy (FT-IR) were used to identify the structure of the films. The thickness of the film was measured by atomic force microscopy (AFM). The interfacial behavior and friction coefficient of the films were measured by ball-on-disk (BOD) tribometer. A constant normal force of 1 N was applied for all the tests, and the force corresponds to an initial Hertzian mean contact pressure of about 343 MPa [7]. The details of BOD test conditions are summarized in Table 2.

3. Results and discussion

3.1. Measurement of the BN films

3.1.1. Elemental composition and structure of the films

Table 3 shows the elemental composition of the films deposited with different filament temperature measured by EDX analysis. The elemental composition of the films was influenced by the different filament temperature. With increasing the filament temperature, the N concentration in the films increased. There is no C contains in the films when the filament temperature increased to 1800 °C. The thicknesses of the films deposited under different filament temperatures were 133 nm, 265 nm and 316 nm, respectively. This suggested that the higher temperature of filament makes the precursor being decomposed completely, and the deposition rate of the films were increased with increasing the filament temperature.

Table 3. Element compositions of the films deposited under different filament temperatures.

Samples	Filament temperature (°C)	Element compositions (at.%)				
		B	N	C	O	Si
No. 1	1500	52.50	14.96	13.46	2.16	16.93
No. 2	1800	61.50	22.24	0	0.57	15.67
No. 3	2000	61.84	25.66	0	0.77	11.73

The structures of the films were analyzed by FT-IR and XRD analysis. The FT-IR analysis is a most important and widely used tool for characterizing BN films, for hBN, the two IR-active phonons have TO frequencies at about ~780 and ~1370 cm^{-1}, respectively. The 1370 cm^{-1} mode is a stretching of the B-N bond within the basal plane, and the 780 cm^{-1} mode is a bending of the B-N-B bond between the basal planes [1]. Figure 2 shows the FT-IR results of the films as shown in Table 3. The FT-IR spectra of the films deposited with different filament temperature were different. At the lower filament temperature, no clear peak was observed, the spectra looks smooth and with some broad peaks. When the filament temperature increased to 1800 °C, two peaks with wide wavenumber were detected. The intensity of the peaks increase and become narrow when the filament temperature increased to 2000 °C. These two peaks correspond well to the peaks of hBN in the FT-IR spectra [1]. It indicated that

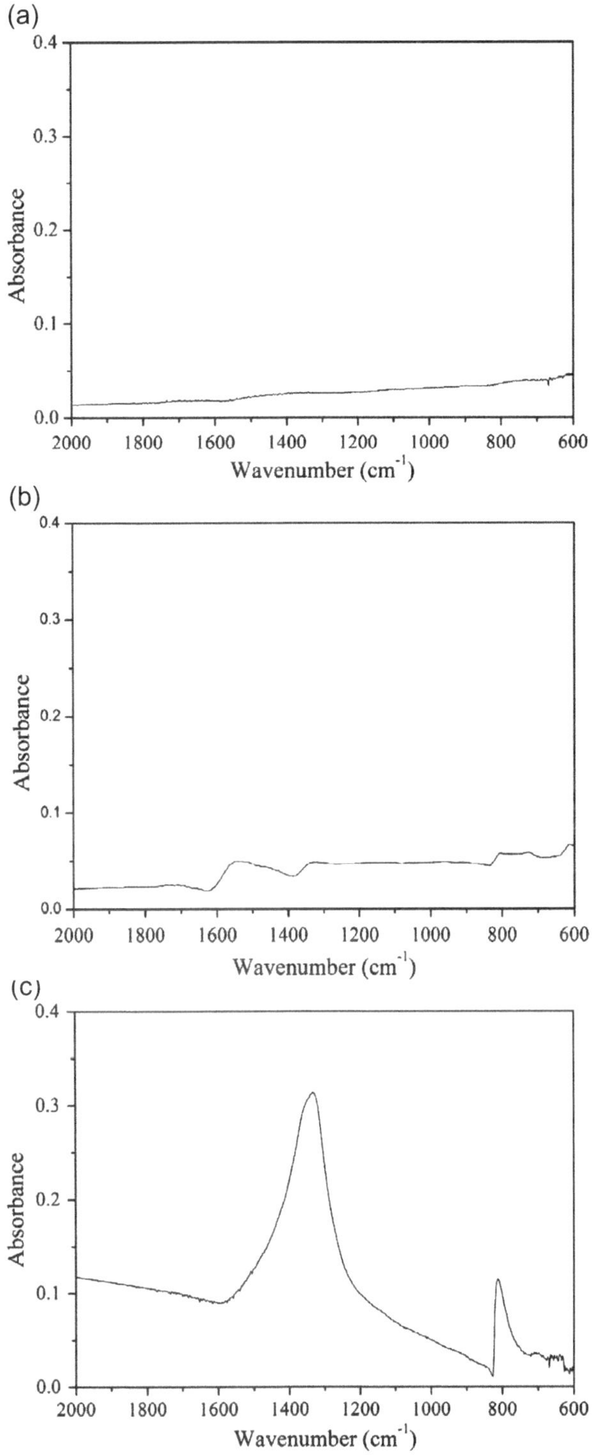

Figure 2. FT-IR results of the films deposited under different filament temperatures: (a) 1500 °C, (b) 1800 °C, (c) 2000 °C.

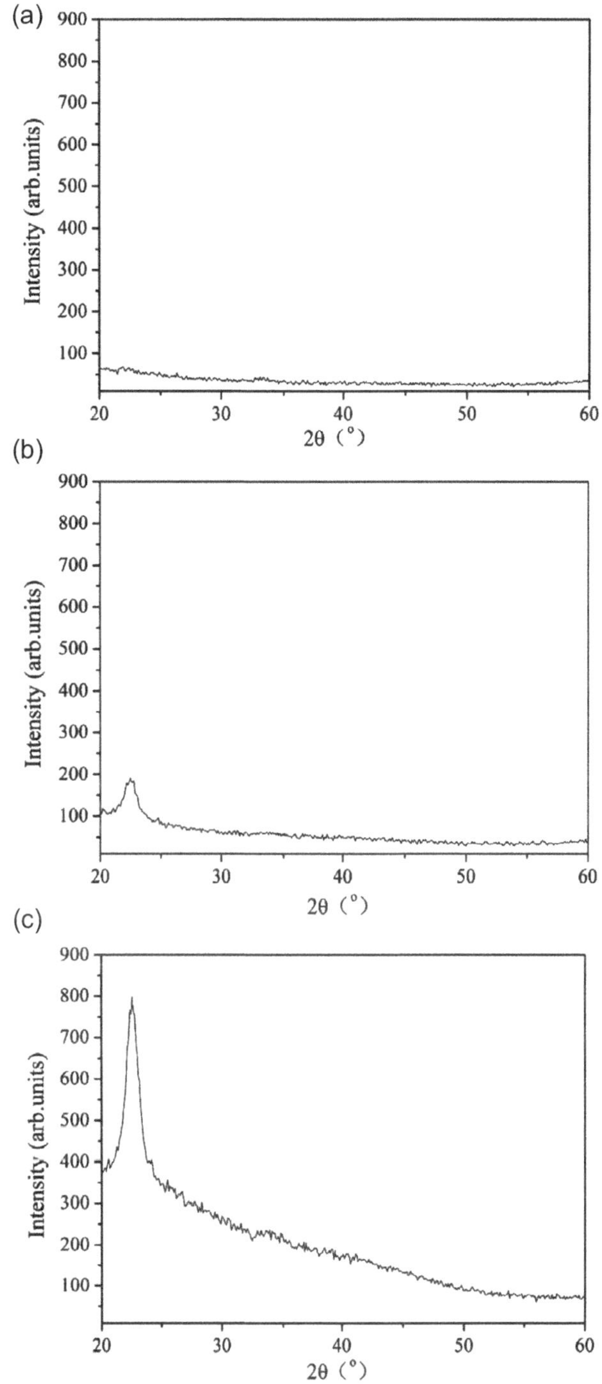

Figure 3. XRD results of the films deposited with different filament temperatures: (a) 1500 °C, (b) 1800 °C, (c) 2000 °C.

the film deposited under lower filament temperature has disordered structures or amorphous, and hBN films could be deposited under higher filament temperatures.

Figure 3 shows the XRD results of the films as shown in Table 3. It can be found that no peak was detected in the XRD spectra when the film was deposited under lower filament temperature. As the filament temperature increased to 1800 °C, a peak around 23° was observed, and the intensity of this peak increases as the filament temperature further increased. It indicated that the crystallization degree of the films was increased with increasing the filament temperature. However, the (0 0 2) diffraction peak position for hBN is reported at $2\theta = 26.8°$ [8], the peak shift to the small angle may be caused by the rich B content in the film. The excessive B atom may incorporate into the structure, and causes the lattice constant increase. According to Bragg's low [9], the angle would be decreased if the lattice constant increase. This needs further investigation in the future.

From the FT-IR and XRD measurements, it suggested that the hBN films were deposited at higher filament temperature, and the crystallization degree of the BN films become better as the filament temperature increased. In order to investigate whether the hBN films can be applied for microforming die for forming pure titanium, the BOD tests were carried out to investigate the tribological properties and interfacial behavior between the BN films and pure titanium.

3.1.2. Tribological properties and interfacial behavior of the film

Figure 4 shows the friction coefficient of the films deposited with different filament temperature. From these results, it can be observed that the films deposited at lower filament temperatures show the higher and big vibration friction coefficient than the film deposited at higher filament temperature. The friction coefficient of the film deposited at higher filament temperature was different compare with the films deposited at lower filament temperature. At the initial stage of the BOD test, the friction coefficient is low and smooth, while it was increased rapidly and vibrated after 20 laps. After BOD test, many wear debris were found on the edge of the wear track. As the hBN film was very soft and it was easily to be removed, the wear debris on the wear track may increase the friction coefficient, and the friction coefficient increased after the film was worn out.

As the life time of the film was too short due to the thin thickness, in order to investigate the wear properties and interfacial behavior of the film sliding against with pure titanium, the thick hBN film was needed. Figure 5 shows the surface morphology of the films deposited for 120 min with the filament temperature at 2000 °C, the other deposition conditions were the same as before. The thickness of the film is around 2 μm measured by AFM. From the SEM image of the hBN film, it was found that the film has regular structures with the size around 1 μm.

Figure 6 shows the friction coefficient of the film measured by BOD test under the same testing condition except the sliding lap increased from 100 to 800. The friction coefficient of the film shows the same tendency as the film with thin thickness. The friction coefficient of the film was low and stable at the initial stage, then it increases as the sliding laps increased. In order to study the interfacial behavior between the film and pure titanium, the elemental composition of the wear track on the film was measured during the initial stage and after the BOD test.

Figure 7 shows the surface morphology and EDX results of the film after 50 and 800 laps. The results show that there was no titanium adhesion on the wear track after 50 laps, when the sliding lap increased to 800 laps, the titanium was detected and adhesion on the wear track. It was considered that the titanium was adhesion on the film only when the film was worn out. In other words, if hBN film was remained, the titanium would not adhesion on the film.

From the BOD test, it was found that the hBN film has low friction coefficient and low reaction with pure titanium, it could be applied to the microforming die for forming pure titanium. However, the adhesion strength or the wear resistance of the film was low, it was easily to be worn out during the BOD test.

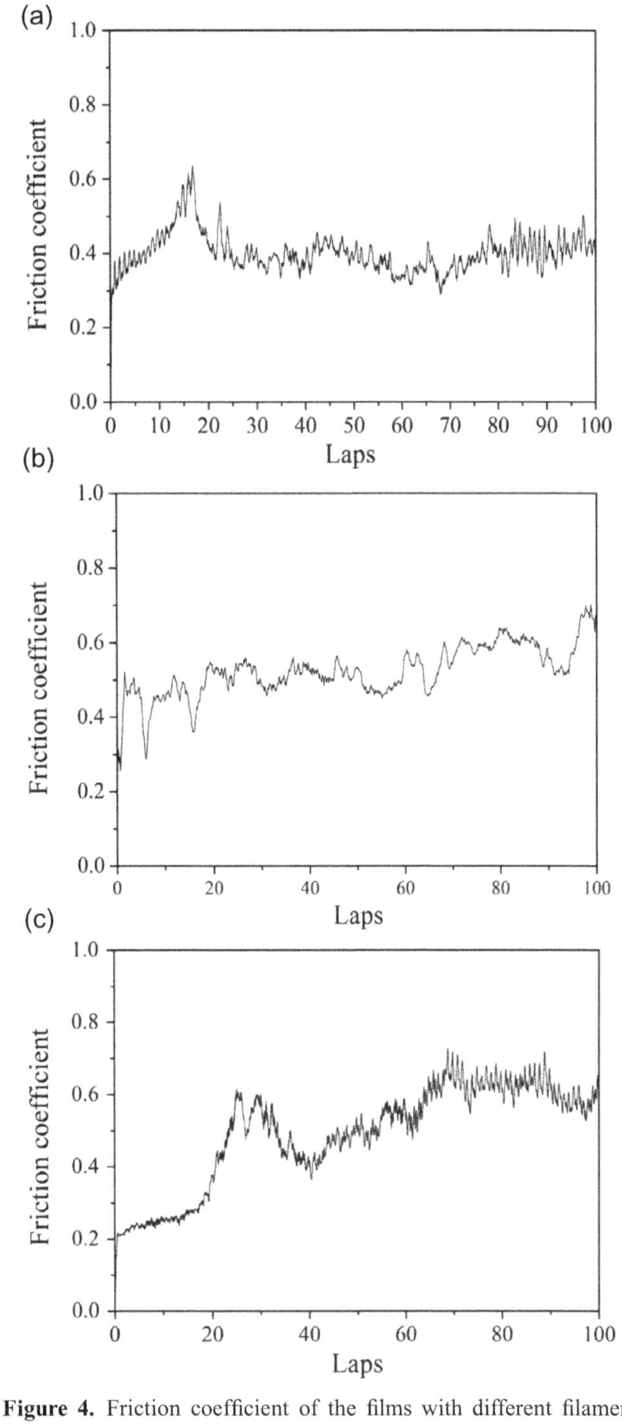

Figure 4. Friction coefficient of the films with different filament temperature: (a) 1500 °C, (b) 1800 °C, (c) 2000 °C.

3.2. Discussion

BN films contain hexagonal structure could be deposited at higher filament temperature by the filament assisted chemical

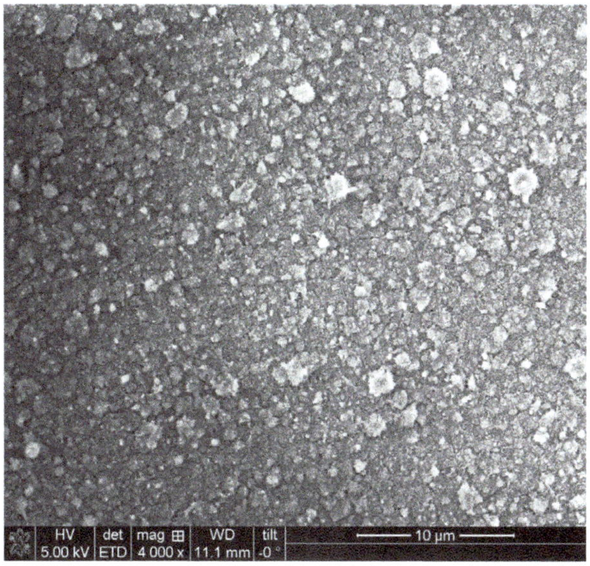

Figure 5. SEM image of the hBN film deposited for 120 min.

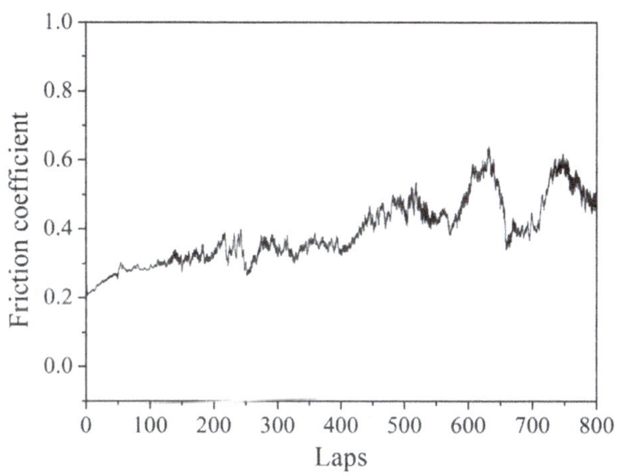

Figure 6. Friction coefficient of the hBN film deposited for 120 min.

vapor deposition, the higher temperature of filament makes the higher electron energy and lead to the hexagonal structure of BN formation. As there was no energetic ion/plasma bombardment during the deposition process, the internal stress in the films was much lower than the films deposited by PVD or PECVD process. The energy for cBN nucleation is higher than the energy for hBN film formation, if the electron energy from filament can be higher enough for cBN nucleation, it was possible to deposit the thick cBN film with lower internal stress by the filament assisted CVD process.

For microforming die application, the coating system should has low friciton coefficient, low wear rate, high adhesion strength on the tool and low reaction with forming

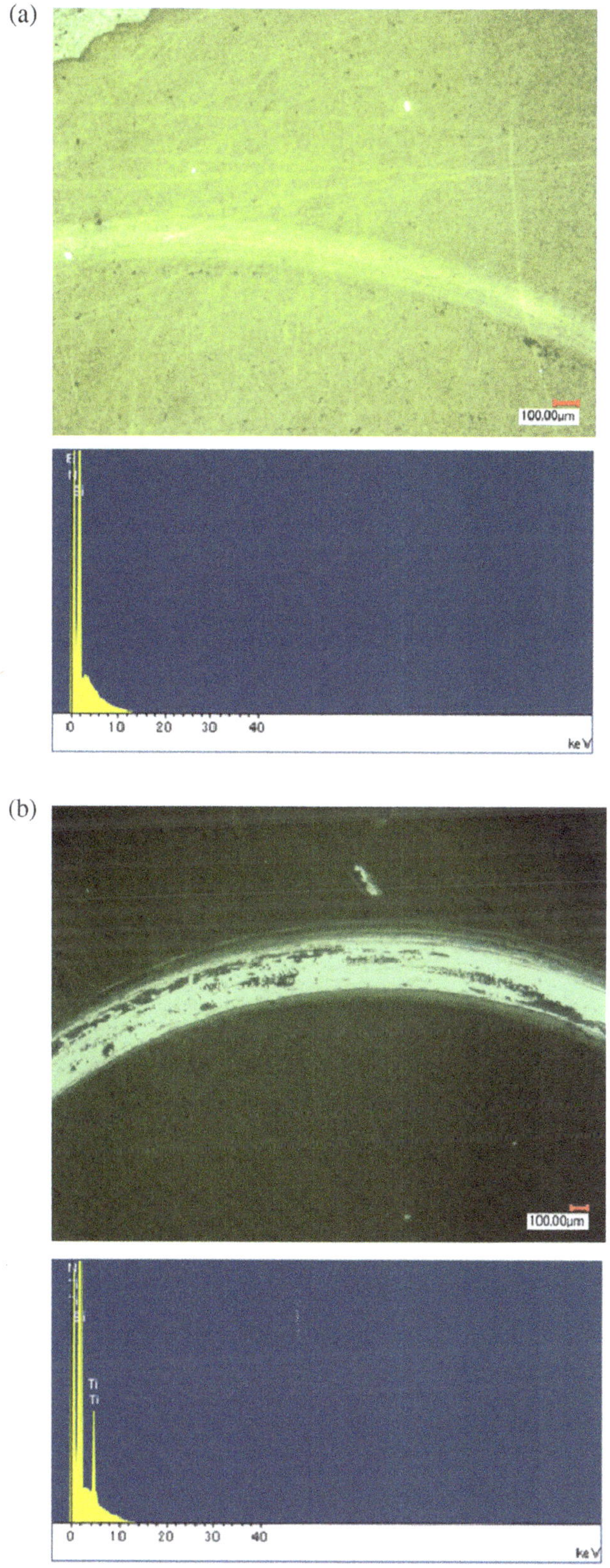

Figure 7. EDX measurement of the wear track on the film after BOD test: (a) wear track of the film after 50 laps, (b) wear track of the film after 800 laps.

materials. The BOD test shows that the hBN films has low friction coefficient and chemical inertness to pure titanium, it could be used as the coatings for microforming tool to prevent pure titanium sticking on the tool surface. However, the wear rate of the hBN films was high and it was easily to be worn out. In order to apply the BN film to microforming die, future work should be done to decrease the wear rate of the film. As is known to all, the cubic boron nitride film has high hardness and low wear rate, if the hBN film contains some cubic structure, the wear resistance of the coating system could be improved. On the other hands, some other substrate material like WC should be used, as this material is a common material for tool application.

4. Conclusion

The BN film was deposited by hot filament CVD. The filament temperature influenced the decomposition rate of single source precursor and lead to different film's elemental composition and structure. The film deposited under lower filament temperature was amorphous, hBN film was deposited under higher filament temperature, and the crystallization degree of the film was improved at the higher filament temperature. The hBN film shows the lower and stable friction coefficient at the initial stage, moreover, there was no titanium adhesion on the film after ball on disk test when the film was remained. It was expected that the hBN film could be used for microforming die to prevent pure titanium adhesion on the tool surface.

References

1. P.B. Mirkarimi, K.F. McCarty, D.L. Medlin, Review of advances in cubic boron nitride film synthesis, Materials Science and Engineering R21 (1997) 47–100.
2. S. Logothetidis, C. Charitidis, P. Patsalas, T. Kehagias, A comparative study of composition, structure and elastic properties of boron nitride films deposited by magnetron and ion beam sputtering, Diamond and Related Materials 8 (1999) 410–414.
3. J.X. Deng, B. Wang, L.W. Tan, H. Yan, G.H. Chen, The growth of cubic boron nitride films by RF reactive sputter, Thin Solid Films 368 (2000) 312–314.
4. G. Schwarz, F. Friess, G.K. Wolf, Deposition of c-BN by ion beam assisted CVD, Surface and Coatings Technology 125 (2000) 106–110.
5. R. Weissenbacher, R. Haubner, HF-CVD of B,C,N containing layers on various substrates by the decomposition of Triethylborazine, Diamond and Related Materials 12 (2003) 884–890.
6. H. Dumont, B. Bayle, B. Bonnetot, J. Bouix, Deposition and characterization of BN/Si (0 0 1) using tris(dimethylamino)borane, Materials Research Bulletin 37 (2002) 1565–1572.
7. J. Etter, Report on hertzian contact stress theory, OPTI 521, College of Optical Science, University of Arizona, Tucson, AZ, 2013.
8. M.A. Mannan, Y. Baba, N. Hirao, T. Kida, M. Nagano, H. Noguchi, Hexagonal nano-crystalline BCN films grown on Si(100) substrate studied by X-ray absorption spectroscopy, Materials Sciences and Application 4 (2013) 11–19.
9. Information on http://en.wikipedia.org/wiki/Bragg's_law

9

Mechanical and tribological studies on nano particles reinforced hybrid aluminum based composite

Aniruddha V. Muley[1,*], S. Aravindan[2], and I.P. Singh[2]

[1] Netaji Subhas Institute of Technology (NSIT), Dwarka, New Delhi 110078, India
[2] Indian Institute of Technology Delhi, Hauz Khas, New Delhi 110016, India

Abstract – Hybrid metal matrix composites are new class of materials due to their better mechanical properties which can be achieved through proper selection and combination of materials. The work reported in this paper is based on fabrication of hybrid composites by using nano particles as reinforcements. The hybrid composites were fabricated by reinforcing them with nano sized SiC and Al_2O_3 particles in order to study mechanical and tribological properties of these enhanced materials. A stir casting method was used to obtain hybrid composites. LM 6 aluminum alloy was used as a matrix material. The results shown increase in hardness as well as in ultimate tensile strength of the composites with small wt.% of nano-sized hybrid reinforcements. The composites produced also exhibit better tribological properties.

Key words: Hybrid composites, Nano-sized reinforcement, Mechanical properties, Tribology

1. Introduction

Aluminum based composite system has been studied over last three decades where micro sized reinforcements were used commonly. Only few research work available so far on nano sized particle reinforcement aluminium matrix composite [1, 7]. Hybrid metal matrix composite are new class of composite materials that gained the attention of scientist and researchers recently [2, 10]. Hybrid metal matrix composites are engineered combination of two or more reinforcements. Hybrid reinforcement gives us high degree of freedom in material design [3]. It was reported that the properties of hybrid composites are superior to those of single reinforcement composite. Their have been very few studies carried out on hybrid composites [4].

It was reported that use of hybrid composites helps to improve strength, stiffness, modulus, wear resistance and decreased the coefficient of thermal expansion [5–11]. Aluminum matrix composites are one of the advanced engineering material developed for variety of applications [4]. Reinforcing the aluminium alloy with nano particle have a significant influence on the overall strengthening [7].

This paper presents the work carried out on nano hybrid aluminium matrix composite to study the effect of nano hybrid reinforcement on mechanical and tribological properties of nano hybrid composite.

*Corresponding author: avmuley2000@yahoo.com

2. Experiment

To develop nano hybrid aluminium matrix composite, procured aluminium alloy billet (LM6) cut into small pieces. The weight of each piece is recorded. The nano particles of SiC and Al_2O_3 in equal ratio are measured on micro balance having wt.% of 0.5, 1.0, 1.5 and 2 with respect to cut Al alloy pieces. A piece of LM6 alloy is placed in graphite crucible, which is heated with the help of induction furnace to melt Al alloy. The temperature of molten Al alloy raised 10–20 °C above melting point (LM6 Al alloy have melting point of 580 °C). The nano particles are preheated to a temperature of around 100 °C to make them free from moisture and to improve its wettability with Al alloy. The nano particle are then feed manually into crucible containing molten Al alloy. Then the stirring is carried out at constant speed of 400 rpm to mix the nano particles into Al alloy for 4–5 min. The stirrer is made of low carbon steel. To avoid oxidation and contamination to hybrid composites argon inert gas envelope is used. After stirring the molten metal mixed with nano particles poured into preheated metallic mould to get rods of circular cross section after solidification. The preheating of metallic mould to a temperature of around

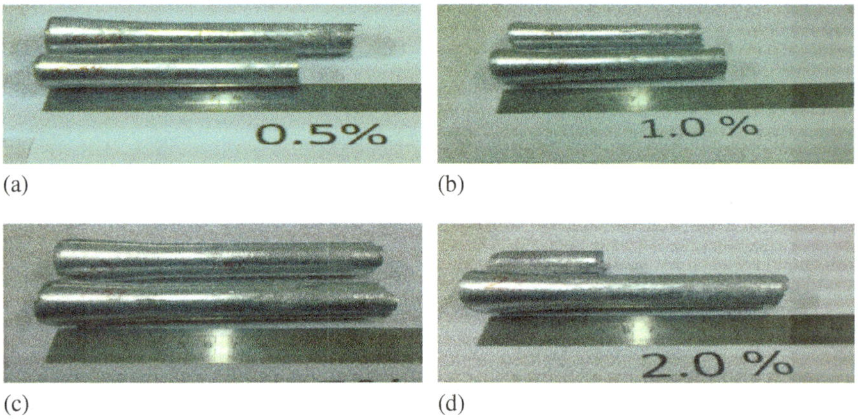

Figure 1. Hybrid composite stir cast rods with different wt.%. (a) 0.5 wt% reinforcement, (b) 1% reinforcement, (c) 1.5 wt% reinforcement, (d) 2 wt% reinforcement.

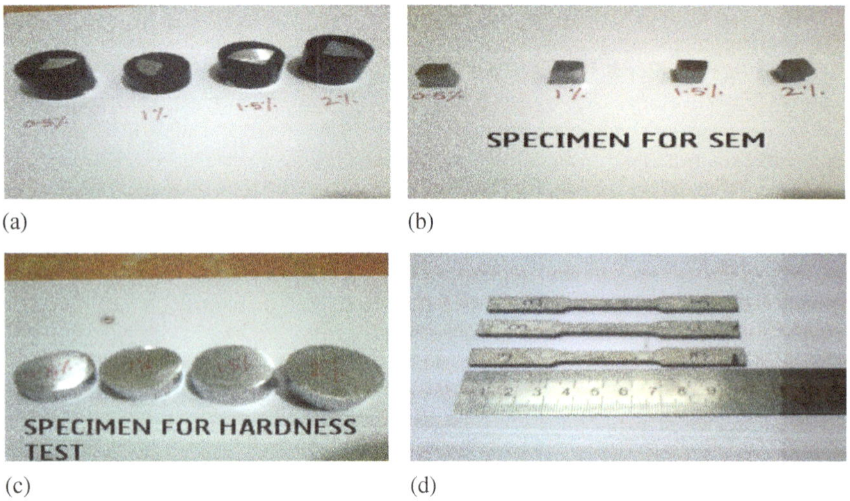

Figure 2. Specimen for (a) optical microscopy, (b) SEM, (c) hardness test and (d) tensile test.

Table 1. Chemical composition of LM6 (Al-12Si alloy) wt.%.

Si	Mg	Cu	Ni	Fe	Mn	Zn	Al
11–13	0.8–1.5	0.5–1.5	0.5–1.5	<0.2	<0.04	<0.02	Balance

100 °C is carried out to reduce humidity effect. Nano hybrid composites cylindrical rods thus obtained consist of 0.0 wt.%, 0.5 wt.%, 1.0 wt.%, 1.5 wt.%, 2.0 wt.% nano sized hybrid reinforcements in equal ratio. Figure 1 shows stir cast rods of hybrid composite with 0.5 wt.% to 2 wt.%. Further samples for characterization are prepared as shown in Figure 2.

Characterization studies included investigation of density, ultimate tensile strength, hardness, microstructure and wear test.

Material: Aluminum alloy LM6 was used as matrix (Table 1).

Nano sized SiC (>98.6% purity) and Al_2O_3 (>99.8% purity) powers were purchased from Reinste Nano Venture Pvt. Ltd. company New Delhi, India. The average particle size is 25–50 nm and 40 nm for SiC and Al_2O_3 nano reinforcement respectively.

Density measurement was carried out by using Archimedes principle technique. To examine microstructure and distribution of nano sized reinforcements in aluminium matrix optical microscopy and SEM were used respectively. All samples were polished to obtained mirror like surface with no visible scratches followed by etching. Micro hardness test was carried out by using Vickers hardness tester with ASTM E384-99 standard. Tensile test was done using ASTM E8 standard. Pin on disc test was used to study wear behaviour of composite.

3. Results and discussion

3.1. Processing

The Figure 1 shows the stir cast rod with different wt.% of hybrid reinforcement (nano sized SiC and Al_2O_3 particles) in

Table 2. Results of density and porosity measurements.

Materials	Nano sized SiC (wt.%)	Nano sized Al_2O_3 (wt.%)	Theoretical density (kg/m^3)	Experimental density (kg/m^3)	Porosity (%)
LM6 alloy	0	0	2700	2663	1.37
LM6 + 0.25% SiC + 0.25% Al_2O_3	0.25	0.25	2703	2667	1.33
LM6 + 0.5% SiC + 0.5% Al_2O_3	0.5	0.5	2706	2570	5.02
LM6 + 0.75% SiC + 0.75% Al_2O_3	0.75	0.75	2710	2610	3.69
LM6 + 1.0% SiC + 1.0% Al_2O_3	1.0	1.0	2713	2573	5.27

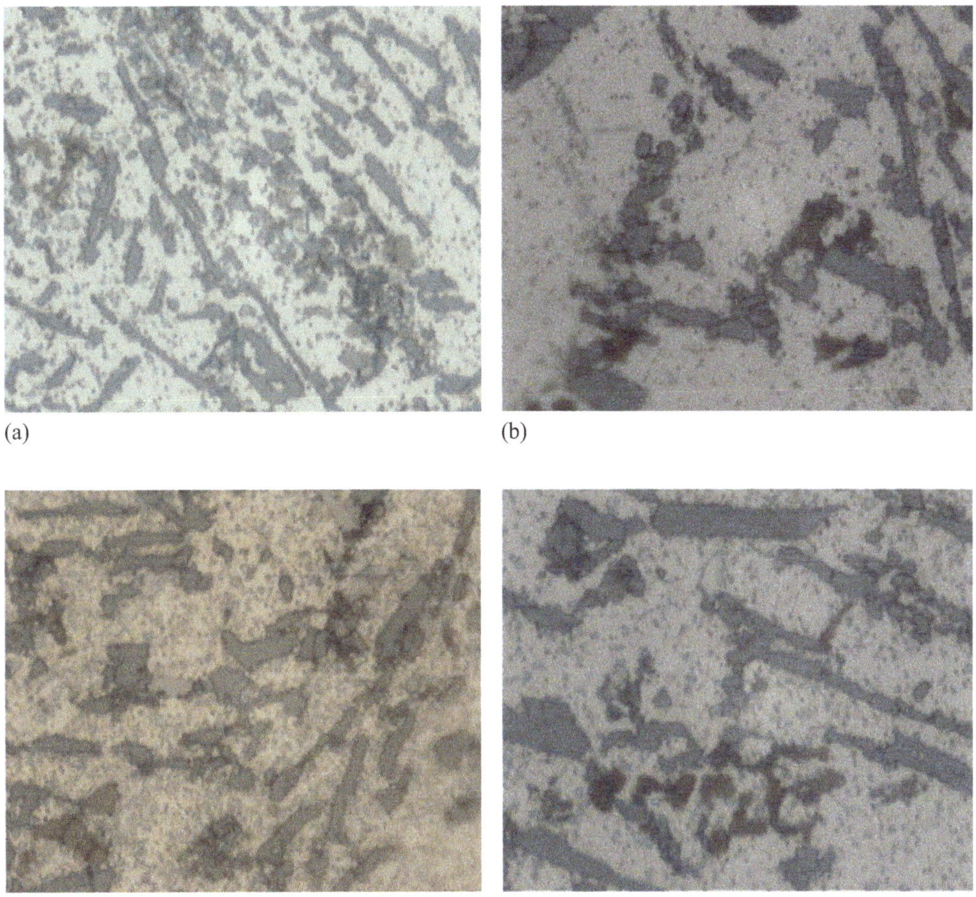

(a) (b) (c) (d)

Figure 3. (a)–(d) show optical microscopic images of hybrid composite specimens. (a) Optical microscopic image of 0.5 wt.% nano particles composite, (b) optical microscopic image of 1.0 wt.% nano particles composite, (c) optical microscopic image 1.5 wt.% nano particles composite, (d) optical microscopic image 2.0 wt.% nano particles composite.

equal ratio. The fabrication of al based hybrid composites was done by using stir casting method. It is one of the most economical techniques available to produce large near net shaped parts of composite materials [12].

3.2. Density measurement

The Table 2 shows the theoretical and experimental density of composite specimens. They are very close to each other showing consistency in density. The samples used in this experiments were randomly cut from the stir cast rods, which confirms uniform distribution of hybrid reinforcement. The low amount of porosity indicates better densification of composites.

3.3. Micro-structural characterization

3.3.1. Optical microscopy

Figures 3a–3d show optical microscopic images of hybrid composites. These shows the microstructure of composites specimens.

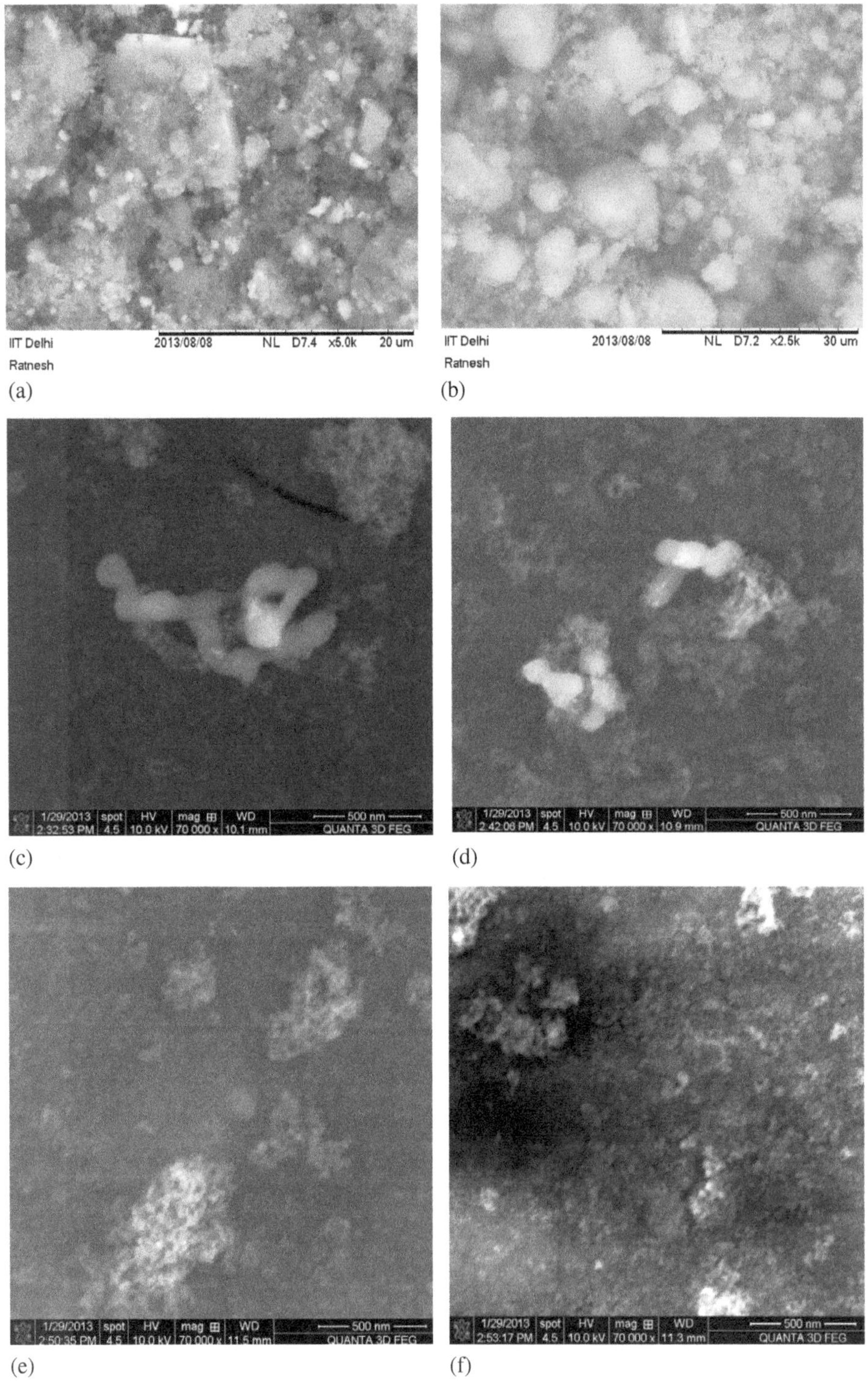

Figure 4. SEM images of nano particles and hybrid composites with different wt.%. (a) SiC nano particles, (b) Al_2O_3 nano particles, (c) 0.5 wt.% nano particles composite, (d) 1.0 wt.% nano particles composite, (e) 1.5 wt.% nano particles composite, (f) 2.0 wt.% nano particles composite.

3.3.2. SEM

Figures 4a and 4b show SEM image of SiC and Al_2O_3 nano particles respectively.

Figures 4b–4f show SEM images of the hybrid composite specimens shows uniform distribution of reinforcement with small amount of clustering. An uniform distribution of reinforcement is necessary in improving mechanical properties of composites. Some amount of clustering and agglomeration of nano-sized particles observed this can be attributed to high surface energy associated with nano-sized particles.

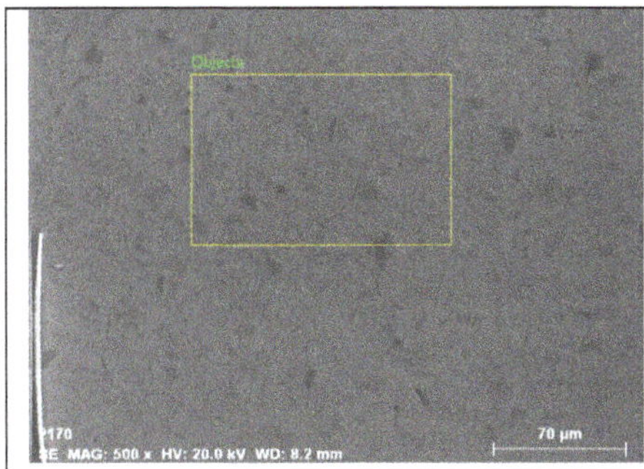

Spectrum: Objects

El	AN	Series	unn. C [wt.%]	norm. C [wt.%]	Atom. C [at.%]	Error [wt.%]
Al	13	K-series	76.81	78.08	70.78	3.7
Si	14	K-series	11.27	11.46	9.98	0.5
C	6	K-series	7.09	7.20	14.67	2.1
O	8	K-series	2.84	2.89	4.42	0.7
Fe	26	K-series	0.36	0.37	0.16	0.1
		Total:	98.37	100.00	100.00	

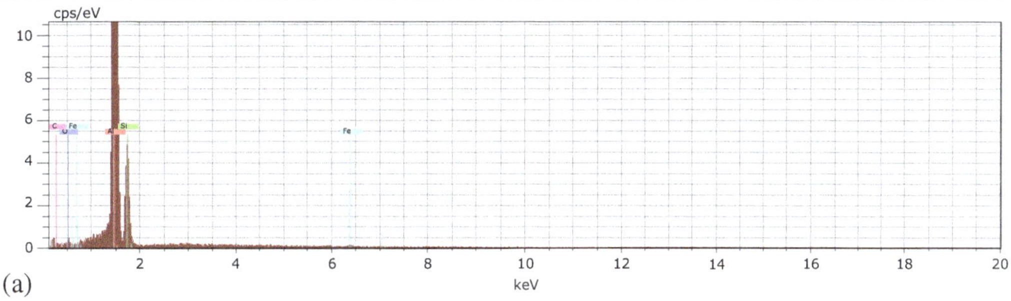

(a)

Spectrum: Objects

El	AN	Series	unn. C [wt.%]	norm. C [wt.%]	Atom. C [at.%]	Error [wt.%]
Al	13	K-series	79.25	77.14	68.07	3.8
Si	14	K-series	10.37	10.09	8.56	0.5
C	6	K-series	9.64	9.38	18.59	2.6
O	8	K-series	3.24	3.15	4.69	0.8
Fe	26	K-series	0.24	0.23	0.10	0.0
		Total:	102.73	100.00	100.00	

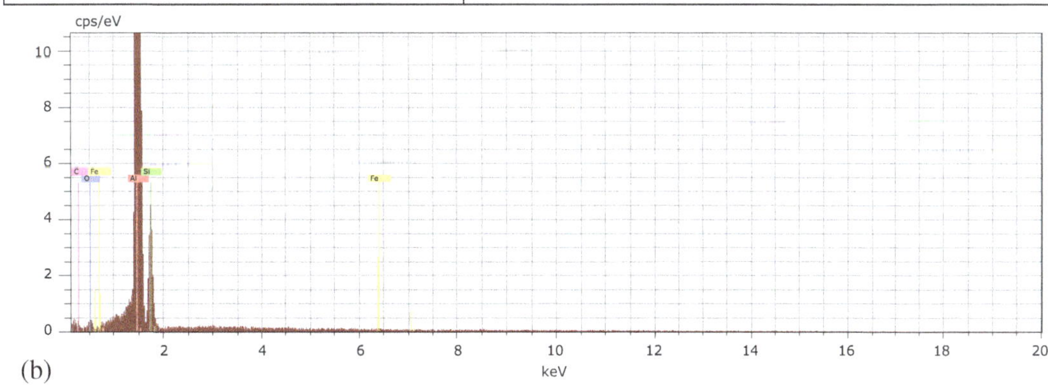

(b)

Figure 5. (a)–(d) show EDAX results of hybrid composites. (a) 0.5% nano hybrid composite, (b) 1.0% nano hybrid composite, (c) 1.5% nano hybrid composite, (d) 2.0% nano hybrid composite.

3.3.3. EDAX

It confirm presence of SiC and Al_2O_3 in hybrid composites. Figures 5a–5d show EDAX of various wt.% hybrid composites.

3.4. Hardness test

Figure 6 shows the results of Vickers micro hardness test. It was observed that the hardness increased with increase in reinforcement quantity. The hardness is higher than LM6 Al alloy. This can be attributed to the presence of hybrid reinforcement particles. The presence of reinforcement imposed higher degree of constraint to the localised matrix deformation during hardness test; which obstructed the motion of dislocations and resisted the deformation of matrix. Increase in the residual stresses induced due to the mismatch of thermal expansion between the matrix and reinforcement resulted in higher dislocation density, increased hardness of composites [1].

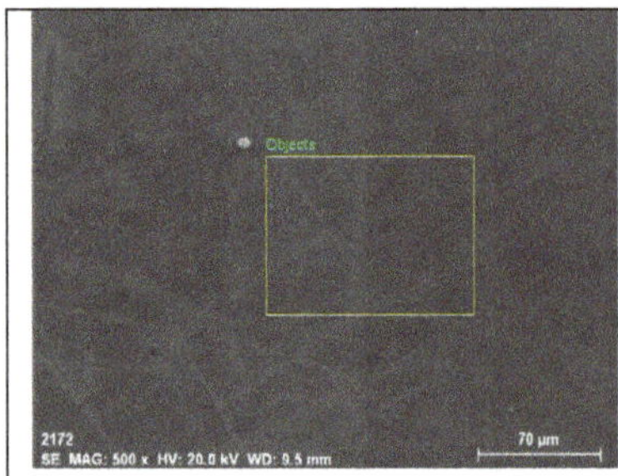

Spectrum: Objects

El	AN	Series	unn. C [wt.%]	norm. C [wt.%]	Atom. C [at.%]	Error [wt.%]
Al	13	K-series	76.17	71.44	60.68	3.7
O	8	K-series	10.17	9.54	13.66	1.8
Si	14	K-series	10.08	9.46	7.72	0.5
C	6	K-series	9.97	9.36	17.85	2.5
Fe	26	K-series	0.22	0.21	0.09	0.0
		Total:	106.62	100.00	100.00	

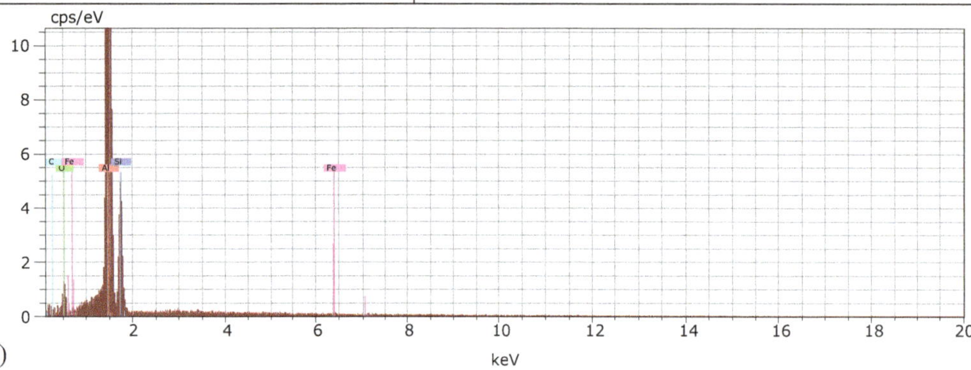

(c)

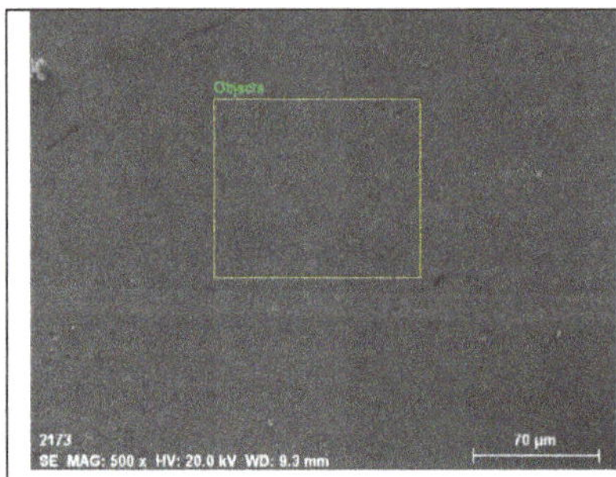

Spectrum: Objects

El	AN	Series	unn. C [wt.%]	norm. C [wt.%]	Atom. C [at.%]	Error [wt.%]
Al	13	K-series	86.36	75.43	67.61	4.2
Si	14	K-series	14.38	12.56	10.82	0.7
C	6	K-series	8.83	7.71	15.52	2.5
O	8	K-series	4.45	3.88	5.87	1.0
Fe	26	K-series	0.47	0.41	0.18	0.1
		Total:	114.49	100.00	100.00	

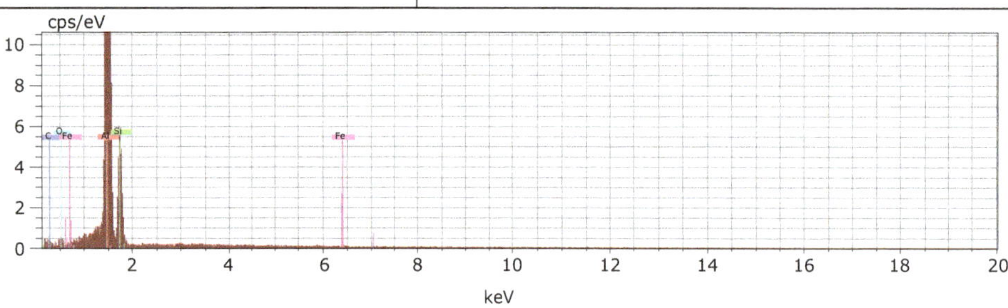

(d)

Figure 5. Continued.

3.5. Tensile Test

Figure 7 shows results of ultimate tensile strength of hybrid composites. The ultimate tensile strength of the hybrid composites increases with increase in nano particle wt.% percentage. The tensile test is carried out as per ASTM standard. The specimens for tensile test (as shown in Figure 2d) were cut along the length of cast rod by wire cut EDM machine. The increase in the tensile strength can be attributed to obstruction to dislocation motion, higher dislocation density, induced internal stresses due to thermal expansion mismatch between matrix and reinforcements and effective load transfer from matrix to better bonded and uniformly distributed reinforcements [1, 7, 9].

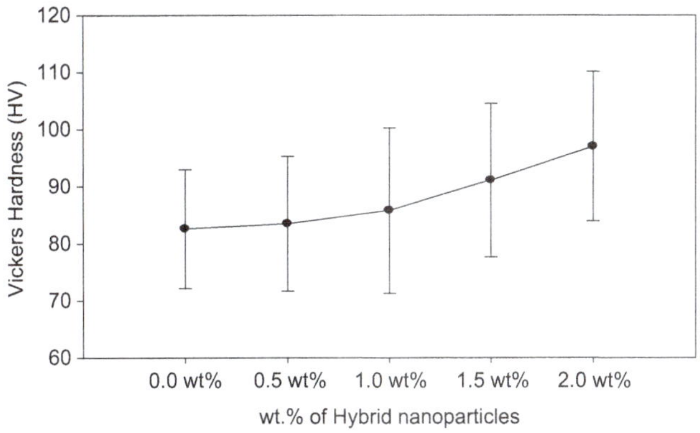

Figure 6. Vickers micro hardness.

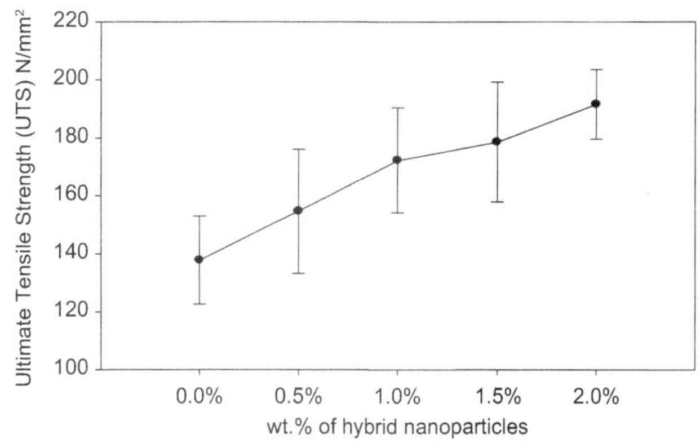

Figure 7. Ultimate Tensile Strength (UTS) of hybrid composites.

3.6. Pin on disc wear test

The pin on disc wear test was performed to study the tribological properties of the hybrid composites. Figure 8 shows the pins used in this study. The diameter of the pin is 10 mm and length is 30 mm. The counter surface disc of EN 31 material was used. The pin and disc surfaces were polished with emery paper. It was ensured that pin surface should be perfectly flat. To know wear loss, the difference between weight of the pin before and after wear test was calculated. The experiments were performed over load range of 10 N–50 N keeping sliding distance 1000 m constant and sliding speed of 1 m/s and 2 m/s was used. As the wt.% of reinforcement increased wear loss decreased (Figures 9 and 10). But with increased load the wear loss increased for 0.5 wt.% hybrid composite. For other composition with higher load there no much increase in wear loss. It is indicated that hybrid nano reinforcement reduced the wear loss. At the higher load (40 N and above) due to higher temperature between pin and disc, pin had tendency of being ceased. In Figure 10 as the sliding speed increased from 1 m/s to 2 m/s the wear loss increased for all wt.% hybrid composites but less than shown in Figure 9. It was reported that addition of SiC particles in matrix reduced the natural tendency for materials flow at the wear surface and formed iron reach (Fe_3O_4) layer on the surface of composite pin and the steel disc track which helps in reducing wear rate [13, 14, 17]. The well distributed SiC and Al_2O_3 particles acted as load bearing elements [14, 15]. In case of Al-Si alloys it was reported that when iron reach surface layers are formed on the composite the wear rate is low [16]. Some studies also shown that SiC particles subjected to tribo-chemical interaction during sliding process which forms SiO_2 which act like a lubricant and helps to reduce the wear. Further with increased volume fraction wear resistance reported to be enhanced [17]. The enhanced wear performance of hybrid composites can be attributed to improved hardness and strength of composite with the addition of nano Al_2O_3 particulates also as composites with improved hardness and strength exhibit better wear resistance [18]. The hybrid composite have better wear resistance than single reinforcement reinforced composites. The SiC particles are more effective than Al_2O_3 particles due to their high hardness. Also it was noted that porosity was mainly located around Al_2O_3 particles while it was less pronounced around SiC particles. This is due wetting behaviour of Al alloy. The SiC particles have better wettability and are more compatible with Al alloy than Al_2O_3 particles. The addition of mg improves wettability and reduces porosity. The SiC particles reacts with Al to produce Al_4C_3 precipitation at Al/SiC interfaces. However any reaction of Al melt and Al_2O_3 has not been reported [19]. It was reported that hybrid aluminium matrix composites are superior to unhybrid aluminium matix composites in wear performance. Mostly studies on hybrid composites are focussed on the form of reinforcements such as mixture of particles and fibers or fibers and whiskers. But information on hybrid composites reinforced with the nano particles is limited. It was reported that as the size of reinforcement increased from nano to micro the rate of wear increased, this is due to reduction in relative density, hardness and inter-particle spacing. The wear mechanism observed for nano particles was abrasion. In micro sized particles dominant wear mechanism is adhesion and particle cracking induced delamination. The tendency of particle cracking decreases with reducing particle size [20].

Figure 8. Pin for wear test.

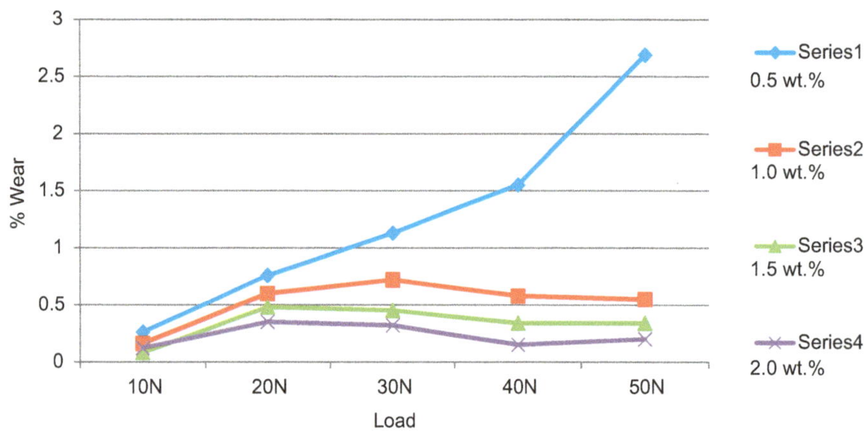

Figure 9. Load vs. wear (sliding speed 1 m/s and sliding distance 1000 m).

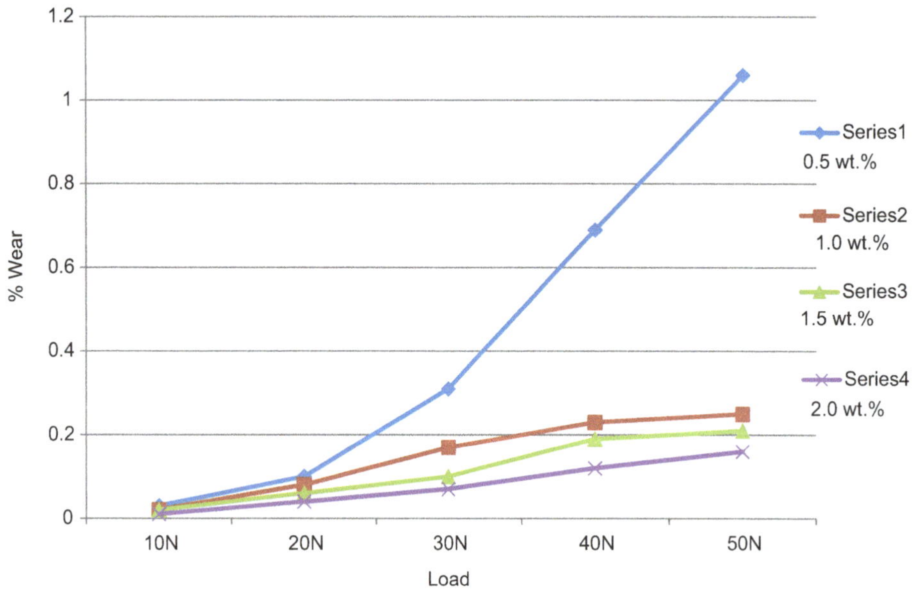

Figure 10. Load vs. wear (sliding speed 2 m/s and sliding distance 1000 m).

4. Conclusions

The following conclusions are drawn from the work reported above:

1. The stir casting method is simple and cheaper method for mixing reinforcement into matrix materials. But it has limitations when used for mixing nano particles, as nano particles have tendency of agglomeration. This results in clustering of nano particles and affect uniform distribution in matrix material. But continuous stirring at 400–600 rpm helps to distribute nano particles to a great extent.
2. The nano hybrid composites shown improved mechanical properties compared to LM6 alloy matrix. Vickers hardness for as cast LM6 matrix with 0 wt.% reinforcement is 84 Hv. With the addition of small wt.% of nano hybrid reinforcement it has shown increasing trend. For 2 wt.% of nano hybrid reinforcement it is increased to to 98 Hv. This increase in hardness is approx. 17%. Similarly the Ultimate Tensile Strength (UTS) also shown increasing trend. It is 138 N/mm^2 for 0 wt.% reinforcement and increased to 193 N/mm^2 for 2 wt.% nano hybrid reinforcement. This increase in UTS is approx. 39% compared to LM6 Al alloy. The further improvement in mechanical properties can be obtained by using suitable manufacturing method which helps to distribute nano particles uniformly in matrix and by using secondary processing techniques like hot extrusion.
3. The pin on disc wear test is carried out with aim to use nano hybrid composite for braking system application in automobile. The results shown that the nano hybrid composite exhibit reduced wear loss with increase in nano

hybrid reinforcement from 0.5 wt.% to 2 wt.%. The reduction in wear loss is approx. 80%. So nano hybrid composites have huge potential to enhance tribological property and can become most sought after material for tribological application.

The hybrid nano-composites exhibit better mechanical and tribological properties, compared to monolithic Al alloy. It has also have better properties than single reinforced composites. Because of its superior properties it can be a preferred material for many applications especially in braking system of automotive vehicles.

References

1. S.K. Thakur, K.S. Tun, M. Gupta, Enhancing uniform, non-uniform and total failure strain of aluminium by using SiC at nanolength scale, Journal of Engineering Materials and Technology 132 (2010) 1–6.
2. J.S.S. Babu, C.G. Kang, Nanoidentation behaviour of aluminium based hybrid composites with graphite nanofiber/alumina short fiber, Materials and Design 31 (2010) 4881–4885.
3. X.N. Zhang, L. Geng, G.S. Wang, Fabrication of Al based hybrid composite reinforced with SiC whiskers and SiC nanoparticles by squeeze casting, Journal of Materials Processing Technology 176 (2006) 146–151.
4. Y.C. Feng, L. Geng, P.Q. Zheng, Z.Z. Zheng, G.S. Wang, Fabrication and characteristic of Al based hybrid composite reinforced with tungsten oxide particle and aluminium borate whisker by squeeze casting, Materials and Design 29 (2008) 2023–2026.
5. G.H. Fan, L. Geng, Z.Z. Zheng, G.S. Wang, P.Q. Zheng, Preparation and characterization of $Al_{18}B4O_{33}$ + $BaPbO_3$/Al hybrid composite, Materials Letters 62 (2008) 2670–2672.
6. B.S.B. Reddy, K. Rajasekhar, M. Venu, J.S.S. Dilip, S. Das, K. Das, Mechanical activation assisted solid state combustion synthesis of in situ aluminium matrix hybrid (Al_3Ni/Al_2O_3) nanocomposites, Journal of Alloys and Compounds 465 (2008) 97–105.
7. L. Geng, X. Zhang, G. Wang, Z. Zeng, B. Xu, Effect of aging treatment on mechanical properties of (SiC_w + SiC_p)/2024 Al hybrid composites, Transactions of Nonferrous Metals Society of China 16 (2006) 387–391.
8. D. Jun, L. Yao-hui, Y. Si-Rong, L. Wen-Fang, Dry sliding friction and wear properties of Al_2O_3 and carbon short fibres reinforced Al-12Si alloy hybrid composites, Wear 257 (2004) 930–940.
9. Y.C. Feng, L. Geng, G.H. Fan, A.B. Li, Z.Z. Zeng, The properties and the microstructure of hybrid composites reinforced with WO3 particles and $Al_{18}B_4O_{33}$ whiskers by squeeze casting, Materials and Design 30 (2009) 3632–3635.
10. L. Guan, L. Geng, H. Zhang, L. Huang, Effect of stirring parameter on microstructure and tensile properties of (ABO_w + SiC_p)/6061 Al composites fabricated by semi-solid stirring technique, Transactions of Nonferrous Metals Society of China 21 (2011) s274–s279.
11. Y. Zhao, S. Zhang, G. Chen, Aluminum matrix composites reinforced by in situ Al_2O_3 and Al_3Zr particles fabricated via magnetochemistry reaction, Transactions of Nonferrous Metals Society of China 20 (2010) 2129–2133.
12. B.F. Schultz, J.B. Ferguson, P.K. Rohtagi, Microstructure and hardness of Al_2O_3 nano particles reinforced Al-Mg composites fabricated by reactive wetting and stir mixing, Materials Science and Engineering A 530 (2011) 87–97.
13. R.L. Deuis, C. Subramanium, J.M. Yellup, Dry sliding wear of alumium composites – a review, Composite Science and Technology 57 (1997) 415–435.
14. R. Chen, A. Iwabuchi, T. Shimizu, S.H. Shin, H. Mifune, The sliding wear behavior of NiAl and SiC particles reinforced aluminum alloy matrix composites, Wear 213 (1997) 175–184.
15. D. Jun, L. Yao-hui, Y. Si-Rong, L. Wen-Fang, Dry sliding friction and wear properties of Al_2O_3 and carbon short fibres reinforced Al-12Si alloy hybrid composites, Wear 257 (2004) 930–940.
16. M. Narayan, M.K. Surappa, B.N. Pramiala Bai, Dry sliding wear of Al alloy 2024-Al_2O_3, Wear 181–183 (1995) 563–570.
17. A. Ravikiran, M.K. Surappa, Effect of sliding speed on wear behaviour of A356 Al-30 wt% SiC_p MMC, Wear 206 (1997) 33–38.
18. M. Shanti, Q.B. Nguyen, M. Gupta, Sliding wear behaviour of calcium containing $AZ_{31}B/Al_2O_3$ nanocomposites, Wear 269 (2010) 473–479.
19. H. Alhatci, T. Kocer, E. Candan, H. Cimenoglu, Wear behaviour of Al/(Al_2O_{3p} + SiC_p) hybrid composites, Tribology International 39 (2006) 213–220.
20. E.M. Sharifi, F. Karimzadeh, Wear behaviour of aluminium matrix hybrid nanocomposites fabricated by powder metallurgy, Wear 271 (2011) 1072–1079.

10

A review of automation in manufacturing illustrated by a case study on mixed-mode hot forging

Colin S. Harrison*

Mechanical and Electrical Engineering, School of Engineering, Glasgow Caledonian University, Cowcaddens Road, Glasgow G4 0BA, UK

Abstract – This paper describes a review of automation techniques as currently practiced in the forging and forming industry. That industry has been relatively resistant to handling automation because of the severe challenges that are faced by having typically small batch sizes, extremely challenging operating conditions in terms of temperatures and pressures and lighting conditions. The paper discusses how aspects of these difficulties can be overcome and it is hoped that the specific solutions and methodology presented will find general applications in other aspects of manufacturing industry. The concept of the "Automated Forging of the Future" is introduced in which advanced handling, rapid tooling-change and flexibility, smart metrology, advanced sensing and intelligent gripping techniques are presented. The paper represents the distillation of work on automated forging based on the authors' implementation of a key theme in forging automation in *Advanced Forming Research Centre* (AFRC), The University of Strathclyde, Glasgow, UK.

Key words: Automation, Robotics, Forging, Forming

1. Introduction

The manufacturing forging industry can be classified across several industries including Automotive, Aerospace and Medical being dominated by relatively high value added operation, much of the reminder having migrated overseas. Continual cost pressures have meant that industries have had to work smarter rather than harder, since lower costs bases attract plug-compatible manufacturing replication, unless there is a high skill, higher value-add differentiator. Smart automation is presented as one of these differentiators and the ability to quickly change over between batches, achieve higher manufacturing tolerances and overall quality, whilst in many cases increasing volume have been key drivers for the implementation of advanced automation.

Typical product families that have been automated include the production in the automotive industry of gears, hubs, crankshafts with the concurrent objective of optimising materials performance by manipulating materials structure for example, hardness and straight by the automated forging processes themselves.

*e-mail: colin.harrison@gcu.ac.uk

The key advantages that are presented for automation include:

- Increased Volume (capacity).
- Improved Quality – via consistency of manufacturing and reduction in variability.
- Reduced Costs.
- Reduced dependency on hard to find labour to work in harsh operating conditions.
- Improved Safety Occupational health statistics continue to illustrate that powered presses are a major source of industrial accidents.

However, the financing of these new systems inevitably must and should be filtered by a cost – justification approach in order to test the economic viability of the particular investment proposition. A widely used system is based on manufacturing systems simulation allied to a cost benefit analysis in order to adopt a relatively low cost approach, one the initial business benefits have been outlined. Systems such as Witness [1] Hocus, and the like can be used to explore discrete event simulation approaches, especially where hard production data exist on Tact time, time in process, batch sizes and other production data can refine the investment proposition. In industrial situations a question raised almost from the first will always be "What are the benefits" and some attempt at clarifying the financial, engineering and any intangible benefits will need to be constructed with increasing rigour. Some large multinationals have their own specific cost modelling software to address financial returns [2], others use a combination of payback period or internal rate of return, perhaps via a spreadsheet. Intangible benefits can often be influential, but present difficulties in taking them above the line and are usually fought off.

Academic work on forging automation is relatively sparse, however case studies can be found which illustrate the public face of successful implementations [3].

These key drivers automatically define an investment route for scarce capital in that fundamentally the automation engineer is faced with the challenge of automating for lower forecast cost. This means that many developments which might have desirable features are often not cost effective dependant on the analysis regime used. Systems simulation for automation can be applied particularly where the implementation includes data on queuing, mean time between failures (MTBF) and mean time to repair (MTTR) though the results tend to be general in nature. One possible approach when considering a large investment in forging/forming particularly due to the high cost, and longevity of press technology, is often to simulate the process in order to clarify in detail how the modified operation will work and this can then be verified in broad brush terms with a cost modelling approach in order to quantify the benefits.

The existing academic work on "Automation in Forging" is not extensive, for example a search of UK theses on the keywords "Forging Automation" (via the on line index) will reveal a relatively weak result. Contrast this with a search on the abstract keyword "Forging" where over 35 relevant theses will appear including studies on lubrication, wear and damage as well as 3D simulations. Similarly in the main sources (journals having an impact factor and rated conferences) a search using these keywords will result in few results, which are mainly concerned with kinematic modelling [4, 5], often without real world data [6], discussions of potential applications [7], thixotropic forging studies [8] and the like. In contrast there has been relatively steady development by suppliers and manufacturers' websites often have useful reporting material, though discussions with robot suppliers (ABB/Fanuc) have revealed that there has been little application development of automation in forging, which reflects the lack of academic output.

This represents an opportunity to fully investigate the integration of automation in forging operations with the intention of cost reduction, increases in quality and consistency and in overall process control.

Forging presents a tremendous challenge to the systems integrator wishing to build in automation to the manufacturing process, combining as it does the effects of varying temperatures, immense potentially destructive forces, and accumulations of hardening lubricant. This combines with the high cost of replacing forging/forming presses where economic lives can be decades, means that retrofitting automatic systems is one feasible industrial approach. Here often the main implementation expertise lies in the area of the industrial supplier or systems integrator. The Forging Industries Association Technology Plan [9] has identified a number of key issues that need to be addressed which prioritises process control and modifications of the "Hoteye" system [10] and other sensing data for Forging.

The methodology of this paper has been to survey the existing, though rather sparse academic literature, interview with industry expert practitioners, and to describe the associated case study. The first task is to try to define what is meant by automation in forging and this description is sufficiently broad that a number of modular topics can be included:

- automated handling,
- automated lubrication,
- automated heating,
- automated process control.

2. Automation for forging applications

2.1. Automated robotic handling

A key factor in automated handling for forges is the use of robots and the following section places this in context.

The British Automation and Robot Association (BARA) [11] publishes an annual review on robot use based on feedback from 40 leading suppliers [12] which reports that the market is dominated by automotive applications, and recent statistics suggest there are 25,000 robots in service though the international federation suggests that only half of these are less than 12 years old. Even though the market is dominated by automotive applications, trends since 2005 indicate that other sectors are becoming progressively more important. However the use of robotics in forging and is one that is increasing from a relatively low level depending on category definition (see Figure 1) and the non-automotive applications have continued to grow.

It is estimated that 75% of installed robots are of the 6 axis type (BARA-2009) figures, although these can also be mounted on rails or hang from a track with over 90% of the 460 installations in 2009 being priced at between £10,000 and £50,000 although there is an increasingly active market in robot refurbishment, which on average is about 50% of the cost of a new installation.

A recent review by Rimrock Corp on forge shop automation [13, 14] describes an approach to adopting automation by subdividing the process (of handling) into the following operations:

- billet loading,
- trim press loading,
- parts transfer robot,
- spray robot.

A conventional approach is to apply these to a pre-existing press situation in order to improve the process and the benefits as listed can be substantial though there is cost associated with redesign of the process and part which is often difficult to quantify at the outset though can provide substantial benefits in successful installations. This approach could be described as

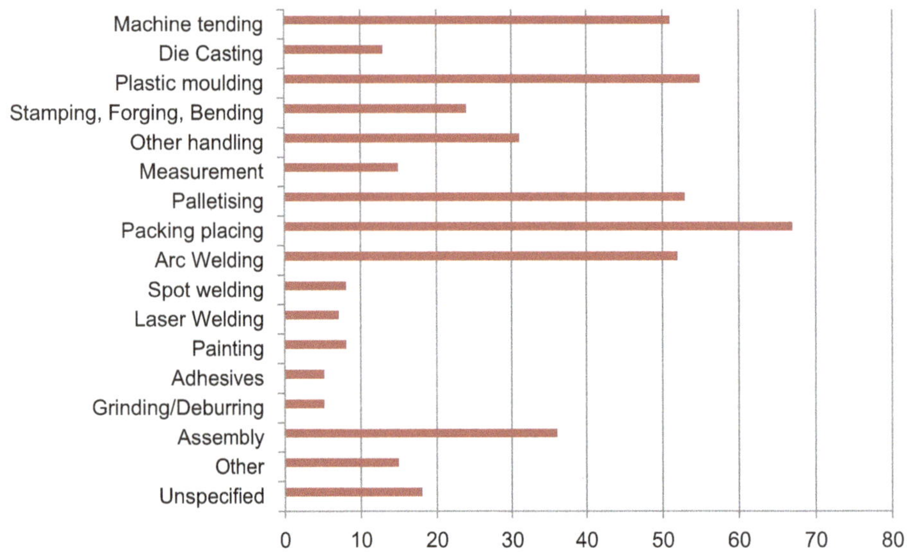

Figure 1. UK Robot application analysis for 2009 (BARA).

Figure 2. High volume "Compact" Automotive crossbar press (Schuler Automation) [17].

"Robotizing the man" [15]. A more fundamental approach which relates more to the state of the art is a system such as that promoted by Schuler Automation [16] involving a completely integrated transfer press line.

This system uses a fully enclosed approach and a central drive to co-ordinate manufacture (see Figure 2). The part handling is achieved internally by using a modular lever transfer arrangement in order to handle formed panels between each individual press station.

A variety of handling systems are available such as cross shuttles, robot loading with conveyors, and independently controllable feeders depending on the client automation needs.

A similar retrofit crossbar approach is possible with a conventional press situation that transfers parts directly between presses without releasing the part. This should result is more accurate and reliable picking since the part does not undergo any intermediate drop convey and pick stage. Here the robot is mounted on a rail system and itself travels between sites whilst the part is still held (see Figure 3).

Offline robotic programming is available via systems such as MasterCam [18] Roboguide [19], Robotmaster [20] or the open source ROS [21], in order to generate path control and servo instructions for the preferred robot system [22].

In robotic handling a key factor is the reliability of the system and integrators aim to show that their systems perform robustly. An example of a reliable system is the use of MTBF and some suppliers claim to have an MTBF of 60,000 h for robots at Jaguar/Land Rover, meaning that a particular robot would only be down once in every 15 years [24] (Figure 4).

Typical advertised benefits of robotic handling include:

- improved consistency,
- greater production volume,
- reduced labour-hours,
- safer operation,
- lowered production costs,
- improved safety,
- environmentally friendly through lower energy costs,
- typical productive life of robot 10–15 years [25].

Figure 3. "Crossbar Robot" transfers parts directly from press to press (Schuler Automation) [23].

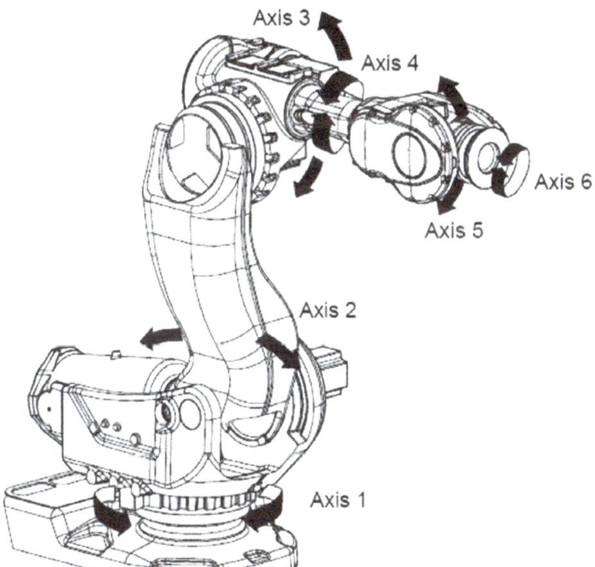

Figure 4. Standard 6 axis Robot model shown IRB 7600 (ABB).

Here the prevalence of high temperatures (up to 1100 °C for some materials) and airborne detritus means that exceptionally robust robots are used, often incorporating specially adapted grippers [26]. Note the use of caged robot enclosure for personnel protection (see Figure 5). Robot suppliers can provide systems with payloads ranging up to 500 kg or more depending on reach and are typically programmed via a teach pendant, though offline programming is often possible. A typical "6-axis" servo has changed little in the basic configuration over 2 decades [27] though have now become extremely reliable, though overall system integrity is more difficult to assure

Figure 5. Robot handling crankshafts for the auto industry (ABB).

because of the many interactions that occur with real world devices in a forging cell, particularly during the setup cycle.

Note that the addition of vision guided packages, including 2D and 3D aspects are slowly becoming more prevalent though 3D vision is not suitable for all applications in forging and infra-red interference provides special difficulties as does lens contamination issues from free floating dag.

An excellent demonstration of a robotic forging cell for turbine blades was described in the "How it's made" series [28] which describes a robotic forge using a two gripper approach to manipulate slugs, and form performs which also integrates furnace heating, deburring and automated lubrication.

Note that Leistritz claim [29] that by using Superplastic *forging* in isothermal conditions (rather than super plastic forming) that a near net shape can be forged, used in conjunction with electrochemical-machining for metal removal. The stated Leistritz philosophy on automation relates to a common user interface and is really a user-centred control systems approach. This presents an alternative to substitutive thinking on automation which has the overall goal of minimising human interaction and is based largely on a machine dominated process. Example Leistritz blades are shown in Figure 6.

2.2. Automated lubrication

Discussions with Houghton Plc and interaction with leading manufacturers reveal that automatic lubrication systems are often a part of the forming process – usually named "dagging". This applies a colloidal graphite based suspension of water and mineral oil [30] and one approach to optimisation is to use a

Figure 6. Forged compressor blades (Leistritz).

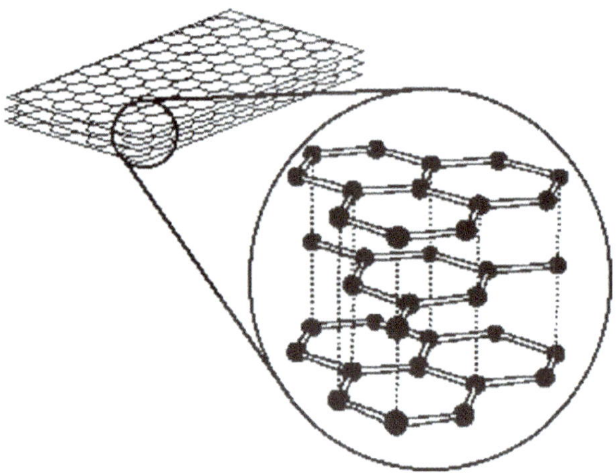

Figure 7. Crystalline graphite structure illustrating lubricious cleavage planes (McIntosh).

Taguchi or design of experiments approach. A diagram illustrating the structure for graphite dag is shown in Figure 7.

Once the parameters of the specific concentration have been determined for a particular parts configuration (e.g. extrusion and heading) and temperature then Statistical Process Control (SPC) can be used to manage the process. An example of dag application for extrusion is shown in Figure 8.

Many presses include auto-lubrication functions however these do not always carry an approved "CE Mark" [31] and for this reason manual dagging is relatively common. Obtaining a CE mark for the particular auto-lubrication press is a costly and time consuming process and one best undertaken by the press manufacturer at the time of the build, therefore manual dagging is likely to continue until an uncertified pre-existing press is replaced by one having the relevant certifications. There also a perception that when it comes to the application of dag that the operator knows best.

Surprisingly in automation applications the dagging function seems to be persistently applied by the human operator. A series of videos on forging illustrate this and very often a manually applied spray process can be seen entering the field of view from the side. The illustrated cell in this video of truck axles forging shows robots moving on a floor mounted rail in a two stage forging process however the dag is still applied by an operator [32].

One novel application for a retrofit is described by East Engineering [33] similar to that of an inkjet printer (see Figure 9). Here the dag is directly programmed via a series of small nozzles to spray particularly on one area and developed by "Jerko Sprue-system [34]". The advantages being very fine atomisation, precise delivery, and cooling effects limited to only where the dag is applied meaning a more stable process (see Figure 10).

This system has been installed by DIC Engineering as part of a crankshaft line for General Motors incorporating an NKMZ 7000T forging press, with piezoelectric load cells to protect against overload.

It should be possible using a *robot spray technique* to generate a relatively sophisticated dag spray pattern that only operates where it is absolutely needed using a modified lance rather than in the general area of the part and tooling. This will result in less dag being applied together with reduction in the associated cooling effects, and could be of use particularly for blade forming.

Some of the ideas presented in the above review were distilled into an example automated cell illustrating the key features, and is presented here as a case study, in the *Advanced Forming Research Centre* (*AFRC*) at the University of Strathclyde.

A key issue to be addressed in forge manufacturing is that of the variability introduced by manually controlled methods. Individual forge operators tend to use individualised technique which means that the output varies according to the speed of the operator, the amount of lubricant applied and leads to results which can be inconsistent. Automation of this process, as well as leading to increases in production volume additionally gives improvements quality and especially in consistency of output [35]. For the research investigator the latter feature (related to stabilisation of experimental conditions) – means that the automated process can then itself be used as a process debugging tool, without the confounding that tends to obscure the process improvements obtainable from any particular change. Therefore the automated system itself can be used to diagnose the process, for example where new consistent defects start to emerge from a stable process. Where manufacturing operations consist of both manual and automated processes improvements can be reflected back from the automated to the manual operations in terms of for example, the application of lubricant.

The introduction of Automation in forging clearly has a number of specific challenges. These are described in the following section.

3. Research challenges in forging automation

There are a series of important challenges that impact on industrial implementation of forging automation including

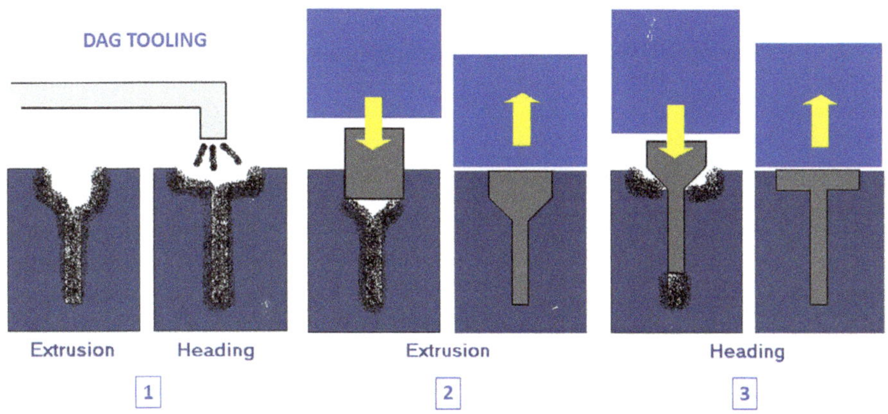

Figure 8. Dag application in extrusion and heading (McIntosh).

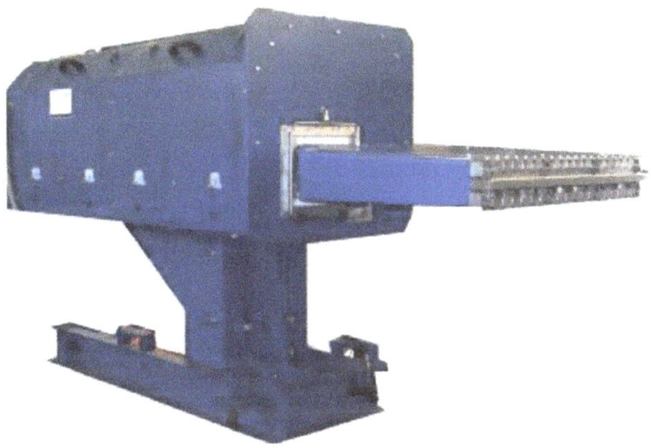

Figure 9. Jerko Dag "printer" (Jerko Spruesystems).

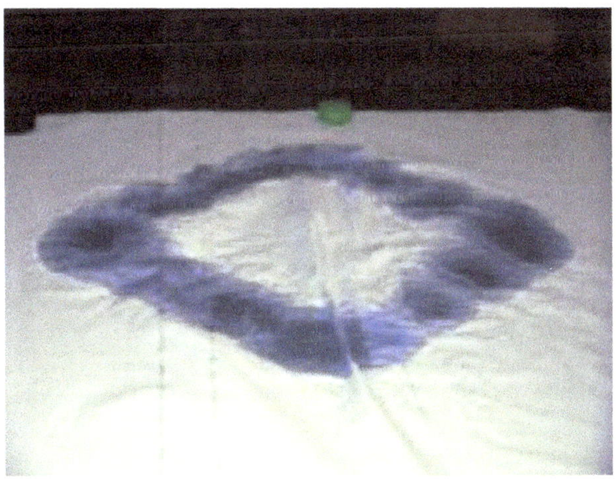

Figure 10. Printed Dag pattern showing local directed spray (Jerko Spruesystems).

potential financial benefits [36] such as return on investment, Payback period, which interact with specific technical ones, which are described here.

Overall these can be grouped into Process Specific Challenges, and No specific challenges applicable to any automation project.

3.1. Non specific automation process challenges

It is critical to any technical project that the basic groundwork required by a long term effort investment is laid by obtaining full backing from senior management. The issue is that if there is insufficient or lukewarm support at senior production levels then when unforeseen difficulties arise, and especially where there is no immediate solution, the danger is that the project will be stopped. It is also important to have an extremely complete understanding of the detailed nature of the task that requires automation in order to be able to specify and implement an automated system. Barriers to implementation can include fear of redundancy [37] and labour displacement – though in practice robotic automation tends to mean that existing staff are redeployed into other areas after an increase in volume and quality has been achieved, though some of those who leave voluntarily may not be replaced. There can also be a process knowledge gap on an operation, and a standard technique is to video the precise manufacturing operations as well as to elicit knowledge form the particular operator, some redesign of the task is often required in order to cope with the limitations of the automation system. This is against the known background that robots are not "accurate" in the sense that CNC machines are accurate, but the repeatability and flexibility means that it these limitations can be handled and still make a productive system.

Nevertheless we focus here on the technical aspects of implementation, assuming that the solid financial and particularly managerial backing exists. Automation in general has a range of challenges but forging presents some that rare relatively uncommon. There is a market led desire to move towards small smaller batch sizes [38] and the case for automation may mean that the technical aspects of implementation cannot easily be demonstrated – technical achievability influences the benefits.

The major long term strategic challenge involved is that of worker substitution, the original source of the word robot [39], particularly applicable in a hostile environment such as a forge or foundry operations involving high temperatures, levels of contamination, and extremely powerful machinery. The next great leap forward is the deployment of the Asimo style self-organising robot-android that has a capability without being micro-managed at every step of the production process, and ideally capable of easy redeployment between processes. Even basic self organising systems indicate that that this is some way off in the future [40], nevertheless the goal is to provide plug compatible replacement, at lowered cost and raised quality. Therefore the benefits of automation can be gained without the inherent disadvantages of bolted down arms with smart sensors, programming and end effectors, slaved to a master controllers which comprise the bulk of automated systems today. Such systems are at least a decade away or more from being deployed in manufacturing.

3.1.1. Teamwork/collaborative approaches

A challenge for the research practitioner is that of robot-human interaction at a number of levels. The majority of robot programming is still done in practice via the traditional teach pendant – Illustration – backed by vendor training courses, heavy course documentation and support channels. The reason for pendant based operation is that the real world is still difficult to model at the sufficiently detailed level in order to allow true offline programming to occur. This means that advanced. Program capability is thus dependant on high levels of familiarity with the various menu trees, detailed co-ordinated systems, Tool coordinate variants, in variable declarations, screen prompts and the like. In short it is still fundamentally a programming task requiring programmer skills, which present challenges for the occasional user/and often it means that external expertise has to be bought in to complete a specific task in a scheduled time, especially when production downtime is involved.

Recent alternative approaches to aspects of robot programming include development by KUKA and others of on passive click and drag and interaction via Mobile phone devices. One approach is simply to emulate the pendant software [41] inside the application, though other emerging methods include physically moving the robot itself [42] into position and selecting a point recall button on the robot [43]. This offers a much more direct way of programming the robot since humans can use intuitive honed skills in our own natural way of controlling motion rather than via an artificial menu system – reducing programming and development time. Here falling hardware costs mean that entry level industrial robots, which start at about £20 k, may be only 20% of the cost of an installation due to the additional value-added services that are required to make an installation actually function correctly. That is the area that key research challenge to simplify the introduction of smart flexible robots.

Accuracy – robots are not "accurate" – or as accurate as a typical CNC machine tool, however they are conventionally classed as "repeatable" which allows advanced systems to be successful.

3.1.2. Human robot collaboration/collision

It is expected that next generation robotics for manufacturing is going to interact more directly with the human operator and advanced safety protocols and ways of working are currently being developed. The idea is to move back from the caged robot process to one where human and robot work co-operatively and using the humans sophisticated sensory and knowledge capability rather than automated substitution. This means that intrinsic safety protocols such as advanced sensing in order to manage torque and speed control are required if the human is going "back in the cage". One approach is to have a set of whisker detectors so that if a human (or other unexpected object) is encountered the robot immediately reduces power – so that the human is not hurt. One system currently operating (without a set of guarding) is described by Kuka where the human works closely with the robot – augmenting the robots ability to perform repetitive tasks with the human ability to sense the world, make decisions and to behave in a non–programmed way. The trend is towards more bottom up rather than top down robot implementation [44] see ISO Tech Specification TS 15066 [45] with robot safety via collaborative interworking [43].

For part or product family opportunities where the investment case is reasonably demonstrated difficulties include the following items.

3.2. Process specific challenges

Hot Forging in particular means that handling equipment must be designed to cope with metallic materials at up to 1100 °C, so replaceable, robust handling systems such as gripper jaws are one possible solution for long term use. Cold and warm forging often means a commensurate increase in forging forces. The idea that fragile dexterous robot hands can be used is not one that finds implementation in practice. Hence smart techniques for advanced handling or sensing must be sufficiently robust as to cope with locally high temperatures. In the case of picking from for example an Electrotherm [46] rotary furnace, the entire structure (end effector/gripper/part) is soaked for up to a second in an oven temperature typically greater than 1100 °C. The level of robustness for any equipment to deal with these temperatures should not be underestimated, and in super-plasticity high temperatures are encountered in combination with saggy, tricky to handle and position materials, often in combination with specific cleaning requirements due to uneven deposition of slag. Additionally the volume rates involved are often relatively low, and require a level of skilled haptic feedback that is not commonly found in industrial automation, though research is moving forward rapidly there. Again the issue relates critically back to the business case for the investment in a manufacturing context. Note that whilst manufacturing accounts for about 10% of employees it accounts for 18% of fatalities, and 18% of seven day injuries [47].

Extremely high pressure/forces – upwards of 100 tonne Force are not unusual. Whilst this means that there is an opportunity to improve operator safety particularly with regards to presses – a tangible benefit – it also means that extremely close attention has to be paid to forge interlocks between the robot and the press. Note that where press injuries occur a large percentage lead to hand amputations [48]. Even smaller presses are easily capable of crushing a handling or lubricating robot which means that the system has to have extremely reliable methods of controlling the operations in the tooling area. A previous research project [49, 50] where a cold forming press routinely operated at 300 strokes/min (and on occasion up to 1000/min) described a programmable method of co-ordinating the motorised feeding cycle with the instantaneous linear encoder position on the press, though commonly in addition limit switches are used which interact with the master cell PLC to co-ordinate the robot. This means that for example simply relying on visual indications given the difficult operating environment, which often consists of poorly lit, and variable ambient conditions, presents high risks to equipment and personnel.

An additional challenge related to automatic forging/forming is that loosely contained objects have the potential to be ejected at very high speed from tooling. One of the major causes of injury related to presses [51] is the secondary damage caused by potential high temperature hard projectiles being ejected from the press. This can occasionally occur in the use of ceramic tooling for superplastic forming (SPF) presses temps 450 °C–950 °C at up to 7 MPa forming pressures for Ti-6AL-4V, though redesign techniques have mitigated the tendency of the tooling to explode. The Boeing Company, [52] reports two occurrences of catastrophic explosive brittle failure in 14 years where serious injury was prevented by reinforced heat shields.

Airborne lubricant – often forming and particularly forging tends to use lubricating/release agents and the result is that equipment is covered in a thin dusty black film of graphite. This detritus also presents difficulties for the picking and placing operations since the compacted/layered material tends to strongly adhere (at high temperatures) to the tooling. In manual situations this is dealt with by intermittent cleaning based on operator knowledge of where and how the build-up occurs, however in an automatic context provision has to be made otherwise toolsets become damaged as the effective cavity volume reduces [3].

Handling/haptic dexterity – this can be a particular research challenge where dexterity and adaptation is required and the handling and automation of SPF is an area where automation has been proven to be difficult [52]. Dealing with non-rigid formed structures (e.g. SPF) presents handling challenges that low sensitivity robots often have difficulty with, allied often to lower volumes of very high value parts – this is often linked to a tricky cleaning cycle due to deposits being left on the tooling. The operator skill level is relatively high and the process is not always completely understood which presents difficulties for the automation engineer, or research challenges. Current handling approaches are most frequently based on the use of high temperature protective "silver" suits whilst the operator stands in front of a 1000 °C oven and uses previous skill and a delicate sense of feel to remove the super-plastically formed sheet with basic long tools.

One potential approach is for active force measurement, or current feedback/monitoring to be used allied to an advanced vision system. Here again the difficulty is based on the high temperatures encountered.

3.2.1. Handling flash/excess material

Forging and forming processes often produce small but important amounts of waste or excess flash material either after clipping or as part of the extrusion process. One objective of the forming process is to optimise the material parameters by using the deformation process and thus the idea is to do the minimum possible machining. If clipping operations are used the build-up of flash can become a serious issue as it the one of the parts simply remain stuck in the tooling so that the handling system cannot remove them. Some detailed and practical thought needs to be put in to work out how the material will be dealt with, either via ejector pins or similar approaches.

3.2.2. Press integration

Integrating with existing manufacturers presses presents major challenges and one approach is to go back to the original manufacturer, or specifically the press controller. Presses have long lives typically 50 years+ but much longer is not unusual – with upgrades in controllers in order to interface with advanced metrology, (displacement, force measurement typically) often via industrial style PLC interfaces. Fortunately specialists exist that can deal with this, skills which are often employed by the systems integrator, including the critical safety elements, safety interlocks, guarding as required [19, 53, 54]. Often once the press project moves forwards snags occur, such as dealing with the waste product (flash) created when parts are reduced in volume – failure to do this will jam the tools, as well as how parts that stick in the tool are to be removed. One method of anticipating problems is to generate a detailed video of exactly (in excruciating detail), what goes on in the typical formation of a part. Experience suggests that even if this is done the depth of the micromanagement of every single element of the process has to be focussed on in extreme detail. One hypothesis is: what systems have to be in place to allow a blind person do this task? Learning gained from that pre-analysis phase – will pay off at the implementation stage however some developments may only emerge later, here the use of a knowledgeable systems integrator is invaluable. Automated tool changing is also possible but tends to be extremely expensive, up to 50% of the cost of a new press in some cases, depending on mass, and here conventional Single Minute Exchange of Die techniques are more likely to be cost effective.

4. Case study – advanced forming centre automation

A forging automation research programme at University of Strathclyde has been developed in conjunction with input from world leading companies such as Boeing, Rolls Royce, Timet, Aubert-Duvall and Barnes Aerospace, and formerly, Mettis Aerospace. The industrial input formed a key part of the research agenda based on a membership model [55]. As part of that process a requirement to pursue advanced automation techniques was identified as part of a strategic road-mapping

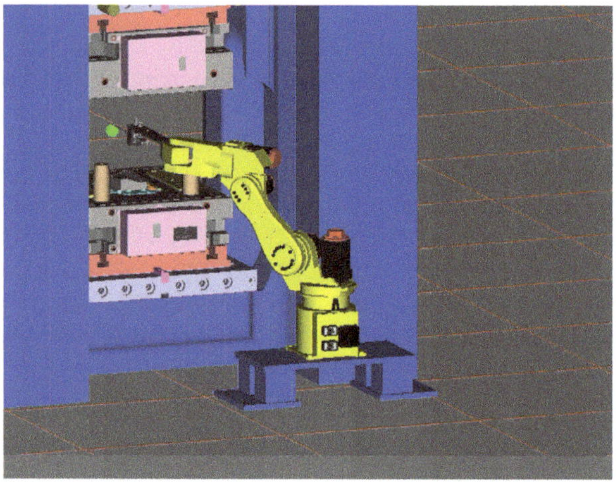

Figure 11. Manual spraying becomes robotically consistent.

exercise and this resulted in a creation by the author of a vision for Automation by the author to encapsulate the main objectives for Automation in AFRC.

It was clear from industrial interaction that manual variability in forging in particular is a key determinant of inconsistency in product quality and for a research organisation attempts to generate process improvements can be masked by the associated indeterminism. Some operators take longer to transport part from furnace to press, even a single operator has variability in the associated processing time dependent on levels of tiredness, and other external factors. Thus part temperature – a key determinant of forging quality and grain size is difficult to control at the point of hit, regardless of how much effort is placed on temperature control at the oven, since the transport time is variable.

Process variability in manually dominated processes has been widely observed by work study exercises and is a feature of simulation exercises in manufacturing [56, 57]. For the process experimenter this variability is anathema in the isolation of improvements so as part of the technological development of advanced forging and forming. Accordingly the key vision of the AFRC automation then evolved into a two-pronged approach, that of developing and researching advanced automation techniques in their own right, and to attempt to expunge, or at least to mitigate, the inherent variability in forging techniques by repeatable, automated, usually robotic approaches.

This vision was then discussed and presented internally so that agreement on the strategic thinking could be embedded in the senior management of the industrial partners, key to assuring the success of the project. Management buy-in is essential from the earliest possible stage in order to ensure the viability of the project when the going gets difficult. A pilot project was selected see Figure 11, and detail associated with the project is presented in the following section.

5. Vision for automation

The author defined the purpose of Automation as:

1. To generate a stable repeatable platform for forging and forming trials/experiments.
2. To generate and prove out automation forming techniques which have a 5–10 year technical benefits lead on existing industrial practice.

These industrial and research benefits include the following key features:

increased production volume,
reduced variability,
improved quality – dimensional accuracy,
Reduced Cost Reduction in scrap,
improved safety "Power presses are among the most dangerous machines used in industry" (PUWER) [58],
process debugging capability via stabilised experimentation.

5.1. Design considerations

As part of the implementation of the Automation Capability a Forge working group was set up comprising forge operators, manufacturing engineers, in order to clarify the key integration strategies as well as the tactical implication of the various technological choices and how they would impact on the project. From this interaction it became clear that here was a desire to operate the main forge – a 2000 tonne max screw press from Schuler (Figure 16) in both manual and automated modes in order to perform relatively short production run research projects on special metals as well as to have an integrated automation cell. This meant that from a design point of view the automation had to be capable of being safely stowed in the short term in order that particular manually orientated projects could be executed. For the Forging automation team his meant a fairly hard think about exactly how to balance the apparently competing requirements of both manual and automated operation.

For example in a typical robotic orientated cell a pick and place robot would be positioned front and centre (or slightly offset due to the geometry of the robot joints) in a permanent floor position, meaning difficulties for both access and interoperability. A possible configuration would be to have the handling robot on a swing arm or similar system to allow it to be deployed when necessary, however this has implications in terms of trailing cables and repositioning and after some consideration an overhead gantry was selected which allowed the handling robot to be safely parked when manual only operation is desired.

As part of the investigative work it was decided to pilot a particular project in the forge handling area and a Tier1 membership poll was taken in order to confirm that this was an acceptable approach, duly ratified by the AFRC technical board. The pilot project selected was automated lubrication & cleaning colloquially known as "dagging" based on the application of a specialised micro-graphite suspension to the tooling. Here consistency of application and lubricant amount on to the tooling was a key objective as well as to develop some learning in the forge on the practicalities of working with industrial scale applications.

The first step was to define what the characteristics of the automated solution was to do which included the area of the tooling to be covered and the reach of the robot – bearing in mind that the recommendation is to reduce exposure of the robot itself to the operation of the 2000 tonne press in case of inadvertent crushing. An automated stirring system was also introduced (agitation) in order to prevent the suspended graphite in the lubricant from settling out into layers. In manual operation an occasional quick stir with a stick does the job but all these small steps have to be automated in an integrated installation.

The first step was to trial various locations based on particular lubricant spray gun choices, and based on this possible selection in conjunction with a possible adaptor to cover the entire area, a design and manufacture of the associated holding fixtures/nozzle assembly.

Once a first estimate of the robot assembly had been created it was then modelled in the Fanuc Software "Roboguide" [59] which contains a detailed model of the each of the Fanuc robots, and can generate motion paths and timing information for the particular type of robot chosen. It is also particularly useful as a communications tool since it can show exactly how a particular robot will work in practice to each member of the team and create supplementary AVI video material (see Figures 11 and 12). Feedback on the simulation was then incorporated into the model and a suitable mounting for the robot was then defined based on the technicalities of positioning a plinth onto a robot pit – defined prior to the automation theme. A student was engaged for this work and having completed the modelling work in ProEngineer can then be exchanged into Roboguide, an approach used in other Fanuc installations [60, 61].

A key part of the design of the manufacturing process was to break the process into a series of automatable tasks which could then be controlled either by the robot controller through conventional 24 V DC I/O or alternatively via a master PLC, to be installed at a later date. In the first instance the development has been focused on controlling I/O via the robot controller since that was convenient and available, though these can migrate to the master PLC later.

Breaking the task down into individual parts means that it was realised that what the forge operator does is more involved than realised and includes a suit of operations so that the following devices have to be slaved to a master control program.

5.1.1. Furnace heating – automation

The AFRC rotary furnace device includes an incrementing hearth which can be moved round to the next position for presentation to the operator/robot.

5.2. Screw press – automation

Here the screw press needs to be controlled remotely by a master PLC in order to execute pre-programmed load conditions and press stroke. Some thought needed to be put into what would happen if a part is stuck in the tooling, as well as into making sure that appropriate communications protocols are put into place and that facilities exist to interact with the master PLC. In the case of the AFRC press it was part of the tender process that the press was automation ready, but often some kind of control retrofit is required on major items of capital equipment like this. Note that large presses in particular can have a useful life commonly 50 years or more, and controls upgrades are a regular feature.

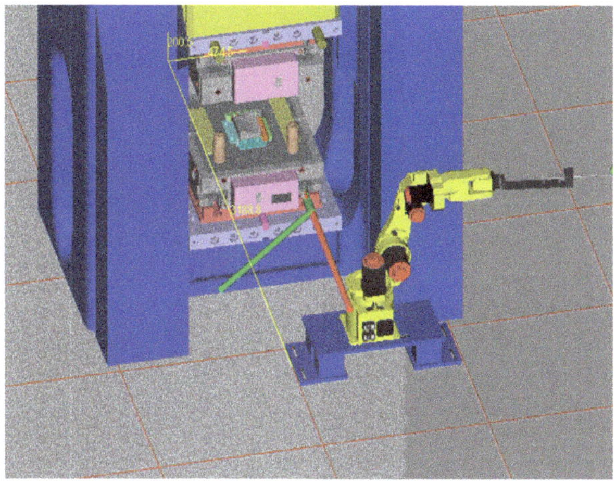

Figure 12. Design of special tool – layout checking.

5.2.1. Presentation devices

Part of the issues of using industrial automation is the level of investment that is required in presentation devices and this aspect needs careful appraisal – and an acceptance that fine tuning will probably be required after the system has been introduced. Since the handling and orientation devices have to strictly aligned on the basis of a pick repeat operation (often without a vision system) then either some sort of vibratory bowl feed system is used or other arrangement to present parts in a consistent location.

In the implementation here since the setup is not for continuous industrial production in the first instance it was decided to manually preload the furnace with billets and then drive the automation from that point. This offers flexibility and does not detract from the benefits of automation since the process only really begins after the billets or preforms have been heated to the correct temperature.

5.2.2. Lubrication robot Fanuc M6iB/6S – basic specifications

951 mm reach,
6 Kg payload at wrist,
upright, Wall or Inverted mounting,
0.08 mm repeatability at full speed,
8 Digital Inputs/Outputs,
mechanical weight 135 kg.

An examination of the setup of the press lubrication at the strategic level allocated the dagging robot and a convenient robot supplied by Fanuc was available the M6iB/6S. Since it was recommended that the robot itself not be placed at any time within the jaws of the press, it was sensible to use one with a relatively long reach since the platens of the AFRC screw press are large and it was desired that the robot be able to lubricate as

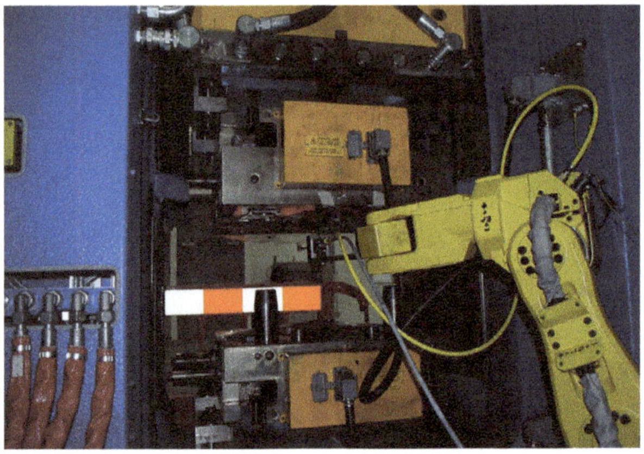

Figure 13. Robot operating inside press tooling.

much of this as possible. The other robot that was initially considered an LR Mate used elsewhere [3] would also have been feasible for smaller forgings but would have provided less coverage of the complete area of the tooling. The M6iB also provides fairly sophisticated Robot I/O that can slave associated devices without the need for a PLC – useful at the prototype/pilot stage in particular. A photo of the setup is shown in Figure 13.

It was necessary to analyse how this would work with the particular choices of lubricant gun available and how that would interact with the screw press and associated obstacles – this is where working between the Software Roboguide/ProEngineer proved invaluable, with CAD data being supplied by subcontractors in order to check the reach and manoeuvrability of the robot, and the required press opening to allow it.

Part of the work that the manual operator does in forging is the application of lubricant, and one popular trade-name is "Dag", a range of technical fluids specially formulated for forming and forging operations. In practical terms the operator usually uses judgement to determine where and how much dag should be applied but a simplified path using the robot – akin to a spraying application is possible in order to obtain complete coverage. It is possible to tune the process – often using compressed air to drive a purpose made spray gun as well as a mixture screw which defines the size of the opening orifice and hence flow rate. Additional debugging of the re-spray pattern is available by close and knowledgeable examination of the parts produced to see if recognisable faults are apparent through the misapplication or over/under-application of lubricant. Here it is important that an experienced operator provided feedback on exactly how specific faults are being produced and what they mean.

In the case of the work being reported here the lubrication robot provided a convenient pilot project since it incorporated the main features of robotic operation without the precision required of a classical pick and place operation, and all the associated feeding supply devices.

5.2.3. Safety protocols/guarding

Legislative requirements and basic consideration for staff safety dictated at an early stage that safety guarding had to be included. Since even the smaller robot (overall mass 140 kg) is capable of moving at up to 600° per second it does have the potential to cause some damage both to staff and itself and auxiliary equipment. Guards were therefore procures and installed internally – having physical interlocks in the first instance, with plans to integrate electrical integration once the final PLC was selected. Final layout is contingent upon placement of the multiforge/horizontal upsetter and integration of the guarding to cope with both the position of that in the one manufacturing cell.

5.2.4. Automation in practice

As the development of the automation work continued it became obvious that the experimental benefits from highly repeatable trials of particular forging situations excited research interest. An additional project of opportunity arose, related to quenching where the advantage of repeatedly quenching an object under programmable control emerged. Despite the fact that the automated handling system was not designed for this from the outset a brief model in Roboguide and clarified that it would be possible to repeatedly quench objects, thus improving the consistency of the experimental procedures. This has been a trend that has been observed elsewhere, robots tend to be redeployed to other factory/R&D functions rather than be replaced.

As part of this work it was decided to add automation capability to a larger furnace presently used for quenching trails based on a forklift. This means that it would be possible to routinely and repeated quench the relatively large (up to 25 kg) objects in a systematic way, since the time to go from the furnace to quenchant by the existing manual (forklift) method was very variable.

5.2.5. Trimming/clipping press

At the output side of the screw press, a clipping press has been placed which removes any final waste from the final stamped part. From the Automation engineers view both presses can be regarded as similar devices – in which program control is handed over to the individual press controller, which then has load displacement parameters required for a particular part sequence.

5.2.6. Robot gantry

It became apparent from the design and review meetings that it would not be possible to have the pick robot at the back of the press for a number of reasons and this posed the issue of access:

the press was fixed in a substantial pit and could not be moved,
Fanuc M710iC,
50 kg Max load capacity at wrist,
repeatability ±0.07 mm,
reach 2050 mm,
mechanical weight 560 kg,
6 axis electric servo robot,
capable of floor wall or ceiling mount capability.

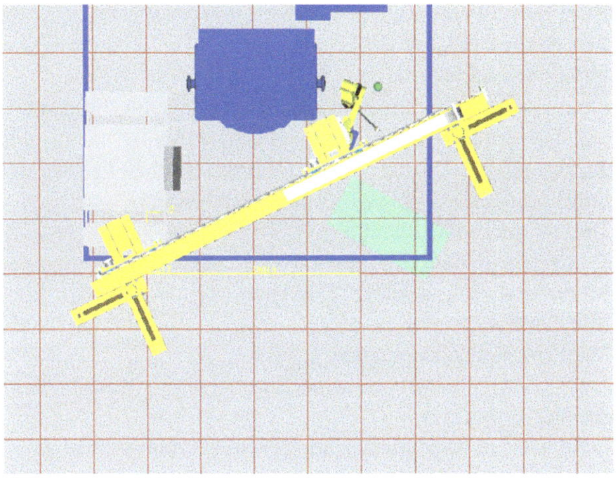

Figure 14. Checking clearance of robot and park position.

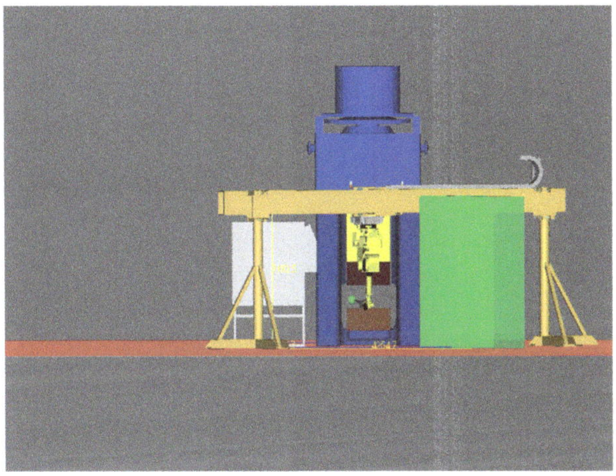

Figure 15. Checking robot interaction with press.

The Electrotherm Furnace was required at the front for manual operations – therefore the robot had to accommodate this too. Therefore the robot had to go at the front. The forklift could not access the rear of the press to change tooling – since that was the case the dagging robot was placed at the rear offset as shown.

The forklift based bolster/tool change mean that any temporary placement of the robot had to also cope with tool-change and no overhead crane was available. This meant that access to the front and the floor area had to be kept clear for both manual operation and for forklift access/egress. Placing the Robot on a swing arm was also considered though this would have meant considerable space implications as well as placing the 500 kg robot mass on the end of a simply supported cantilevered beam. Fanuc Roboguide was used to define the layout (see Figures 14 and 15).

Possibilities of mounting the robot on temporary plinths, on a fixed overhead structure were considered but a useful approach was that of inverting the robot and mounting it on a motorised gantry. This has a number of advantages.

The standard workspace is clear and the robot could be parked safely well away from the press when not in use.

Figure 16. Fanuc M710 robot mounted inverted on Gantry.

The robot could be repeatedly placed in a centralised position when required. The working envelope of the robot and its flexibility would be enhanced considerably.

The disadvantages were the increased cost and increased complexity in programming (7th axis).

However these disadvantages were greatly enhanced by the increased flexibility – especially important in an R&D environment where the ability to change over from task to another and to implement small batch sizes is critically important. Having considered these alternatives the decision was taken to source a motorised overhead rail/Gantry via Fanuc robotics and the installation of this was also managed by Fanuc robotics, with simulation done in Roboguide and laid out using ProEngineer as shown in Figures 15 and 16 and basic configuration in Figure 16. This allowed the location and handling arrangement to be checked so that the robot could be placed in an ideal park position, and deployed in front of both the Programmable furnace, screw press and clipping press. Motion between these was able to be checked and elementary off line programming ideas looked at top check that the robot would be able to pick parts and place them correctly.

6. Conclusions

This paper has described a review of manufacturing automation as related to forging and an outline of a process to design and implement an automated robotic forging cell, based on a large press. The process requires clear management support from the beginning as well as a realisation that for certain applications smart design can accommodate both manual forging and automated forging operation using the same installation. A micro-management approach to the process is required that includes all the corrective actions that intelligent operators do in order to produce smart forgings and the process of laying out the cell has been described, using CAD tools including Roboguide, and ProEngineer. At the time of writing the cell is still undergoing development but the fundamentals have been laid down based on the original authors' vision and are expected to produce stable experimental results as well as a basic structure for flexible technology development.

Acknowledgements. The layout and conception of this cell was assisted greatly by the following staff: Mr I. McLarnon, Mr J. McBain with additional contributions by Mr S. Cunniffe and Dr. M. Rosochowska.

References

1. C.S. Harrison, I.G. Ross, M. Bell-Jack, P.W. McEwan, Manual on manufacturing systems simulation for NCR Dundee, University of Dundee, 1989, p. 180 (unpublished).
2. Industrially orientated software to address cost benefits, http://www.vanguardsw.com/products/vanguard-system/cost-modeling.htm
3. Automated forging of Gas turbine Compressor Blades from billet to finished forging, http://www.fanucrobotics.co.uk/en/countries/fruk/news/rolls%20royce
4. G. Chen, H. Wang, K. Zhao, Z. Lin, Modular Calculation of the Jacobian Matrix and Its application to the performance analyses of a Forging robot, Advanced Robotics 23 (2009) 1271–1279.
5. C. Yan, F. Gao, W. Guo, Co-ordinated kinematic modelling for motion planning of heavy duty manipulators in an integrated open die forging centre, Proceedings IMechE Part B: Journal of Engineering Manufacture 223 (2009) 1299–1313.
6. K.W. Lilly, A.S. Melligeri, Dynamic simulation and neural network compliance control of an intelligent forging center, Journal of Intelligent and Robotic Systems 17 (1996) 81–99.
7. X. Wang, M. Yukawa, Y. Yoshita, T. Sukeda, T. Ishikawa, Research on some basic deformations in free forging with robot and servo-press, Journal of Materials Processing Technology 209 (2009) 3030–3038.
8. R. Badadou, H. Shimahara, G. Hirt, Automated semisolid forging of steel components by means of thixojoining, Solid State Phenomena 116–117 (2006) 383–386.
9. http://www.forging.org/pdf/TechPlan2006-2007Rev.2.pdf
10. http://www.ogtechnologies.com/hoteye.html
11. http://www.bara.org.uk/
12. http://www.bara.org.uk/info/statistics/Industrial_Robot_Facts_2010.pdf
13. Jan Hutson, Forge Shop Automation, part 1, April 2009, pp. 20–23, www.forgemag.com
14. Jan Hutson, Forge Shop Automation, part 2, July 2009, pp. 23–26, www.forgemag.com
15. Robot – from Rossums Universal Robots – Capek, translates as "hard worker" or even "serf labour".
16. http://www.schulergroup.com/major/documents/downloads/download_technologien_branchen_produkte/download_technology_forging/massivumformung_broschuere_e.pdf
17. http://www.schulergroup.com/major/us/11_Highlights/20_Compact_Crossbar_Pressen/index.html
18. http://www.mastercam.com/
19. http://www.fanucrobotics.co.uk/en/products/software
20. http://www.robotmaster.com/
21. http://www.ros.org/
22. Private conversation, G. Dr G. Pierce Centre for Ultrasonic Engineering, Univ. Strathclyde.
23. http://www.schulergroup.com/major/us/04_Presse/010_Pressemitteilung/2009/2009_02_05_crossbar_roboter/index.html
24. ABB Robotics "Foundry Magazine", November 2008.
25. Retraining Robots, Ben Hargreaves, professional Engineering, 21st April, 2010, pp. 29.
26. http://www.abb.com/
27. http://www.rimrockcorp.com/files/News/ForgeShop_pt_2.pdf
28. http://www.youtube.com/watch?v=WIMzH87RWJM. Channel Series 9 Episode 3 segment D, Jet Compressor blades Episode 107.
29. http://www.leistritz.com/extrusion/en/04_products/steuerungstechnik_2.html
30. L. McIntosh, Internal Rolls-Royce report, Department of Mechanical Engineering, University of Strathclyde, 2009 (unpublished).
31. http://en.wikipedia.org/wiki/CE_mark
32. http://www.youtube.com/watch?v=xHH24_LCPR8&feature=related
33. http://www.east-engineering.eu/Segment%20Spray%20eng.pdf
34. http://www.jerko-spruehsysteme.com/index.htm
35. http://forgingmagazine.com/feature/2x-robots-doubles-productivity
36. P.S. Adler, Managing Flexible Automation, California Management Review Spring 30, 3 (1988) 34–57.
37. Professional Engineering, Institution of Mechanical Engineers 27, 5 (2014) 27–28.
38. http://www.ibef.org/download/corporatebharatforge_june06.pdf
39. K. Capek, Rosumovi Univerzální Roboti, Science Fiction Play (1920).
40. N. Ay, N. Bertschinger, R. Der, F. Guttler, E. Olbrich, Predictive information and explorative behaviour of autonomous robots, European Physical Journal B 63 (2008) 329–339.
41. https://www.youtube.com/watch?v=yFi7UL70zTo
42. T. Hulin, C. Preushe, E. Yechiam, A. Telpaz, V. Schmirgel, U.E. Zimmerman, Haptic and Visual Training of system behavior a case study for Robotic Programming-by-Demonstration, BIO Web of Conferences 1 (2011) 00039.
43. https://www.youtube.com/watch?v=r7gU74Yv9Es&list=UU6HrPPoLdjNynZCvUrJbmBw
44. S. Calinon, Challenges in adapting imitation and reinforcement learning to compliant robots, Bio Web of Conferences Skills 1 (2011) 00013.
45. R.N. Shea, Robot Safety standard Update, RIA meeting, 20–23 Feb 2013, Orlando World Marriot Centre, Orlando Florida.
46. http://www.electrothermindustry.com/Small%20Rotary%20Hearth %20Furnace
47. http://www.hse.gov.uk/statistics/industry/manufacturing/manufacturing.pdf
48. Durosay et al., Journal of Hand Surgery (European) 36, 5, 383–391.
49. A. Razali, Y. Qin, J. Zhao, C.S. Harrison, R. Smith, Development of a new high precision feeder for micro-sheet forming, Journal of Manufacturing Science and Engineering, Transactions of the ASME 133, 6 (2011) 061025, DOI: 10.1115/1.4005046.
50. Y. Qin, A. Brockett, Y. Ma, A. Razali, J. Zhao, C.S. Harrison, W. Pan, X. Dai, D. Loziak, Micro-manufacturing research, technology outcomes and development issues, International Journal of Advanced Manufacturing Technology 47, 9–12 (2009) 821–837, DOI: 10.1007/s00170-009-2411-2.
51. F.R. Spellman, Nancy E. Whiting, Machine Guarding Handbook, Rowan and Littlefield, 1999.
52. D.G. Sanders, Reinforced ceramic dies for superplastic forming operations, Journal of Materials Engineering and Performance 13, 6 (2004) 753–757.
53. http://www.forgex.co.uk

54. http://www.CorePD.co.uk
55. www.strath.ac.uk/afrc
56. C.S. Harrison, R.J. Twist, K.K.B. Hon, Simulation of the effects of uncertainty and its effects on production planning lead times, in 6th International Conference on Computer Aided Production Engineering, University of Edinburgh, 1990, pp. 237–244.
57. A. Carrie, Simulation of Manufacturing Systems, John Wiley, 1988.
58. http://www.hse.gov.uk/pubns/books/l112.htm
59. http://www.fanucrobotics.co.uk/
60. S. Sakabara, The latest Robot systems which reinforce the manufacturing sector, in Proceedings of the 2003 IEEE International Conference on Robotics and Automation, Taipei, September (2003).
61. Z. Haiming, Y. Xichen, L. Lifeng, Path planning and off-line programming in laser remanufacturing robot system, Mechanic Automation and Control Engineering (MACE) (2010) 3796–3798.

11

Review on powder-based electron beam additive manufacturing technology

Xibing Gong[1], Ted Anderson[2], and Kevin Chou[1,*]

[1] Mechanical Engineering Department, The University of Alabama, Tuscaloosa, AL 35487, USA
[2] Advanced Manufacturing Team, Marshall Space Flight Center, Huntsville, AL 35811, USA

Abstract – This paper presents a thorough literature review of the powder-based electron beam additive manufacturing (EBAM) technology. EBAM, a relatively new additive manufacturing (AM) process, can produce full-density metallic parts directly from the electronic data of the designed part geometry. EBAM has gained broad attentions from different industries such as aerospace and biomedical, with great potential in a variety of applications. The paper first introduces the general aspects of EBAM. The unique characteristics, advantages and challenges of EBAM are then presented. Moreover, the hub of this paper includes extensive discussions of microstructures, mechanical properties, geometric attributes, which impact the application ranges of EBAM parts, with focus on commonly used titanium alloys (in particular, Ti-6Al-4V). In the end, modeling efforts and process metrology of the EBAM process are discussed as well.

Key words: Electron beam additive manufacturing (EBAM), Geometric attributes, Mechanical properties, Microstructures, Metallic powder, Process modeling, Process metrology.

Introduction

Additive manufacturing (AM), also known by various terms, e.g., direct digital manufacturing, based on "layer-by-layer" fabrications is an emerging technology, by which physical solid parts are made directly from electronic data, generally files from computer-aided design (CAD) software. This group of technologies offers many design and manufacturing advantages such as short lead time, complex geometry capability, and tooling free. Electron beam additive manufacturing (EBAM) is a relatively new AM technology [1]. Similar to electron-beam welding, EBAM utilizes a high-energy electron beam, as a moving heat source, to melt and fuse, by rapid self-cooling, metal powder and produce parts in a layer-building fashion. Moreover, EBAM is one of a few AM technologies capable of making full-density functional metallic parts, drastically extending AM applications. In particular, the ability of direct fabrications of metallic parts can significantly accelerate product designs and developments in a wide variety of metallic-part applications, especially for complex components, e.g., fine network structures, internal cavities and channels, which are difficult to make by conventional manufacturing means [1, 2].

EBAM machines were first commercialized, around 1997, by Arcam AB Corporation in Sweden. Because EBAM has many unique characteristics such as high energy efficiency, high scan speed, and moderate operation cost, the technology has attracted, in recent years, increased interests from different industries. The use of an electron beam offers extensive features such as higher build rates due to increased penetration depths and rapid scanning speeds. Since then, many research groups have been studying the EBAM technology from different aspects and for various applications. Despite the potential benefits over conventional manufacturing technologies, EBAM still has a few process deficiencies, such as process stability, part defects and quality variations [3], etc. As EBAM technology is relatively new, there have not been detailed reviews. The objective of this paper is to offer a survey of various investigations into EBAM, especially for the study of the titanium (Ti) alloy parts, along with the EBAM part microstructures and associated mechanical properties. In addition, modeling and simulation of the EBAM process, though rare, for process understanding and advancements are also discussed.

EBAM details

Process principle

A conceptual schematic of an EBAM machine is shown in Figure 1 [2]. The principle is very similar to that of a scanning

*e-mail: kchou@eng.ua.edu

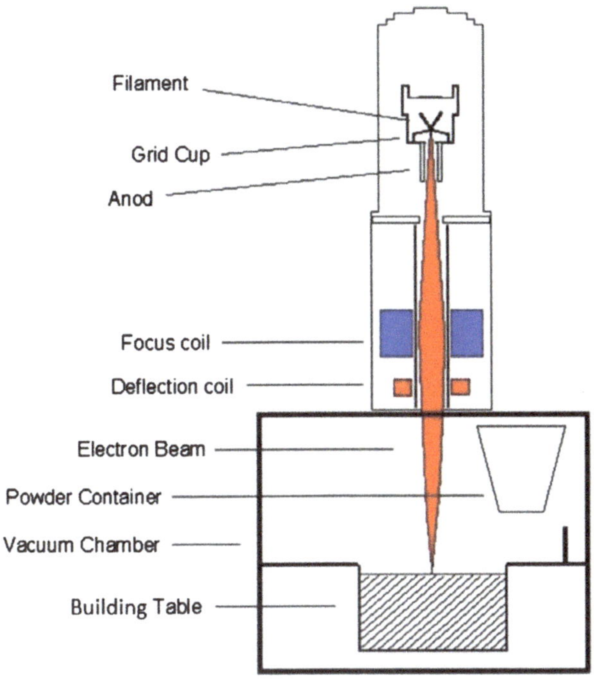

Figure 1. Schematic drawing of an Arcam EBAM machine [2].

electron microscope. A heated tungsten filament, in the upper column, emits electrons which are collimated and accelerated to a kinetic energy of about 60 keV. The electron beam is controlled by two magnetic coils, which are housed in the lower column. The first one is a magnetic lens which focuses the beam to the desired diameter, and the second one deflects the focused beam to the desired point on a build platform. The electron-beam gun itself is fixed, no moving mechanical parts involved in beam deflections. The beam current is controlled in the range 1–50 mA and the beam diameter can be focused down to about 0.1 mm. In the chamber of the middle part of the machine, fine metal powder, on the order of 10–100 μm, is supplied from two hoppers and forms a thin layer by a raking mechanism before each layer build. The typical layer thickness is in the range 0.05–0.2 mm. The computer-controlled electron beam scans over the powder layer in a predefined pattern and consolidates the desired areas into solid and dense metals. The beam has to first scan at a high speed (order of 10 m/s) in multiple passes to preheat powder to a sintered state, while a beam scan on the order of ~0.5 m/s is used during the melting cycle. Then, a new powder layer is laid on top and the scanning process is repeated until all layers are completed.

The entire process takes place under a high vacuum. The typical pressure of residual gases in an EBAM machine is 10^{-1} Pa in the vacuum chamber and 10^{-3} Pa in the electron gun [2]. During the melting process, a low pressure of inert helium gas (10^{-1} Pa) is added to the vacuum chamber to avoid build-up of electrical charges in powder. When all layers have been completed, the built part is allowed to cool inside the process chamber, which is then filled up with helium as to assist cooling. Because of radiation from electrons, the process observation is not as accessible as other AM technologies, only through a leaded-glass viewport. Therefore, what exactly happens inside the build chamfer is not as well perceived as other AM processes. Recently, Oak Ridge National Laboratory published a video animation of the EBAM process [4] that offers good illustrations of the process details, especially to those without access to an EBAM machine.

Applications and challenges

Due to EBAM's unique capability, it is especially beneficial to such industries as the aerospace sector, creating new opportunities for both prototyping and low volume productions. The time, cost, and challenges of machining or other processes are eliminated, which makes the components readily available for functional testing or installation on a system, e.g., aircraft [5]. Additionally, the additive process opens a door to new design configurations (e.g., cellular structures) and weight-reduction alternatives.

The energy density of the electron beam is high enough to melt a wide variety of metals and alloys. EBAM processes have the potential to work with many material classes, for example, aluminum alloys [6], stool steel (H13) [7], and cobalt-based superalloys [8], etc. However, Ti alloys, in particular, Ti-6Al-4V, were the first material extensively researched, also widely used in EBAM technologies. Ti alloys have numerous potential applications as a consequence of their superior properties: low density, high mechanical strengths, corrosion resistance, human allergic response, and good biocompatibility [9–11]. The industrial processing routes for titanium alloys include ingot casting, powder processing, ingot forging, and sheet production by hot-rolling. Using traditional processes makes it difficult to manufacture highly complex and functional Ti-alloy parts, such as artificial knee joints, hip joints, and bone plates, etc. EBAM has been used for fabrications of specific net-shaped parts for use within both aerospace and medical implant industries. The fabrication of implants from patient specific data with adaptation to the region of implantation is made possible with EBAM (e.g., temporomandibular joint prosthesis in Figure 2a), eliminating expensive secondary processing such as machining or forming and related lead times. Moreover, with geometric freedom, EBAM has enabled one step fabrications of intricate architecture (meshed, porous, cellular), e.g., porous custom implants with controlled porosity to meet the requirements of the anatomy and functions at the region of implantation. In addition, EBAM has been used to assist the Environmental Control and Life Support System (ECLSS) group of NASA for a new design of hardware, Figure 2b, to pack zeolite materials for improving CO_2 removal from the International Space Station environment [12]. Using EBAM, the total production time is 1 week including machining the matted surfaces in contrast to 13 weeks when using traditional processes. In addition, the ability to produce a negative Poisson's ratio structure, the so-called "auxetic behavior" is another unique capability of EBAM. A negative Poisson' ratio leads to higher impact and shear resistances, and fracture toughness. Schwerdtfeger et al. [13] found negative Poisson's ratios in the range of −0.2 to −0.4 depending on the orientation.

Since Ti-6Al-4V is the most widely used material in EBAM, the major effort of this review is focused on this group

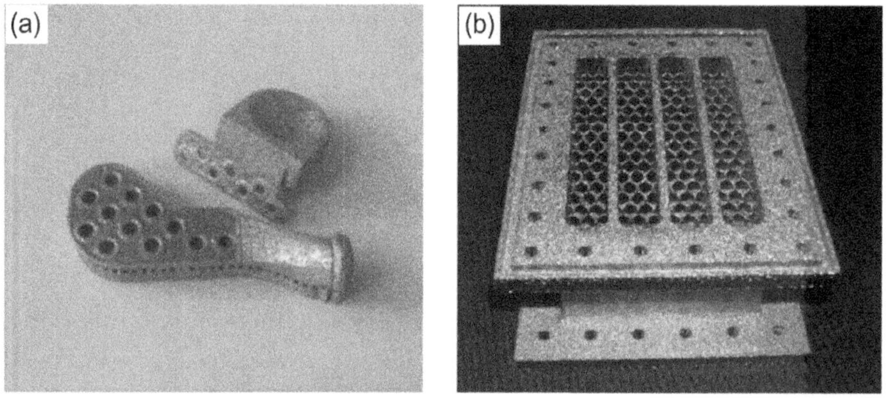

Figure 2. (a) Prototype of temporomandibular joint prosthesis, and (b) ECLSS test chamber [12], both produced by EBAM.

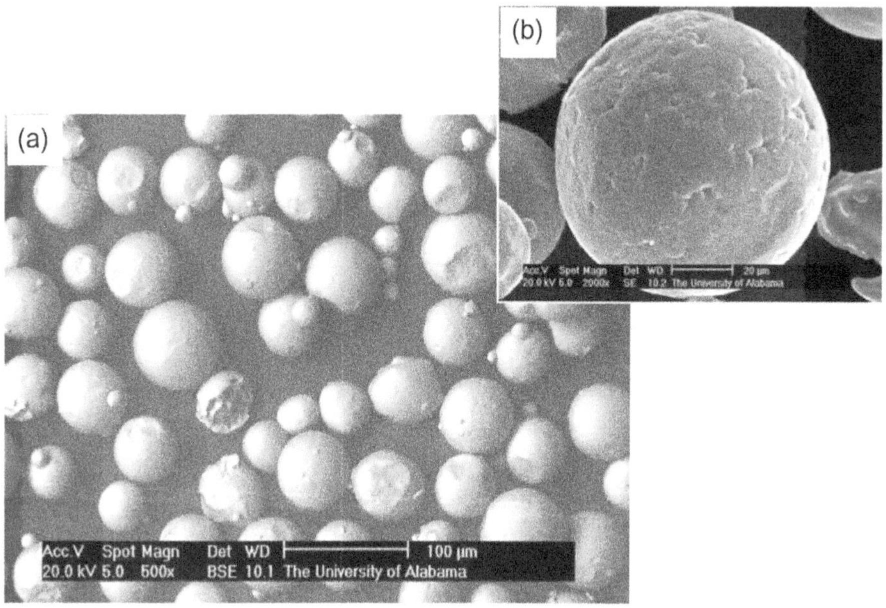

Figure 3. SEM images of Ti-6Al-4V powder: (a) particles with different sizes and (b) an individual particle.

of materials, its usages, applications and EBAM part characteristics.

Powder characteristics

Raw powder

Raw materials used in EBAM are metallic particles from powder metallurgy, and the characteristics and quality of powder strongly affect the process performance. The powder morphology is one important factor in EBAM affecting processing conditions such as flowability, powder packing, and ultimately, heat transfer process phenomena [14]. Neira-Arce observed that the powder used in EBAM is spherical in shape. The spherical shape may contribute to improved flowability, and thus, may ensure high build rates and part accuracy [14].

In general, fine powder is used in EBAM. Figure 3 shows scanning electron microscopic (SEM) images of Ti-6Al-4V powder: (a) low magnification showing a population of particle sizes and (b) high magnification of an individual powder. The powder size distribution also has a significant effect on the build part density, surface finish and mechanical properties [14]. Syam et al. [15] studied the size distribution of Ti-6Al-4V powder in EBAM and reported the size range as from 45 to 100 µm. Murr et al. [9] also investigated the distribution of Ti-6Al-4V powder diameters. The spherical powder diameters were measured and plotted in a histogram, shown in Figure 4. It is noted that the powder size distribution is bimodal. The average overall powder diameter in Figure 4 is 30 µm. The average sizes of large particles and small attached particles are around 60 µm and 10 µm, respectively. For the chemical composition, according to a study by Murr et al. [9], Ti-6Al-4V powder used in EBAM has a nominal composition of 6.04% Al, 4.05% V, 0.013% C, 0.07% Fe, 0.13% O,

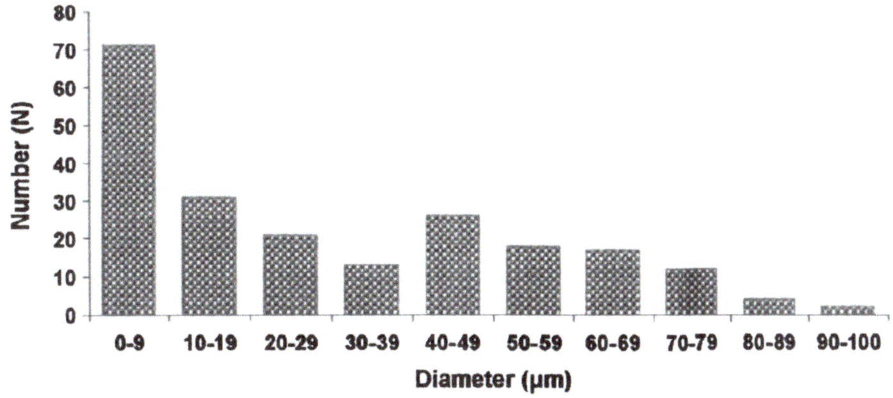

Figure 4. Histogram of raw Ti-6Al-4V particle size distribution [9].

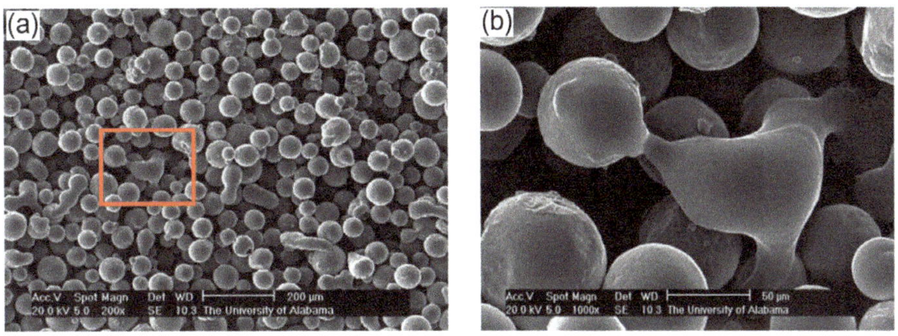

Figure 5. SEM images of sintered Ti-6Al-4V particles from preheating.

<0.005% N and H; the balance is Ti (in weight percent), comparable to the common Ti-6Al-4V specification.

The preheating process contributes to the metallurgical bonds or even partial melting of the powder.

Sintered powder

The thermal cycle in EBAM, which includes the preheating, subsequent melting and solidification, is critical to determine the microstructure and mechanical properties of the EBAM parts. Different from the laser additive manufacturing process, the EBAM process applies the preheating to lightly sinter the precursor powder layer by using electron beam at a low power and a high scanning speed. According to the study by Cormier et al. [7] and Rodriguez et al. [16], the preheating process serves two purposes: holding the metal powder in place during the subsequent melting scan and reducing the thermal gradient in the build part.

He et al. [17] studied preheating of Ti-6Al-4V powder in EBAM. According to the authors, the sintering mechanism is that small particles partially or completely melt and they play an important role as a binder to bond the majority of large particles together, which will not only hold the particles and withstanding the impact from electrons, but also prevent the spheroidization effect in the part surface. Gong and Chou [18] also investigated the morphology of the sintered powder in EBAM. The phenomenon of necking could be noted on both the build and side surfaces. Figure 5 illustrates the SEM images of the aggregated powder on the scanning surface. The diameter of the necks is on the order of 1–10 μm.

Microstructures

Typically, Ti-6Al-4V samples from EBAM show an ordered lamellar microstructure, consisting of extremely fine grains, as can be expected by the thermal characteristics of the EBAM process: small melt pool and rapid cooling. EBAM components possess a columnar shaped morphology of the prior β phase with a growing direction parallel to the build direction [19], which is a consequence of primary thermal gradients that exist in the build direction. Antonysamy et al. [20] studied the prior β-grain texture of EBAM components. It is shown that the columnar structure shows strong fiber texture of $< 001 > \beta$, which is normal to the deposited powder layers. The favored texture could be attributed to the highly elongated shape of the moving melt pool and the raster pattern used in the EBAM process. The phases of EBAM Ti-6Al-4V samples include some α phase at the β grain boundaries, which is finer than that obtained by metal casting [21]. Facchini et al. [22] investigated microstructures by X-ray diffratometry (XRD) and reported that the main constituent is α HCP phase with only a small contribution of β phase in EBAM Ti-6Al-4V. The HCP pattern can be attributed to both α phase and the α'-martensite, which is spatially smaller than the α-phase platelets. Safdar et al. [19] also reported that a typical ($\alpha + \beta$) structure, namely Widmanstätten α platelets with rod-like β phase, formed

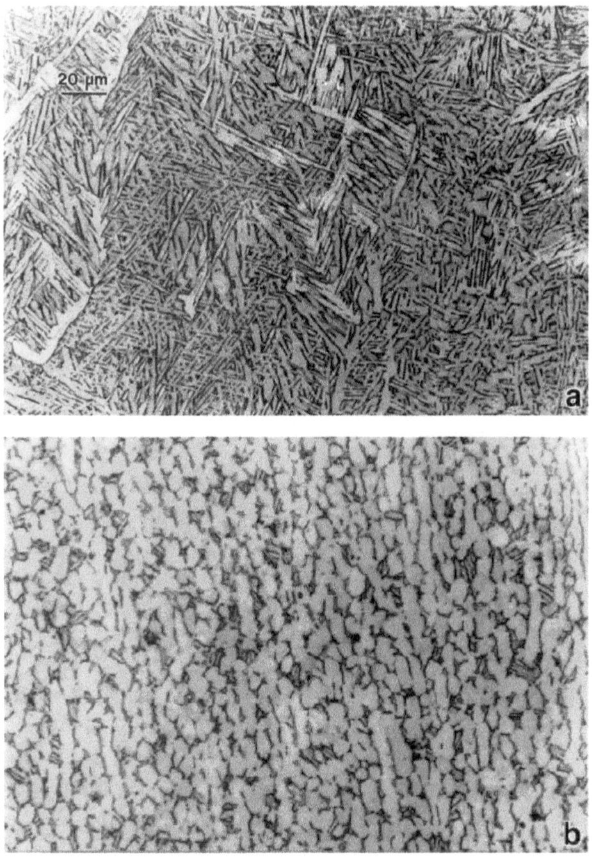

Figure 6. Microstructures of Ti-6Al-4V specimens made by (a) EBAM vs. (b) casting [9].

at the interfaces of the fine α grains, have been observed in the columnar prior β grains. Grain boundary α phase was found to be formed around the boundaries of the columnar prior β grains. Koike et al. [23] compared the EBAM microstructures to counterparts from cast and wrought Ti specimens. For the cast specimen, a typical α-case microstructure consisting of columnar α-crystals can be clearly observed. The microstructure of the wrought Ti-6Al-4V specimen consisted of slightly elongated α grains and intergranular β grains which is a typical microstructure from heat treatment conditions. On the other hand, no α-case was observed near the surface of the EBAM specimens. However, Murr et al. [9] observed that the EBAM Ti-6Al-4V prototypes exhibit the α-phase, acicular platelet microstructures similar to commercial wrought products, while the mesh and foam prototypes exhibit mainly α'-martensitic platelets or mixtures of α and α' platelets, giving rise to harder structures and consequently greater strengths. Figure 6 compares the microstructures of Ti-6Al-4V specimens from (a) EBAM and (b) casting [9]. The authors of this paper have attempted to study the microstructure of Ti-6Al-4V parts processed by EBAM. The microstructure of Ti-6Al-4V consists of columnar grains of prior β growing along the build direction, as can be seen from Figure 7a. The width of the columnar structure is about 25–30 μm. After the solid phase transformation in EBAM, the prior β grain structure transforms to a bulk microstructure of fine Widmanstätten ($\alpha + \beta$) structure and martensite α' owing to the rapid cooling process. Figure 7b illustrates the microstructure of the scanning surface of an EBAM Ti-6Al-4V sample, with "equiaxed" grains shown. It demonstrates that the prior columnar structure is generally rod-shaped. There are also studies reporting different microstructures (grain size and morphology) obtained from the EBAM, attributed to noticeably different cooling rates [24, 25] and re-melting [26]. As to the chemistry specification, Heinl et al. [27] tested the element composition of the EBAM parts. The authors claimed that the composition of the EBAM parts conforms to the standard specifications for Ti-6Al-4V alloy castings for surgical implants. However, Gayten et al. reported that the chemistry of EBAM made Ti-6Al-4V parts may have 10 to 15% reduction in Al content [28] possibly due to a higher vapor pressure of Al.

The gas voids or porosities are a typical defect in EBAM parts, which is exemplified in Figure 8 below. Gaytan et al. [28] investigated porosity defects and control during EBAM and attempted using hot-isostatic-pressing (HIP) to eliminate gas voids in built samples. The result showed that while the voids can be largely eliminated by a single standard HIP cycle, remnants sometimes persist. In addition, some gas bubbles actually came from recycled powder and stay in EBAM-built parts. Because of the melt and liquid phase surface tension as well as the low gas pressure, it is essentially impossible to eliminate the intrinsic gas bubbles in EBAM parts. Reported by Gaytan et al., however, because of the small void size (order of 10 μm), they may not impact the mechanical properties of EBAM-built parts.

In addition, electron beam-material interactions reduce the EBAM process stability. The melt pool instability is affected possibly by an inadequate energy density transmitted from the electron beam into the powder, and it may result in the balling effect, when the surface tension of molten liquid exceeds the wetting ability of the previously solidified layer [29]. The melt balls further prevent the process from continuation due to rough textures of the top layer. Moreover, as EBAM relies on selective solidification of the top powder layer, energy is inserted into the material in a non-uniform way. Large temperature gradients may emerge due to selective heating of powder areas and thus, residual stresses may be induced. If the residual stresses exceed the bonding abilities between layers, it results in delamination, which depends on the scanning strategy [30]. Specifically, the orientation of the scan vectors has a considerable influence to delamination [3].

It has been reported that operating parameters have significant effects on the part characteristics [24, 25], quality consistency and process performance. Murr et al. reported that variations in melt scan, beam current, and scan speed affect the EBAM built defects such as porosity [24], and may cause significant property-performance variations [31]. In general, the beam power, diameter, and speed, as well as the pre-heat temperature are four major process parameters; the first three are tied to the thermal cycle variables, temperatures and cooling rates, and the pre-heat temperature governs the sintering state of powder prior to the melting scans. Puebla et al. [32] studied the effect of the scan speed on microstructure and macrostructure of EBAM Ti-6Al-4V. When the scan speed is increased from 100 to 1000 mm·s^{-1}, the cooling rate and solidification speed both increase, which results in a decreasing α-phase acicular grain width as well as an increasing proportion of martensite plate production. Correspondingly, increasing the scan speed

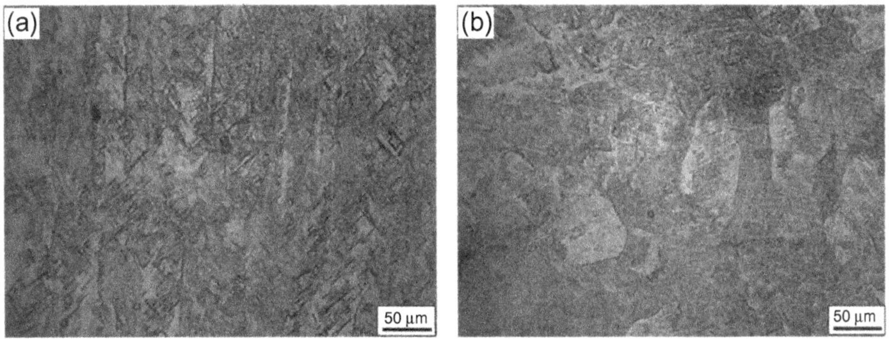

Figure 7. Optical metallograph of EBAM Ti-6Al-4V: (a) side surface and (b) scanning surface.

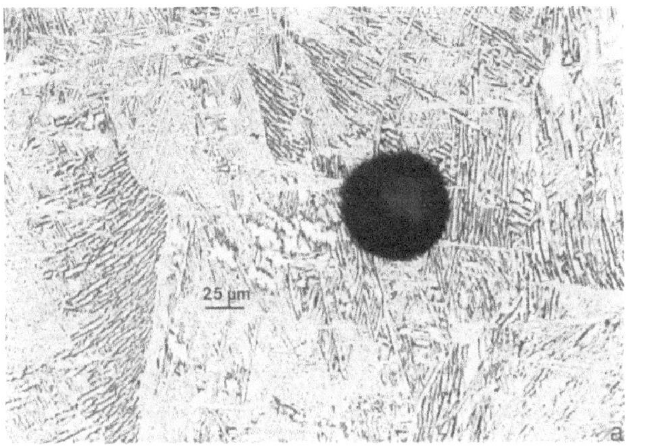

Figure 8. An example of gas void: optical metallograph of a polished and etched section [28].

increases the porosity by creating unsintered powder volumes within the layers. In addition, Hrabe and Quinn [33] investigated the effect of energy input on the microstructure of the built parts. It was found that the melt pool size, α lath thickness and prior-β grain size increased with the increase of energy input.

Rapid self-cooling, which results in the EBAM layer building/bonding mechanism, has been noted to be different upon comparing different melt-scan parameters. Gaytan et al. reported that in Ti-6Al-4V, the differences in melt-scan parameters create microstructural variations characterized by sizes or dimensions as well as phase differences and dislocation density variations [28]. Bontha et al. [25] investigated the effects of process parameters on solidification microstructures in beam-based fabrications. It shows that variations in the beam power and speed can alter both solidification cooling rates and thermal gradients by several orders of magnitude, which have a significant effect on resulting microstructures. Results specifically for Ti-6Al-4V suggest that process size-scale can have a significant effect on microstructures, including a transition from columnar to equiaxed microstructure at higher powers.

Mechanical properties

EBAM part properties have been frequently investigated. Some studies indicated that properties of EBAM parts are comparable to those from conventional processes (wrought) [9, 34]. However, other research indicated improved hardness of EBAM parts [35]. Changes in local chemistry and different microstructures have been suggested as possible causes.

Tensile testing

Tensile testing has been widely used to characterize the mechanical properties of EBAM parts. Some researchers [22] found that the ultimate tensile strength (UTS) of EBAM built specimens is higher than the wrought or annealed ones, with a lower ductility. However, others [23] presented that the UTS and ductility of the cast and wrought Ti-6Al-4V specimens were higher than those of EBAM counterparts. The reason for the difference could be attributed to the variation in the built parameters, which result in different structures such as composition, structures, pore size, and porosity distribution, etc. Hrabe and Quinn [33] also studied the effect of energy input on the mechanical properties of the EBAM built parts. The results show that UTS, yield strength (YS) decreased with the increase of energy input. However, the change in UTS (2% change) and YS (3% change) was small. Ladani and Roy [36] investigated the anisotropic effect on the mechanical behavior of Ti-6Al-4V manufactured by EBAM. Tensile testing indicated that YS and UTS for flat build samples have distinguishably higher values than those of the side-build and top-build samples.

Koike et al. [23] investigated and reported that the YS and UTS of the EBAM specimens were 735 and 775 MPa, respectively. The ductility was 2.3% elongation. These values were comparable to those of cast specimens, which exhibited slightly higher UTS and ductility. These differences were considered to be due to rippled, rough specimen surfaces and possibly higher oxygen content in the EBAM specimens (0.34 vs. 0.22%), originated from the alloy powder. Other researchers showed better ductility (elongation) in EBAM parts. Murr et al. [9] studied that the elongation of the EBAM samples shown is about 16–25% compared to the elongation of 12–14% for wrought alloys. Table 1 below summarizes the mechanical properties of EBAM vs. wrought Ti-6Al-4V specimens.

Compressive strength

Compressive testing has also been used to evaluate EBAM parts, mainly for meshed, porous, or cellular structures

Table 1. Comparisons between EBAM and wrought Ti-6Al-4V.

	YS (GPa)	UTS (GPa)	Ductility (%)	*H* (GPa)	Ref.
EBAM	1.10–1.15	1.15–1.20	16–25	3.8–4.1	[9]
	0.83	0.915	–	3.2	[22]
	0.735	0.775	2.3	3.619	[23]
Wrought	1.17–1.22	1.23–1.29	12–14	3.8–4.3	[9]
	0.79	0.87	–	–	[22]
	0.860	0.931	14	3.207	[23]

in biomedical applications. Li et al. [37] utilized EBAM to fabricate porous Ti-6Al-4V parts with fully interconnected, controlled internal pore architecture. The compressive test showed that a linear elastic deformation stage, followed by a long plateau stage with a nearly constant flow stress to large strains, in which cells collapse due to buckling and plastic-yielding, and the final stage with the stress reaching the maximum value. The authors reported that 66% porosity Ti-6Al-4V parts has a compressive strength approximately 116 MPa and the Young's modulus of 2.5 GPa, close to the human cancellous bone.

Parthasarathy et al. [11] reported the stiffness and elastic modulus for the open cellular Ti-6Al-4V foams varied with density consistent with the Gibson-Ashby foam model [38]. Moreover, the stiffness varies inversely with porosity or pore density consistent with literature values for a number of metal and alloy systems, especially aluminum. The authors also reported that both compressive stiffness and strength decrease with an increase in the porosity. Heinl et al. [39] investigated cellular Ti-6Al-4V structures with a controllable interconnected porosity by EBAM. The mechanical properties of the structures were examined under compression load and compared with the properties of human bone. It is found that cellular solids have an important influence to their stiffness and strength.

Other testing

Other types of mechanical testing such as hardness tests and flexural tests have also been used in studying mechanical properties of Ti alloys processed by EBAM. Koike et al. [23] investigated EBAM Ti-6Al-4V specimens by Vickers microhardness testing. The hardness of EBAM specimens were higher than that of the cast or wrought specimens, which is probably attributed to finer α/β lamellar microstructures according to Hall-Petch relation [40, 41]. Karlsson et al. [42] conducted the nanoindentation test of EBAM components with two different Ti-6Al-4V powder sizes (diameters of 45–100 μm and 25–45 μm). The results show that the Young's modules and hardness were not significantly affected by the powder size or the layer thickness within the range of studied process parameters. However, the part surface appearance was noted to be different with the different powder sizes. Additionally, Cansizoglu et al. [29] investigated EBAM Ti-6Al-4V bars using a three-point bending tester. The results of the flexure tests showed that the elastic properties of the structures are relatively consistent between builds.

Geometric attributes

Even with impressive advantages over conventional manufacturing technologies, EBAM still exhibits several process challenges, such as dimensional accuracy and surface finish. Despite an intense interest in attainable accuracy and strengths by EBAM [43], few studies have emphasized the geometric aspects in EBAM. Cooke and Soons [44] studied the geometric accuracy of metallic test part manufactured by EBAM and other powder-based processes. The model was a circle-diamond-square test part with an inverted cone that is used to evaluate the performance of five-axis milling machines. The authors reported that overall, the observed errors of EBAM parts are significantly larger than those of typical machined parts by at least an order of magnitude. The errors seem to be repeatable, providing opportunities for compensation strategies. In addition, errors of parts may be due to the process; cyclic thermal effects, including deformations due to residual stresses, are most likely the cause.

In research conducted by Koike et al. [23], an enlarged view of the gauge section of the EBAMTi-6Al-4V specimen shows a rough exterior appearance compared to as-cast counterparts. The surface of the EBAM specimens are covered by rippled layers. On the other hand, the exterior surfaces of cast specimens are much smoother. Figure 9 below shows the surfaces of EBAM Ti-6Al-4V parts (tensile specimens), exhibiting the scanned pattern (Figure 9b) and surface morphology from a white-light interferometer (Figure 9c). The surface has a roughness about 20 μm of R_a. On the other hand, it has also been suggested that as-fabricated rippled surfaces of EBAM processed alloys may be beneficial in some biomedical applications where an irregular surface is desired.

Other materials used in EBAM

In addition to widely used Ti alloys, EBAM has also been used with other metal-based materials including intermetallics, tool steels, superalloys and copper, etc.

Titanium aluminide, a TiAl-based intermetallic, has been used for propulsion exhaust system components and other aerospace applications. Murr et al. [45] characterized the microstructures of TiAl powder and solid TiAl components fabricated by EBAM. The EBAM process results in a phase transformation: the powder was α_2-phase-rich (HCP), while the EBAM components were largely γ-TiAl (FCC). The EBAM specimens exhibited an equiaxed γ-TiAl grain structure with a

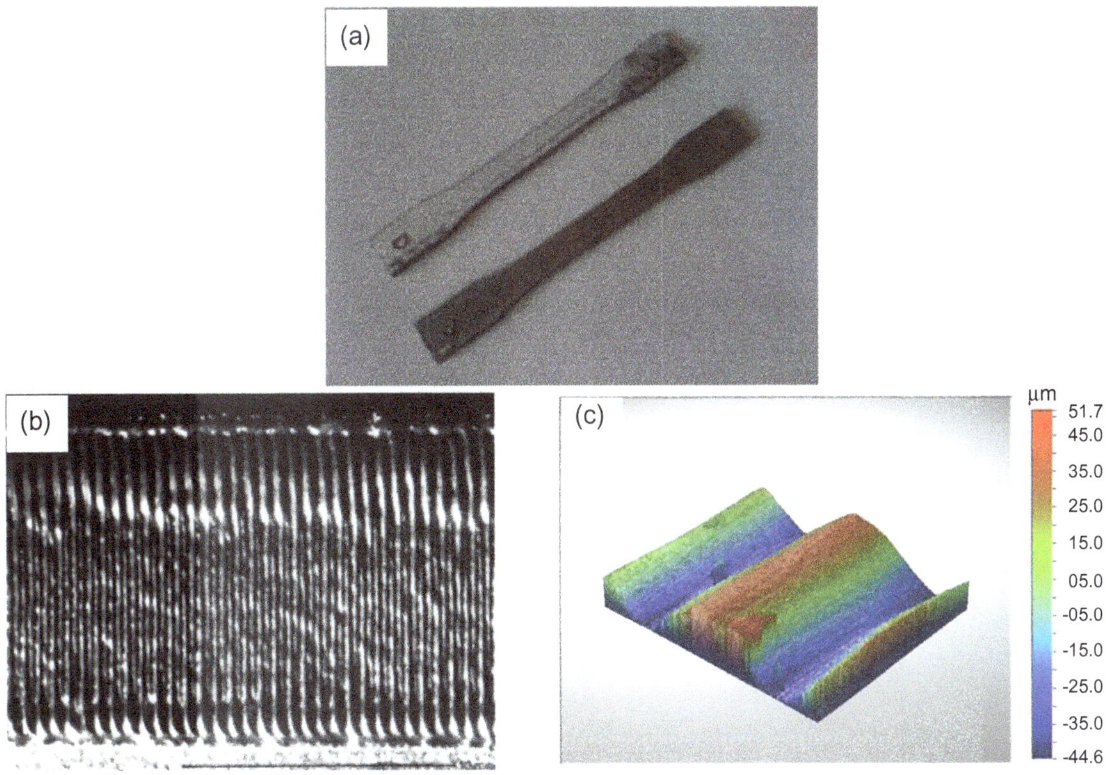

Figure 9. Exterior surfaces of (a) EBAM Ti-6Al-4V samples showing: (b) the scan path and (c) surface morphology.

lamellar γ/α_2 colony structure within the γ-grains. A relatively high dislocation density contributed to residual hardness, due in part to rapid cooling, associated with EBAM process. Heat treatment is generally required to adjust the microstructure and increase mechanical properties of EBAM TiAl parts. Sabbadini et al. [46] studied the effects of HIP treatments. It was shown that HIP produces a tempered microstructure, with near γ equiaxed morphology, resulted from an almost full recrystallization and little grain growth. Biamino et al. [2] also investigated the mechanical properties of both as-fabricated EBAM TiAl specimens and after HIP. After HIP, YS appears to be fairly temperature independent up to 815 °C.

H13 steel is a popular and versatile, hot-work steel, providing good balance of toughness, thermal crack resistance, and high temperature strength needed for forming tooling. Cormier et al. [7] studied the microstructure and properties of H13 steel produced by EBAM. The authors reported that EBAM H13 parts exhibit full interlayer bonding with virtually no porosity. The as-fabricated material was martensite having a hardness of 48–50 HRC. After annealing, HRC value was under 20. A small number of isolated shrinkage cracks confined within specific layers could be observed.

Co-based alloys are a group of bone implant materials. Murr et al. [47] investigated the microstructure and mechanical properties of EBAM processed Co-29Cr-6Mo alloy. The EBAM fabricated solid, mesh, and foam Co alloy prototypes all exhibited a directional, columnar $Cr_{23}C_6$ precipitate architecture parallel to the EBAM build direction intermixed with some stacking faults in the FCC matrix. The columnar precipitates were spaced within textured, directional grains. Following HIP-annealing, the columnar precipitates dissolved and were replaced by a higher density of intrinsic stacking faults, resulting in hardness essentially constant, while there was a slight drop in YS, but UTS increased by 20%. Gaytan et al. [48] also investigated the EBAM process on Co-Cr-Mo alloys. The tensile testing of specimens fabricated by EBAM produced average UTS of 1.45 GPa, YS of 0.51 GPa, and an elongation of 3.6%, which are better than wrought or cast Co-Cr-Mo alloys. Sun et al. [49] studied microstructure and high temperature tensile properties of the Co-28Cr-6Mo-0.23C-0.17N alloy fabricated by EBAM. The authors found that the preferential crystal orientations of the γ phase in the as-EBAM-built 0°-, 45°-, 55°- and 90°-samples along the cylindrical axial directions were near [0 0 1], [1 1 0], [1 1 1] and [1 0 0], respectively. In addition, $M_{23}C_6$ precipitates were observed to align along the build direction with intervals of around 3 µm in all of the samples. The sample built with 55° deviation from the *Z* axis had the highest UTS of 806 MPa at 700 °C.

Ramirez et al. [50] investigated the effects of precipitate (Cu_2O) on the microstructure and mechanical properties of EBAM Cu parts. These precipitate-dislocation architectures create increased hardness, ranging from a base-plate hardness of HV 57 to an EBAM part hardness of HV 88, referenced to the precursor powder containing Cu_2O precipitates that has a hardness of HV 72.

Nickel-based superalloys present prominent high-temperature, corrosion and oxidation resistance applications including jet engine components. Murr et al. [51] investigated Ni-based superalloys regarding to the effects of heat treatment on the mechanical properties of EBAM built parts. Tensile properties

Table 2. Property comparisons between EBAM and wrought Ti-6Al-4V.

Material	Application	Mechanical properties	Ref.
TiAl	High temperature, propulsion exhaust system	YS/UTS (700 °C): ~300 MPa, 360 MPa	[2, 45, 46]
H13	Tool, die/mold	As-processed hardness of 48-50 HRC	[7]
Co-Cr based alloys	High temperature bearings, biomedical implants	Micro-hardness: 4.4 to 5.9 GPa	[47, 48]
Cu	Electrical	Micro-hardness: 88 HV	[50]
Ni based alloys	Jet engine parts	YS/UTS (538 °C): 300 MPa, 610 MPa	[51]

of EBAM cylinders at 538 °C didn't significantly alter micro-hardness from as-fabricated cylinders tested at room temperature, but YS dropped from 0.33 to 0.30 GPa, and UTS decreased by 23%. The corresponding elongation increased from 44 to 53%. Similarly, the fabricated and HIP-ed cylinders tested at 538 °C also illustrated no significant change in micro-hardness, but both YS and UTS decreased marginally with little increase in ductility comparing to tests at room temperature. By comparison, wrought Ni superalloy tested at 538 °C exhibited a YS of 280 MPa and a UTS of 830 MPa.

Table 2 below briefly summarizes different materials that have been tested in EBAM, their corresponding applications and properties.

Process simulations

Despite the potential benefits of EBAM technologies and increasing studies/reports of this process, mostly applications, case studies, microstructures/properties, there has been little literature in process modeling/simulations of EBAM. EBAM is a rather complicated process and fundamental understanding of process physics is a key to improve the process performance and part quality consistency. Due to temperature dependent material properties and a moving heat source, an analytical solution of the thermal process model is considerably demanding. Zäh and Lutzman developed a simplified heat transfer model of EBAM to study the scan speed and beam power effects on the weld-pool geometry [3]. Therefore, the mathematical-physical model is being transferred into a simulation software based on the finite-element (FE) method and thus, melt pool instabilities as a function of various input parameters can be predicted. Three different melt pool conditions are compared with each other. The melt pool shapes at the beam power (PB) of 150 W vary significantly with the applied scanning speed, V_S. Within the compared results, the parameter combination V_S = 50 mm/s and PB = 150 W yields the minimum length-to-width ratio of approximately 1.3. The authors were able to define a suitable process window, however, acknowledged a more comprehensive model is needed if new materials are to be utilized.

Shen and Chou recently initiated a study to investigate the thermal phenomenon of the EBAM process. A finite element (FE) model incorporating Gaussian heat flux distribution, fusion latent heat, temperature-dependent thermal properties was developed to study the thermal response when subject to a moving source of heat of high intensity [52]. Figure 10 below shows an example of temperature contours in the part during the process. The results are temperature contours of a half part modeled (because of symmetry) around the beam center at the

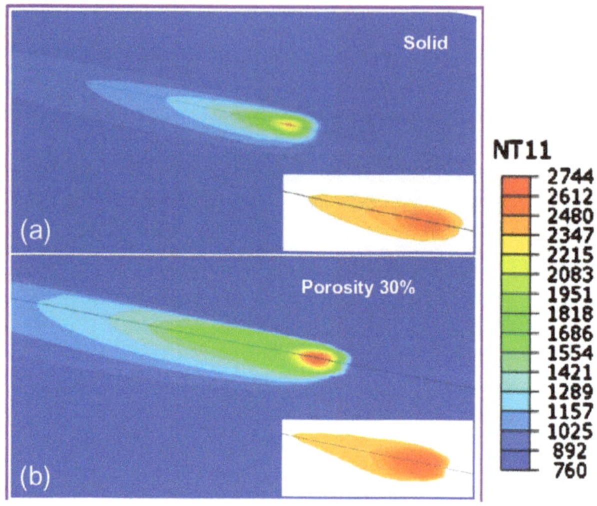

Figure 10. Temperature fields and melt pool geometries of (a) solid vs. (b) powder top layer [52].

quasi-steady state. High temperatures with large gradients around the maximum temperature area are noted. It is also noted because the powder porosity strongly affects the thermal conductivity, inclusion of powder as the top layer in the model is required for modeling accuracy. Further, since this is a moving heat source condition, the temperature spatial distributions can be translated to temporal responses to evaluate the corresponding cooling rates. The model, once validated, may be applied to systematically study various issues such as the process parameter effects and the raw material effects.

Jamshidinia et al. [53] applied a coupled computational fluid dynamic (CFD)-FE model to study the heat and thermal stress distribution in EBAM with Ti-6Al-4V powder. According to the numerical results, the negative temperature coefficient of surface tension is responsible for the formation of an outward flow in the molten pool on the top surface during the EBAM. Also, the influence of electron beam scanning speed on the thermal stresses developed in the domain is studied. The heat input and the cooling rate are the dominant factors influencing the level of thermal stresses developed during melting and cooling steps, respectively. The minimum electron beam scanning speed of 100 mm/s resulted in the maximum and minimum thermal stresses during melting and cooling steps, respectively.

Körner et al. [54] conducted the mesoscopic simulation of the EBAM process. A 2D lattice Boltzmann model is developed to investigate melting and re-solidification of a randomly packed powder bed under the irradiation of a Gaussian beam during EBAM. Numerical experiments demonstrate that the packing density of the powder bed has the most significant effect on the melt pool characteristics.

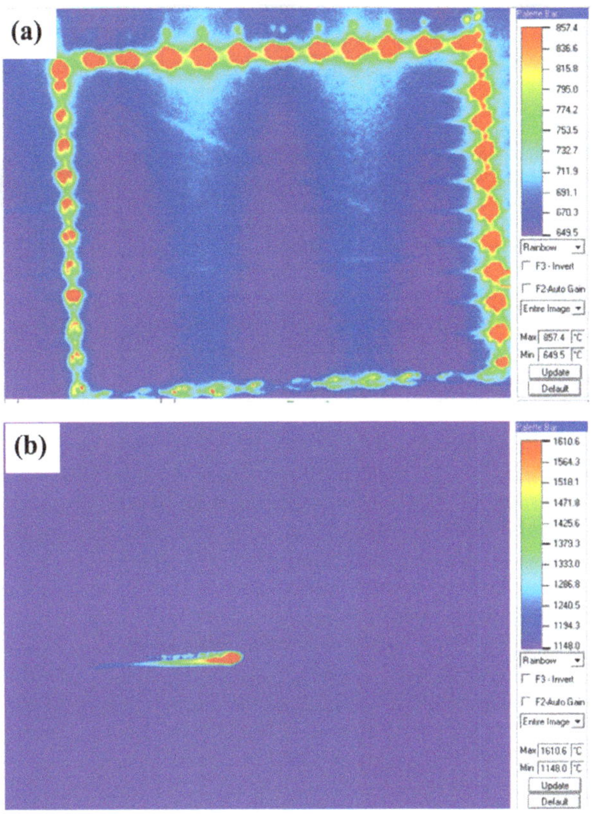

Figure 11. Example of near-infrared thermal images taken during an EBAM building process: (a) during contour melting and (b) during hatch melting.

Moreover, process variable measurements such as temperatures have seldom been reported for EBAM despite the need to validate the simulation model and to monitor the process. Zäh and Lutzman applied thermocouples, attached to the build plate, to evaluate temperature response during EBAM to evaluate the model developed by this research group [3]. However, because of high temperatures and steep gradients, thermocouple techniques are limited in accuracy and resolutions. On the other hand, the authors of this paper have attempted to use a near-infrared (IR) thermal camera for process temperature measurements in EBAM [55, 56]. A methodology has also been developed to analyze temperature distributions and history around the melting scan area. A preliminary result is shown in Figure 11 below, displaying examples of thermal images at particular frames collected during the EBAM experiment that built a 25.4 mm square block. Figure 11a was taken during the contour melting and the square boundary being heated can be observed. Figure 11b was during the hatch melting; at this particular time frame, the electron beam was moving from the left side toward the right. A localized high temperature zone (over 1600 °C) with a rapid temperature diminishing tail, similar to, qualitatively, the simulation result (Figure 10) can be noted. In addition, IR-imaging was applied for flaw detection during EBAM by Schwerdtfeger et al. [57]. The method applied demonstrated a good correspondence between the flaws visible in images recorded of an EBAM build using an IR camera and the flaws recorded by traditional metallographic sectioning of the samples after the process was finished.

Conclusions

EBAM, an enabling technology for design and manufacturing integration, can efficiently assist developments and modifications of products, especially for complex, difficult-to-fabricate components. However, the EBAM process physics is complex, and the part characteristics seem to be sensitive to the process parameters, the effects of which may not be well understood. There has been increased literature of EBAM, and this study intends to provide an overview of the EBAM technology, process principles, applications, part microstructures and mechanical properties, and process simulations. The collected and analyzed information from this review can be summarized as below.

1. Though Ti-6Al-4V is the most widely used materials for EBAM, other materials such as intermetallics, tool steels, Co-Cr alloys and Ni-based superalloys, and Cu, etc., have been attempted for various applications popularly in the aerospace and biomedical industries, some with unique geometry such as cellular structures.
2. EBAM parts of different raw alloys have been characterized in microstructures and mechanical properties. In general, EBAM parts have very fine microstructures with some porosity, which can be mostly eliminated by HIP. EBAM parts typically have comparable (to wrought counterparts) or superior mechanical properties: high YS and UTS and higher hardness, but a lower ductility.
3. Dimensional errors of EBAM parts are significantly larger than those of typical machined parts by at least an order of magnitude, and yet, surface finish is poorer than cast parts.
4. A thorough literature survey shows limited published studies of EBAM process simulations. To effectively use the EBAM technology, it is necessary to model the process phenomenon and correlate part characteristics with process parameters and variables.

Acknowledgements. The materials presented in this paper are supported by NASA, under Award No. NNX11AM11A.

References

1. http://www.arcam.com, ArCam AB, accessed in December 2013
2. S. Biamino, A. Penna, U. Ackelid, S. Sabbadini, O. Tassa, P. Fino, M. Pavese, P. Gennaro, C. Badini, Electron beam melting of Ti-48Al-2Cr-2Nb alloy: microstructure and mechanical properties investigation, Intermetallics 19 (2010) 776–781.
3. M.F. Zäh, S. Lutzmann, Modelling and simulation of electron beam melting, Production Engineering 4 (2010) 15–23.
4. http://www.youtube.com/watch?v=BxxIVLnAbLw, Oak Ridge National Laboratory, accessed in December 2013.

5. J. Hiemenz, Electron beam melting, Advanced Materials & Processes 165 (2007) 45–46.
6. P. Yu, M. Qian, D. Tomus, C.A. Brice, G.B. Schaffer, B.C. Muddle, Electron beam processing of aluminium alloys, Materials Science Forum 618–619 (2009) 621–626.
7. D. Cormier, O. Harrysson, H. West, Characterization of H13 steel produced via electron beam melting, Rapid Prototyping Journal 10 (2004) 35–40.
8. S.M. Gaytan, L.E. Murr, E. Martinez, J.L. Martinez, B.I. Machado, D.A. Ramirez, F. Medina, S. Collins, R.B. Wicker, Comparison of microstructures and mechanical properties for solid and mesh cobalt-base alloy prototypes fabricated by electron beam melting, Metallurgical and Materials Transactions A 41 (2010) 3216–3227.
9. L.E. Murr, E.V. Esquivel, S.A. Quinones, S.M. Gaytan, M.I. Lopez, E.Y. Martinez, F. Medina, D.H. Hernandez, E. Martinez, J.L. Martinez, S.W. Stafford, D.K. Brown, T. Hoppe, W. Meyers, U. Lindhe, R.B. Wicker, Microstructures and mechanical properties of electron beam-rapid manufactured Ti-6Al-4V biomedical prototypes compared to wrought Ti-6Al-4V, Materials Characterization 60 (2009) 96–105.
10. S.M. Gaytan, L.E. Murr, F. Medina, E. Martinez, L. Martinez, R.B. Wicker, Fabrication and characterization of reticulated, porous mesh arrays and foams for aerospace applications by additive manufacturing using electron beam melting, TMS 2010-139th Annual Meeting and Exhibition-Supplemental Proceeding, Seattle, WA, February 14–18, 2010, pp. 283–290.
11. J. Parthasarathy, B. Starly, S. Raman, A. Christensen, Mechanical evaluation of porous titanium (Ti6Al4V) structures with electron beam melting (EBAM), Journal of the Mechanical Behavior of Biomedical Materials 3 (2010) 249–259.
12. J. Goods, MSFC, private communication, February 2008.
13. J. Schwerdtfeger, P. Heinl, R.F. Singer, C. Körner, Auxetic cellular structures through selective electron-beam melting, Physica Status Solidi B 247 (2010) 269–272.
14. A. Neira-Arce, Thermal modeling and simulation of electron beam melting for rapid prototyping on Ti6Al4V alloys, Ph.D. Dissertation, North Carolina State University, Raleigh, NC, 2012.
15. W.P. Syam, A.M. Al-Ahmari, M.A. Mannan, H.A. Al-Shehri, K.A. Al-Wazzan, Metallurgical, accuracy and cost analysis of Ti6Al4V dental coping fabricated by electron beam melting process, Proceedings of the 5th International Conference on Advanced Research in Virtual and Rapid Prototyping, Leiria, Portugal, September 28–October 1, 2011, pp. 375–383.
16. E. Rodriguez, F. Medina, D. Espalin, C. Terrazas, D. Muse, C. Henry, E. MacDonald, R.B. Wicker, Integration of a thermal imaging feedback control system in electron beam melting, WM Keck Center for 3D Innovation, University of Texas at El Paso, pp. 945–961.
17. W. He, W. Jia, H. Lin, H. Tang, X. Kang, H. Yu, Research on preheating of titanium alloy powder in electron beam melting Technology, Rare Metal Materials and Engineering 40 (2011) 2072–2075.
18. X. Gong, K. Chou, Characterizations of sintered Ti-6Al-4V powders in electron beam additive manufacturing, Proceedings of the ASME 2013 International Manufacturing Science and Engineering Conference, Madison, WI, June 10–14, 2013, MSEC 2013-1131.
19. A. Safdar, L.Y. Wei, A. Snis, Z. Lai, Evaluation of microstructural development in electron beam melted Ti-6Al-4V, Materials Characterization 65 (2012) 8–15.
20. A.A. Antonysamy, J. Meyer, P.B. Prangnell, Effect of build geometry on the β-grain structure and texture in additive manufacture of Ti6Al4V by selective electron beam melting, Materials Characterization 84 (2013) 153–168.
21. A. Gulzar, J.I. Akhter, M. Ahmad, G. Ali, M. Mahmood, M. Ajmal, Microstructure evolution during surface alloying of ductile iron and austempered ductile iron by electron beam melting, Applied Surface Science 255 (2009) 8527–8532.
22. L. Facchini, E. Magalini, P. Robotti, A. Molinari, Microstructure and mechanical properties of Ti-6Al-4V produced by electron beam melting of pre-alloyed powders, Rapid Prototyping Journal 15 (2009) 171–178.
23. M. Koike, K. Martinez, L. Guo, G. Chahine, R. Kovacevic, T. Okabe, Evaluation of titanium alloy fabricated using electron beam melting system for dental applications, Journal of Materials Processing Technology 211 (2011) 1400–1408.
24. L.E. Murr, S.M. Gaytan, F. Medina, E. Martinez, D.H. Hernandez, L. Martinez, M.I. Lopez, R.B. Wicker, S. Collins, Effect of build parameters and build geometries on residual microstructures and mechanical properties of Ti-6Al-4V components built by electron beam melting (EBM), Proceedings of 20th Annual International Solid Freeform Fabrication Proceedings, Austin, TX, August 3–5, 2009, pp. 374–397.
25. S. Bontha, N.W. Klingbeil, P.A. Kobryn, H.L. Fraser, Effects of process variables and size-scale on solidification microstructure in beam-based fabrication of bulky 3D structures, Materials Science and Engineering A 513–514 (2009) 311–318.
26. P. Wanjara, M. Brochu, S. Girard, M. Jahazi, Electron beam freeforming on type 321 stainless steel using BNi-2 brazing paste, Materials Science and Technology 21 (2005) 613–618.
27. P. Heinl, A. Rottmair, C. Körner, R.F. Singer, Cellular titanium by selective electron beam melting, Advanced Engineering Materials 9 (2007) 360–364.
28. S.M. Gaytan, L.E. Murr, F. Medina, E. Martinez, M.I. Lopez, R.B. Wicker, Advanced metal powder based manufacturing of complex components by electron beam melting, Materials Technology 24 (2009) 181–190.
29. O. Cansizoglu, O. Harrysson, D. Cormier, H. West, T. Mahale, Properties of Ti-6Al-4V non-stochastic lattice structures fabricated via electron beam melting, Materials Science and Engineering A 492 (2008) 468–474.
30. J. Schwerdtfeger, P. Heinl, R.F. Singer, C. Körner, Selective electron beam melting: a new way to auxetic cellular structures, Proceedings of 20th Annual International Solid Freeform Fabrication Proceedings, Austin, TX, August 3–5, 2009, pp. 724–729.
31. L.E. Murr, S.M. Gaytan, M.I. Lopez, E. Martinez, F. Medina, R.B. Wicker, Metallographic characterization of additive-layer manufactured products by electron beam melting of Ti-6Al-4V powder, Practical Metallography 46 (2009) 442–453.
32. K. Puebla, L.E. Murr, S.M. Gaytan, E. Martinez, F. Medina, R.B. Wicker, Effect of melt scan rate on microstructure and macrostructure for electron beam melting of Ti-6Al-4V, Materials Sciences and Applications 3 (2012) 259–264.
33. N. Hrabe, T. Quinn, Effects of processing on microstructure and mechanical properties of Ti-6Al-4V fabricated using electron beam melting (EBM), Part 1: Distance from build plate and part size, Materials Science & Engineering A 573 (2013) 264–270.
34. M. Svensson, U. Ackelid, Titanium alloys manufactured with electron beam melting mechanical and chemical properties, Medical Device Materials V-Proceedings of the Materials and

Processes for Medical Devices Conference, Minneapolis, MN, August 10–12, 2009, pp. 189–194.
35. S.M. Gaytan, L.E. Murr, D.H. Hernandez, E. Martinez, S.A. Quinones, F. Medina, R.B. Wicker, Structure-property-process optimization in the rapid-layer manufacturing of Ti-6A1-4V components by electron beam melting, TMS 2009-138th Annual Meeting and Exhibition, San Francisco, CA, February 15–19, 2009, pp. 363–369.
36. L. Ladani, L. Roy, Mechanical behavior of Ti-6Al-4V manufactured by electron beam additive fabrication, Proceedings of the ASME 2013 International Manufacturing Science and Engineering Conference, Madison, WI, June 10–14, 2013, MSEC 2013-1105.
37. X. Li, C. Wang, W. Zhang, Y. Li, Fabrication and characterization of porous Ti6Al4V parts for biomedical applications using electron beam melting process, Materials Letters 63 (2009) 403–405.
38. L.J. Gibson, M.F. Ashby, Cellular solids: structures and properties, Pergamon Press, 1988.
39. P. Heinl, L. Müller, C. Körner, R.F. Singer, F.A. Müller, Cellular Ti-6Al-4V structures with interconnected macro porosity for bone implants fabricated by selective electron beam melting, Acta Biomaterialia 4 (2008) 1536–1544.
40. X. Gong, S.B. Kang, S. Li, J.H. Cho, Enhanced plasticity of twin-roll cast ZK60 magnesium alloy through differential speed rolling, Materials and Design 30 (2009) 3345–3350.
41. X. Gong, H. Li, S.B. Kang, J.H. Cho, S. Li, Microstructure and mechanical properties of twin-roll cast Mg-4.5Al-1.0Zn alloy sheets processed by differential speed rolling, Materials and Design 31 (2010) 1581–1587.
42. J. Karlsson, A. Snis, H. Engqvist, J. Lausmaa, Characterization and comparison of materials produced by electron beam melting (EBM) of two different Ti-6Al-4V powder fractions, Journal of Materials Processing Technology 213 (2013) 2109–2118.
43. W. Liu, L. Li, K. Kochhar, A method for assessing geometrical errors in layered manufacturing. Part 2: Mathematical modeling and numerical evaluation, The International Journal of Advanced Manufacturing Technology 14 (1998) 644–650.
44. A.L. Cooke, J.A. Soons, Variability in the geometric accuracy of additively manufactured test parts, Proceedings of 21st Annual International Solid Freeform Fabrication Symposium, Austin, TX, August 9–11, 2010, pp. 1–12.
45. L.E. Murr, S.M. Gaytan, A. Ceylan, E. Martinez, J.L. Martinez, D.H. Hernandez, B.I. Machado, D.A. Ramirez, F. Medina, S. Collins, R.B. Wicker, Characterization of titanium aluminide alloy components fabricated by additive manufacturing using electron beam melting, Acta Materialia 58 (2010) 1887–1894.
46. S. Sabbadini, O. Tassa, P. Gennaro, U.R. Ackelid, Additive manufacturing of gamma titanium aluminide parts by electron beam melting, 2010 TMS Annual Meeting & Exhibition-Global Innovations in Manufacturing of Aerospace Materials: The 11th MPMD Global Innovations Symposium, Seattle, WA, February 14–18, 2010, pp. 267–274.
47. L.E. Murr, K.N. Amato, S.J. Li, Y.X. Tian, X.Y. Cheng, S.M. Gaytan, E. Martinez, P.W. Shindo, F. Medina, R.B. Wicker, Microstructure and mechanical properties of open-cellular biomaterials prototypes for total knee replacement implants fabricated by electron beam melting, Journal of the Mechanical Behavior of Biomedical Materials 4 (2011) 1396–1411.
48. S.M. Gaytan, L.E. Murr, E. Martinez, J.L. Martinez, B.I. Machado, D.A. Ramirez, F. Medina, S. Collins, R.B. Wicker, Comparison of microstructures and mechanical properties for solid and mesh cobalt-base alloy prototypes fabricated by electron beam melting, Metallurgical and Materials Transactions A 41 (2010) 3216–3227.
49. S.H. Sun, Y. Koizumi, S. Kurosu, Y.P. Li, H. Matsumoto, A. Chiba, Build direction dependence of microstructure and high-temperature tensile property of Co-Cr-Mo alloy fabricated by electron beam melting, Acta Materialia 64 (2014) 154–168.
50. D.A. Ramirez, L.E. Murr, E. Martinez, D.H. Hernandez, J.L. Martinez, B.I. Machado, F. Medina, P. Frigola, R.B. Wicker, Novel precipitate-microstructural architecture developed in the fabrication of solid copper components by additive manufacturing using electron beam melting, Acta Materialia 59 (2011) 4088–4099.
51. L.E. Murr, E. Martinez, S.M. Gaytan, D.A. Ramirez, B.I. Machado, P.W. Shindo, J.L. Martinez, F. Medina, J. Wooten, D. Ciscel, U. Ackelid, R.B. Wicker, Microstructural architecture, microstructures, and mechanical properties for a nickel-base superalloy fabricated by electron beam melting, Metallurgical and Materials Transactions A 42 (2011) 3491–3508.
52. N. Shen, K. Chou, Thermal modeling of electron beam additive manufacturing process – powder sintering effect, ASME International Manufacturing Science and Engineering Conference, Notre Dame, IN, June 4–8, 2012, MSEC 2012-7253.
53. M. Jamshidinia, F. Kong, R. Kovacevic, The coupled CFD-FEM model of electron beam melting (EBM), ASME District F-Early Career Technical Conference, Birmingham, AL, November 2–3, 2013, pp. 163–171.
54. C. Körner, E. Attar, P. Heinl, Mesoscopic simulation of selective beam melting processes, Journal of Materials Processing Technology 211 (2011) 978–987.
55. S. Price, J. Lydon, K. Cooper, K. Chou, Experimental temperature analysis of powder-based electron-beam additive manufacturing, 24th Annual International Solid Freeform Fabrication Symposium, Austin, TX, August 12–14, 2013, pp. 162–173.
56. X. Gong, B. Cheng, S. Price, K. Chou, Powder-bed electron-beam-melting additive manufacturing: powder characterization, process simulation and metrology, Early Career Technical Conference, Birmingham, AL, November 2–3, 2013, pp. 55–66.
57. J. Schwerdtfeger, R.F. Singer, C. Körner, In situ flaw detection by IR-imaging during electron beam melting, Rapid Prototyping Journal 18 (2012) 259–263.

12

Material modelling and its application to creep-age forming of aluminium alloy 7B04

Aaron C.L. Lam[1], Zhusheng Shi[1,*], Xia Huang[2], Yo-Lun Yang[1], Yuansong Zeng[2], and Jianguo Lin[1]

[1] Department of Mechanical Engineering, Imperial College London, London SW7 2AZ, UK
[2] AVIC Beijing Aeronautical Manufacturing Technology Research Institute, Beijing 100024, PR China

Abstract – Creep-ageing behaviour of aluminium alloy 7B04-T651 at 115 °C under a range of tensile stress levels has been experimentally investigated and numerically modelled for creep-age forming (CAF) applications. Creep strain, yield strength evolution and precipitate growth of creep-aged specimens were investigated. The alloy was modelled using a set of unified constitutive equations, which captures its creep deformation and takes into account yield strength contributions from three creep-age hardening mechanisms. Applications of the present work are demonstrated by implementing the determined material model into a commercial finite element analysis solver to analyse CAF operations carried out in a novel flexible CAF tool. Stress relaxation, yield strength, precipitate size and springback were predicted for the creep-age formed plates. The predicted springback were further quantified and compared with experimental measurements and a good agreement of 2.5% deviation was achieved. This material model now enables further investigations of 7B04 under various CAF scenarios to be conducted inexpensively via computational modelling.

Key words: Creep-age forming, Material modelling, Process modelling, Aluminium alloy 7B04

1. Introduction

Creep-age forming (CAF) is a metal forming technique that utilises the collective behaviour of viscoplastic deformation (creep) and microstructural change (ageing) of aluminium alloys under external loading at elevated temperature [1]. Manufacturing using this technique, in addition to better formed part quality and industrial environment, can reduce the number of manufacturing operations to form a part and thus the associated costs as well [2, 3]. As current deformation of a metallic alloy also depends on its microstructure, which changes dynamically during CAF, accurate prediction of formed part quality cannot be achieved solely using conventional viscoplastic creep equations. Attention has been drawn towards developing mechanism-based creep-ageing constitutive equations that consider evolutions of microstructure and mechanical properties, and efforts have been made towards the concurrent modelling of both the shaping and material processing functions of CAF as a result [3, 4].

In this work, the creep-ageing behaviour of a peak-aged aluminium alloy 7B04 is experimentally investigated and then modelled using a set of unified constitutive equations. The equation set was employed in finite element analysis to simulate an actual CAF process of the alloy in the form of approximately 8-mm-thick plates. Predictions were made for the yield strength evolution, precipitate growth and springback of the physically formed plates. A comparison is made between the simulated and experimentally measured springback values and the springback predictive ability using the material model is briefly discussed. The implications and influences of this work are given at the end of this article.

*e-mail: zhusheng.shi@imperial.ac.uk

2. Experimental procedure

2.1. Material

The material used was aluminium alloy 7B04 whose chemical composition is presented in Table 1. Creep-ageing tensile test pieces of geometry and dimensions shown in Figure 1 and plates of dimensions 600 mm × 350 mm × 8.2 mm had been machined from the same bulk material in the longitudinal rolling orientation and provided in T651 temper by AVIC BAMTRI (Beijing, PR China).

2.2. Material testing

Figure 2 illustrates a schematic of the material's heat treatment cycle from solutionising to the subsequent material testing operations. The creep-ageing test procedure used by

Table 1. Chemical composition (wt.%) of 7B04.

Zn	Mg	Cu	Mn	Fe	Cr	Si	Ni	Ti	Al
5.97	2.48	1.51	0.33	0.16	0.16	0.07	<0.05	<0.05	Bal.

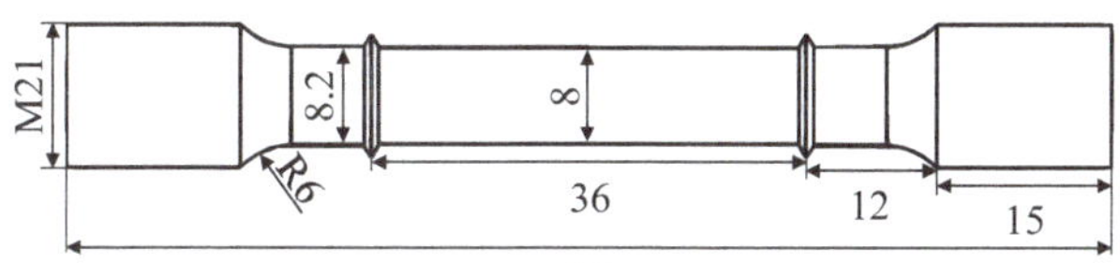

Figure 1. Geometry and key dimensions of the test piece (units in mm).

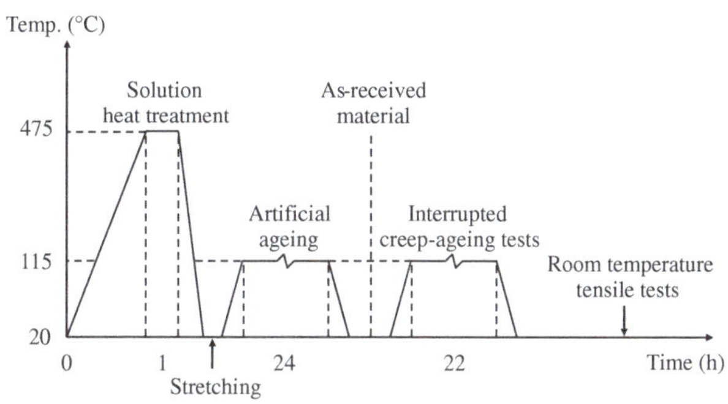

Figure 2. Heat treatment history of the material.

Ho et al. [4] was adopted and the tests were carried out at 115 °C under four tensile stress levels: 200, 240, 260 and 280 MPa. Room temperature tensile tests were carried out on selected creep-aged specimens. In addition, samples of the as-received material and those that had received a further 22 h stress-free-ageing were analysed using transmission electron microscopy (TEM) to enable subsequent calibration of the alloy's precipitate growth.

2.3. Creep-age forming tests

A novel flexible CAF tool was used for the CAF tests. The tooling setup, as shown in Figure 3, comprised of 25 sparsely spaced forming pins on each side of the top and bottom die holders. Flexible forming surfaces were on both sides of the plate with an additional elastomeric sheet layer between the plate and lower flexible surface. Further details about the method of CAF tool design and test setup can be found in [5] and [6] respectively. The main steps of the forming test include:

1. Setup the forming tool surface by adjusting the heights of forming pins;
2. Load the top die holder to deform the plate to the required shape;
3. Lock the top and bottom die holders using nuts and bolts;
4. Run the creep-ageing thermal cycle;
5. Release the load by loosening the nuts to let springback occur.

Figure 3. Experimental setup for the creep-age forming tests.

CAF tests were carried out on two plates at a single curvature bending radius of 1200 mm. After unloading, profiles of the unloaded plates were measured and springback of the plates (S) were calculated using $S = 1 - (\delta_f/\delta_l)$, where δ_f and δ_l are the final and loaded centre deflections of each plate respectively. S is a dimensionless quantity and varies from 0 (no springback) to 1 (full springback). Repeated deflection measurements were obtained from each side of the formed plates and the final average value was taken.

3. Material modelling

3.1. Unified creep-ageing constitutive equations

7B04 is an age-hardenable aluminium alloy. During ageing of 7000-series aluminium alloys, microstructure evolves from supersaturated solid solution and precipitation occurs in the following sequence [7, 8]: Solid solution → GP zones → η' → η ($MgZn_2$).

The strength of the alloy changes as precipitation takes places. At the early stage, coherent GP zones form and the yield strength increases as precipitates nucleate and grow. GP zones are then gradually replaced by semi-coherent η' until a peak-age condition is reached. This is when the alloy attains its peak strength with an optimal precipitate size and density (spacing) combination. Afterwards, η' are replaced by incoherent and stable η whilst precipitates coarsen, leading to a gradual decrease in strength (over-ageing). Throughout this process, contribution from solid solution hardening gradually decreases and leads to a gradual softening of the alloy. During creep-ageing, dislocation hardening also contributes to the overall strength of the material and shows rapid increase in the early stage.

In order to model the described behaviour of 7B04, the multiaxial unified constitutive equations proposed by Zhan et al. [9] were adapted:

$$\dot{\varepsilon}_e^{cr} = A_1 \sinh\{B_1[\sigma_e(1-\overline{\rho}_d) - k_{ec}\sigma_y]\} \quad (1)$$

Table 2. Initial conditions used for CAF of 7B04-T651 (115 °C, 22 h).

σ_{sol}^0 (MPa)	σ_{age}^0 (MPa)	σ_y^0 (MPa)	$\bar{r}_p^0$	$\bar{\rho}_d^0$
190	347	537	1	1E-5

Table 3. Material constants determined for CAF of 7B04-T651 (115 °C, 22 h).

A_1 (h^{-1})	B_1 (MPa^{-1})	$k_{\varepsilon c}$	C_a (MPa)	m_1	C_s (MPa)	m_2	A_2 (MPa)	n_d
1.95E-6	0.042	0.16	4	0.9	0.2	0.45	260	0.7
C_r (h^{-1})	Q_r	m_3	γ_r	m_4	A_3	C_ρ (h^{-1})	m_5	
0.098	1.8	2.2	3	2	150	0.145	1.5	

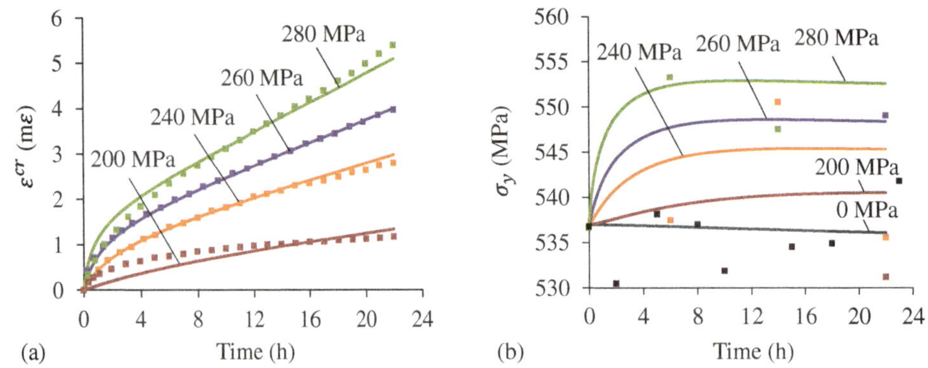

Figure 4. Comparisons of the experimentally obtained (symbols) and calibrated (solid lines) (a) creep-ageing curves and (b) yield strength evolutions for creep-ageing at 115 °C under the indicated stress levels.

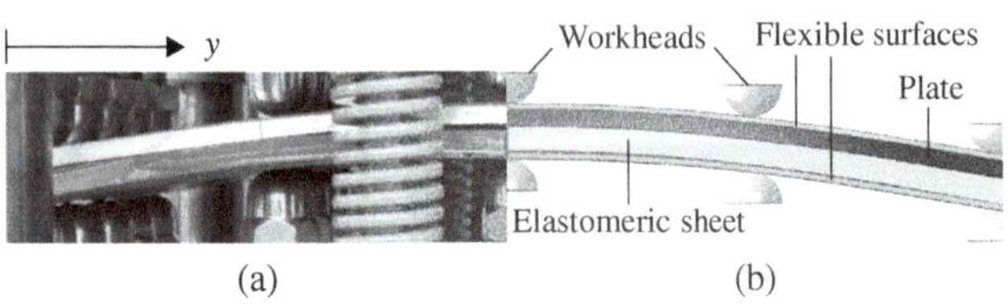

Figure 5. Side-view images taken at full loading of (a) a physical forming test and (b) finite element simulation with to-scale rendering of shell thicknesses.

$$\dot{\sigma}_{age} = C_a \dot{\bar{r}}_p^{\,m_1}(1 - \bar{r}_p) \tag{2}$$

$$\dot{\sigma}_{sol} = -C_s \dot{\bar{r}}_p^{\,m_2}\left|\bar{r}_p - 1\right| \tag{3}$$

$$\dot{\sigma}_{dis} = A_2 \cdot n_d \cdot \bar{\rho}_d^{\,n_d - 1}\dot{\bar{\rho}}_d \tag{4}$$

$$\sigma_y = \sigma_{sol} + \sqrt{\sigma_{age}^2 + \sigma_{dis}^2} \tag{5}$$

$$\dot{\bar{r}}_p = C_r(Q_r - \bar{r}_p)^{m_3}(1 + \gamma_r \bar{\rho}_d^{\,m_4}) \tag{6}$$

$$\dot{\bar{\rho}}_d = A_3(1 - \bar{\rho}_d)\dot{\varepsilon}_e^{cr} - C_\rho \bar{\rho}_d^{\,m_5} \tag{7}$$

where $A_1, B_1, k_{\varepsilon c}, C_a, m_1, C_s, m_2, A_2, n_d, C_r, Q_r, m_3, \gamma_r, m_4, A_3, C_\rho, m_5$ are material constants. The normalised dislocation density is given by $\bar{\rho}_d = (\rho - \rho_i)/\rho_m$, where ρ_i and ρ_m are respectively the dislocation density of the virgin material and the maximum (saturated) dislocation density. The normalised radius of precipitate is defined as $\bar{r}_p = r/r_c$, where r and r_c are the current and peak-aged precipitate size respectively. Equations (1)–(7) describe the inter-relationship amongst the equivalent stress (σ_e), normalised dislocation density ($\bar{\rho}_d$), normalised radius of precipitate ($\bar{r}_p$) and effective creep strain (ε_e^{cr}), as well as the strength contributions from solid solution hardening (σ_{sol}), age hardening (σ_{age}) and dislocation hardening (σ_{dis}). Equation (1) describes the rate of creep strain accumulation, which is not only a function of stress but also of dislocation density and yield strength (σ_y). Readers are referred to [9] for detailed descriptions of Equations (1)–(7).

3.2. Determination of material constants

An evolutionary-algorithm-based optimisation software developed by Zhang et al. [10] was used for material constants determination. The alloy's yield strength evolutions and creep-ageing curves obtained from the tensile creep-ageing tests were used for model calibration. This was coupled with manual constants adjustments carried out in MATLAB to calibrate for the TEM analysis result.

Tables 2 and 3 present respectively the initial conditions used and the material constants determined. Referring to the

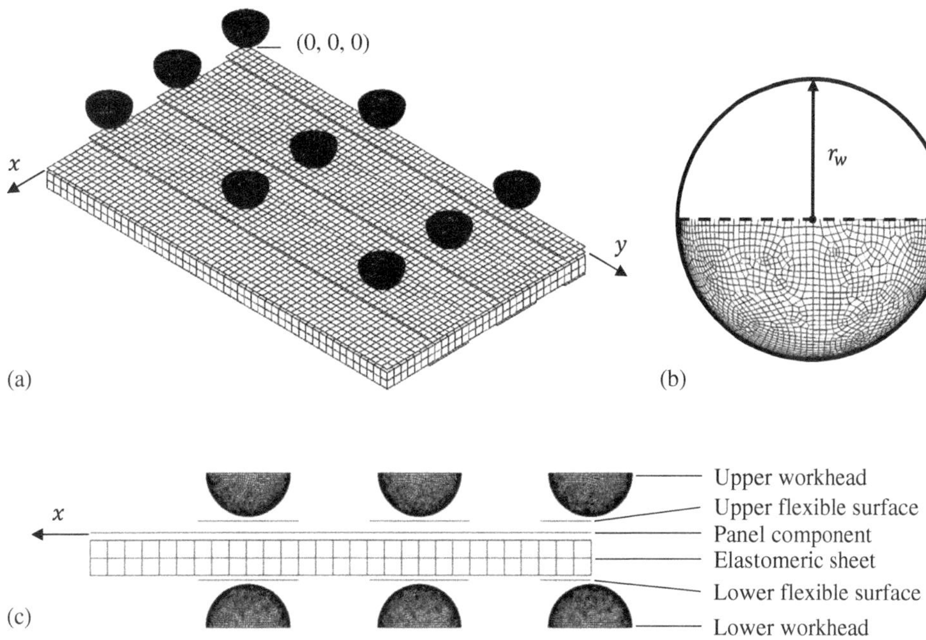

Figure 6. (a) Isometric view of the finite element model; (b) Finite element mesh of a workhead with an arbitrary revolving axis (dashed line) and the radius of revolvement; (c) *xz* plane view of the finite element model with the different parts illustrated. Adapted from [6].

initial conditions, the as-quenched yield strength of 7B04 was 206 MPa [11]. A value of σ^0_{sol} = 190 MPa was chosen for the as-recieved T651 condition. Before any creep-ageing takes place, zero dislocation hardening was assumed and thus $\sigma^0_{\text{age}} = \sigma^0_y - \sigma^0_{\text{sol}}$, where $\sigma^0_y = 537$ MPa was obtained from the tensile tests. $\bar{r}^0_p = 1$ was set as the material is at its peak-aged condition. The initial dislocation density is significantly smaller than the maximum dislocation density and $\bar{\rho}^0_d = $ 1E-5 was used.

Figure 4 illustrates a comparison for each of the calibrated creep-ageing curves and yield strength evolutions with the corresponding experimental data. After 22 h of creep-ageing, maximum deviations of 0.29 mε and 9.35 MPa are observed which represent a 5.6 and 1.8% difference respectively. Thus, characteristics of both the viscoplastic deformation and yield strength evolution of 7B04-T651 during creep-ageing have been captured with close agreements. In addition to the creep and strength evolutions, the alloy's precipitate growth was also calibrated. This was based on the TEM analysis result which estimated an average growth in precipitate radius from 8.5 to 12.6 nm after 22 h of stress-free-ageing.

4. Application of the material model to finite element modelling of creep-age forming

4.1. Finite element model

In order to demonstrate the application of the material model to CAF process, a finite element (FE) model that represents a quarter of the flexible CAF tool was constructed in Abaqus. As shown in Figure 5, only workheads of the forming pins were modelled. Detailed descriptions of the FE model and modelling procedure are available in [6], whilst the influences of residual stresses and initial distortion of plates on springback prediction are not considered in this study.

For completion, the FE model used in this study is briefly described here. Mesh convergence studies were carried out following a one-mesh-at-a-time principle whereby the convergences of the flexible surfaces, elastomeric sheet and panel component were individually assessed. Figure 6 illustrates the final condition of the FE meshes. The workheads were assumed to be perfectly spherical with a radius of r_w = 14.65 mm and modelled as rigid bodies. The plates were modelled using initially flat shell elements with their actual plate thicknesses, a Young's modulus (E) of 67 GPa, a Poisson's ratio (ν) of 0.33 and a thermal expansion coefficient (α) of 2.3E-5 K^{-1} assigned. Equations (1)–(7) were implemented into Abaqus via the user-defined subroutine CREEP with the initial conditions and material constants in Tables 2 and 3 assigned. The flexible surfaces were modelled as spring-tempered steel with E = 210 GPa, ν = 0.3 and α = 1.3E-5 K^{-1}. The elastomeric sheet has α = 2.5E-5 K^{-1} and was modelled as an isotropic hyperelastic material and assigned with parameters equivalent to those of a 60 Shore A hardness elastomer. The following six-step modelling procedure was employed:

1. Initiate the simulation with temperature, boundary conditions and contact scheme defined;
2. Load the upper workheads to deform the plate whilst the lower workheads remain fixed;
3. Heat the entire configuration up to the CAF temperature;
4. Propagate into a visco step for creep-ageing to take place;
5. Cool the entire configuration down to room temperature;
6. Unload the upper workheads to allow springback to occur.

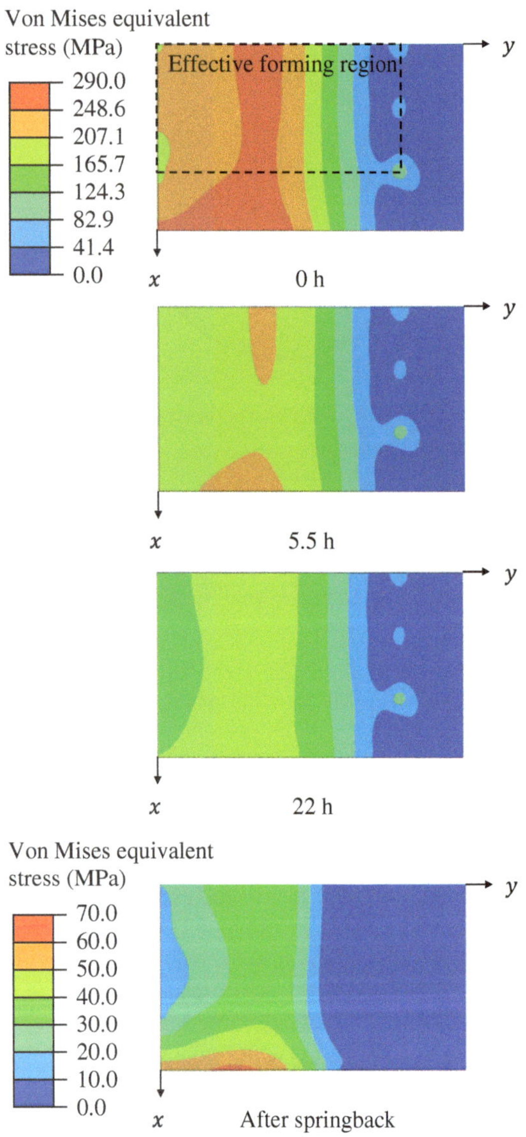

Figure 7. Evolution of von Mises equivalent stress at 0, 5.5 and 22 h of the creep-ageing time and the final stress distribution after springback. Distributions are plotted for the topmost tensile surface of a quarter 8.2-mm-thick plate at a single curvature bending radius of 1200 mm under 115 °C. The dashed line illustrates the effective forming region of the flexible form tool.

Simulations of the CAF tests were carried out using the FE model under conditions that match those of the forming tests. Stress relaxation, yield strength evolution, precipitate growth and springback of the plates were modelled and analysed, with the springback further quantified and compared with those of the physically formed plates.

4.2. Stress relaxation

Figure 7 shows the stress relaxation that occurred on the topmost tensile surface of a quarter (300 mm × 175 mm) 8.2-mm-thick plate during CAF in the flexible form tool. Three field plots were recorded at the 0 (at an infinisimal first numerical time increment), 5.5 and 22 h (the total creep-ageing time) time points of the creep-ageing period. The final plot illustrates the stress distribution after springback.

The physical flexible form tool has an effective forming region (EFR) of 480 mm × 240 mm, which translates to a 240 mm × 120 mm region from the centre point of symmetries of the quarter model. When the plate is fully loaded at 0 h, a higher overall stress state is observed in the region between the $y = 0$ and 120 mm workheads. The maximum stress level is observed outside the EFR at $x = 175$ mm. This is due to the act of Poisson's effect on the plate region where it is unconstrained (outside the EFR), which as a result introduces a greater presence of transverse strain to the region, increasing the overall stress level there. In that region, a maximum stress relaxation of 102 MPa was recorded, reducing the stress level there from 288.2 to 186.2 MPa over the 22 h period. Sixty-eight percent of the 102 MPa stress relaxation took place within the first 5.5 h of creep-ageing.

In general, the overall stress state decreases gradually with increasing y and the minimum stress level is observed outside the EFR. Local stress concentrations can be observed at the x-border of the EFR, somewhere between $y = 200$ and 300 mm, which is at the proximity of where the $y = 240$ mm workheads are located. Because internal stresses of the plate were not fully relaxed by the 22 h of creep-ageing, the plate springs back upon tool unloading due to elastic stress recovery. After springback, the region of high stress relaxation outside the EFR has the highest amount of residual stress on the plate surface; about four times lower than the stress level when the plate was fully loaded. Within the EFR, distribution of the stress field is similar to that of the plate immediately before unloading, but with approximately five times less in the magnitude of stress.

4.3. Yield strength prediction

Figure 8a shows the field plots of yield strength evolution for the topmost tensile surface of the same plate during CAF. At 0 h, the plate is deformed elastically and the yield strength throughout the plate has an expected initial value of 537 MPa. Due to the different stress levels experienced by the plate at different regions, differences in the rate of yield strength evolution over the area of the plate begin to emerge with time. A higher strength region is observed at approximately $y = 90$ mm, which correlates with the high stress region in Figure 7. Within this region, the yield strength of the material has grown from 537 MPa to a median value of 540 MPa after 5.5 h. At 5.5 h, the yield strength gradually decreases in a y-outward manner, agreeing with the stress distributions and their relationship defined by the material model. Outside the EFR in the y direction is a region of the plate where stress-free-ageing occurs most and a 0.2 MPa minor softening of the material is observed.

Under the CAF condition investigated and of the three discrete time points captured, the overall material's yield strength is highest at 5.5 h, leading to an overall trend of softening over the whole plate area from then onwards.

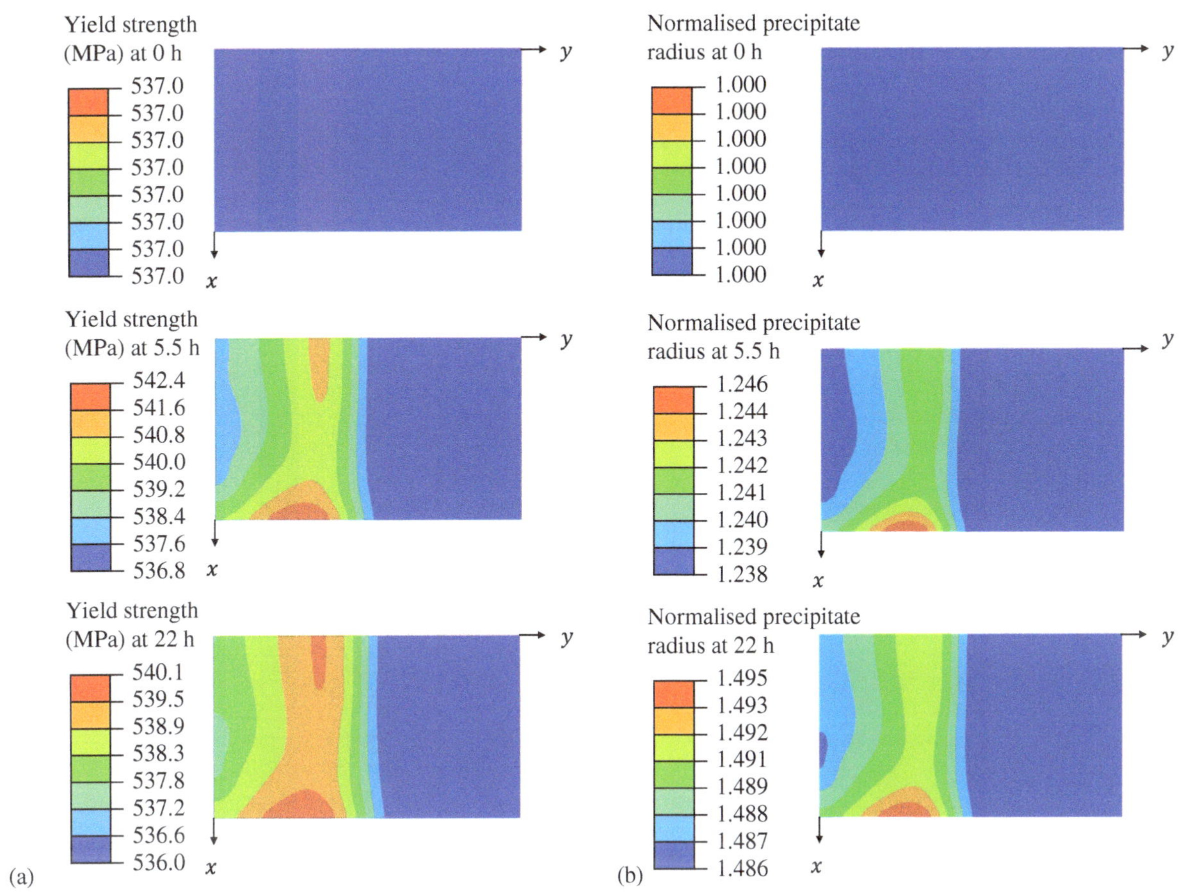

Figure 8. Evolutions of (a) yield strength and (b) normalised precipitate radius at the topmost tensile surface of a quarter 8.2-mm-thick plate at a single curvature bending radius of 1200 mm during CAF under 115 °C.

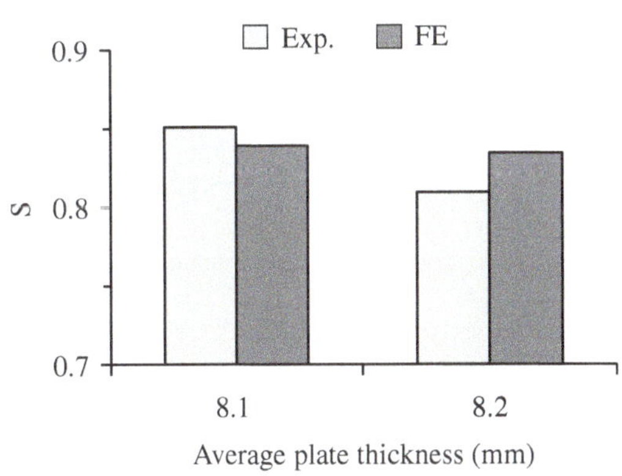

Figure 9. Comparison of the experimentally measured and FE-simulated springback of 7B04-T651 plates. The forming tests were conducted at a single curvature bending radius of 1200 mm under 115 °C for 22 h.

At the end of the CAF process, the median strength value of the plate in the high strength region has reduced from the 540 MPa at 5.5 h to 538.3 MPa at 22 h. For the outer y region where the stress levels are negligible, the yield strength of the plate has softened further by 0.8 MPa, to a final value of 536 MPa.

4.4. Precipitate size prediction

Figure 8b shows the field plots of normalised precipitate radius for the same plate and arranged in the same format as those illustrated in Figure 8a. In general, the field plots show a similar precipitate growth trend as those recorded for the yield strength evolution in Figure 8a. This is because the hardening mechanisms of CAF (Eqs. (2)–(5)) are dependent on precipitate size, dislocation density and their inter-relationship, which in turn, are influenced by the stress levels experienced and therefore the creep strain rate of the plate. On the other hand, unlike yield strength which has mechanisms that activate the softening of the material and led to a decrease in its value, the precipitate radius of the alloy has grown steadily over the entire course of creep-ageing time. Within the first 5.5 h of strength growth period, the median $\bar{r}_p$ value has increased by 24% from 1.000 to 1.241. Over-creep-ageing of the alloy from its peak strength value occurred at some point within the next 16.5 h period, during which a further 20% increase in the median precipitate radius value, from 1.241 to 1.489, is observed.

4.5. Springback prediction and comparison

Figure 9 illustrates a comparison between the measured and FE-simulated springback values. The FE-simulated springback value is lower than the experimental value for the 8.1-mm-thick plate whilst the opposite is observed for the 8.2-mm-thick plate. Based on the FE-simulated results which show a 0.4% difference between the two plates, it can be concluded that the springback sensitivity to a 0.1 mm difference in plate thickness is close to negligible under the current CAF condition. Therefore, the 4.2% difference between the experimental springback values of the two plates is likely to be due to experimental uncertainties, which may include their difference in residual stress level and initial plate distortion prior to CAF submission [6]. Nonetheless, the measured and FE-simulated results both show a decrease in springback as the plate thickness increases from 8.1 to 8.2 mm. The FE-simulated springback values deviate slightly from the experimental values but they are within an acceptable range of −1.2 and +2.5% difference in springback or +0.3 and −0.6 mm difference in vertical deflection. Thus, a good prediction of springback was achieved using this material model.

5. Summary

A material model was presented with its constants determined for creep-age forming of aluminium alloy 7B04-T651. In addition to creep deformation, the model also predicts the precipitate growth and yield strength evolution of the alloy. Its applications to creep-age forming have been demonstrated through the analyses of stress relaxation, yield strength evolution, precipitate growth and springback using finite element modelling techniques. A good agreement of 2.5% deviation from the physical forming test results was achieved for the predicted springback values.

Implications and influences

Due to the long processing hours required for creep-age forming, the technique is often found associated with the production of extra-large panel components such as aircraft wing skin panels. As a result, the quality of a creep-age formed part, largely determined by its final shape, strength specification and condition of microstructure, can vary greatly within the large area covered by the component. Full experimental investigations are time consuming to perform and also highly costly. The test programme, material modelling techniques and their applications summarised in this article enable creep-age forming of aerospace-grade alloys, under various initial, boundary and loading conditions, to be investigated inexpensively via computational modelling. Cost-efficient and accurate investigations can now be performed for creep-age forming of aluminium alloy 7B04-T651, enabling further advancement of the forming technique and its associated technologies.

Acknowledgements. The strong support from Aviation Industry Corporation of China (AVIC) Beijing Aeronautical Manufacturing Technology Research Institute (BAMTRI) for this funded research is much appreciated. The research was performed at the AVIC Centre for Structural Design and Manufacture at Imperial College London.

References

1. M.C. Holman, Autoclave age forming large aluminium aircraft panels, Journal of Mechanical Working Technology 20 (1989) 477–488.
2. F. Eberl, S. Gardiner, G. Campanile, G. Surdon, M. Venmans, P. Prangnell, Ageformable panels for commercial aircraft, Journal of Aerospace Engineering 222 (2008) 873–886.
3. L. Zhan, J. Lin, T.A. Dean, A review of the development of creep age forming: Experimentation, modelling and applications, International Journal of Machine Tools and Manufacture 51 (2011) 1–17.
4. K.C. Ho, J. Lin, T.A. Dean, Constitutive modelling of primary creep for age forming an aluminium alloy, Journal of Materials Processing Technology 153–154 (2004) 122–127.
5. A.C.L. Lam, Z. Shi, J. Lin, X. Huang, Y. Zeng, T.A. Dean, A method for designing lightweight and flexible creep-age forming tools using mechanical splines and sparse controlling points, International Journal of Advanced Manufacturing Technology 80 (2015) 361–372.
6. A.C.L. Lam, Z. Shi, J. Lin, X. Huang, Influences of residual stresses and initial distortion on springback prediction of 7B04-T651 aluminium plates in creep-age forming, International Journal of Mechanical Sciences 103 (2015) 115–126.
7. R. Ferragut, A. Somoza, A. Tolley, Microstructural evolution of 7012 alloy during the early stages of artificial ageing, Acta Materialia 47 (1999) 4355–4364.
8. G. Sha, A. Cerezo, Early-stage precipitation in Al-Zn-Mg-Cu alloy (7050), Acta Materialia 52 (2004) 4503–4516.
9. L. Zhan, J. Lin, T.A. Dean, M. Huang, Experimental studies and constitutive modelling of the hardening of aluminium alloy 7055 under creep age forming conditions, International Journal of Mechanical Sciences 53 (2011) 595–605.
10. P. Zhang, J. Lin, D. Balint, OPT-CAF: the optimisation tool for calibrating creep constitutive equations, Project report, Imperial College London, 2012.
11. Z. Li, B. Xiong, Y. Zhang, B. Zhu, F. Wang, H. Liu, Microstructural evolution of aluminum alloy 7B04 thick plate by various thermal treatments, Transactions of Nonferrous Metals Society of China 18 (2008) 40–45.

13

Development of microforming process combined with thin film transfer printing

Kazushi Koshimizu*, Qiu Zheng, Tetsuhide Shimizu, and Ming Yang

Tokyo Metropolitan University, 6-6 Asahigaoka, Hino-shi, Tokyo 191-0065, Japan

Abstract – Microforming receives a lot of attentions in the recent years due to the increased use of microparts in electronics and medical sectors. For the further functionalization of these micro devices, high functional surface with noble metals are strongly required for the devices in bio- and medical fields, such as bio-sensors. To realize the sub-millimeter structure of metal foils and micro to nanometer structures in one forming process, the present study proposes a combined process of microforming for metal foils and transfer printing of gold (Au) thin films. To clarify the availability of the proposed combined process, transferability of Au thin films under micro bulging deformation are investigated. 0.1 mm-thick pure titanium (Ti) foils and 100 nm-thick Au films were used as blank and functional materials, respectively. The forming tests of the proposed process were conducted. With increasing strain of Ti foils, Au TP areas increase. By this experiment, it's confirmed that the hydrogen reduction of oxidation layers and the strain of Ti foil are significant factor for Au TP on Ti foils.

Key words: Microforming, Transfer printing, Ti foils, Au thin films, Micro bulging

1. Introduction

Because of low cost, high productivity and stability of process, the demand of micro press forming is increasing to realize small devices made by metallic parts taking advantage of good properties of metals, such as mechanical strengths, electrical conductivity and chemical resistance. Especially for medical devices, biocompatibility of metal has been highly required, and its miniaturization technology is needed. Additionally bio-devices with noble metal, such as gold (Au) on the surface, are often used in bio-technology by exerting good functions like detecting antibody [1] or catalytic reaction [2]. Therefore, the process to fabricate the functional surface, which is combined with the biocompatible bulk metals is importance in the fabrication of bio- and medical devices.

In the present study, a process with the combination of micro press forming and fabrication of the functional surface for metal foils is proposed. This process can realize the submillimeter structure of metal foils and micro to nanometer structures of noble metals in one step. In this process, transfer printing (TP) [3] is selected to fabricate the functional surfaces. As shown in Figure 1, the noble metal films of Au is first deposited on the punch, and then micro press forming process is conducted to transfer Au to the processed material (Ti).

However, the availability of this combined process is not reported and the significant factors for this TP are not clarified yet. Thus, the aim of this study is to clarify the availability of the proposed process and to investigate the factors influencing the TP of Au thin films to Ti foils after press forming.

*e-mail: koshimizu-kazushi@ed.tmu.ac.jp

2. Experiment

2.1. Au TP on Ti thin films

To realize Au TP on Ti foils, basic parameter for Au TP on Ti have to be investigated. In recent studies, Hur et al. [4] reported that temperature, pressure and time have influence on the TP rate. Taniguchi et al. [5] reported that the atmosphere of TP and released layer under thin Au films also influenced the TP rate. According to these previous reports, the present study investigates the transferability of Au TP on Ti substrate under these parameters.

Thin Ti film and Au film were deposited by electron cyclotron resonance sputtering machine (ELIONIX Corp.) on Silicon (Si) substrate, respectively. The thickness of Ti is 500 nm, while the thickness of Au is 100 nm. As shown in Figure 2, Thin Ti film was deposited on the Si substrate uniformly and Au films were patterned on the Si substrate with a width of 50 μm and a pitch of 100 μm. But, in the case of any condition, the released layer of Cr_2O_3 was deposited under Au films.

After preparing two substrates, Ti film was first placed on the surface of Au film, and then was put in a chamber to be heated under the atmosphere of pure argon gas or mixed gas of hydrogen and argon with load for TP. TP conditions are

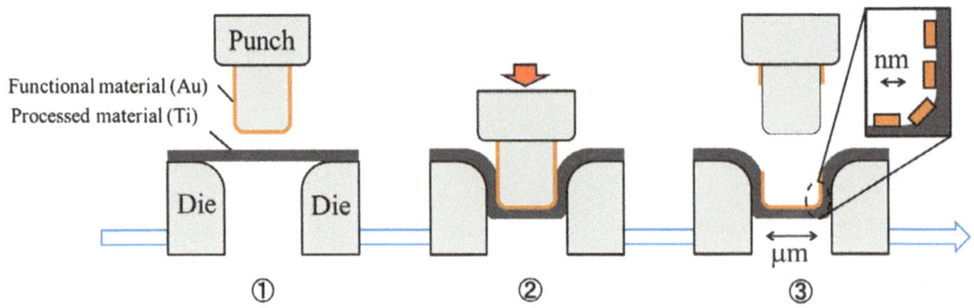

Figure 1. Schematic illustration of the proposed process with the combination of micro press forming and transfer printing.

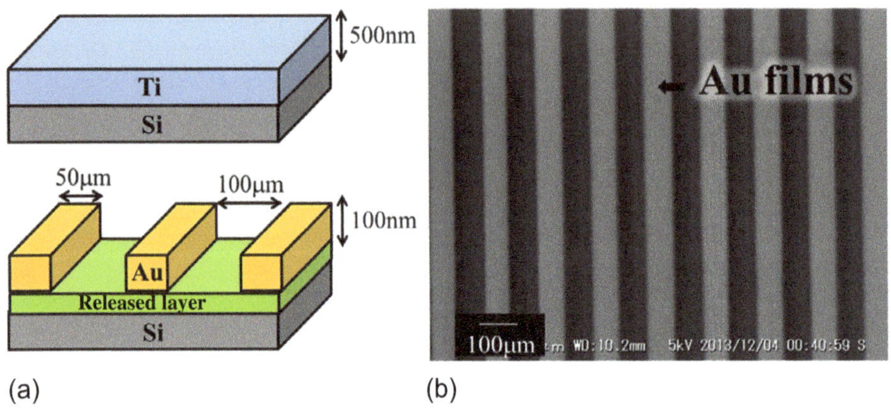

Figure 2. (a) Schematic illustration of Si substrate deposited Ti and Au respectively, (b) SEM image of Au.

Table 1. Conditions of Au transfer printing (TP).

	Atmosphere	Temperature (°C)	TP pressure (kPa)	TP time (min)	Release layer
A	Ar	700	150	30	SiO_2
B	H_2/Ar	700	150	30	SiO_2
C	H_2/Ar	800	450	30	SiO_2
D	H_2/Ar	800	450	30	Cr_2O_3

shown in Table 1. To evaluate the feasibility of functional surface made by this process, TP rate defined as the ratio of the area of printed Au to deposited Au in a square area of 1 mm × 0.6 mm is calculated (see Figure 3).

2.2. Au TP on the deformed Ti foil

To investigate the influencing factors of TP of Au on deformed Ti foils, press forming system was developed (see Figure 4). This system mainly consists of a servo press machine, vacuum chamber, rotary pump, die for bulging process and resistance heating equipment. This system can realize the controlling of atmosphere and temperatures of Ti foil.

As shown in Figure 5, back plates of SUS deposited Au thin films were put under 0.1 mm-thick Ti foils. The thickness of back plate is also 0.1 mm. The patterns of thin Au films deposited on the back plate are the same as shown in Figure 2 and the released layer was Cr_2O_3 in this experiment. To reduce the oxidation layer of the Ti foils, the specimens were annealed at 700 °C with the atmosphere of H_2 for 30 min before bulging. After bulging, the keeping time of punch for TP was 30 min. Then the Au films can be transferred to the deformed Ti foils. Other experimental conditions, such as the atmosphere for bulging, strain of deformed area and TP pressures, are shown in Table 2. To investigate the strain of the deformed area of Ti foil and back plate, the bulging process was analyzed by finite element method (FEM) using ABAQUS. The schematic illustration of the model is shown in Figure 6. As shown in Table 3, where, the mesh size of specimens was 25 µm × 25 µm, the punch stroke was 1 or 2 mm from the punch contacted with Ti foil, the mass density of Ti foil and back plate were 4.5×10^{-9} and 7.9×10^{-9} (mg/µm^3), the young's modulus of them were 110 and 171 (GPa), the yield stress of them were 330 and 292 (MPa), the poisson's ratio of them were 0.34 and 0.28, and the friction coefficient was 0.1.

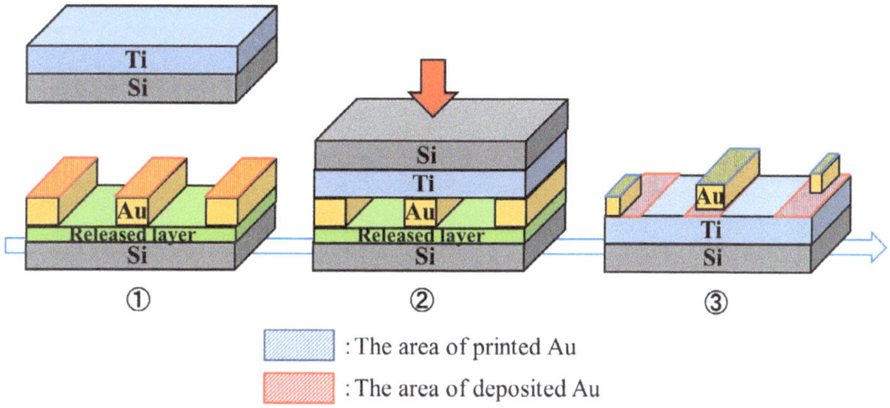

Figure 3. Schematic illustration of transfer printing.

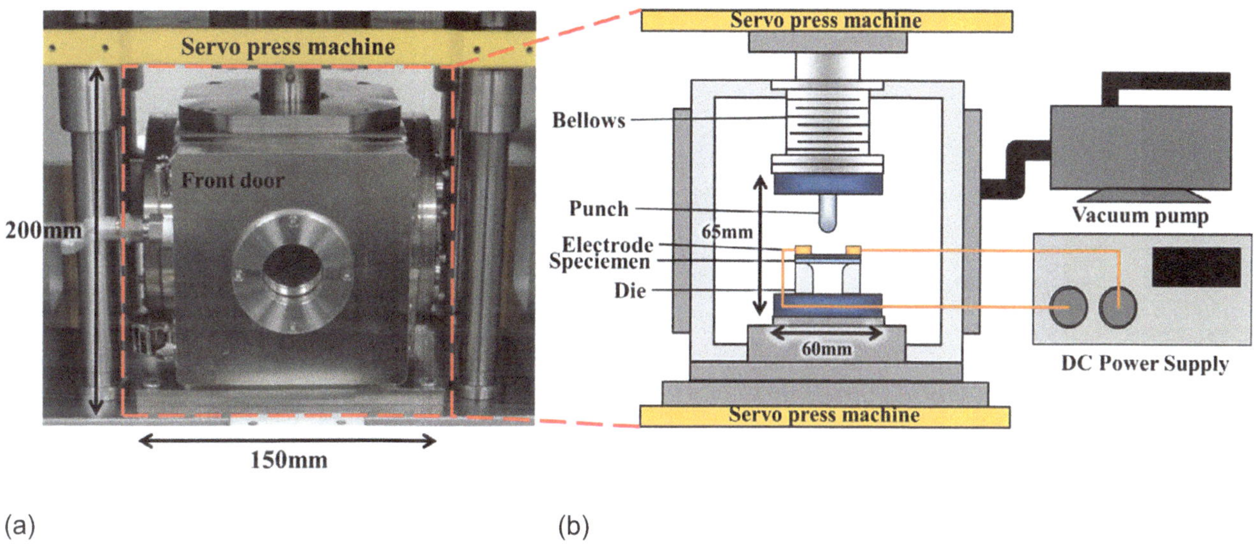

Figure 4. Developed press forming system (a) appearance of chamber, (b) schematic illustration.

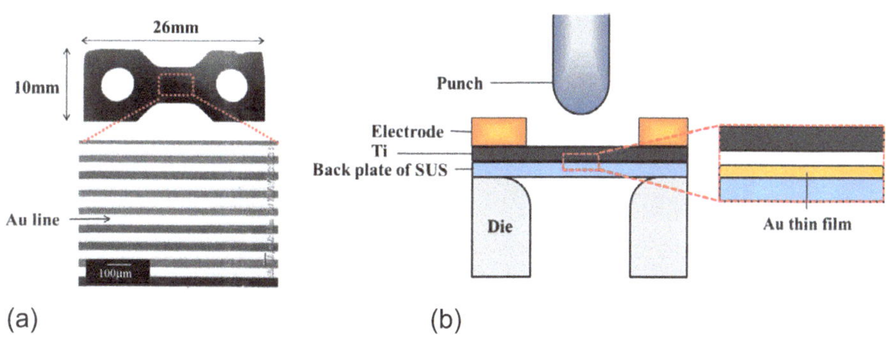

Figure 5. (a) Shape of back plate and SEM image of surface, (b) schematic illustration of die for bulging process.

3. Results and discussion

3.1. Au TP on Ti thin films

Figure 7 shows the images of surfaces for Ti thin film after Au TP taken by scanning electron microscope (SEM) in the conditions of A–D. The qualitative analysis for these surfaces was taken by energy dispersive X-ray analyser (EDX). In condition A, no Au thin films were printed on Ti foils. However Au films can be found to be printed on Ti for the conditions B–D. The TP rate for conditions B–D is shown in Figure 8. As the conditions change from B to C and D, the TP rate

Table 2. Conditions of Au TP on deformed Ti foil.

	Atmosphere	Strain (%)	TP pressure (N)
E	Ar	8.4	390
F	H_2/Ar	8.4	250
G	H_2/Ar	2.6	390
H	H_2/Ar	8.4	390

Table 3. Input mechanical properties of specimens and FE simulation conditions.

Specimen	Ti	Back plate (SUS)
Mass density (mg/μm)	4.5×10^{-9}	7.9×10^{-9}
Young's modulus (GPa)	110	171
Yield stress (MPa)	330	292
Poisson's ratio	0.34	0.28
Friction coefficient	0.1	

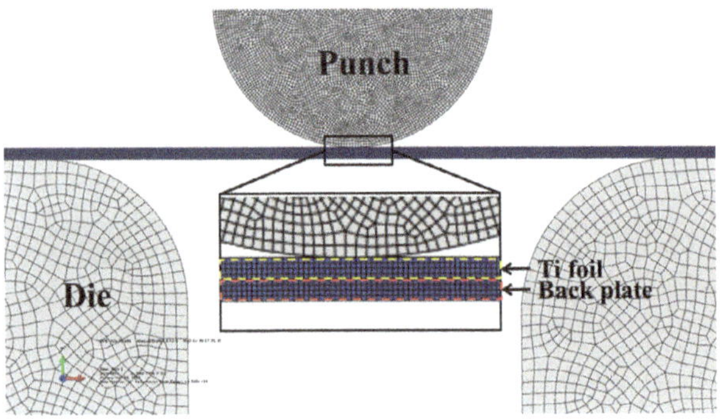

Figure 6. Schematic illustration of FEM model.

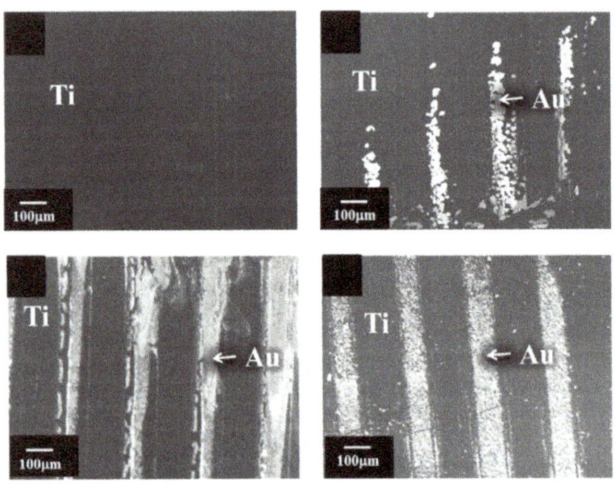

Figure 7. SEM image of Au thin films transfer printed on Ti thin films.

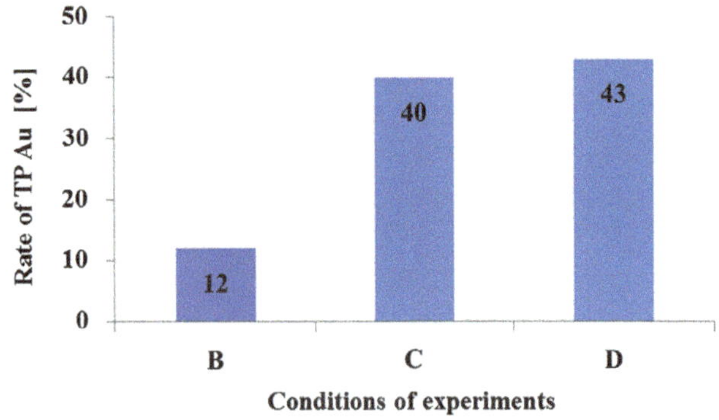

Figure 8. Rate of transfer printed Au thin films for three different conditions shown in Table 1.

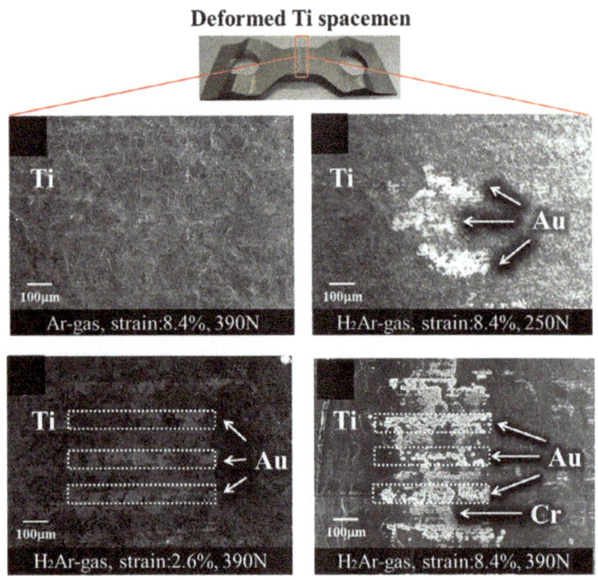

Figure 9. SEM image of Au thin films transfer printed on deformed Ti foils.

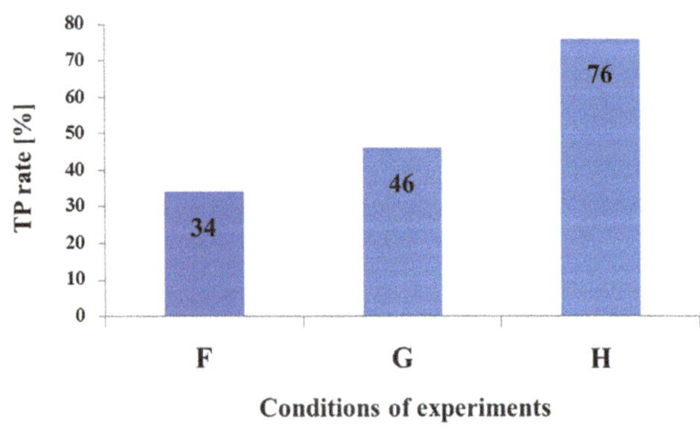

Figure 10. Rate of transfer printed Au thin films on deformed Ti foil s for three different conditions shown in Table 2.

increases from 12% to 40% and to 43%. From this measurement, it was confirmed that the atmosphere of H_2 gas, high temperature of Ti film, high pressure of TP and release layer of Cr_2O_3 can increase Au TP rate. The effect of H_2 gas can be explained by the reduction of oxidation layer of Ti due to the reaction with H_2 gas under high temperature. Thus, Ti surface becomes active and the chemical reaction between Ti and Au is promoted. As TP temperature increases, this reaction is also promoted. The effect of TP pressure can be explained by

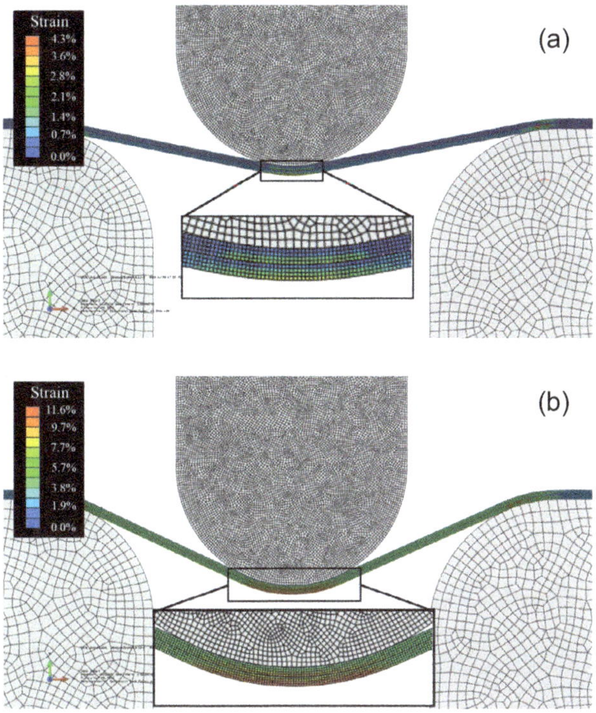

Figure 11. Distribution of strain (a) punch stroke was 1 mm and strain of middle of Ti foil was 2.6%, (b) punch stroke was 2 mm and strain of middle of Ti foil was 8.4%.

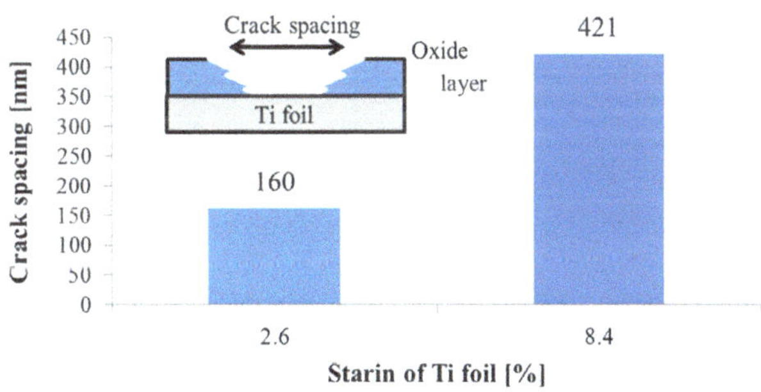

Figure 12. Crack spacing of oxidation layer for different strain of Ti foil.

increasing the true contact area between Ti and Au, which result in the increase of TP rate. The effect of release layer of Cr_2O_3 can be explained by the decrease in the binding energy of substrate with Au. Taniguchi et al. [5] reported that the adhesive force between two surfaces relates to these surface energy. This energy of Cr_2O_3 is smaller than SiO_2. Therefore, Au thin films are easily released and transferred to Ti films.

3.2. Au TP on the deformed Ti foil

Figure 9 shows the images of surfaces for Ti foil after Au TP taken by SEM and EDX in the conditions of E–H. From these results, Au thin films were printed in deformed area of Ti foils. Figure 10 shows the Au TP rate on Ti foils. Here, TP rate is defined as the ratio of the area for printed Au to deposited Au in the square area of 500 μm × 50 nm. It was confirmed that Au thin films were printed in H_2/Ar gas. This indicates that the effect of the H_2 reduction of oxidation layer has significant influence on TP. In addition, with increasing strain and TP pressure, Au TP rate increases. It was considered that high TP pressure enlarges true contact area between Ti and Au, therefore Au TP rate increases from condition F to H. However, in condition H, some Cr thin films were also printed on Ti surface. This effect of Cr TP can be explained by elastic deformation of Au thin films, which result in the contacting two surfaces of Cr and activated Ti in the condition of high TP pressure. With the further increase of strain, active surfaces of Ti foil seem to be exposed by the cracks of oxidation layer. As shown in Figure 11, the strain of Ti foil was concentrated around the radius of punch. Latella et al. [6] reported that the crack space of oxidation layer can be observed in the area with distribution of strain. These crack spacing were considered to expose the active surface of specimen. The crack spacing of oxidation layer is calculated by equation (1) [6]:

$$Cs = \frac{\sigma_c t}{\tau} \tag{1}$$

where, Cs is the crack spacing, σ_c is the fracture strength, t is the thickness of oxidation layer, and τ is the interfacial shear strength. As shown in Figure 12, the crack spacing became from 160 nm to 421 nm as strain increased from 2.6% to 8.4%. Therefore, the amount of exposed surface has the same effect as the reduction of the oxide surface under H_2 atmosphere.

4. Conclusions

In this study, we proposed a process with the combination of micro bulging and transfer printing. The factors influence for the Au TP on Ti film and deformed Ti foil were investigated. The results of Au TP on Ti thin film showed that the atmosphere of H_2, increasing TP temperatures, increasing TP pressure and release layer of Cr_2O_3 could increase the Au TP rate. In addition, the results of Au TP on deformed Ti foil showed that the strain of Ti foil influenced the Au TP rate significantly. It was considered that active surfaces exposed by the cracks of oxidation layer increased the reaction area for TP of Au on Ti foil.

In the future, the detail of relation between strain of Ti foil and Au TP rate and the influence of process parameters for the quality of Au films such as TP rate or feature will be considered in order to define, in detail, the process scope of this combining process.

References

1. X. Cao, Y. Ye, S. Liu, Gold nanoparticle-based signal amplification for biosensing, Anal. Biochem. 417 (2011) 1–16.
2. H. Zhao, Y. Wang, Q.H. Tang, et al., Pt catalyst supported on titanium suboxide for formic acid electrooxidation reaction, Int. J. Hydrogen Energy 39 (2014) 9621–9627.
3. J.W. Kim, K.Y. Yang, S.H. Hong, H. Lee, Formation of Au nanopatterns on various substrates using simplified nano-transfer printing method, Appl. Surf. Sci. 254 (2008) 5607–5611.

4. S.H. Hur, D.Y. Khang, C. Kocabas, J.A. Rogers, Nanotransfer printing by use of noncovalent surface forces: applications to thin-film transistors that use single-walled carbon nanotube networks and semiconducting polymers, Appl. Phys. Lett. 85 (2004) 5730–5732.
5. J. Taniguchi, S. Ide, N. Unno, H. Sakaguchi, Nanoprint lithography of gold nanopatterns on polyethylene terephthalate, Microelectron. Eng. 86 (2009) 590–595.
6. B.A. Latella, B.K. Gan, H. Li, Fracture toughness and adhesion of thermally grown titanium oxide on medical grade pure titanium, Surf. Coat. Technol. 201 (2007) 6325–6331.

14

Approaches for modelling and simulation of metal machining

Jürgen Leopold*

TBZ-PARIV GmbH, 09126 Chemnitz, Germany

Abstract – Manufacturing is as old as the human race. Modelling and Simulation of Machining Operations are mainly based on the development of the Natural Sciences. Within the last 40 years there have been dramatically improvements of both: the scientific foundations of models and simulation techniques on one hand side and a strong industrial application on the other hand. The main focus of this paper is a critical assessment of the relevant single modelling methods for the simulation of machining, as well as providing some roadmaps for further directions towards the closed-loop modelling of machining. Limitations are discussed and challenges for the future are highlighted.

Key words: Manufacturing, Machining, Modelling, Numerical Methods

Nomenclature

u_x, v_y, w_z	Displacements (respectively velocities)
$\dot{\varepsilon}$	Strain rate components
σ, τ	Stress components
ε_{eff}	Effective strain
k	Yield shear stress

1. Introduction

To understand the progress in manufacturing and the associated modelling and simulation methods, a short look back in to the history will identify the next important steps of development and application.

Starting with very simple cutting tools and hand driven technology, the industrial revolution opened the chance to use mechanical energy in addition to man-power with the development and design of machine tools, applied up to the present century. To the classical design, a new concept – Parallel Kinematic Machine Tools (PKM) – has been developed and introduced into industry. Compared with Henry Ford's principle of Mass Production, the main step forward in general manufacturing technology from "Conventional machining" processes was started with the "Flexible Machine Tool System" (FMS) concept. This development, in conventional machine tools, PKM and FMS has been driven by the objectives of increasing product quality whilst retaining higher manufacturing productivity, cost reduction and environment protection. The development and application of relevant models and simulation techniques has a very simple background: Machining is the most predominant manufacturing operation in terms of volume and expenditure. Machined components are used in almost every type of manufactured product. It has been estimated that machining expenditure contributes to approximately 5% of the GDP in developed countries, while in the US alone it translates into approximately $250B per year [1]. In the face of worldwide competition, the time-to-market of new products can be responsible for economic survival. Advanced modelling and simulation techniques can be a competitive advantage for companies. Returning to the short history of manufacturing (Figure 1), the challenge of further modelling and simulation techniques application are in the increasing of the accuracy for the single model on one hand side but more and more on the application of the so called "Closed-loop-principle" [2] in modelling. This will be pointed out within the next section.

*email: jleopold@tbz-pariv.de

2. Closed-loop principle of modelling

The manufacturing of a workpiece can be a very complex situation.

The main goal of machining is the manufacturing of a workpiece with specified properties: geometrical properties (length, diameter, concentricity, etc.); surface properties (roughness, waviness, polish, etc.); sub-surface properties (residual stress; crack-free, etc.), mechanical stability, etc.

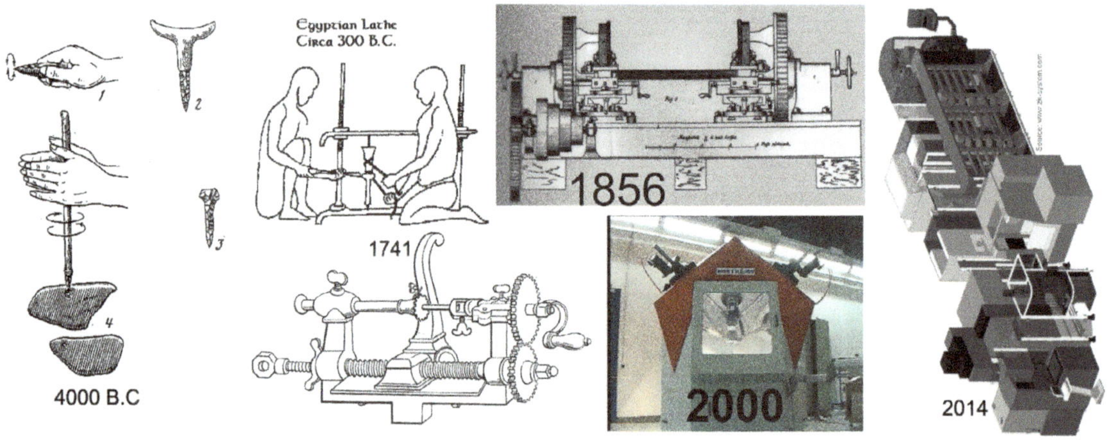

Figure 1. Main steps in manufacturing from past to present.

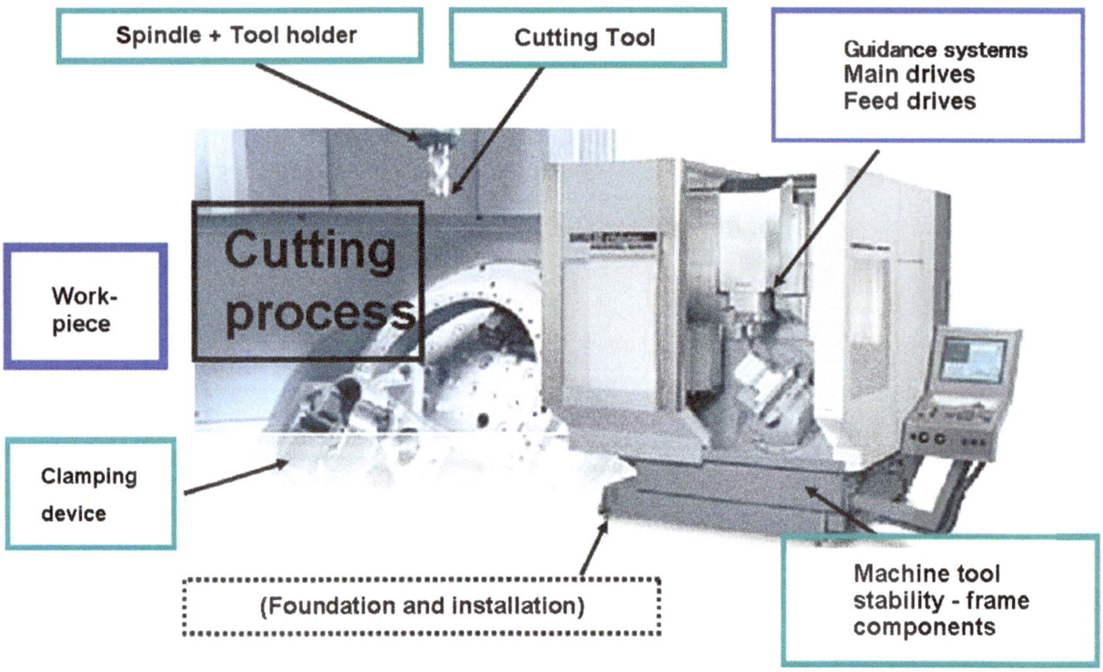

Figure 2. Machining of pre-defined workpieces.

To manufacture a product with such properties, a complex system (Figure 2) must be applied.

Following Georg Schlesinger's postulate, that the "profit will be determined at the cutting tool tip" – the cutting tool is the most important component part of the closed-loop modelling principle. Geometrically properties such as length, flute geometry, tool-tip geometry (rake; chip-former), micro-geometry, cutting tool material and coatings; coolant bore holes and design; mechanical stability, clamping areas of the tool within the tool holder, etc. The interaction of the cutting tool with the workpiece is responsible for the generation of the new workpiece, which is why the workpiece is one of the next main modelling part in this system. Typically single modelling systems are focussing on only this interaction – with no deformable cutting tool and also no deformable workpiece. This is commonly used – but did not represent the real situation in manufacturing. The workpiece geometry can be realized in different steps (roughing and finishing) – but in the airplane industry for instance, there are often thin-walled parts manufactured. These parts are commonly produced by casting and can include previously-induced residual stresses. The physical based machining residual stresses must be superposed with the residual stressed caused by previously processes and can deform the final part into an undesirable condition. A similar situation can be found for parts coming from previously forming operations. In addition to this serious problem, the simulation techniques require the workpiece properties and the material law for this workpiece (which depend on strain, strain-rate and temperature for the contemporary homogeneously and isotropic modelling world).

Both the cutting tool and the workpiece are fixed within the machine, which is why on one hand machine tool properties

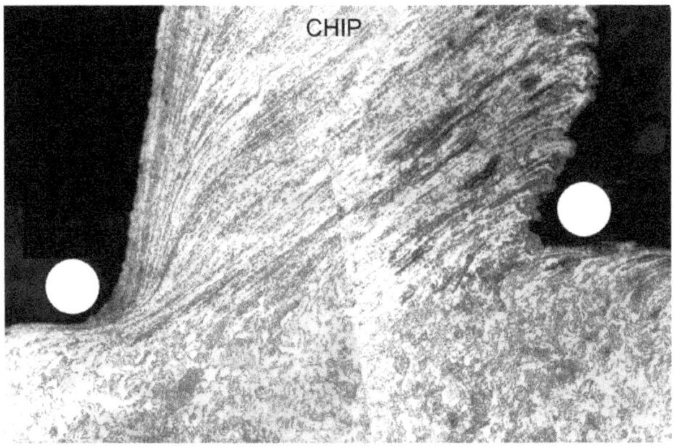

Figure 3. Metallographic view of the chip formation zone.

such as damping and stiffness and also the same properties for the spindle and tool holder within the machine tool, both being influenced by the interaction of the chip-formation process at the cutting tool tip with the workpiece. On the other hand, the clamping device with similar properties (stability; stiffness; damping) also influences considerable the machining process itself. In addition, most of the machining operations are based on cooling and lubrication processes, so this liquid situation must be included with in model.

Following the above very short discussion of the simulation, the final focus for modelling and simulation must be on the development of closed-loop models. The next section will give an overview of contemporary modelling and simulation methods, applied in machining and will be followed by the "white" areas in modelling and simulation.

3. Modelling and simulation techniques

3.1 Slip-line method

Consider the physically based machining situation of the tool-tip interaction with the workpiece [3]. The surface of the "old" workpiece is on the left-side and that of the new one on the right side of Figure 3.

The originally structure of the previous workpiece surface on the left side is dramatically deformed into the chip area. The line (or area) from tool tip Point P1 to the free surface Point P2 is called the shear line (or shear plane) and has been defined by the modelling pioneers about 130 years ago (Time – first shear line equation [4]; Merchant 1944 – the so-called Merchant circle [5]) concerned with previous experimental investigation by Tresca [6], Mallok [7] and Svorykin [8].

With the definition of the shear-line, the well known slip-line method in chip-formation was born. The principles are summarized briefly below [3].

In addition to the usual assumptions that the metal is isotropic and homogeneous, the common approach involves the following:

- the metal is rigid-perfectly plastic; this implies the neglect of elastic strains and treats the flow stress as a constant,
- deformation is by plane strain,
- possible effects of temperature, strain rate and time are not considered,
- there is a constant shear stress at the interfacial boundary. Usually, either a frictionless condition or sticking friction is assumed.

The principal ways in which slip-line field theory fails to take account of the behaviour of real materials are:

- It deals only with *non-strain-hardening materials*.
- There is no allowance for *creep* or *strain-rate effects*. The rate of deformation at each given point in space and in the deforming body is generally different, and any effect this may have on the yield stress is ignored.
- All inertia forces are neglected and the problems treated as *quasi-static (a series of static states)*.
- In the cutting operations which impose *heavy deformations*, most of the work done is dissipated as heat; the temperatures attained may affect the material properties of the body or certain physical characteristics in the surroundings, e.g. lubrication.

The following equations are the basics for the slip-line method.

3.1.1 Plane plastic strain

Since elastic strains are neglected, the plastic strain increments (or strain-rates) may be written in terms of the displacements (or velocities) $u_x(x, y)$, $v_y(x, y)$, $w_z = 0$, as below:

$$\begin{aligned} \dot{\varepsilon}_x &= \frac{\partial u_x}{\partial x} & \dot{\gamma}_{xy} &= \frac{1}{2}\left(\frac{\partial u_x}{\partial y} + \frac{\partial v_y}{\partial x}\right), \\ \dot{\varepsilon}_y &= \frac{\partial v_y}{\partial y} & \dot{\gamma}_{yz} &= \frac{1}{2}\left(\frac{\partial v_y}{\partial z} + \frac{\partial w_z}{\partial y}\right) = 0, \\ \dot{\varepsilon}_z &= \frac{\partial w_z}{\partial z} = 0 & \dot{\gamma}_{zx} &= \frac{1}{2}\left(\frac{\partial w_z}{\partial x} + \frac{\partial u_x}{\partial z}\right) = 0. \end{aligned} \tag{1}$$

3.1.2 State of stress

It follows from the Levy-Mises relation that τ_{xy} and τ_{yz} are zero and therefore that σ_z is a principal stress. The Levy-Mises relationship between stress and strain for an ideal plastic solid where the elastic strain are negligible is as below:

$$\sigma = \begin{pmatrix} \sigma_{xx} & \tau_{xy} & 0 \\ \tau_{xy} & \sigma_{yy} & 0 \\ 0 & 0 & \sigma_{zz} \end{pmatrix} = \begin{pmatrix} \sigma_{xx} & \tau_{xy} & 0 \\ \tau_{xy} & \sigma_{yy} & 0 \\ 0 & 0 & \frac{1}{2}(\sigma_{xx} + \sigma_{yy}) \end{pmatrix}. \tag{2}$$

It follows that the plastic strain increment $\partial\varepsilon_{zz} = 0$.

The material is incompressible, $\dot{\varepsilon}'_x = -\dot{\varepsilon}'_y$, and each incremental distortion is pure shear. The state of stress throughout the deforming material is represented by a constant yield shear stress k, and a hydrostatic stress $-p$ which in general varies

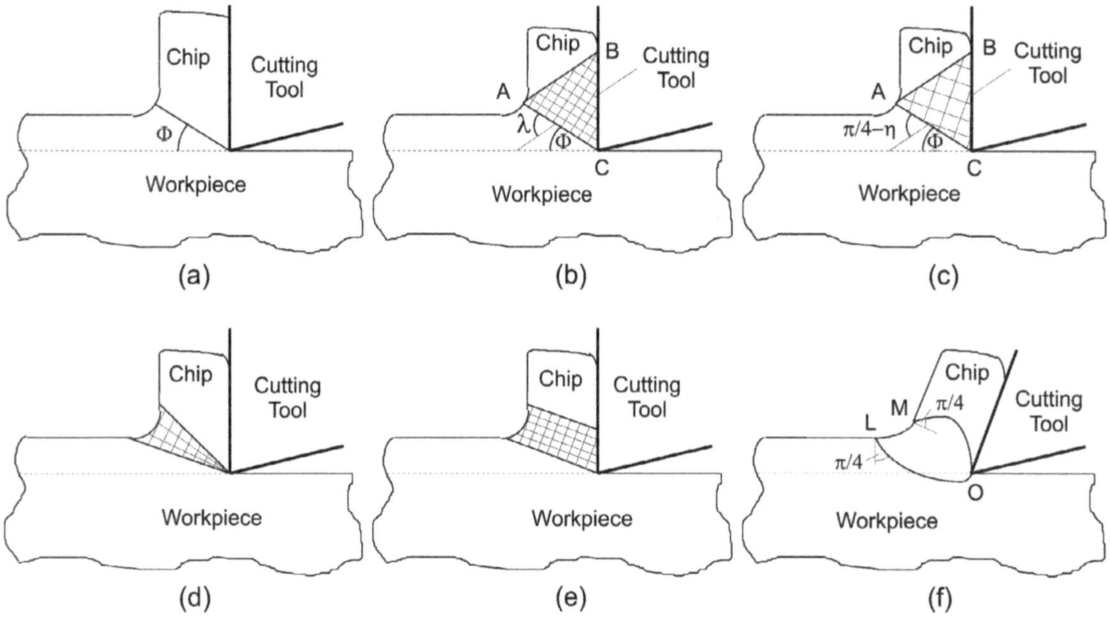

Figure 4. Selected examples for slip-line field solutions in machining.

from point to point throughout the material. k is the yield shear stress in plane strain and the yield criterion for this condition is:

$$\tau_{xy}^2 + \left(\sigma_x - \sigma_y\right)^2/4 = k^2, \tag{3}$$

where $k = Y/2$ for the Tresca criterion and $k = Y/\sqrt{3}$ for the Mises criterion.

The state of stress at any point in the deforming material may be represented in the Mohr circle diagram. Together with Hencky's hydrostatic stress conditions along slip-lines and Geiringer velocity field equation the initial equations can be solved graphically or analytically.

Over the last 70 years this method has been used in basic research by hundreds of scientists (Merchant 1945; Lee and Shaffer 1951; Shaw 1955; Palmer and Oxley 1959; Weber 1969; Stevenson and Oxley 1970; Makino and Usui 1973; Fang and Jawahir 2002). For analytically solution, the computation time compared with numerically methods extreme short. This is responsible for the ongoing investigations, despite the main disadvantages (Figure 4).

3.2 Finite Element Modelling (FEM)

With the design and manufacturing of the computer [9] new numerical methods by mathematicians have been developed. Figure 5 gives a short history of FEM application with respect to metal cutting simulation.

There are two main FEM concepts: the EULERIAN flow formulation and LANGRANCE's formulation.

In a very earlier stage of chip formation simulation, the Eulerian FEM principle [10] was applied to steady-state machining [3, 11–15].

The plastic metal flow can be considered as an incompressible non-Newtonian flow problem with the following unknown quantities:

v – flow velocity and
p – hydrostatic pressure

defined by the differential equations:

$$\begin{aligned} \operatorname{div} v &= 00 \quad \text{(incompressibility condition)} \\ -\operatorname{div}[\sigma] &= \rho g \quad \text{(equilibrium equation, i.e. balance of internal and external forces)} \end{aligned}$$

where $[\sigma]$ is the stress tensor and ρg represents the external forces.

For incompressible materials only the deviatoric stress tensor $[\sigma'] = [\sigma] + p[I]$ ($[I]$ is the unity tensor) is influenced by the effective strain ε_{eff} and the strain rate tensor $[\dot{\varepsilon}]$ in the form

$$[\sigma'] = 2\eta(\dot{\varepsilon}_{\text{eff}}, \varepsilon_{\text{eff}})\,[\dot{\varepsilon}],$$

$$\varepsilon_{\text{eff}} = \frac{2}{\sqrt{3}}\sqrt{J_2([\varepsilon])},\ \dot{\varepsilon}_{\text{eff}} = \frac{2}{\sqrt{3}}\sqrt{J_2([\dot{\varepsilon}])},$$

J_2 – 2nd tensor invariant.

The uniaxial flow stress σ_{eff} of the workpiece metal is assumed to be governed by the general flow rule

$$\sigma_{\text{eff}} = \eta\left(\dot{\varepsilon}_{\text{eff}}\right)\dot{\varepsilon}_{\text{eff}}, \tag{4}$$

$$\text{with } \eta(\dot{\varepsilon}_{\text{eff}}) = \frac{\sigma_y + \left(3\eta_0\dot{\varepsilon}_{\text{eff}}\right)^{1/m}}{3\dot{\varepsilon}_{\text{eff}}}, \tag{5}$$

and the material parameter are according to the temperature θ

$$\begin{aligned} \sigma_y(\theta) &= \sigma_y^{(1)} + \sigma_y^{(2)} \cdot \exp\left(\sigma_y^{(3)} \cdot \theta\right), \\ \eta_0(\theta) &= \eta_0^{(1)} + \eta_0^{(2)} \cdot \exp\left(\eta_0^{(3)} \cdot \theta\right), \\ m(\theta) &= m^{(1)} + m^{(2)} \cdot \theta. \end{aligned} \tag{6}$$

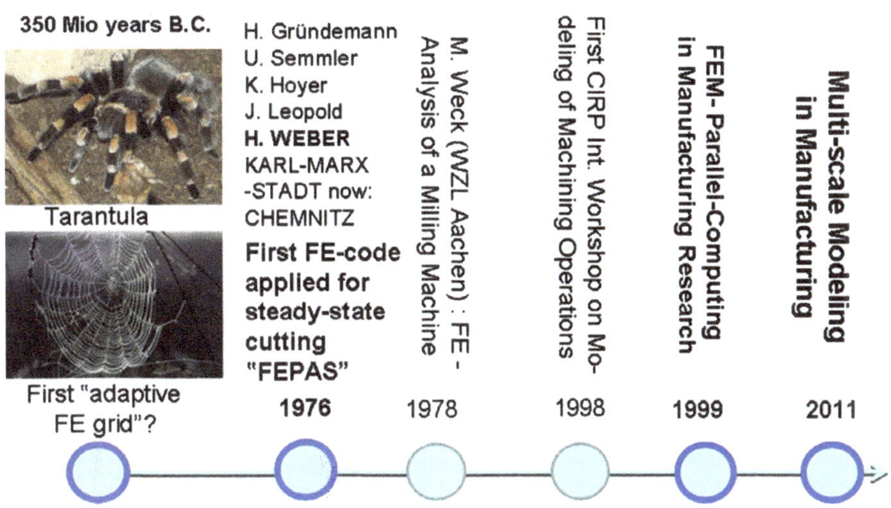

Figure 5. Finite element application in the modelling and simulation of metal cutting.

This general law includes linear Newtonian and the non-linear Bingham and Norton-Hoff laws as special cases. The parameters in the flow rule may depend on the strains. Instead of the flow rule of type, other formulations for the thermo-visco-plastic behaviour of the material may be used. In general, the flow rules give a functional relation of the dependence of the flow stress on the actual effective strain rate, the actual effective strain and the actual temperature. Typical material laws are the Johnson-Cook [16] and the Zerilli-Armstrong [17] law.

Frictional stress τ_F on the tool rake face may be defined using a modified model of Usui and Hoshi [18]:

$$\tau_F = k \cdot \sigma_{\text{eff}}\left(1 - e^{-\mu_F \frac{\sigma_n}{\sigma_{\text{eff}}}}\right), \tag{7}$$

where k is the friction factor (≤ 1.0) and μ_F is the friction number.

Heat flux within the workpiece is modelled by the stationary heat conduction-convection equation

$$\rho c_p v\nabla\theta = \nabla(\lambda\nabla\theta) + \dot{w} \tag{8}$$

where ρc_p represents the heat capacity, λ is the thermal conductivity and $\dot{w}$ is the heat dissipation rate. The parameters λ and c_p depend on the temperature. It may be assumed that the heat dissipated is equal to the plastic work rate $\dot{w} = \sigma_{\text{eff}} \cdot \dot{\varepsilon}_{\text{eff}}$.

In comparison to the steady-state solution with the Eulerian approach, Lagrange's solution can be used for a non-stationary chip-formation application. This approach must be used for the most typical industrial applications such as: tool-workpiece entry or exit; burr-formation simulation; any chip thickness modifications during the machining operation, etc. This approach is more general – but is also much more time consuming in computation. The leading industrial software code [21], is an explicit dynamic, thermo-mechanically coupled finite element modeling package specialized for metal cutting. The principle is pointed out in the following equations [21].

The *balance of linear momentum* is written as:

$$\sigma_{ij,j} + \rho b_i = \rho\ddot{u}_i. \tag{9}$$

3.2.1 Thermal equations

Heat generation and transfer are handled via the second law of thermodynamics. A discretized weak form of the first law is given by

$$C\dot{T}_{n+1} + KT_{n+1} = Q_{n+1}, \tag{10}$$

where T is the array of nodal temperatures,

$$C_{ab} = \int_{Bt} c\rho N_a N_b \mathrm{d}V_o \tag{11}$$

is the heat capacity matrix

$$K_{ab} = \int_{B0} D_{ij} N_{a,i} N_{b,j} \mathrm{d}V \tag{12}$$

is the conductivity matrix

$$Q_a = \int_{Bt} sN_a \mathrm{d}V + \int_{B\tau q} hN_a \mathrm{d}S \tag{13}$$

is the heat source array, with h having the appropriate value for the chip or tool. In machining applications, the main sources of heat are plastic deformation in the bulk and frictional sliding at the tool-workpiece interface.

3.2.2 Constitutive model and material characterization

In order to model chip formation, constitutive modeling for metal cutting requires determination of material properties at high strain rates, large strains and short heating times and is quintessential for the prediction of segmented chips due to shear-localization. The increase in flow stress due to strain rate sensitivity is accounted for with the relation

$$\left(1 + \frac{\dot{\varepsilon}^p}{\dot{\varepsilon}_0^p}\right) = \left(\frac{\bar{\sigma}}{g(\varepsilon^p)}\right)^{m_1} \tag{15}$$

where $\bar{\sigma}$ is the effective Mises stress, g the flow stress, ε^p the accumulated plastic strain, $\dot{\varepsilon}_0^p$ a reference plastic strain rate, and m_1 is the strain rate sensitivity exponent. A power

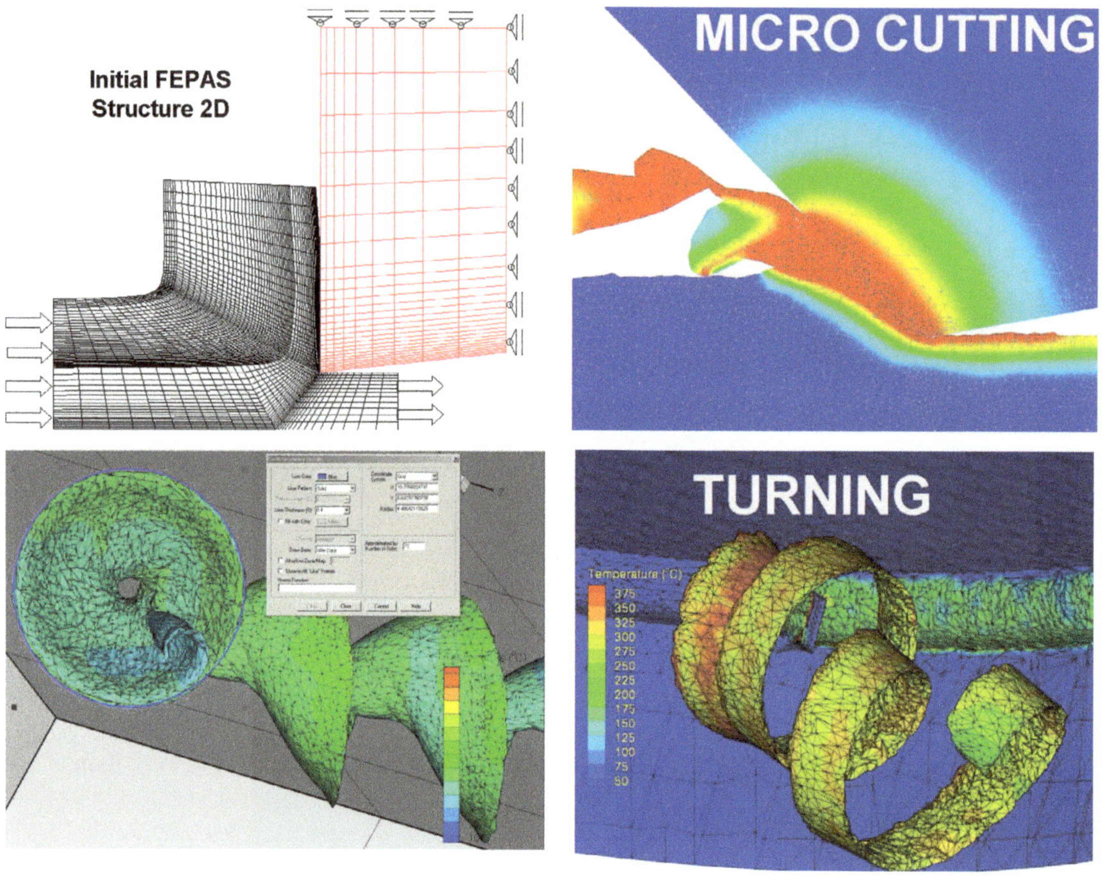

Figure 6. From EULER 2D up to LAGRANGE 2D/3D simulation – examples.

hardening law model is adopted with thermal softening which gives

$$g = \sigma_0 \Theta(T)\left(1+\frac{\varepsilon^p}{\varepsilon_0^p}\right)^{1/n}, \tag{16}$$

where n is the hardening exponent, T the current temperature, σ_0 is the initial yield stress at the reference temperature T_0, ε_0^p the reference plastic strain, and Θ (T) is a thermal softening factor ranging from 1 at room temperature to 0 at melt and having the appropriate variation in between.

Next Figure 6 indicates the grand step forward in industrially driven machining simulations and application.

In addition to this "single" numerical philosophy (EULERIAN or LAGRANGE), the Arbitrary Lagrangian – Eulerian (ALE) technique combines the unique features of Lagrangian and Eulerian formulations, and adopts an explicit solution technique for fast convergence [19]. While the ALE technique with Eulerian boundaries requires a predefined chip geometry assumption, chip formation is free of any geometrical assumption in ALE with Lagrangian boundary conditions [20].

Considering industrial demands, typical 3D-computation times for a drilling operation of more than three month (in the year 2000) have been changed to a few hours by a new numerical algorithm and, above all, due to parallel computing.

Recently [21] the parallel computing capability in chip-formation simulation has been extended to up to 38 processors. In the case of time consuming drilling simulation, the total elapsed time can be reduced from about 119 h to about 7 h. Thus, batch jobs can run overnight with the results being available on the next morning (Figure 7).

For coating-substrate simulations, this method has been developed and applied by [22–27] in the last century. The most time and memory consuming part of the numerical program is the solver for the linear FEM equation systems. Iterative methods should be preferred, because direct solvers have bad error propagation and, due to the unavoidable fill-in, require too much storage when applied to large equation systems. Moreover, the FEM matrices' large condition number, mainly caused by the inevitable high mesh graduation, requires special pre-conditioning techniques, which is why the Pre-Conditioned Conjugate Gradient Method (PCCG) has been applied for the FEM solver. Additional coarse-mesh pre-conditioning has substantially reduced the necessary number of iterations. The accuracy of FEM simulations on standard hardware is basically limited by the available storage and the acceptable computing time. The application of parallelised software on a distributed memory computer with N processors reduces the computing time to less than the Nth part and increases the available storage by almost N times. The software developed for the simulation of coating-substrate-systems on cutting tools is scalable to any number of processors and portable to a variety of hardware and operating systems. The dominating programming model for parallel FEM applications is the Domain Decomposition Method (DDC; Figure 8). For N processors, the simulation

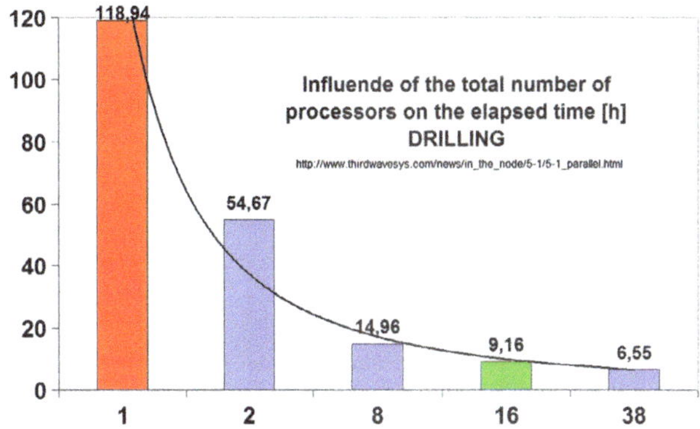

Figure 7. Example for the elapsed time reduction in parallel computing.

domain e.g. the tool's geometrical model, is divided into N disjunctive sub domains and each processor does all the computational work related to "it's" sub domain, with some internal communication for data exchange and control purposes. Based on this parallel FEM software, deformations, strains and stresses caused in coated tools by external loads can be computed on a microscopic scale and become available inside and between the coating layers. Based on computation results with this parallel FEM software, a cutting insert test (workpiece material: GGG-60) with an adopted new coating-substrate system [28] increased the lifetime of the tool by up to 146%.

3.3 Complementary methods

Supplementary methods that are used successfully in other scientific applications have been transferred to machining simulations. Figure 9 point out the commonly used complementary methods.

Boundary Element Method (BEM), Finite Variation Method (FVM) and Finite Difference Method (FDM) are used for linear elastically and heat conductivity problems.

The boundary element method is a numerical technique to solve partial differential equations (PDEs) of a variety of physical problems with well defined boundary conditions. The PDE over a problem domain is transformed into a surface integral equation over the surfaces that enclosed the domain. This integral equation can be solved by discretization of the surfaces into small patches – the boundary elements.

The transformation into a surface integral is not possible generally. There is not known a global way to transform the complex thermo-mechanical equations describing the chip and burr formation process to a surface integral equation. Therefore the BEM has been used only for sub problems in machining simulation e.g. for the thermal problem in steady state chip formation [29].

New mathematical methods for the coupling of finite elements and boundary elements for problems in elastoplasticity and for contact problems with friction which yield variational inequalities were developed and applied at the Institute of Applied Mathematics in Hannover [30].

A few years ago [31], a previously developed numerically code based on the Finite Point Set Method (FPM) (or sometimes called the Grid Free Method or Smoothed Particle Hydrodynamics – SPM) [32], has been applied for a simple 2D chip formation simulation (Figure 10). The main advantage could be to delete the second numerically step in Figure 10 – normally used in FEM. For a "relatively good accuracy" [33] the CPU time can be reduced up to approximately 2.75 times less cost compared to the Eulerian model. The basic equations follow the Hydrodynamic's:

$$\frac{\mathrm{d}}{\mathrm{dt}}\rho = 0 \quad \Rightarrow \quad \nabla_V = 0,$$
$$\frac{\mathrm{d}}{\mathrm{dt}}v = -\frac{1}{\rho}\nabla p + \frac{\eta}{\rho}\cdot \Delta v + g,$$
$$\frac{\mathrm{d}}{\mathrm{dt}}T = -\frac{1}{\rho \cdot c}\cdot \nabla(k \cdot \nabla T).$$

Due to missing friction conditions between chip and cutting tool and a very simple material law, the development was stopped [31]. A recently published new paper [34] did not present any substantial progress in this direction. The authors focused the further development in: "The FPM for the simulation of chip formation still requires considerable development effort however. Thus the algorithms for the calculation of the material behaviour and for the consideration of various boundary conditions such as friction, inertia and heat transfer will be developed further" [34]. In addition, other interesting papers have been published recently [35, 36].

Molecular Dynamics simulation applications for cutting has been developed in the earlier 90th of the last century [37–39] and are commonly used for special material removal, such as mono crystalline silicon [40].

4. Modelling and simulation in the industry – selected examples

4.1 Cutting tool optimisation

State-of-the art in the leading cutting tool companies is the daily job for tool design and tool optimization [41–44]. Investigations are already focused on: the tool tip geometry, the coolant bore design, cutting tool stability, to the chip-geometry design, mainly for deep-hole drilling, the influence of cutting tool material in relation to the workpiece material, etc.

4.2 High-speed cutting or high-performance cutting

The influence of high speed or high loads on the stability of the cutting process and the quality of the new surface is widely investigated. There are interesting solutions available due to the holding of special International Conferences [45–47].

4.3 Dry machining

From the modelling and simulation point of view is the manufacturing without coolants or lubricants a very welcome

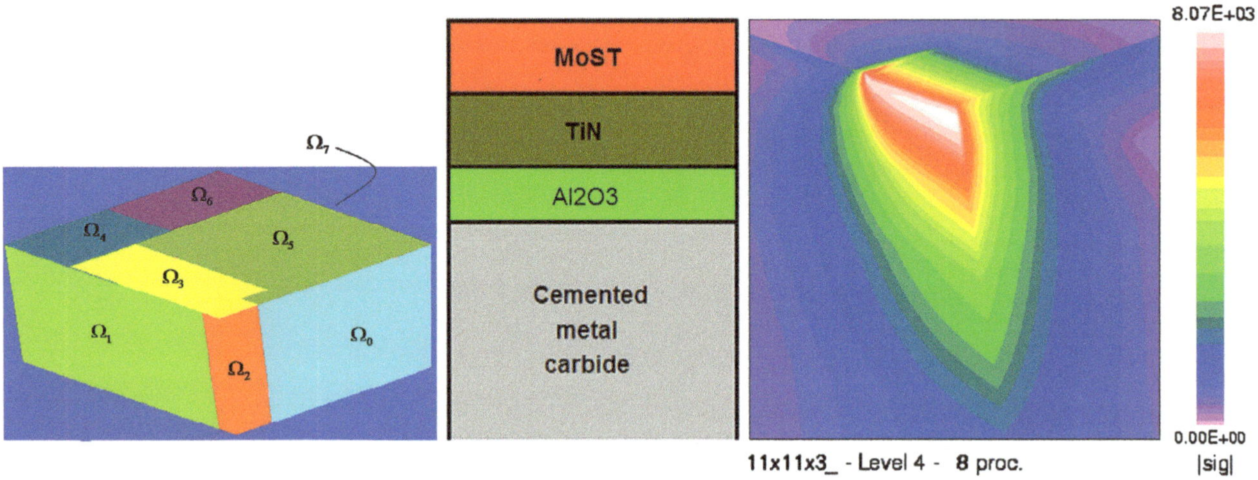

Figure 8. DDC Method applied to coating-substrate optimisation.

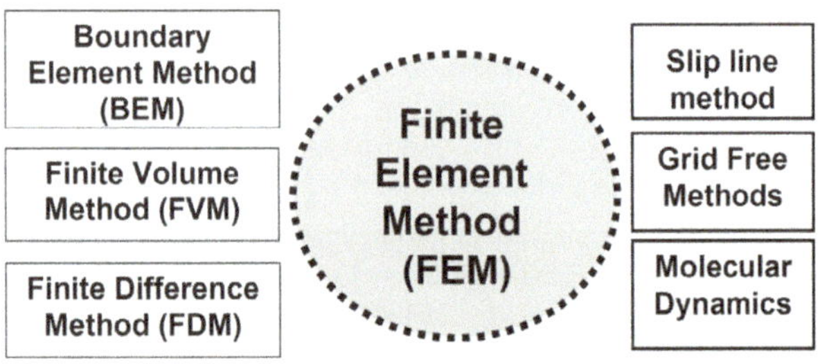

Figure 9. Methods for machining simulations.

task. The boundary conditions of the cutting model are simpler, because no special friction conditions between the chip and cutting tool (rake face, etc.) must be included in to the calculation. Due to the fact, that a lot of cutting operations are not stable enough in dry conditions, leading software tools has included cooling options for coolants and lubricants.

4.4 Burr-formation and clean manufacturing

The best choice is: *If you do not make the burr, you do not have to remove it.* In most cases this philosophy is not applicable in the industrial environment. With modelling and simulation, the generation of burrs can be investigated and good cutting conditions (including cutting tool, coatings, speed, feed, tool geometry, etc.) can be find out to prevent burrs.

In the earlier 90th a quantitative model of burr formation for ductile materials which does not include fracture during orthogonal machining has been proposed [48]. The burr formation mechanism is divided into three parts – initiation, development and formation of the burr with appropriate assumptions. A good overview to the state-of-the art was published recently [49]. Based on the new mechanical principle of the "hydrostatic bowl" [50], the mechanism of burr formation can be better understood.

The hydrostatic bowl as a highly negative hydro-static pressure region immediately below the cutting edge in the workpiece (Figure 11), which acts as a slipstream, may have a positive effect on burr formation as well as on the burr shear zone, assuming that chip and burr will tear off there. A bigger hydrostatic bowl leads to a better burr. The downscaling influences the size and shape of the hydrostatic bowl [51]. The size of the pressure region increases in the course of the time up to the moment when the tool crosses the left edge of the workpiece. The hydrostatic bowl bigger the smaller is the rake angle. Hence there is an obvious influence of the rake angle on the size and depth of the pressure region [52]. In addition, there are additional investigations, concerning the failure criteria [53] to simulate burr-formation, Ultrasonic assisted cutting [54], Modification of the local material properties to prevent burrs [55]. Burr formation and clean ability are closely connected. In addition to prevent burrs in manufacturing, some new methods to increase the clean less within the machine tool has been developed [56]. With standard coatings (modified DLC) as well as new coatings based on nanoscale metallic fluoride powders, the amount of stuck-on chips at the sheet metal surface in a machine tool after the machining of steel and aluminium alloy was reduced by up to 45% and in dry machining with Nanocer Coatings by up to 94%. Geometrical cutting tool design influences burr formation [57–59].

4.5 Cryogenic machining

The most emerging needs of the modern metal cutting operation are to increase the material removal rate with better surface finish, high machining accuracy and acceptable tool life. These objectives can be achieved by reducing the cutting temperature in the cutting zone. In metal cutting, high cutting temperatures are generally reduced by proper selection of process parameters, proper selection and application of coolants and lubricants and using heat- and wear-resistant cutting tool materials – including special coatings. Conventional coolants fail to provide desirable control of cutting temperature in the cutting zone and also they create some techno-environmental problems such as disposal and dermatological problems to the user. These problems were eradicated using liquid nitrogen as a coolant in

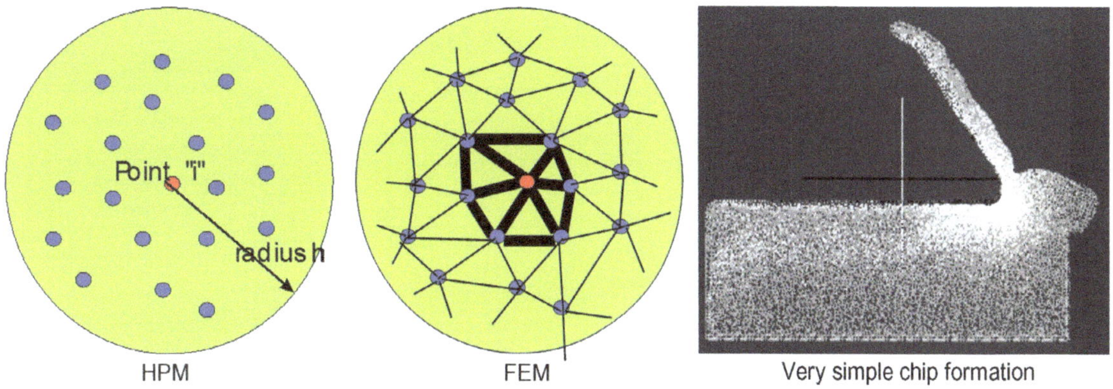

Figure 10. Principles of FPM and a simple example for cutting.

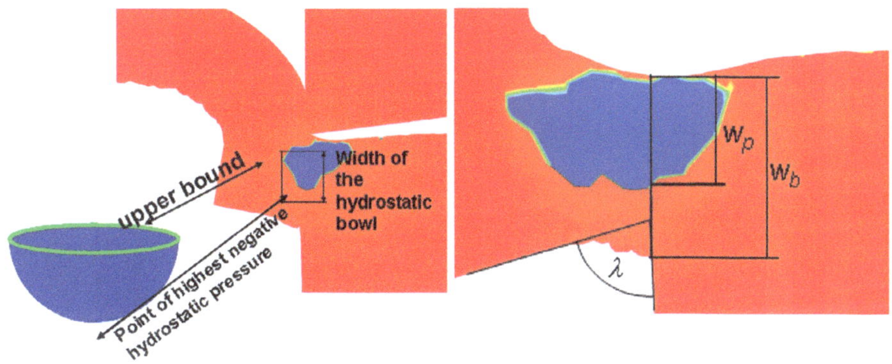

Figure 11. Definition of the "hydrostatic bowl" – applied to burr formation.

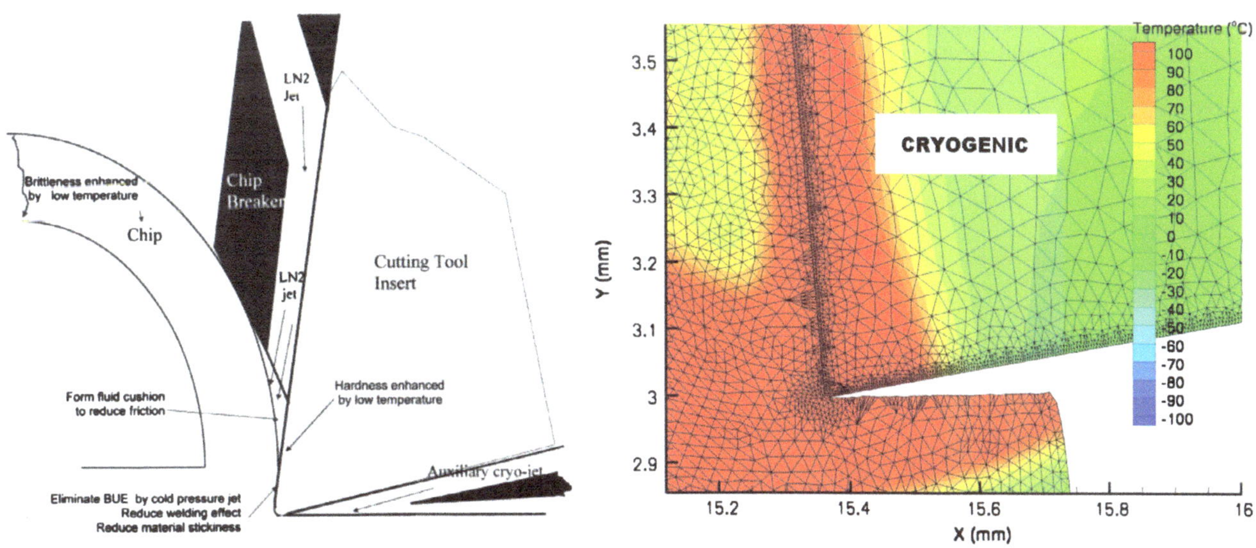

Figure 12. Left the model [60] and right the temperature distribution around the cutting edge with cryogenic cooling.

cryogenic machining due its excellent cooling ability along with environmental friendliness [60–66]. Cryogenic titanium machining can increase cutting-tool life by up to a factor of 10 and double the material-removal rate, compared to conventional machining methods in certain applications [67]. Modelling and Simulation of the chip formation process helps to identify the best available parameters and help to reduce the number of shop floor tests (Figure 12).

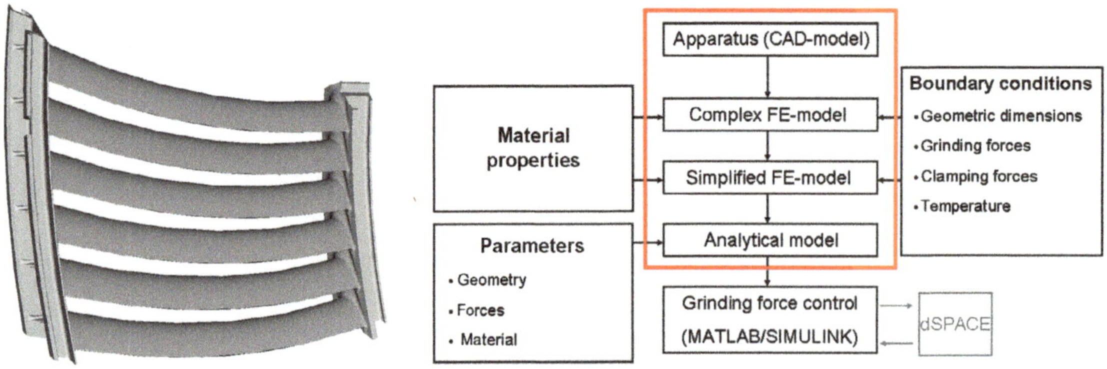

Figure 13. Application of closed-loop-simulation on a nozzle-guide-vane part.

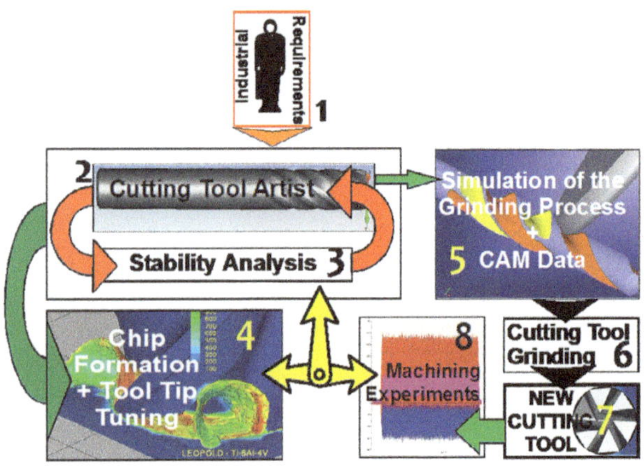

Figure 14. Closed-loop design of cutting tools.

5. Open issues

5.1. Hybrid modelling and closed-loop-design

Hybrid Modelling is the combination (or extension) for any combination of analytical, numerical, experimental or communication engineering methods.

Well known approach is the combination of the slip-line approach with FEM simulations [68], the hybrid method FEM simulation with the visoplasticity [69–73] and previously discussed modelling techniques together with AI-based methods [74].

This philosophy has been introduced as "house of models" [2]. The most important deficit of knowledge is already in the system approach of modelling and simulation. As discussed in the previous sections, there is an exceptionally great step-forward in physically based modelling and simulation for single solutions. The next step forward is expected to be for a closed-loop-design. There are a few examples, as follow.

5.1.1 Clamping devices and machining

The reduction of overall-costs, lead-times and improvement of product quality is one of the main goals in manufacturing today. At the same time innovative materials, including titanium alloys, composites and metal-composite sandwiches, are being used for assembling lighter and increasingly more complex parts. In manufacturing processes, fixturing is used for accurate location and tight holding of the workpiece. The ability to establish and secure desired positions is significant in assessing the effectiveness of a fixture system. Further, product differentiation increases the need of highly efficient, flexible, accurate and automatic fixing systems in industry. Especially in the aerospace and automotive sector the number of low volume, high value and difficult-to-handle products is increasing. Specialised fixtures are economic only within mass production and cannot be reconfigured easily, otherwise the design and construction of modular fixtures are very time consuming and expensive, with problems in repeatability and positioning accuracy. Furthermore, quality and precision cannot be controlled and influenced actively during the manufacturing process and during loading/unloading. Thus there is a need of active, efficient, flexible and precise fixturing systems, which are fields of research in many sectors of industry nowadays. As a basis of all development processes, the forces acting upon the fixing components during the manufacturing process have to be analysed. An iterative improvement of fixture prototype designs is still common and causes high development costs.

Consequently, an effective modelling and simulation method for the prediction of the workpiece-fixture behaviour during machining is required, which would also enable the possibility of fixture optimisation. Therefore, the finite element method (FEM) is an adequate tool to be used for the simulations.

In [75–77] a complex FEM-model with a huge number of elements has been created followed by a parametric model with similar mechanical properties but fewer elements. The usability of this closed-loop method has been revealed by forecasting the transverse reaction forces acting on knee levers fixing a Nozzle-Guide-Vane (NGV) during shape grinding. With continuum-mechanical principles and a parametric FEM model the calculation time has be reduced dramatically (Figure 13).

5.1.2 Closed-loop cutting tool design

The current engineering of tools often proceeds as follows: a model of the tool is developed and a prototype is ground and analyzed. Results are then evaluated by development engineers together with costumers who make appropriate changes to the tool geometry. A new prototype is developed, and the cycle is repeated. Each single iteration demands high costs in terms

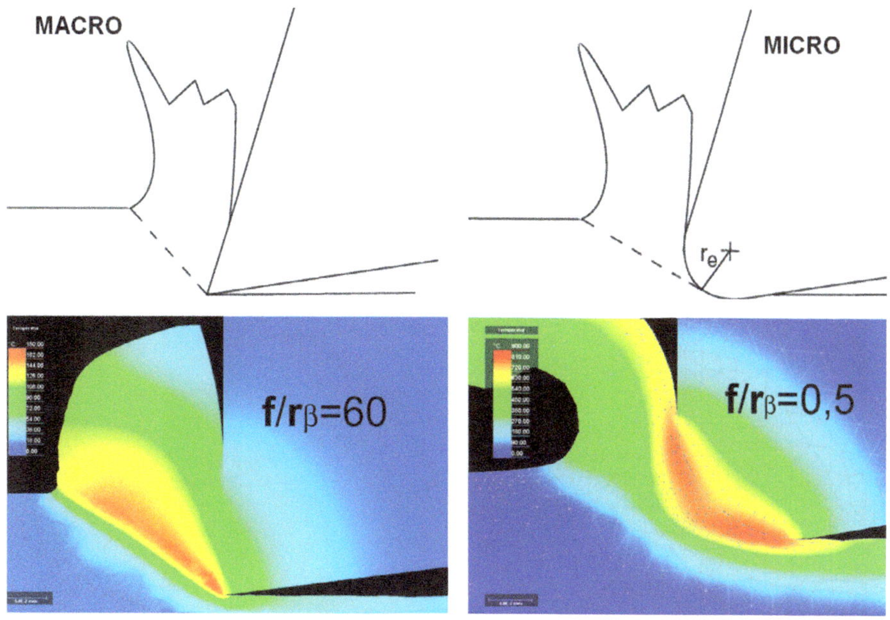

Figure 15. Geometrical differences.

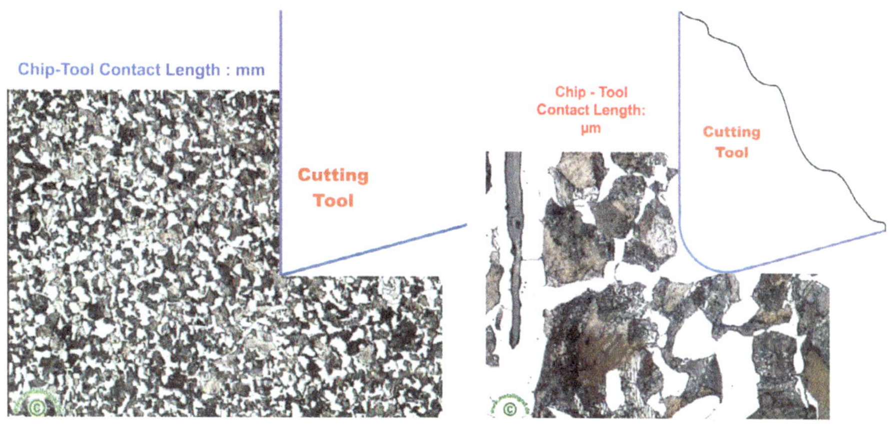

Figure 16. Material differences.

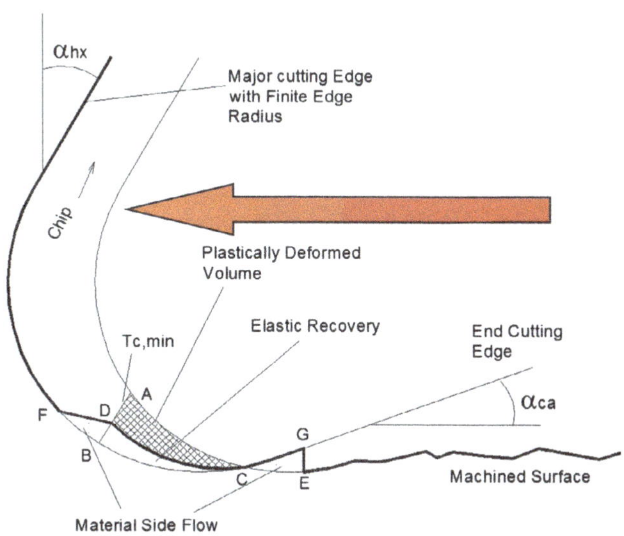

Rake face	Roughness R_z [µm]
Polished surface	0,05
Delivery quality	3,09 ... 3,89
TiAlN - coating	4,23 ... 4,80
TiN - coating	4,42 ... 5,47
Grinded surface	3,30 ... 3,91

Figure 17. Differences in frictional influences near the cutting edge.

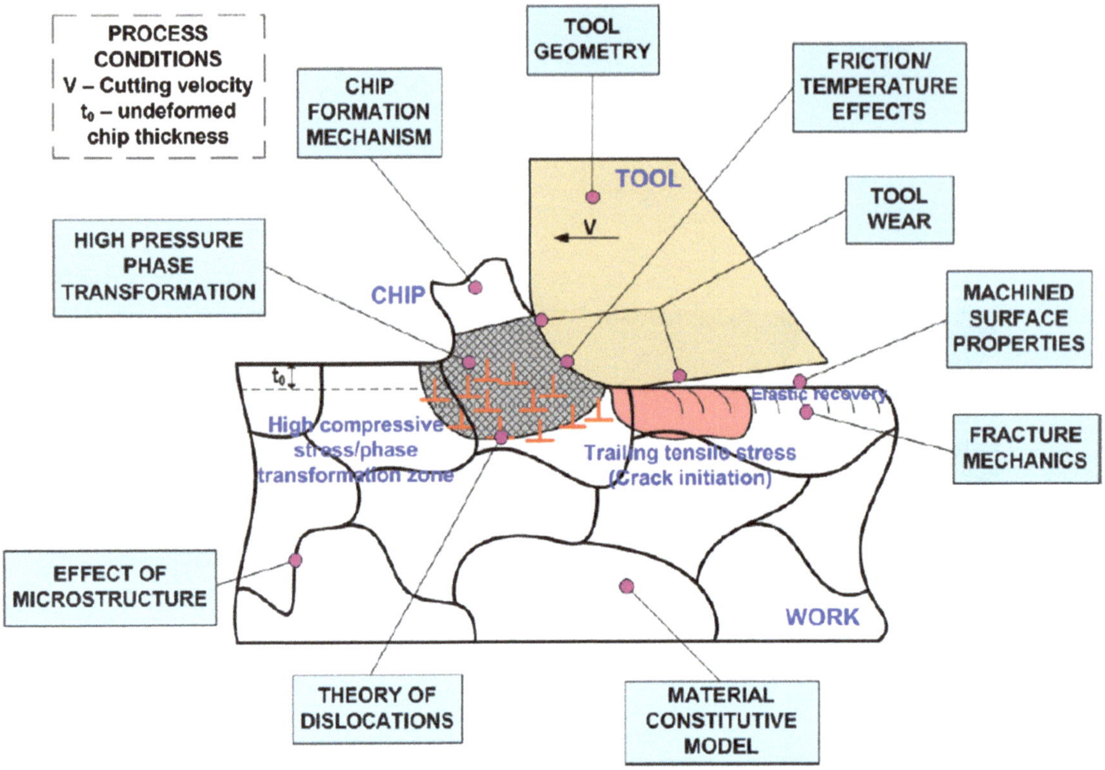

Figure 18. Overall modelling task in micro cutting.

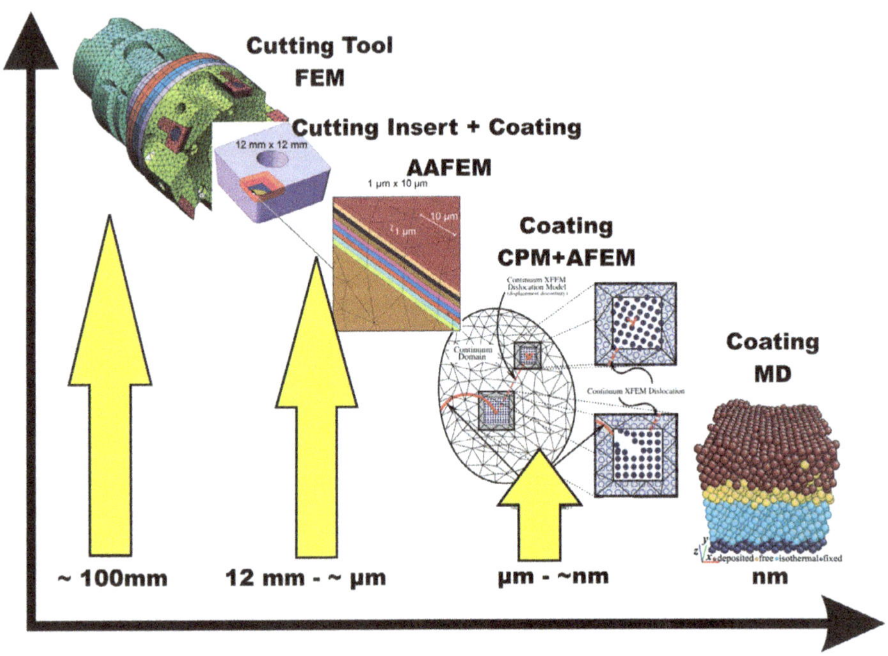

Figure 19. Top-down concept for a multiscale modelling of coating-substrate systems [94].

of personnel, materials and machine time. For a special type of cutting tool applied in high-performance cutting operations, an integrated simulation system has been developed and tested (Figure 14).

The closed-loop system enables users to relocate the prototyping process from the machine tool onto a personal computer, keeping machine tests to a minimum. This system has been tested for the multivalent applicable ratio milling tool.

The time-to-market was strongly reduced and the same methodology can be used for other types of cutting tools [79].

5.2 Multiscale modelling in machining

Going back into the history, micromachining was and is an important topic within manufacturing [80–87].

The modelling task differs from other metal cutting simulation methods in the following main areas (geometry, material structure; friction; dislocations, etc., Figures 15–17).

It must be pointed out, that up to now there has been no final model or simulation code available to investigate micro cutting with the topics presented in Figure 18. Investigations into the material flow with the micro visioplasticity method have been used to verify numerical investigations – but they are very time consuming [88]. In addition, there are a lot of investigations dealing with the application of molecular dynamics – in combination with FEM simulation [89].

5.3 Multiscale modelling in coating-substrate simulation

Following Figure 19, the geometrical differences between the well known boundary conditions at the cutting tool clamping area, in the machine tool spindle and at the chip-tool interaction area at the coating-substrate-system are about 100 mm to 1 nm.

The main four levels for investigation of Coating-Substrate-Simulation (CCS) are:

- the macroscopically level – continuum mechanics > Macro FEM,
- the microscopically level – continuum mechanics > Micro FEM,
- the material science based level > Quasi Continuum (QC) method Crystal plasticity and Atomistic FEM,
- the atomic-scale level > Molecular Dynamics and "First-principle Methods".

The commonly used method is the well known finite-element-method for the macroscopically and microscopically coating-substrate investigation. The basic research at this time is focused to develop the methods separately on one hand and finding methods to close the gap between these methods on the other hand (Figure 19).

First steps to close the gap between the four levels of simulation have been presented [8].

The goal of the European Project [90, 91] was to develop an integrated multiscale modelling approach to link molecular dynamics crystal plasticity and continuum mechanics modelling activities for the applications.

A new Advanced Adaptive Finite Element code [92] has been used to investigate the continuum mechanics modelling of typical cutting tools.

Molecular Dynamics (MD) techniques have been applied to model atom deposition processes and the atomic structure and interfaces to achieve optimal coating microstructures. The MD method was used to simulate the formation of lattice structures of coatings and layer interfaces in thermo-chemical processes and the results were used as input for nano- and micro-mechanics modelling for defining material structures.

Continuum diffusion models are being used to simulate such processes as plasma nitriding, plasma carburising and plasma nitro carburising. These surface modification processes have been widely used in the design of multilayered surface systems to improve their load supporting capacity. The outcomes have been used to define material structures as inputs for mechanics analysis.

New physically based constitutive equations based on evolving micro structural state variables (e.g. dislocation density) were developed to predict damage. These, incorporated with the newly developed controlled Poisson Voronoi tessellation technique for generating virtual grain structures, enable the load bearing capacity of polycrystalline films and multilayers to be predicted.

In parallel, mesoscopic (called atomic) and crystal plasticity FEM modelling capabilities and facilities for predicting deformation and failure of individual layers of nano-/micro-coatings have been applied. The "atomic" lattice model is solved by finite element methods. Elements are used like bars or trusses.

The behaviour of the bar is described by an interatomic potential. This enables an atomic lattice to be modelled by a truss-like assemblage of members that deform according to a prescribed interatomic potential, which governs their relative separations under applied loading. A key feature of the atomic finite element method is that it results in a finite element equation with displacement degrees of freedom, hence an atomic finite element mesh can be directly coupled to other element types and larger scales. Discrete dislocation models have also been used to model at scales between the atomic and crystal scales, with concurrent coupling. The lattice structure of coatings, atomic mismatch and dislocations at layer interfaces modelled at the atomistic level is used for the FEM models. Therefore, an integrated technique to determine mechanical properties of nano- and micro-scale coatings have been developed [93–98].

6. Summary

This paper presents a comprehensive summary of the state-of-the-art developments in the modelling of machining operations with special consideration of the application of models and simulation techniques in the daily-job in an industrial environment as well as an outlook to next steps in modelling. As seen, significant advances have been made in developing advanced computational tools and analytical methods for a physically based modelling at the single process level and its application in the industry. There is a lack in modelling of closed-loop design, which could be one of the main interesting topics within the next decade. Some other interesting aspects such as friction in cutting, residual stress topics, part distortion due to cutting and initial residual stresses due to previous manufacturing steps, cutting tool wear task, material law design and aspects of machining stability has been excluded in this review – but are actually existing in the literature [20, 99, 100].

References

1. R.W. Ivester, et al., Assessment of machining models: Progress report, Mach. Sci. Technol. 4 (2000) 511–538.
2. J. Leopold, Proceeding's of the 3rd Int. Workshop on Modelling of Machining Operations, Sydney, 2000.
3. H. Weber, T.N. Loladze, Grundlagen des Spanens, VEB Verlag Technik, Berlin, 1986.
4. I. Time, Soprotivlenie Metallov i Dereva Rezanyu, St. Petersburg, 1870.
5. M.W. Merchant, Mechanics of metal cutting process. I. Orthogonal cutting and a type 2 chip, J. Appl. Phys. 16 (1945) 267–275.
6. H. Tresca, Mémoire sur le rabotage des métaux, Paris, 1877.
7. A. Mallock, The action of cutting tools, Proc. Roy. Soc. London 33 (1881) 127–139.
8. K.A. Svorykin, Rabota i Usilie Neobkhodimyya dlya Oteleniiya Metallichcskikh Struzhek, Moscow, 1883.
9. K. Zuse, http://zuse.zib.de/collection/aseFhOEB5XjjvE2c, 1938.
10. O.C. Zienkiewicz, Y.K. Cheung, The Finite Element Method in Structural Mechanics, McGraw-Hill, London, 1967.
11. H. Weber, et al., Wiss. Schriftenreihe der TH Karl-Marx-Stadt, 1986, 11, ISSN 0323–6374.
12. J. Leopold, U. Semmler, K. Hoyer, Applicability, robustness and stability of the Finite Element analysis in metal cutting operations, Proc. CIRP Workshop on Modelling of Machining, Nantes, 1999.
13. J. Leopold, Wiss. Schriftenreihe der TH Karl-Marx-Stadt, 1, 1980, ISSN 0323-6374.
14. H. Weber, J. Leopold, Wiss. Schriftenreihe der TH Karl-Marx-Stadt, 1981, ISSN 0323-6374.
15. H. Weber, J. Leopold, Acta Technica Academia Scientiarum Hungaricae 86 (1978) 287–300.
16. G.R. Johnson, W.H. Cook, Proc. of the 7th International Symposium on Ballistics, Netherlands, 1983, pp. 541–547.
17. F.J. Zerilli, R.W. Armstrong, Dislocation-mechanics-based constitutive relations for material dynamics calculations, J. Appl. Phys. 61 (1987) 1816–1825.
18. E. Usui, K. Hoshi, Slip-line fields in metal machining which involve centered fans, Proc. of Int. Production Engineering Research Conference, Pittsburgh, ASME, 1963, pp. 61–71.
19. P.J. Arrazola, T. Özel, Investigations on the effects of friction modeling in finite element simulation of machining, Int. J. Mech. Sci. 52 (2010) 31–42.
20. P.J. Arrazola, et al., Recent advances in modelling of metal machining processes, CIRP Ann. – Manuf. Technol., 2013.
21. http://www.thirdwavesys.com/
22. J. Leopold, et al., 8th Int. Workshop on Computational Mechanics of Materials, Stuttgart, October, 1998, pp. 8–9.
23. J. Leopold, M. Meisel, 9th Int. Workshop Computational Mechanics and Computer Aided Design of Materials, Berlin, October, 1999, pp. 4–5.
24. J. Leopold, M. Meisel, Comput. Mater. Sci. 19 (2000) 205–212.
25. J. Leopold, et al., Surf. Coat. Technol. 142–144 (2001) 916–922.
26. J. Leopold, M. Meisel, Report on HPS-CSS (HPS-CSS has been supported by the ESPRIT HPCN programme from the European Commission), 2001 (unpublished).
27. H. Oosterling, et al., Low friction, $M0S_2$-composite coated cutting tools for dry, high speed machining of steel; LoFriCo Final technical report – BRST-CT98-5361, 2001 (unpublished).
28. J. Leopold, Mechanical and thermal behaviour of coating-substrate-systems investigated with parallel FEM, Proc. Int. Conf. on Metallurgical Coatings and Thin Films – ICMCTF 2002, San Diego/USA, April 22–26, 2002.
29. L.C. Cho, C. Abhijit, A boundary element method analysis of the thermal aspects of metal cutting processes, Trans. ASME, J. Eng. Ind. 113 (1991) 311–319.
30. E.P. Stephan, Coupling of boundary methods and finite element methods, E. Stein, R. de Borst, T.J.R. Hughes (Eds.), Encyclopedia of Computational Mechanics, Chapter 13: Fundamentals, Vol. 1, John Wiley & Sons, 2004.
31. J. Kuhnert, A. Mattes, J. Leopold, Internal report, FhG IWU; FhG ITWS and TU Berlin, 2005.
32. J. Kuhnert, A. Tramecon, P. Ullrich, Proc. of the EUROPAM Conf., 2000.
33. S.S. Akarca, W.J. Altenhof, A.T. Alpas, Proceedings of the 10th International LS-DYNA Users Conference, 2008.
34. E. Uhlmann, et al., Proc. of the CIRP Conf. on Modelling of Machining Operations, Donostia-San Sebastian, pp. 145–151, (2009).
35. V. Gyliene, V. Ostasevicus, M. Ubartas, Proc. of the 9th European LS-DYNA Conf. 2013 and Proceedings of the 12th CIRP Conference on Modelling of Machining Operations, 145–151, 2009.
36. C. Espinosa, et al., Proc. of the 10th Int. LS-DYNA Users Conference, 2008.
37. N. Ikawa, et al., Ann. CIRP, 40 (1991) 551–554.
38. S. Shimada, et al., Feasibility study on ultimate accuracy in microcutting using molecular dynamics simulation, Ann. CIRP 42 (1993) 91–94.
39. R. Rentsch, I. Inasaki, Ann. CIRP 44 (1995) 295–298.
40. H. Tanaka, S. Shimada, L. Anthony, Requirements for ductile-mode machining based on deformation analysis of mono-crystalline silicon by molecular dynamics simulation, Ann. CIRP 56 (2007) 53–56.
41. J. Leopold, Werkzeuge für die Hochgeschwindigkeitsbearbeitung, HANSER, 1999.
42. J. Leopold, et al., Fliehkraftverhalten von Feinbohrwerkzeugen bei hohen Drehzahlen, VDI-Z III (1996) 48–50.
43. J. Leopold, et al., Festigkeits- und Verformungsanalyse von Werkzeugen für die Hochgeschwindigkeitsbearbeitung, DIMA 12 (1997) 33–39.
44. J. Leopold, G. Schmidt, A. Kieninger, FEM-Analyse modular aufgebauter HSC-Werkzeuge mit 3-D einstellbaren Schneiden, DIMA 3 (2000).
45. http://www.hsm.tu-darmstadt.de/index_hsm/conference_program_hsm/conferenceprogram_hsm.de.jsp
46. http://www.rcmt.cvut.cz/hsm2014/en/
47. http://hpc2014.berkeley.edu/
48. D.A. Dornfeld, S.L. Ko, A study on burr formation mechanism, Trans. ASME. J. Eng. Mat. Technol. 113 (1991) 75–87.
49. J. Leopold, Prediction and verification of models of burr formation, Int. J. Mater. Product Technol. 35 (2009) 89–117.
50. A. Freitag, C. Sohrmann, J. Leopold, Simulation of burr formation, Proc. of the 8th CIRP Int. Workshop on Modeling and Machining Operations (2005) 641–650.

51. A. Stoll, J. Leopold, R. Neugebauer, Hybrid methods for analysing burr formation in 2D-orthogonal cutting, Proc. of the 9th CIRP Int. Workshop on Modelling of Machining Operations, Bled, Slovenia, May 11–12, 2006.
52. J. Leopold, G. Schmidt, K. Hoyer, A. Stoll, Modelling and simulation of burr formation – State-of-the-art and future trends, Proc. of the 8th CIRP Int. Workshop on Modelling of Machining Operations, Chemnitz, Germany, May 11–12, 2005.
53. J. Regel, A. Stoll, J. Leopold, Numerical analysis of crack propagation during the burr formation process of metals, Int. J. Machining and Machinability of Materials, 54–68.
54. A. Stoll, N. Ahmed, A.V. Mitrofanov, V. Silberschmidt, J. Leopold, Influence of ultrasonically assisted cutting on burr formation, Proc. of the 9th CIRP Int. Workshop on Modelling of Machining Operations, Bled, Slovenia, May 11–12, 2006.
55. M. Dix, R. Leopold, Investigations on the influence of local material properties of burr formation, Proc. of the 10th CIRP Int. Workshop on Modeling of Machining Operations, Reggio Calabria, Italy, August 27–28, 2007.
56. J. Leopold, A. Mucha, Non-stick coating for clean manufacturing – cleanability in high-performance cutting, Proc. of the Conf. NANOFAIR, Karlsruhe, 2006.
57. J. Leopold, R. Neugebauer, M. Löffler, M. Schwenck, P. Hänle, Influence of coating-substrate-systems on chip and burr formation in precision manufacturing, Proc. IMechE Part B: J. Eng. Manuf., 219 (2005) 1–8.
58. J. Leopold, T. Matsumura, Modelling of burr formation of coated-cutting tools for clean manufacturing, Proc. of the 5th Int. Conf. on Leading Edge Manufacturing in 21st Century – LEM21, Osaka, 2009.
59. T. Matsumura, J. Leopold, Simulation of drilling process for control of burr formation, Journal of Advanced Mechanical Design, Systems and Manufacturing 4 (2010) 966–975.
60. S.Y. Hong, I. Markus, W. Jeong, New cooling approach and tool life improvement in cryogenic machining of titanium alloy Ti-6Al-4V, Int. J. Mach. Tools Manuf. 41 (2001) 2245–2260.
61. R. Ghosh, Z. Zurecki, J.H. Frey, Cryogenic machining with brittle tools and effects on tool life, Proc. of IMECE'03, Paper No.: ICMECE2003-42232.
62. M. Dhananchezian, M. Pradeep Kumar, A. Rajadurai, Experimental investigation of cryogenic cooling by liquid nitrogen in the orthogonal machining process, International Journal of Recent Trends in Engineering 1 (May 2009) 55–59.
63. S.C. Jun, Lubrication effect of liquid nitrogen in cryogenic machining friction on the tool-chip interface, Journal of Mechanic, Science and Technology (KSME Int. J.) 19 (2005) 936–946.
64. F. Pušavec, A. Stoić, J. Kopač, The role of cryogenics in machining processes, Technical Gazette 16 (2009) 3–10.
65. F. Pušavec, J. Kopač, Sustainability assessment: Cryogenic machining of inconel 718, Strojniški vestnik, J. Mech. Eng. 57 (2011) 637–647.
66. T. Lu, O.W. Dillon, Jr., I.S. Jawahir, A thermal analysis framework for cryogenic machining and its contribution to product and process sustainability, Proc. of the 11th Global Conference on Sustainable Manufacturing – Innovative Solutions, Berlin, pp. 262–267, 2013.
67. http://www.mag-ias.com/en/mag/technologies.html
68. J. Leopold, W. Arnold, H. Gründemann, Ein integriertes Gleitlinien-Finite-Elemente-Modell zur Spannungsberechnung vor der Scherlinie, Wiss. Z. d. Techn. Hochschule Karl-Marx-Stadt, 21 (1979) 185–189.
69. G. Spur, J. Leopold, G. Schmidt, Ermittlung des Spannungs-, Verformungs- und Temperaturverhaltens spanned bearbeiteter Werkstücke mit Hilfe der Visioplastizität und der Finiten Elemente Methode, Proceedings of the DFG Priority Prpogramm "Wirkflächenreibung bei inelastischer Verformung metallischer Werkstoffe, Hannover, 1995.
70. J. Leopold, et al., Wiss. Z. d. Techn. Hochschule Karl-Marx-Stadt 21 (1979) 185–189.
71. J. Leopold, FEM modelling and simulation of 3D-chip formation, Proc. of the CIRP Int. Workshop on Modelling of Machining Operations, Atlanta, 1998.
72. J. Leopold, G. Schmidt, Proc. of the second CIRP international workshop on modelling of machining operations, Nantes, January 25–26, 1999.
73. J. Leopold, Proc. of the 3rd International Workshop on Modelling of Machining Operations, Sydney, 2000.
74. X.P. Li, K. Lynkaran, A.Y.C. Nee, A hybrid machining simulator based on predictive machining theory and neuronal network modelling, Journal of Material Processing Technology 89–90 (1999) 224–230.
75. M. Klärner, J. Leopold, L. Kroll, Analysis of clamping within a fixing system, Lecture Notes in Computer Science 5315 (2008) 356–367.
76. J. Leopold, et al., Investigations to new fixturing principles for aerospace structures, Proc. of the Int. Conf. on Applied Production Technology, Bremen, 2007.
77. J. Leopold, et al., Interaction between machining and new fixturing principles for aerospace structures, Proc. of ESAFORM, 2008.
78. AFFIX consortium. Affix – aligning, holding and fixing flexible and difficult to handle components, http://www.affix-ip.eu, 2006.
79. J. Leopold, et al., High performance cutting with optimized cutting tools, Proc. of the 4th CIRP Int. Conf. on High Performance Cutting, Gifu, 201.
80. M. Sato, K. Kato, K. Tuchiya, Effect of material and anisotropy upon the cutting mechanism, Trans. JIM. 9 (1978) 530–536.
81. W. König, N. Spenrath, The influence of the crystallographic structure of the substrate material on surface quality and cutting forces in micromachining, Proc. 6th Int. Precision Engng. Seminar, Braunschweig, Germany, pp. 141, 1991.
82. J.D. Kim, D.S. Kim, Theoretical analysis of micro-cutting characteristics in ultra-precision machining, J. Mater. Process. Technol. 49 (1995) 387–398.
83. W.B. Lee, C.F. Cheung, S. To, Characteristics of micro-cutting force variation in ultraprecision diamond turning, Mater. Manuf. Process. 16 (2001) 177–193.
84. W.B. Lee, C.F. Cheung, S. To, A microplasticity analysis of micro-cutting force variation in ultra-precision diamond turning, Transactions of the ASME 124, May (2002), 170–177.
85. A. Simoneau, E. Ng, M.A. Elbestawi, Grain size and orientation effects when microcutting AISI 1045 steel, CIRP Ann. 56 (2007) 57–60.
86. V. Schulze, J. Michna, F. Zanger, R. Pabst, Modelling the process-induced modifications of the microstructure of work piece surface zones in cutting processes, Advanced Materials Research 223 (2011) 371–380.
87. M. Abouridouane, F. Klocke, D. Lung, O. Adams, A new 3D multiphase FE model for micro cutting ferritic-pearlitic carbon steels, CIRP Ann. 61 (2012) 71–74.

88. J.W. Erben, Mikrovisioplastische Untersuchungen an ein- und zweiphasigen metallischen Werkstoffen als Bindeglied zur numerischen Modellierung der Mikrostruktur, Abschlussbericht DFG Le746/17, 1999.
89. G. Schmidt, R. Leopold, R. Neugebauer, FE-simulation of nonlinear dynamical effects in coating-substrate-systems, 4th Int. Symposium: Investigations of Non-Linear Dynamic Effects in Production Systems, Chemnitz, Germany, April 8–9, 2003.
90. FP7 project, "Multiscale Modelling for Multilayered Surface Systems" (M3-2S), Grant No: NMP3- SL-2008- 213600.
91. http://ec.europa.eu/research/industrial_technologies/pdf/modelling-brochure_en.pdf
92. J. Leopold, et al., An advanced adaptive finite element code for coating-substrate simulation, J. Multiscale Modelling 03 (2011) 91.
93. A.V. Byakova, J. Leopold, Effect of stress state on failure resistance of brittle high-strength coatings, Unpublished report, 1996.
94. J. Leopold, R. Wohlgemuth, D. Shan, Y. Qin, Modelling and simulation of coating-substrate-systems: state-of-the-art and future trends, Proc. of the Conference "THEA Coatings", Thessaloniki, 2011.
95. J. Leopold, R. Wohlgemuth, J. Lin, S.V. Subramanian, T. Matsumura, New concepts for micro-structural simulations of coating-substrate-systems, Proc. of the 12th CIRP Conf. on Modelling of Machining Operations, Donostia, San Sebastián, Spain, 1 (2009) 117–124.
96. S. Wang, J. Lin, D. Balint, Modelling of failure features for TiN Coatings with different substrate materials, J. Multiscale Modelling 03 (2011) 49.
97. D.Q. Qin, et al., Prediction of residual stress in multilayered coatings with a linear elastic model incorporating density functional theory calculations, J. Multiscale Modelling 03 (2011) 65.
98. R. Neugebauer, R. Wertheim, U. Semmler, The atomic finite element method as a bridge between molecular dynamics and continuum mechanics, J. Multiscale Modelling 03 (2011) 39.
99. J. Leopold, H. Gründemann, W. Totzauer, Kontinuumsmechanische Methoden zur Modellierung des Spanbildungsprozesses, Sitzungsberichte der AdW der DDR, 12N/1979, 51–81.
100. C.A. Luttervelt, et al., Present situation and future trends in modelling of machining operations, CIRP Ann. 47 (1998) 587–626.
101. E.H. Lee, B.W. Shaffer, The theory of plasticity applied to a problem of machining, Transaction of ASME 73 (1951) 405–413.
102. M.C. Shaw, Metal cutting principles. 3rd ed., MIT Press, Cambridge, 1954.
103. W.B. Palmer, P.L.B. Oxley, Mechanics of orthogonal machining, Proceedings of the Institute of Mechanical Engineers 173 (1959) 623–654.
104. H. Weber, Mechanik der Spanbildung, Wiss. Z. der TH Karl-Marx-Stadt 11 (1969) 597–629.
105. M.G. Stevenson, P.L.B. Oxley, An experimental investigations of the influence of strain rate and temperatures on the flow stress properties of a low carbon steel using machining test, Proc. Inst. Mech. Eng. 185 (1970) 741–754.
106. R. Makino, E. Usui, An analysis of stress and strain distributions in the plastic region of slow speed, steady-state machining, Bull. Japan Soc. of Prec. Engg. 7 (1973) 43–50.
107. N. Fang, I.S. Jawahir, An analytical predictive model and experimental validation for machining with grooved tools incorporating effects of strains, strain-rates and temperature, CIRP Annals – Manufacturing Technology 51 (2002) 83–87.

15

Digitalization of automotive industry – scenarios for future manufacturing

Steven Peters[1,*], Jung-Hoon Chun[2], and Gisela Lanza[1]

[1] Institute of Production Science (wbk), Karlsruhe Institute of Technology (KIT), Karlsruhe, 76131 Germany
[2] Laboratory for Manufacturing and Productivity, Massachusetts Institute of Technology (MIT), Cambridge, MA, 02139 USA

Abstract – Among the current challenges to the global automotive industry are changes in global markets leading to product variety, regulation leading to pressure for new technologies in body and powertrain, and competition by new players such as huge information and communication technology companies. Automotive original equipment manufacturers (OEMs) deal with these issues in different ways. This paper uses the scenario technique to illustrate possible answers to the question, how future value chains of automotive industry will look like. In almost all cases, information and communication technology (ICT) plays a major role in future strategies to cope with the aforementioned challenges. On the one hand, ICT can boost the way to more efficient production of variants by utilizing smart manufacturing approaches, on the other hand ICT enables new features such as autonomous driving.

Key words: Smart Manufacturing, Digitalization, Scenario Technique

1. Motivation

Globally, the automotive industry has sold more than 71 million cars in 2014, up 3.5% compared to 2013 and the fifth record sales year in a row [1]. Nevertheless, the global automotive industry faces multiple challenges such as changes to the market, regulation, and the emergence of new competitors.

1.1. Diversification of regionalized markets

While most European automotive markets remain weak, the US market has recovered from crisis, the Chinese market is slowly growing [2], the Indian market is inconsistent, while Russian and South American market shares are decreasing [2]. Car companies try to succeed in promising markets by applying different global product and production strategies. For example, Volkswagen offers the VW Lavida exclusively for the Chinese market [3] and General Motors developed its brand Buick to successfully appeal to Chinese customers [4]. Some premium brands offer the same worldwide portfolio but have introduced long wheelbase derivatives (even for mid-size cars) and adjusted interior options to reflect regional preferences. Global markets and different customers' needs have led to a huge growth in automotive derivatives leading to complexity in manufacturing.

*Corresponding author: `steven.sp.peters@outlook.com`

1.2. Strict and dynamic regulations

By 2021 the average of the portfolio of an automotive original equipment manufacturer (OEM) has to stay below EU's limit of 95 g CO_2 emission per kilometer [5]. Similar regulations have been imposed in all major markets including China. In Elmau, Germany in June 2015, the G7 states agreed to a zero-carbon emissions economy by 2100. However, car makers are already working very hard to stay below current emissions standards [6]. Besides the ongoing optimization of internal combustion engines, companies' R&D activities focus on alternative powertrains like plug in-hybrids, battery electric vehicles and fuel-cells. Because of uncertainty of future regulations and customers' acceptance of the new technologies, flexibility in the existing car plants becomes crucial.

As important as the powertrain itself are improved aerodynamics and lightweight construction which are also influenced by emissions regulations leading to additional complexity in design, material and production phases. For example BMW has introduced a CFRP-intensive (CFRP = carbon fiber reinforced plastics) body in its BMW i3 (for details see [7] and for sales figures [8]), Ford uses aluminum in extensive ways [9] and several OEMs show complex multi-material-bodies. Again, the influences of those approaches on the production are significant. OEMs have to build up new know-how on processes and materials e.g. for the usage of CFRP [10] and multi-material-bodies change the requirements of the used joining

Table 1. Impacts of the challenges on car and production technology.

Challenge	Impact on cars	Impact on automotive production
Individualization	Many derivatives; platform approaches	Complexity; new processes such as 3D-printing
Regulation	Improved combustion engines; alternative (electric) powertrains; advanced lightweight materials and multi-material-approaches; improved aerodynamics; assistance systems	Micro-structured surfaces and micro-drilling; new processes for manufacturing of batteries, fuel-cells; joining of new materials and material-mixtures; flexible approaches in body-in-white production
ICT	Autonomous driving; connectivity in all dimensions	Smart manufacturing/Industrie 4.0: scalable model-mix-lines, operational excellence, assistance systems for worker

technologies and thus leading to new approaches in the body-in-white production and its automation. Those examples show how manufacturing has changed its role from restrictor to an enabler and how production related issues have to be integrated in the very early phase of design (compare [11]).

1.3. New players and competitors from other industries

The current content of electronics and software of a car is 35% ([2] referring on M. Broy, TU Munich) and electronics systems comprise of more than 90% of innovations and new features of cars [2] such as semiautonomous driving like automatic parking, lane-keeping assistance or sensor-based reporting [2]. Moreover, information and communication technology (ICT) becomes a crucial enabler for more efficient automation and operational excellence inside the manufacturing plants [12]. The authors call this ongoing rise of the importance of ICT in the automotive industry – in both products and production – the "digitalization" of the automotive industry.

Well-known ICT companies from Silicon Valley or China have been entering the car market. Today they are mainly partnering with the automotive OEMs in the fields of in-car infotainment [13, 14]. In addition to those systems, Google has been testing its own self-driving car [15] and rumors about an Apple car have been circulating [16]. Recently there have even been speculations on a joint approach of Apple, BMW and Magna [17]. However, it is questionable whether the ICT giants are really going to enter the automotive OEM business which has a smaller profit margin compared to consumer electronics and Internet business. Regardless of the ultimate business model, ICTs will be a tremendous part of the value of any future premium car which could lead to tough competition, new alliances and business models among established automotive suppliers and the carmakers with ICT companies.

In addition, ICT approaches, namely the "Internet of Things" have become one of the strongest enablers for robust, flexible and efficient manufacturing especially when it comes to increased complexity and variety (compare Sects. 1.1 and 1.2). "Industrie 4.0" [12] or the Industrial Internet [18] is a major trend in worldwide manufacturing. Bosch even expects to achieve 30% efficiency increases by implementing Industrie 4.0 in their manufacturing [19]. With respect to the above mentioned challenges of different markets, customers, and technological complexity, Industrie 4.0 and its core technologies such as secure plug & work-solutions for reconfigurations of machines, augmented reality based assistance devices for workers, cyber-physical systems with inexpensive sensors to automatically collect data in value streams, machines and components as well as machine learning and big data algorithms can be seen as enablers for car makers towards flexible automation, scalable model-mix-factories as well as a boost for the next step of operational excellence and lean production [20].

Table 1 shows the aforementioned challenges and collects their discussed impacts on cars and their production plants. These impacts have the potential to change the value chain of automotive industry in general, which will be analyzed in this paper. On the one hand side there is a need for innovation in almost all fields of the established car business e.g. in engines, in materials and in manufacturing and on the other hand, ICT companies become competitors and question the importance of these technological fields as well as the importance of own value creation. The later point is closely related to the scope of this paper.

2. Literature review

Before introducing and applying the methodology of scenario technique to illustrate potential pictures of the value chains of the future, a short literature review on the importance of an own value creation in established industries is given. The role of own manufacturing will be treated as a major influencing factor in the illustrated scenarios later on.

2.1. Importance of the production depth in literature

Decisions related to the question of which components or products a company should manufacture with own resources or buy on the market are called "make-or-buy"-decisions and are fundamentally important for a company as a whole [21]. More generally speaking, it is the strategic decision on the production depth of a company. The OECD defines the production depth of a nation as an "indicator [which] reflects the share of national production that is created in the country itself" [22]. Here, the production depth of a company (or of one of its products) is defined as the value added content created by a company itself (compare definition of Dürr [23]). Shifting parts of the value creation (or services) which have previously been done inside a company to external suppliers or partners is called outsourcing, while bringing it back is called "vertical integration" (or insourcing).

There are multiple reasons for outsourcing and industrial cases have shown that it can be especially helpful in:

- reducing costs by outsourcing selected activities (catering services or production of parts of lower priority like normed screws) by benefitting from the suppliers' economies of scale,
- speeding up development by creating a close cooperation with specialists benefitting from the suppliers' economies of scope [24],
- enabling innovative features by cross-industry cooperation,
- reducing complexity inside the company and focus own limited resources on market-/brand- and quality-relevant components and features,
- fulfilling local-content requirements (regulated by governments),
- hedging against currency exchange rate fluctuations (by using suppliers in the corresponding regions),
- increasing flexibility (e.g. to keep job-shopper trained as capacity extenders in volatile markets).

Outsourcing has been a major topic since the 1980s [25] and consequently, a lot of research has been carried out to understand the complex influences it might have on a company's (long-term) performance. To get an overview of the focused issues, an analysis was done of the articles listed in Scopus. Figure 1 shows the breakdown of related articles by major effects: cost, innovation, flexibility and quality. Articles were counted which contain in their abstracts the term "outsourcing" and the terms "manufacturing" or "production" and in addition one of the four named effects: cost, quality, innovation and flexibility.

While well-known sources like Stuckey and White [26] warned companies only to vertically integrate "unless it is absolutely necessary to create or protect value", Lacity et al. found that empirical findings on outsourcing are conflicting [27]. Some authors identified negative relationships between outsourcing and indicators like market share when outsourcing reaches a level over a certain optimum [28]. Other studies showed that outsourcing really reduces operational costs [29]. Fuchs and Kirchain showed how wages, yields, downtimes and other parameters are dependent on the location of a factory and demonstrated that with real case data from the optoelectronics industry [30]. They discovered that a geographical separation of design activities and production is difficult in this innovative industry and potentially others using immature processes. Steven et al. presented an empirical study of links between outsourcing and product recalls, in which they used data from different industries and found outsourcing has a negative impact on quality performance, particularly in case of offshoring, outsourcing to a partner abroad [31].

From the perspective of the authors of this paper, literature and industrial cases both have proven that there are good reasons for outsourcing, but the more important the interlinkage between product and production gets, the more difficult it gets to find an optimal balance in terms of quality performance [31], speed to innovate or time-to-market [30, 32].

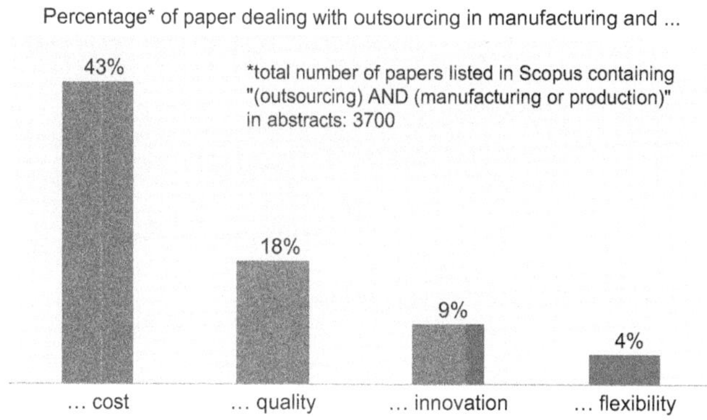

Figure 1. Overview of focused issues in outsourcing literature (data: Scopus).

2.2. Fundamentals of scenario technique

Scenario technique, e.g. implemented by Gausemeier et al. [33], is a well-established method to find feasible perceptions of the future in order to be able to prepare for it – either to avoid risks or to take advantage of chances [34]. It is an instrument used to think about the future in a structured way [33]. Therefore this method takes into consideration an understanding of influencing factors, collects their potential future realizations and finally combines them in multiple futures. Figure 2 illustrates scenario creation steps using Gausemeier et al.'s description. After defining the scenario base, also known as the decision field in its current situation, the methodology is as follows:

- Step 1: Scenario-Field Analysis. In this step, internal and external influencing factors of the decision field, such as that of an automotive OEM are collected using various methods. After brainstorming influencing factors they are filtered to gain a manageable amount of factors (e.g. below 20).
- Step 2: Scenario-Prognostic. For each remaining factor the potential future realizations are described. These projections should define a "window of opportunity" containing the most likely and most extreme scenarios.
- Step 3: Development Scenario. The obtained realizations are combined with each other leading to several thousand (due to combinatorics) "pictures" of the future. Eliminating those with contradictory assumptions and summarizing similar ones to clusters, leads to a manageable size of consistent pictures, which can be described into prosaic descriptions of the future.

3. Scenarios for the future of automotive industry

Selected core competences and current business model approaches of the automotive and the ICT industry are used to describe the effects "digitalization" may have on the automotive industry and their value chain. The focus is on premium

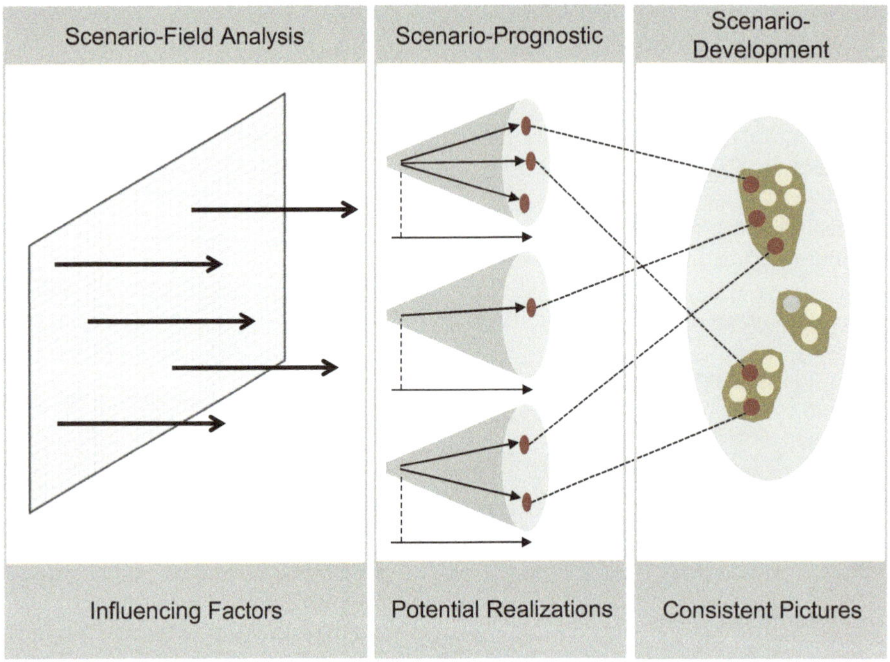

Figure 2. Process of scenario creation [33].

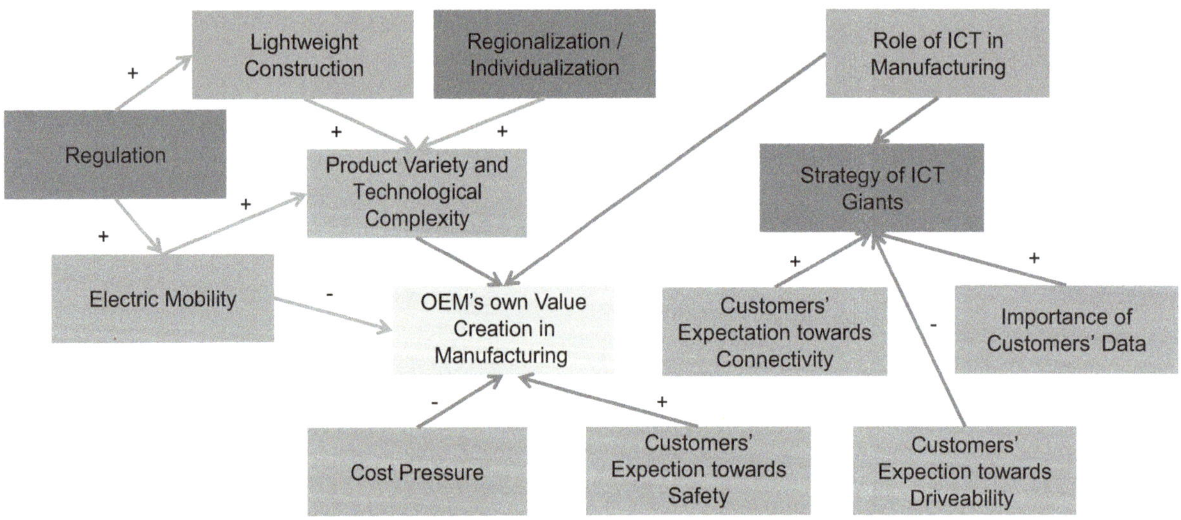

Figure 3. Influencing factors and their interdependencies.

automotive companies. This chapter is structured with Gausemeier et al.'s phases of scenario technique [35].

3.1. Influencing factors

Automotive OEMs' strategy towards their own value creation with respect to the automotive strategy of ICT giants is the core of this observation. Therefore, when identifying influencing factors, the current enablers of the business models of ICT giants in the new economy and today's exclusive core competencies of the automotive OEMs are observed at first. The former focuses on benefitting from (personalized) advertisements enabled by collected customer data [36, 37] and the later on holistic system knowledge leading to driveability, safety and quality. The term "quality" is used to describe a complex system like a car which is very reliable and has well-balanced characteristics. Figure 3 illustrates the collected factors and their potential interdependencies.

It is assumed that increasing cost pressure (alone) hints at decreasing own value creation of the OEMs. Product variety and technological complexity among others caused by the introduced aspects of shifts in markets, alternative powertrains and lightweight construction approaches influences the OEM's production depth, too – however, this influence is not necessarily heading into one clear direction. Some companies outsource to reduce complexity while others want to handle it internally expecting to gain synergies out of platform approaches more easily. In general it is assumed that electric

Table 2. Influencing factors and potential realizations in the future.

	Option 1	Option 2	Option 3
Cost pressure	Increasing e.g. due to new (global) competitors	As it is: high	Decreasing due to higher willingness to pay for premium
Product variety and technological complexity	Increasing due to regulation and regionalization	As it is: high	Decreasing due to restrictive platform approaches
Customers' expectation towards driveability	Increasing – even in developing countries	As it is: very high in developed countries	Decreasing
Customers' expectation towards safety	Increasing – even in developing countries	As it is: very high in developed countries	Decreasing due to lower willingness to pay for it
Customers' expectation towards connectivity	Main criterium	As it is: one of the important criteria	Just nice to have
Strategy of ICT giants	Become automotive OEM	Become major tier 1 system supplier	Stay out of automotive business
Importance of customers' data	Very high: data is the money of the 21st century	As it is: important mainly in the new economy	Low: e.g. reduced willingness to pay with data
Role of ICT in manufacturing	ICT makes production commodity	ICT boosts efficiency, does not change structure	Decreasing (e.g. forced by unions)
OEM's own value creation in manufacturing	Increasing to gain technology multiply effect	As it is: only major core components	Decreasing by using commodity to save costs

mobility leads to a decrease in production depth [38]. ICT companies strategies in the field of automotive are seen as dependent on the customers willingness to share their data which in turn is used for advertising and offering of new services. Moreover, their strategies are supposed to be affected also by customers' expectations towards driveability and (passive) crash safety of a future car – as fulfilling these requirements is usually seen as a major competence of the established automotive OEMs [17].

The role of ICT in manufacturing like the Industrial Internet affects strategies of automotive OEM and ICT companies. The effect of ICT in manufacturing is discussed from improvements in efficiency up to an enabler of simplification of manufacturing so that manufacturing knowledge might not be a crucial success factor anymore – in extreme terms: production might become commodity. However, in this analysis, companies offering ICT-solutions for manufacturing are seen as separated and decoupled from the ICT giants in the B2C-business (business to customer) which are potential candidates for entering the car market. Moreover, influencing factors like the value of brands, customers' psychology and the importance of styling are not covered by the analysis in this paper but have major impact on product development strategies of all players.

3.2. Potential realizations

In accordance with scenario technique, for each factor several options have been defined as potential realizations in the future as given in Table 2.

In order to manage the complexity of the data, some options were excluded such as the greyed out boxes in Table 2. Most of them are seen as unlikely, such as a decrease of cost pressure or variety (compare Sects. 1.1 and 1.2), while others were combined whenever they did not lead to significantly different implications (e.g. to the term "at least as high as today").

3.3. Consistent pictures

In the last phase, the remaining options were put together into four pictures describing a future industrial structure (Table 3). The major goal was to define consistent paths for ICT companies and automotive OEMs. It has been assumed that ICT companies are willing to become automotive OEMs only in those scenarios where customers' data has the utmost importance in their business models. One might argue that is not necessary to develop a whole car to collect customers' data, but, this approach assumes that the data sovereignty first and foremost stays with the manufacturer. All these assumptions may change as business models continue to develop, but seem consistent with today's perspectives and expectations. However, the value of customers' data might change suddenly e.g. in case of privacy violations followed by a loss of confidence by customers.

Picture 1 in Table 3 could be entitled "ICT companies overrule established automotive industry" and describes a state in which only some of today's automotive OEMs can survive. Due to "Industrie 4.0" production becomes commodity and almost anybody is able to manufacture a car. The survivors of the automotive industry will focus either on the design e.g. of specialized crash-structures for cars or become contract-manufacturers for the cars styled and sold by ICT companies.

The second picture is highlighted as "ICT companies become suppliers and competitors" and is a scenario, in which manufacturing knowledge remains important. Today's ICT companies become first tier system suppliers but might also conquer some specific markets (like urban cars) with own cars as well. However, the major goal of these companies in this scenario is to have direct access to the passengers inside the car during fully- or semi-autonomous driving to offer their medial content. Other ICT approaches like "Industrie 4.0" lead to significant increases of efficiency in manufacturing (but not to commoditization). However, companies with a significant

Table 3. Consistent pictures of the future.

	Picture 1	Picture 2	Picture 3	Picture 4
Customers' expectation towards driveability	Decreasing or high only in some (developed) markets	At least as high as today	Decreasing	At least as high as today
Customers' expectation towards connectivity	Main criterium	Main criterium	One of the important or main criteria	One of the important criteria
Importance of customers' data	Very high: data is the money of the 21st century	Very high: data is the money of the 21st century	As it is or lower	As it is or lower
Strategy of ICT giants	Become automotive OEM	Become automotive OEM or tier 1 system supplier	Become major tier 1 system supplier	ICT players stay out of automotive business
Role of ICT in manufacturing	ICT makes production commodity	ICT boosts efficiency, but does not make production commodity	ICT makes production commodity	ICT boosts efficiency, but does not make production commodity
OEM's own value creation in manufacturing	Decreasing by using commodity to save costs	Increasing to gain technology multiply effect	Decreasing by using commodity to save costs	As it is

amount of own value creation benefit from a deeper understanding and knowledge leading to faster ramp-ups of new cars and better quality. In this picture, automotive OEMs use these effects as well as their competences in term of driveability to further boost innovation, defend profit margins and to keep cars made by the ICT companies in niches.

Picture 3 is called "ICT companies become suppliers but define major product-differentiation" and describes again a situation in which almost anybody is able to manufacture a car in "Industrie 4.0". As customers expect high connectivity more than anything else, the diversification on the market is driven mainly by the ICT companies and their brands. That is why automotive OEMs focus on highly standardized, cost-efficient platform approaches. The ICT companies become tier 1 suppliers and gain the highest share of profit margins in the value chain. However, they do not see an additional potential by offering own cars themselves.

The last picture could be called "ICT companies stay outside the automotive business". It is based on advantages of ICT in terms of efficiency in manufacturing (e.g. by Industrie 4.0) but not in the car itself. That is why the established automotive OEMs are in a competition among each other but do not face new competitors from ICT.

4. Implications on production depth

The authors expect picture 2 to be most likely. In this section major aspects and reasons of this opinion are discussed starting with a view on ICT inside a car, followed by a statement on the importance of manufacturing know-how which seems to be supported by latest adjustments of OEM's production depths.

The business model of the ICT companies is based on collecting and managing customers' data in order to offer services and personalized advertising on a massive scale. It is quite obvious that there will be a huge potential for such business models interacting with the passengers when driving autonomously. From today's perspective it is very likely that the ICT companies with their infrastructure and methodology will try to take advantage of this. An open question is the data sovereignty and the business case for sharing customers' data between customer, automotive OEM and ICT companies.

As stated in Table 2, one potentially new role of ICT in manufacturing might lead to a commoditization of production by making operational excellence almost trivial when using the right ICT systems. In contrast to that, the second aspect of the discussion in this section focuses on an observation which is called "technology-multiplier" by the authors. This effect deals with new manufacturing technologies (compare Table 1) which cannot be substituted by any ICT solution. Its multiplying effect is based on the assumption that companies with a high share of their own value creation (deep production depth) are able to learn faster and as a consequence are able to innovate faster, to shorten the time-to-market and to develop and produce complex products with higher accuracy or at a lower cost. This assumption should not be generalized or misunderstood, as, when outsourcing components to suppliers – in the right selection and balance – there are proven potentials enabled by specialization and economies of scale among others, as stated in the literature review. However, there are some indications that (own) manufacturing in general seems to get more important:

- Production-Push-Approach: the view on manufacturing technologies has changed from a restrictor to an enabler of new product features. Examples of such enabling manufacturing technologies are laser technologies for micro-structuring of bearings, aluminum-aluminum resistance spot welding (RSW) in body-in-white manufacturing [39], additive manufacturing or resin transfer molding (RTM) to produce CFRP-parts [40] in series.
- Simultaneous Research: the mentioned complexity and variety of products as well as the fact that almost every new premium car requires at least one first-time application of a new manufacturing technology necessitate a close interaction between product design and production, and in doing so, a new dimension of simultaneous engineering is created [41].

Table 4. Synergies between Mobility 4.0 and Industrie 4.0.

	Mobility 4.0	Industrie 4.0
Data acquisition	Innovative sensors inside and outside the vehicle (Car-to-Infrastructure, Car-to-Car, etc.)	Additional sensors in industrial equipment e.g. machines and logistic systems or collected by a human operator
Data preparation/algorithms	In the vehicle	In the machine control unit, in the factory or in computers at the site of user or equipment builder (e.g. via web communication)
Enablers	Advanced optical sensors, data fusion, online evaluating systems, voice- and gesture control, car-to-X-communication	Physical sensors: temperature, vibration, current; data fusion; voice control
Business models	Out-of-the-box approaches to use collected (big) data recorded by the sensors in the cars in the field (e.g. for traffic forecast)	Improved services by gaining knowledge out of decentral recorded data (e.g. robust Just-In-Time deliveries, predictive maintenance)
Impact on humans	Increased freedom in a range from active participation as a driver to passenger (in fully automated driving mode)	Expanded range of work content instead of Taylorism – enabled by complexity reduction by new assistance systems (e.g. digital Poka Yoke, Human-Machine-Interactions)

- Point of Industrialization: to integrate a new manufacturing technology into series production (known as "industrialization") at the right time in terms of a balanced cost-benefit-ratio is of crucial importance [41, 42], being too late might lead to lost sales or lost cost savings but being too early might lead to increased costs due to the application of a pre-mature technology leading to higher scrap rates, lower speed or bad reliability.

Besides those suggestions made by the authors, there are some hints that OEMs might increase their own value creation. Already in 2007, BMW decided to increase their production depth which was at that time between 25% and 30% [43]. Recently, the importance of OEM's own value creation has been underlined impressively by Porsche AG: for approximately 20 years Porsche was famous for a very low production depth (e.g. 12% for Porsche Cayenne [44]). But in early 2015 the company announced to increase its own value creation for at least the 10–20 most important parts related to quality or brand/image [45]. Japanese OEM, Toyota, itself has a quite low production depth but the depth of Toyota's company grouping ("Kairetsu") is about 76% [46]. Similarly, Korea's Hyundai is also known for a high degree of value creation within the company grouping [47]. In 2012, Wyman forecasted an average global production depth of OEMs of 29% by the year 2025 [38].

5. Additional implications

Besides these findings on future value chains an aspect related to the question of gaining money out of the digitalization in products and production, should be discussed. Therefore, product and production shall not be treated as decoupled anymore. "Industrie 4.0" and autonomous driving (which could be called Mobility 4.0 in this analogy) use quite similar technologies such as various types of (optical) sensors, data fusion systems and decentralized decision making algorithms. Table 4 collects key enablers and potential approaches of Industrie 4.0 and Mobility 4.0. However, autonomous systems like automated guided vehicles in logistics, augmented reality solutions in technical services or condition monitoring systems for predictive maintenance in manufacturing plants are far behind the achievements of autonomous driving. Consequently there might be a chance for ICT companies in addressing both, cars and car factories, with their technologies and solutions in the future.

Today, the usage of advanced product life cycle management systems linked to manifold simulation tools and Big Data algorithms is state of the art during the product development of a car. However, the automotive development processes of today are still static and inspired by pre-defined quality gates. There might be a chance for even more efficient product development in the automotive industry by adapting ICT companies' agile development processes especially when facing volatile markets. Of course, the adaptation of agile methods to automotive has to ensure the same high-quality outcome of today's automotive processes. Successful approaches have already been presented in reference [48]. As a consequence ICT companies might be able to learn how to increase the robustness of their products without losing speed during development.

6. Closing remarks

This paper has presented potential outcomes of the current convergence of the old and new economy as represented by the automotive industry and ICT companies. Digitalization in terms of the rising importance of ICT in production and products might have a strong impact on the automotive industry. However, the authors expect that manufacturing will not become "commodity" as manufacturing knowledge not only leads to improvements in operational efficiency but also drives innovations in fields, which ICT solutions cannot cover, e.g. CFRP-intensive bodies. Therefore automotive OEMs should carefully adjust their own value creation and continuously assess where their individual optimum is – taking into account the ability to innovate quickly. Moreover, global customers' expectations in terms of driveability, quality and safety as well as their behavior towards sharing private data must be observed continuously and very carefully.

Acknowledgements. Steven Peters is indebted to the Baden-Württemberg Stiftung for the financial support of this research project by the "Eliteprogramm für Postdocs".

References

1. Scotiabank (2015) http://www.gbm.scotiabank.com/English/bns_econ/bns_auto.pdf
2. PricewaterhouseCoopers AG (PwC) (2015) http://www.strategyand.pwc.com/perspectives/2015-auto-trends
3. CarNewsChina (2013) http://www.carnewschina.com/tag/volkswagen-lavida/page/2/
4. Business Insider (2013) http://www.businessinsider.com/why-chinese-buyers-love-buick-2013-4?IR=T
5. European Commission (2015) http://ec.europa.eu/clima/policies/transport/vehicles/cars/index_en.htm
6. Reuters (2014) http://www.reuters.com/article/2014/10/02/autoshow-paris-carbon-idUSL6N0RX5S520141002
7. BMW AG (2013) http://www.bmwgroup.com/bmwgroup_prod/d/0_0_www_bmwgroup_com/investor_relations/_pdf/Workshop-Presenation_BMW_i_Innovation_Days.pdf
8. Automobil Produktion (2015) http://www.automobil-produktion.de/2015/06/bmw-seit-produktionsstart-3-500-i8-abgesetzt/
9. Ducker Worldwide LLC (2014) 2015 North American Light Vehicle Aluminum Content Study: Executive Summary. http://www.alcoa.com/car_truck/en/pdf/2014_Ducker_Executive_Summary.pdf
10. New Scientist, http://www.newscientist.com/roadtothefuture/casestudy
11. G. Schuh, K. Korthals, J. Arnoscht, Advanced Materials Research 907 (2014) p. 329–347.
12. National Academy of Science and Engineering (achatech), Recommendations for implementing the strategic initiative INDUSTRIE 4.0, 2013
13. Volkswagen AG (2015) http://www.volkswagenag.com/content/vwcorp/info_center/de/news/2015/05/baidu.html
14. Daimler AG (2015) http://media.daimler.com/dcmedia/0-921-658901-1-1678242-1-0-1-0-0-1-0-0-0-1-0-0-0-0-0.html?TS=1435327318437
15. NBC News (2015) http://www.nbcnews.com/tech/innovation/googles-self-driving-car-prototypes-hit-public-roads-first-time-n381941
16. Macworld (2015) http://www.macworld.co.uk/news/apple/will-apple-make-icar-rumour-roundup-3425394/
17. Süddeutsche Zeitung (2015) http://www.sueddeutsche.de/auto/autoprojekt-icar-buendnispartner-fuer-apple-gesucht-1.2398276-3
18. Industrial Internet Consortium (2015) http://www.iiconsortium.org/vertical-markets/manufacturing.htm
19. Robert Bosch GmbH, 2015, http://www.bosch-presse.de/presseforum/details.htm?txtID=7070&tk_id=107
20. S. Peters, et al., Manufacturing Review 1 (2014) 10.
21. L.R. Jauch, H.K. Wilson, Long Range Planning 12 (1979) 06.
22. OECD, Measuring Globalisation: OECD Economic Globalisation Indicators 2010, OECD Publishing, 2010, p. 208
23. Dürr AG (2015) http://www.durr.com/investor/service-faqs-contact/glossary/technical-glossary/manufacturing-depth/
24. M.E. Porter, Harvard Business Review 74 (1996) 06.
25. M.C.J. Caniels, A. Roeleveld, European Management Journal 27 (2009) 402–417.
26. J. Stuckey, D. White, Sloan Management Review 34 (1993) 71.
27. M.C. Lacity, et al., Journal of Information Technology 25 (2010) 04.
28. M. Kotabe, et al., Journal of the Academy of Marketing Science 40 (2012) 02.
29. B. Jiang, J.A. Belohlav, S.T. Young, Journal of Operations Management 25 (2007) 04.
30. E. Fuchs, R. Kirchain, Design for Location, Management Science 56 (2010) 12.
31. A.B. Steven, Y. Dong, T. Corsic, Journal of Operations Management 32 (2014) 05.
32. A. Gunasekaran, et al., International Journal of Production Economics 161 (2015) 153–166.
33. J. Gausemeier, A. Fink, O. Schlake, in Szenario-Management – Planen und Führen mit Szenarien, Carl Hanser Verlag, München, 1995.
34. H. Bullinger, D. Spath, H. Warnecke, Handbuch Unternehmensorganisation – Strategien, Planung, Umsetzung, edited by E. Westkämper, 3rd edn., Springer-Verlag, Berlin/Heidelberg, 2009
35. J. Gausemeier, A. Fink, O. Schlake, Technological Forecasting and Social Change 59 (1998) 02.
36. Entrepreneurial Insights (2014) http://www.entrepreneurial-insights.com/google-business-model/
37. ZEIT (2013) http://www.zeit.de/2013/41/privatsphaere-internet-datenschutz/seite-2
38. Oliver Wyman, VDA, FAST 2025 – Massiver Wandel in der automobilen Wertschöpfungsstruktur, 2012
39. Automotive News (2015) http://www.autonews.com/article/20141124/OEM06/311249975/as-ford-spends-big-gm-joins-aluminum-with-simple-welds
40. CompositesWorld (2015) http://www.compositesworld.com/blog/post/the-making-of-the-bmw-i3
41. S. Peters, International Journal of Production Research 53 (2014) 13.
42. S. Peters, wt Werkstattstechnik online 5 (2015).
43. Pressebox (2007) http://www.pressebox.de/pressemitteilung/bmw-ag-0/BMW-will-in-diesem-Jahr-die-Produktivitaet-um-zehn-Prozent-steigern/boxid/99347
44. Prognos, Die Veränderung der europäischen Wertschöpfungsstrukturen im Zuge der Vollendung des europäischen Binnenmarktes und der EU- Erweiterungen, 2007
45. Wirtschaftswoche (2015) http://www.wiwo.de/unternehmen/auto/porsche-chef-matthias-mueller-das-apple-modell-ist-nicht-auf-uns-uebertragbar/11498816-2.html
46. INGENIEUR.de (2009) http://www.ingenieur.de/Branchen/Fahrzeugbau/Innovationsfaehigkeit-wichtiger-Lohnkosten
47. Automobil Industrie (2012) http://www.automobil-industrie.vogel.de/oems/articles/367094/index3.html
48. Robert Bosch GmbH (2015) http://www.scrum-day.de/files/scrum-day/content/upload/c4p-vortraege-2015/G1-Volkmar-Denner_Keynote_Scrum_Day_Gn_150615.compressed.pdf

16

A critical review on special forming processes and associated research for lightweight components based on sheet and tube materials

Lang Lihui, Liu Kangning*, Gaoshen Cai, Xiying Yang, Chan Guo, and Guoliang Bu

School of Mechanical Engineering and Automation, Beihang University, Beijing 100191, China

Abstract – Lightweight materials, such as ultra-high strength steel, titanium alloys and aluminum alloys, are used widely in aerospace and automobile industries leading to increasing demands for advanced forming technologies. Recently many innovative sheet- and tube-forming methods have been proposed by researchers and R&D facilities around the word. These methods can greatly improve the formability of materials with low plasticity and can produce complex-shaped parts with good surface qualities. This review aims at providing an insight into the latest developments of particular sheet- or tubular-metal forming process and shows basic applications in modern industry.

Key words: Review, Lightweight material, Sheet forming, Tube forming

1. Introduction

A current widespread trend is to use lightweight materials with the advantage of producing lightweight components to achieve the goal of reduction of emissions and energy. However, the poor formability of advanced lightweight materials at room temperature has seriously hampered their wider application. Lightweight materials, such as ultra-high-strength steel [1], titanium alloys [2, 3] and aluminum alloys [4] are used extensively in aerospace and automobile industries leading to increasing demands for advanced forming technologies, as conventional forming methods lack the ability to meet the need for fabricating lightweight structures. In the 21st century, in metal forming area, many researchers and business R&D facilities have proposed many advanced innovative forming methods dealing with this problem. These particular forming methods have both strengths and weaknesses in their suitability for different types of sheet metal forming processes.

This paper first categories three kinds of lightweight materials and then gives a detailed introduction to some of the particular sheet forming methods, including hydroforming, viscous pressure forming, solid granular-medium forming, superplastic forming, impact hydroforming and asymmetric incremental sheet forming. The state-of-the-art and current applications of these processes are also discussed in addition to some current developments on the forming mechanisms.

*e-mail: lkn_buaa@163.com

2. Lightweight materials

2.1. Aluminum alloys

Aluminum alloys are effective materials for the reduction of vehicle weight and are expanding their applications over years (see Figure 1). Naka et al. [6, 7] investigated the effects of forming speed and temperature on the experimental forming limit diagram (FLD) for a fine grain Al-Mg alloy sheet by performing stretch-forming tests at various forming speeds at several temperatures. Li et al. [8, 9] investigated the uniaxial tensile deformation behavior of three aluminum sheet alloys including Al 5182, Al 5754 and Al 6111-T4 in the warm-forming temperature range using the torsion test method, Cavaliere [10] studied the hot and warm formability of 2618 aluminum alloy, precipitation being found during deformation.

2.2. Titanium alloys

Titanium alloys possess the beneficial properties of high specific strength and excellent corrosion resistance. Therefore, titanium alloys are found in aerospace applications where the

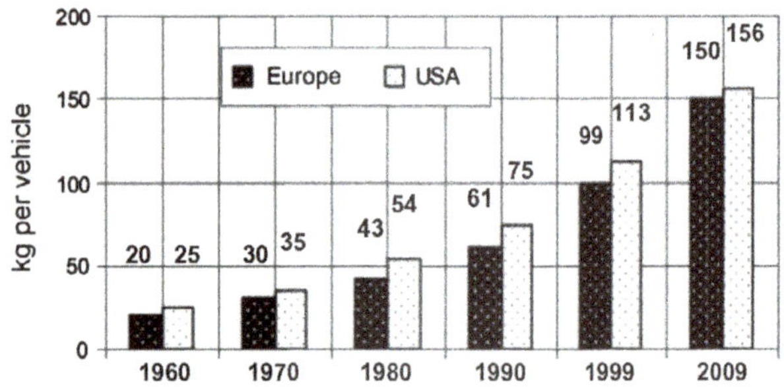

Figure 1. Average use of aluminum [5].

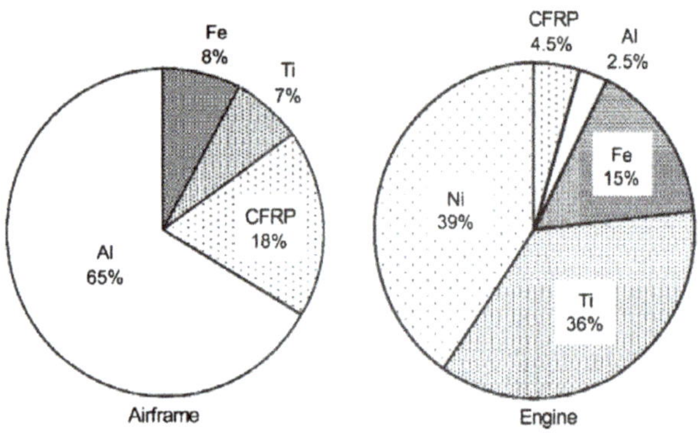

Figure 2. Percentage of aluminum, titanium, and steel alloys and carbon fiber reinforce plastic (CFRP) of the structural weight of modern large commercial aircraft and gas turbine engines [12].

combination of weight, strength, corrosion resistance and high temperature stability is highly valued [11].

Figure 2 shows the percentage of structural weight for various material classes in modern large commercial aircraft, distinguished between airframe and engine materials. The fuselage of the Airbus A330/340, for example, is manufactured of nearly two thirds aluminum. At about 7%, titanium alloys have a similar share of the structural weight to that of steels. However, at over a third of the structural weight, titanium is the second most common material in the jet engine, following Ni-based superalloys; and by volume, titanium alloys are the most abundant material in the engine [12].

2.3. High-strength steels

In order to reduce vehicle weight and improve security, the increasing use of ultra-high-strength steels such as 22MnB5, 27MnCrB5, and 37MnB4 in the automotive industry has become an inevitable trend [13], Within this trend, advanced high-strength steel (AHSS) alloys have seen particularly fast growth [14], being used mainly in chassis components, such as the A-pillar, B-pillar, bumper, roof rail, rocker rail and tunnel (see Figure 3). Such prominent AHSS materials include dual phase, martensitic and boron steels [15]. Individual vehicle models and some companies have incorporated these advanced steels much more quickly than the fleet average: the Mercedes C-class jumped from 38% to 74% HSS in its body redesign,

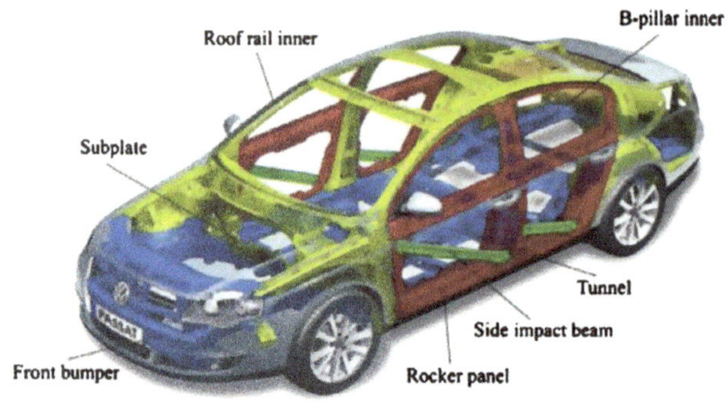

Figure 3. High strength steel used in a typical middle class car [13].

and the BMW X6 has 32% of its body and closure structures composed of AHSS [16]. Estimates from Ducker Worldwide indicate that the automobile industry will see an annual increase in AHSS of about 10% through to 2020 [17].

3. Particular forming methods

3.1. Hydroforming process (HF)

Hydroforming is a well-known long established process which was employed first to form sheets by the use of fluid medium as a soft punch before World War II. In recent years, with the reduction of cycle time and improvement of process controls, the fields of application of hydroforming have become broader. Hydroforming technology has gained increased interest in the world because of its many advantages.

Basically, the hydroforming process can be classified into two categories, sheet hydroforming and tube hydroforming [18], according to the different process features and the usage of different original blanks.

As shown in Figure 4a, a typical method in sheet hydroforming is hydraulic deep drawing (HDD). When the punch pushes the sheet metal into the die cavity, where oil or other liquids are contained, the established high pressure can press the sheet metal tightly onto the punch and a complex part will be formed. Another method is conventional sheet hydroforming (see Figure 4b), where, on the contrary, a rigid punch is replaced by pressurized liquids and the blank is bulged against a die.

Sheet hydroforming attracted widespread attention in the automobile industry when it was first applied to automotive parts as it provides products with relatively good surface quality compared with that of traditional deep drawing. Figure 5 [18] shows some automobile panels fabricated by the sheet hydroforming (SHF) process.

Recently, the enthusiasm generated by HDD technology has cooled down, mainly due to the reasons discussed below [18], miscalculations on these two aspects keeping the SHF process away from series production:

1. The cycle time of the SHF process is too long compared to that of conventional deep drawing or stretch drawing, restricting the SHF process to niche applications.

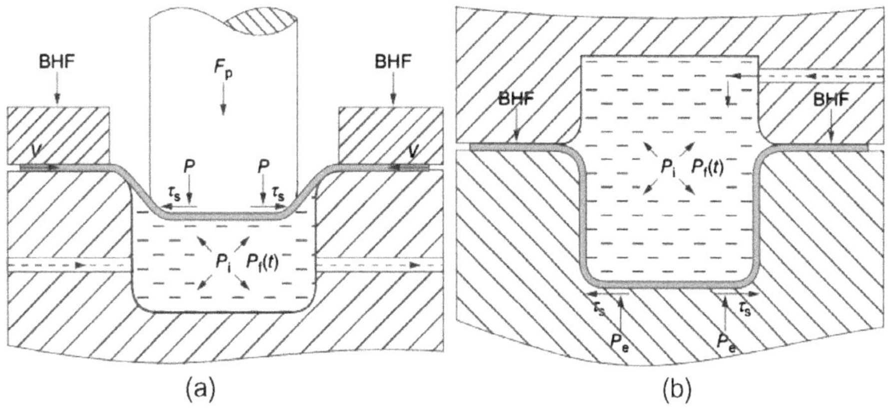

Figure 4. Sheet hydroforming process [19]. (a) hydraulic deep drawing, (b) conventional sheet hydroforming.

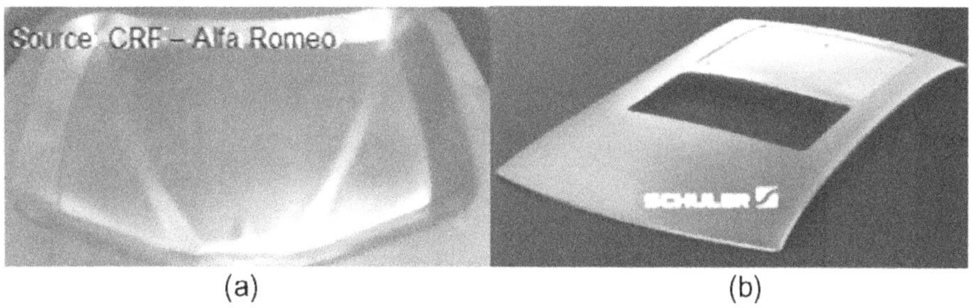

Figure 5. Application of sheet hydroforming (SHF) [18]. (a) Alfa Romeo hood, (b) Maybach roof.

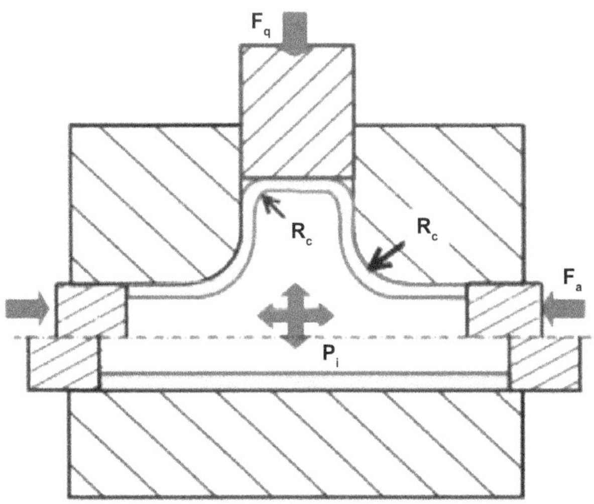

Figure 6. Typical THF process [20].

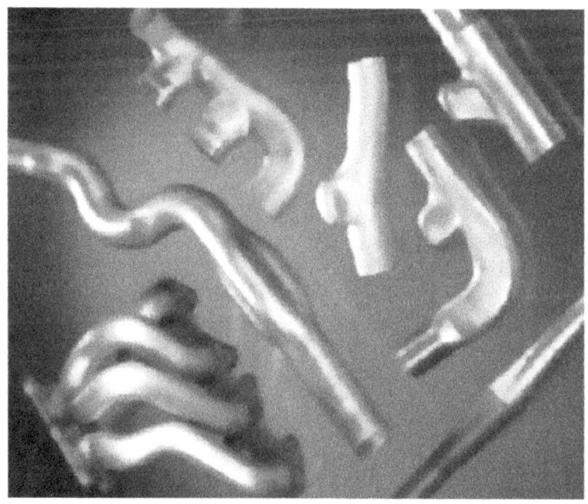

Figure 7. Exhaust parts produced by hydroforming (Schuler Hydroforming, Germany) [23].

2. Special controlling parameters are needed in the process and these parameters have to be acquired by car manufacturers before application in commercialized production. However, the optimized data is not easy to obtain, this being the key point in making full use of the technique.

Figure 6 presents a schematic diagram of a typical tube hydroforming process. A seamless tube or a welded tube is used to form the shape of the die cavity by internal pressure with the corresponding movements of the side punches.

Tube hydroforming has many advantages [21, 22] such as part consolidation, weight reduction, improved part strength. The process was first used in the automobile field for non-structural components, such as exhaust and intake systems or cooling pipes (see Figure 7), since hydroforming is able to fabricate very complex part in a single operation, therefore reducing assembly cost and weight. Applying this process to structural components is a big challenge, but due to the R&D investment of the automobile industry, many parts, such as suspension frames, A-pillars and engine cradles, have been

Figure 8. Engine cradle [23].

Figure 9. An aluminum lower rail for the Volvo car (AP&T, Sweden) [23].

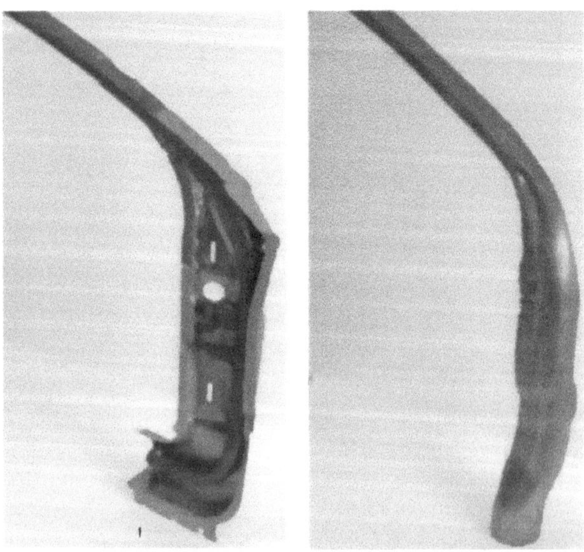

Figure 10. Comparison between a conventional A-pillar (left) and a hydroformed A-pillar (right) [18].

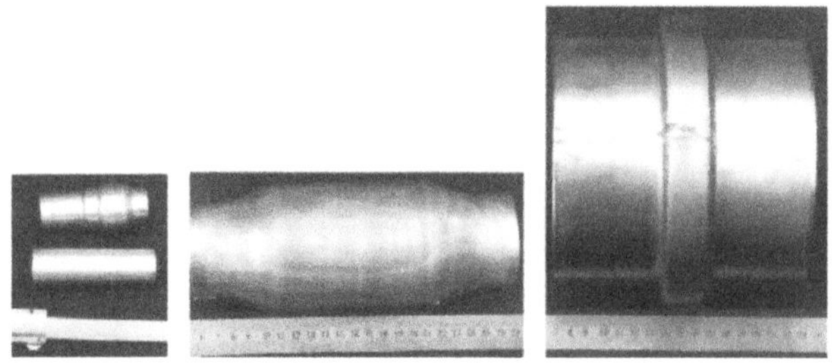

Figure 11. Hydroformed parts used in the aerospace industry (HIT, China) [23].

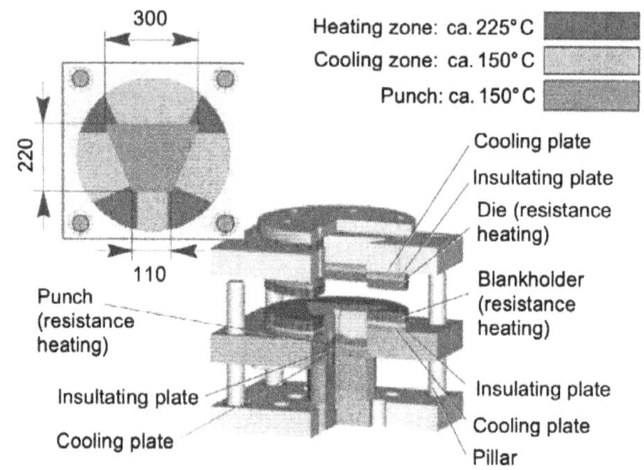

Figure 12. Deep drawing tool at elevated temperature [24].

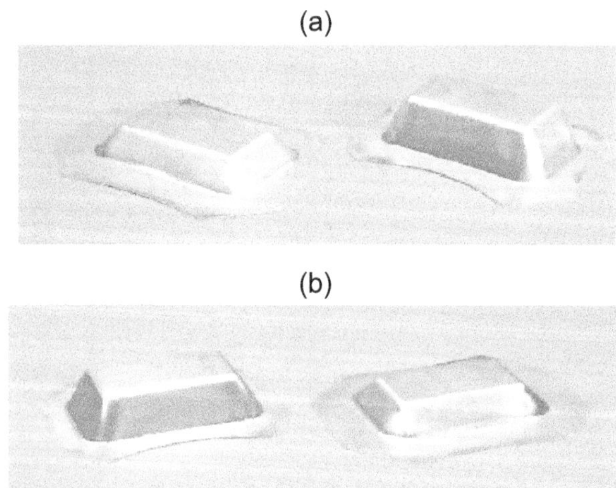

Figure 13. Formed aluminum boxes [25]. (a) Using 5754-O at room temperature (left) and 250 °C (right). (b) Using 6016-T4 at room temperature (left) and 175 °C (right).

successfully developed (see Figures 8 and 9) which not only have the characteristic of being of light weight but also maintain the same structural performance.

Among these, one of the first research programs on this application was carried out in the early 2000 s as an EU-funded project and successfully produced the prototype of an A-pillar reinforcement realized by bending and hydroforming a conical tube (see Figure 10).

In addition to the automotive industry, hydroforming has also gained growing interest in the aerospace industry in the

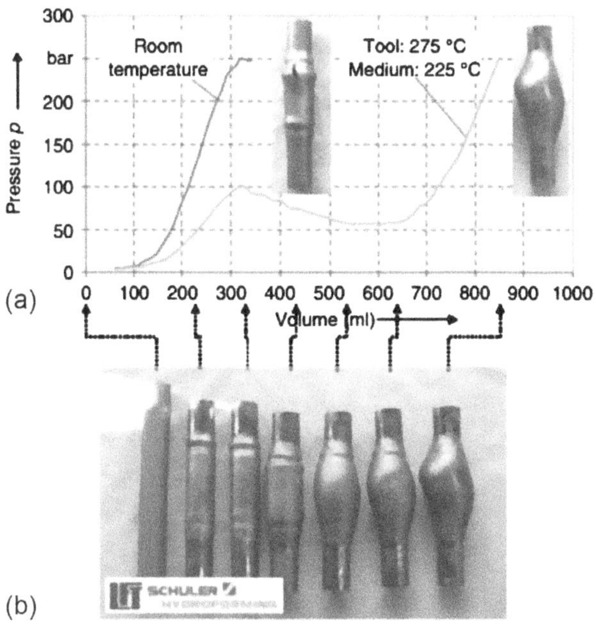

Figure 14. tube warm hydroforming test at room temperature and 220 °C [24].

production of some complicated structural parts. The group led by Wang and Yuan in HIT (Harbin Institute of Technology) [23] has done many relevant researches on the application of hydroforming aircraft parts (see Figure 11), and many parts have been formed out of aluminum alloys or welded stainless tube blanks.

Another new trend is to combine warm forming with sheet/tube hydroforming: the resultant process is called warm hydroforming, and can improve the formability of materials significantly. The tooling for conventional sheet metal forming at elevated temperature has been studied and developed for many years, all of which tooling is almost the same as that shown in Figure 12 [24].

Using 1050, 5754 and 6016 series aluminum sheets, Bolt et al. [25] investigated the formability at elevated temperatures of from 100 to 250 °C. Both box and conical rectangular products were made with a tool with a heated die and blank holder. It was shown that forming at elevated temperature can yield a significant increase in product height, especially for conical products. The minimum die temperature for a significant effect on the process limit is lower for the Al6016 alloy than for the Al5751 alloy. The hardness of the warm-formed products does not differ significantly from that of products shaped at room

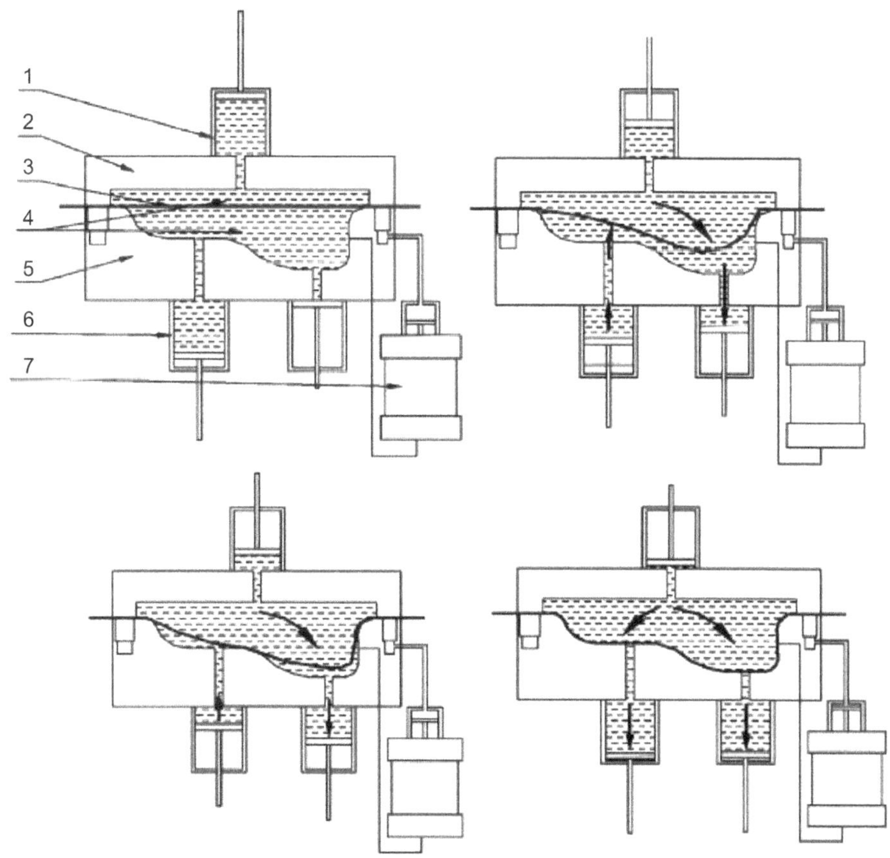

Figure 15. A schematic diagram of VPF [26]: (1) medium injection cylinder; (2) upper die; (3) blank sheet; (4) viscous medium; (5) die; (6) medium outlet cylinder; (7) blank holder cylinder.

(a) (b) (c)

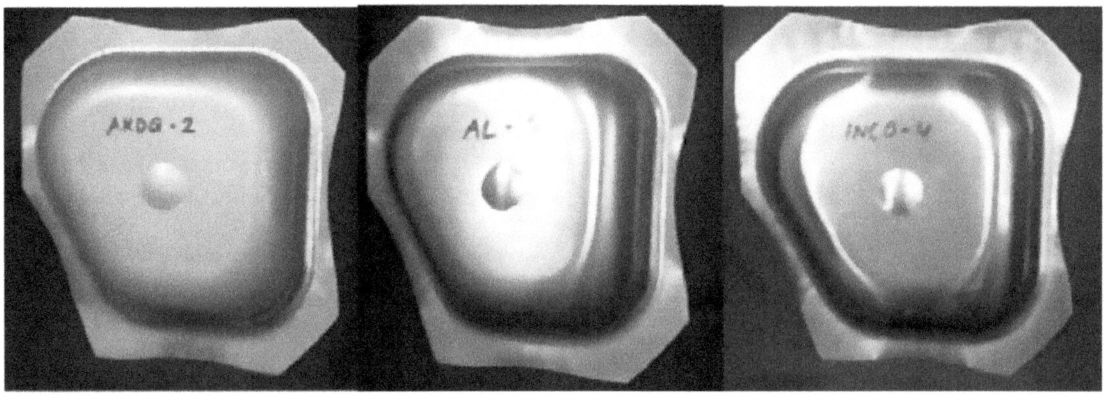

Figure 16. SPF parts formed in VPF tooling [30]. (a) AKDQ steel, (b) Aluminum alloy 6111-T4, (c) Inco 718.

(a) (b) (c)

Figure 17. Parts obtained in the bulging experiment [32]. (a) Viscous medium bulging part, (b) polyurethane bulging part, (c) steel bulging part.

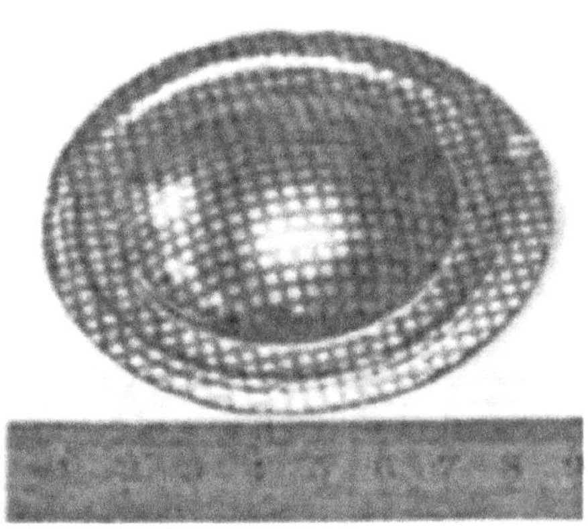

Figure 18. VPF specimens at elevated temperature [33].

temperature. Figure 13 shows the deepest-formed conical boxes using Al5754 and Al6016.

Kleiner et al. [24] conducted bulging tests of tube hydroforming at elevated temperatures. The tube material was an Al alloy (AlMg3.5Mn) and was welded and annealed before the tests in order to secure good formability. Figure 14 shows the pressure curve versus the injected volume for warm tube hydroforming at room and elevated temperature. At room temperature, the pressure increased very quickly and failure of the tube occurred very soon in the process. By increasing the tool temperature to 275 °C and the temperature of the pressurized medium within the tube to 225 °C, the formability of the tube increased significantly.

3.2. Viscous pressure forming (VPF)

The VPF process was invented in 1992 in the form of a patent and it is differed from conventional deep-drawing processes by applying viscous materials as the force-transmitting medium. Specifically, in the VPF process, high pressure exerted by viscous medium is applied on one side of a sheet metal blank to replace rigid punches, whilst back-pressure is also exerted to improve formabilities (see Figure 15). The viscous-medium forming method can make complex-shaped components with good surface quality and high dimensional accuracy, and at the same time reduce leakage [27]. Furthermore, the VPF process also has the advantage of producing parts made of high strength and in other light metals which are difficult to form by conventional sheet metal forming methods [28, 29].

Ahmetoglu et al. [30] conducted VPF experiments on a non-symmetrical part at the ERC/NSM laboratory using three different materials, namely AKDQ steel, aluminum alloy 6111-T4 and Inco 718 SPF and studied the influence of process parameters, such as clamping load, forming speed and sheet

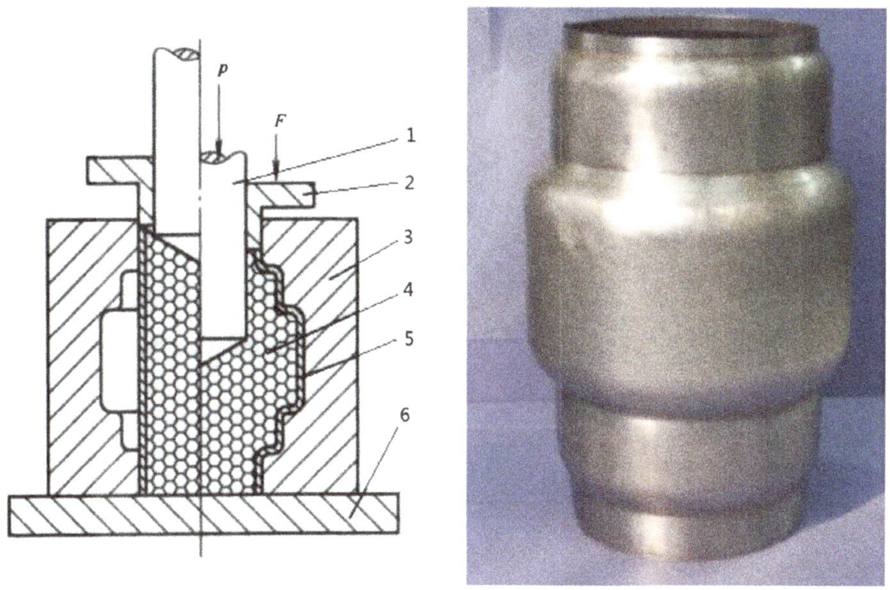

Figure 19. Photo of deformed typical parts [34].

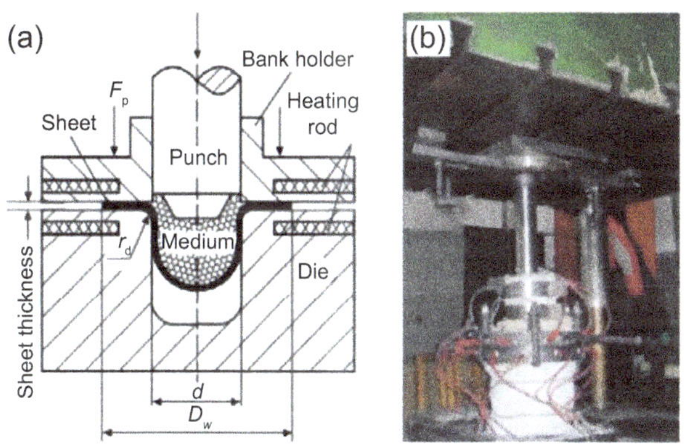

Figure 20. Schematic diagram of deep drawing and experimental equipment of AZ31B magnesium alloy [36]: (a) Diagram of deep drawing, (b) experimental equipment of deep drawing.

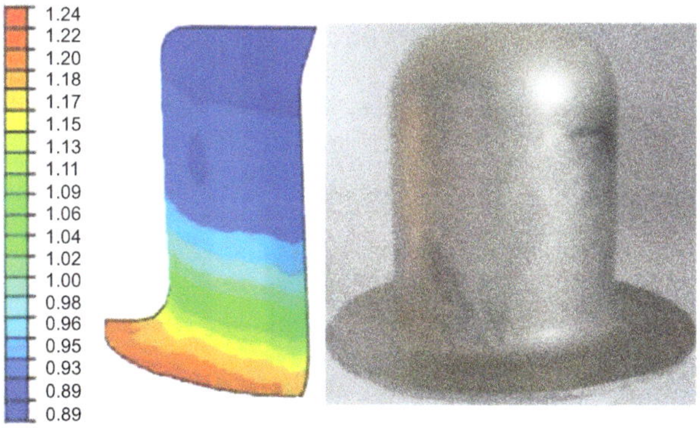

Figure 21. Deep drawing workpiece and contours band of thickness [36].

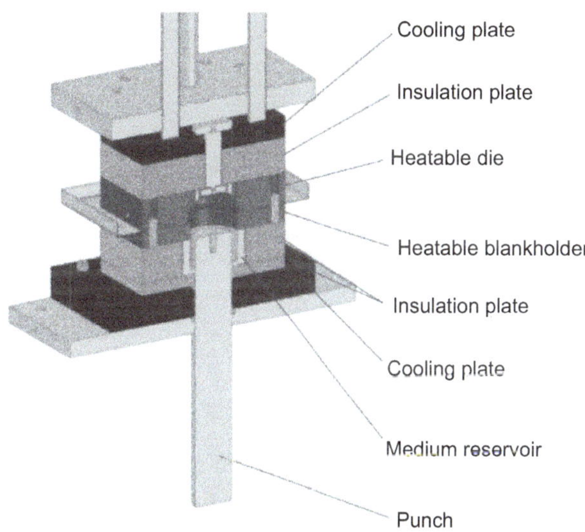

Figure 22. Experimental tool for hydroforming using granular material as a medium at elevated temperatures [37].

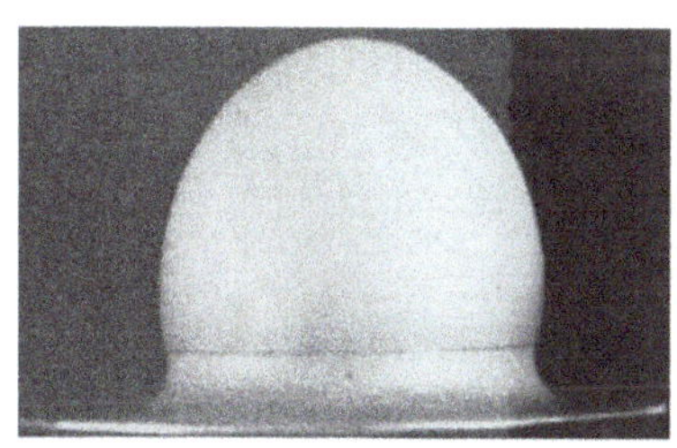

Figure 23. The first superplastically formed bubble [41].

metal blank material, on the technique (see Figure 16). Shulkin et al. [31] introduced an eight-point BHF control system to an experimental VPF machine and studied the formability by both simulation and experiment. Contrast experiments of bulging tests with three kinds of pressure-carrying media, viscous medium, polyurethane and steel, was studied by Wang et al. [32],

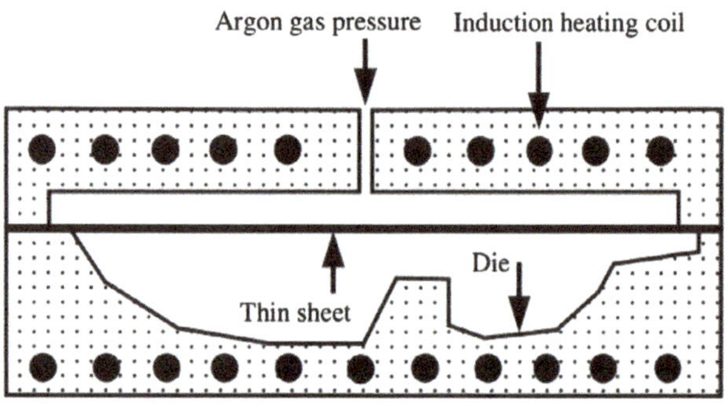

Figure 24. Schematic of SPF process [42].

Figure 25. Non-planar clamp line SPF forming [43].

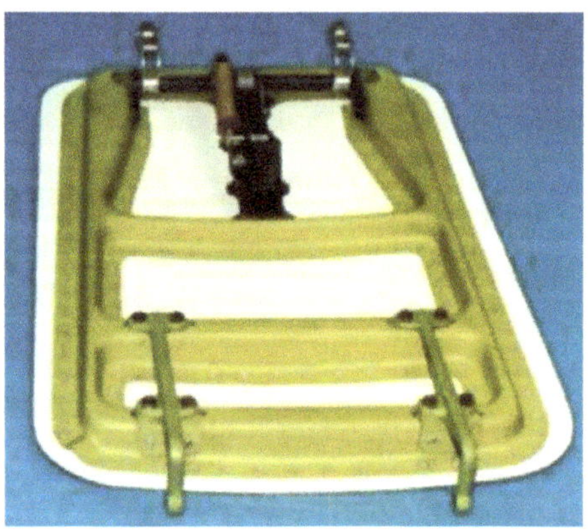

Figure 26. 737 blowout door [45].

the results showing that specimens formed with VPF have less wall-thinning and consequently have a more uniform wall thickness distribution compared with those formed with polyurethane and steel (see Figure 17).

Liu et al. [33] improved VPF technology by using the process at elevated temperature to form an aluminum alloy sheet metal (see Figure 18). The limiting dome height (LDH) of test specimens was measured to evaluate the formability of VPF at different temperatures. Results showed that the LDH of bulging specimen at 200 °C attained 17.4 mm, 10.3% higher than that at room temperature. The influence of forming temperature on thickness distribution, forming load and failure location of sheet metal was also investigated. To the present day, the VPF process has basically remained in the research stage and has not been used extensively in the aerospace or automotive industries.

3.3. Solid granular medium forming (SGMF)

The innovative technology of Solid Granular Medium Forming (SGMF) was proposed to overcome the disadvantages of leakage problems and poor heat resistibility of hydroforming by Zhao et al. [34–36]. In this process, a granular medium replaces water or oil or viscous materials as the force-transmitting medium. Considering that fluids normally used at room temperature are only temperature stable up to about 350 °C, using solid granular medium allow high temperatures and can totally reduce the risk of leakage. Similarly to the hydroforming method, this technology can fabricate both tubular parts (see Figure 19) and sheet parts (see Figure 20). Moreover, the non-uniform distribution of the internal pressure of solid granules can significantly improve the formability of the process, in addition, the interaction between the solid granular medium and the sheet metal surface can also greatly reduce the thinning of the workpiece [35].

By conducting SGMF tests at warm temperatures, Cao et al. studied the granular medium forming performance of magnesium alloy sheet (see Figure 21) and determined that AZ31B magnesium alloy sheet achieves the best deep drawing performance under the conditions of a forming temperature of from 250 °C to 300 °C. These authors also found that the granular medium can apply axial friction on the cylinder wall to further improve the forming performance of metal sheet [36].

Merklein et al. [37, 38] used ceramic beads as a forming medium and designed and assembled an experimental apparatus (see Figure 22) to conduct the process at elevated temperatures. The apparatus can make SGMF parts at temperatures of up to 600 °C, when the lightweight materials such as high- and ultra-high-strength steel or titanium are suitable for forming.

3.4. Superplastic forming (SPF/DB)

Superplasticity is a characteristic of some fine-grained (3–5 μm) alloys when specimens under low stress conditions can exhibit neck-free elongations of very large values [39, 40]. Since the first superplastic part was made pneumatically from a sheet of superplastic Al-Zn eutectoid alloy 50 years ago [41] (see in Figure 23), this technology has been actively pursued in academic, research organizations and industrial R&D facilities around the world due to its special advantages of meeting some unique requirements in the aerospace and automobile industry.

A schematic diagram of the SPF process can be seen in Figure 24, where a thin piece of sheet is put into the tools on a press and sealed at the edges, then the chamber is filled with high pressure gas at elevated temperature. The sheet is gradually deformed in a like manner to the blowing of bubble gum. The commercially available alloys for the SPF process

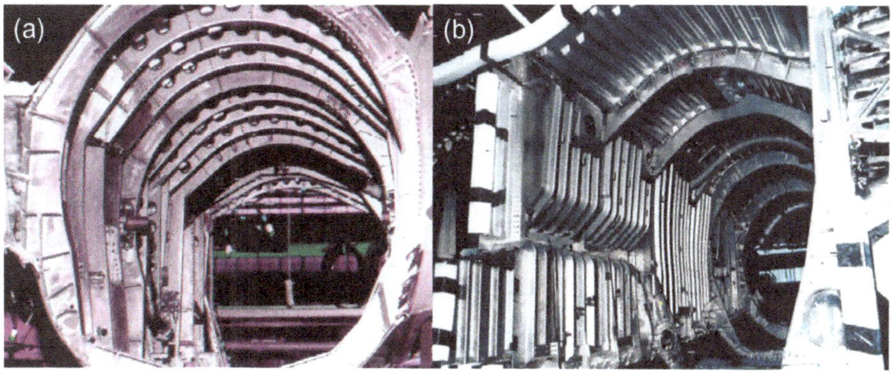

Figure 27. The use of superplastic forming-diffusion bonding to reduce the fuselage part count in F-15E aft fuselage (blats SPF-DB) showing (a-left) before and (b-right) after SPF-DF [46].

Figure 28. The titanium fan blade for Trent 900 engine.

Figure 29. The 2006 limited series reproduction of the 1954 classic Jaguar C-Type [43].

are usually non-ferrous metals such as: aluminum alloy, titanium alloy and magnesium alloy [43].

In general, the SPF process can produce parts with complex geometry and thus replace multi-piece assemblies with one

Figure 30. First bubbles superplastically formed from ceramic sheet.

monolithic component, without the need for welding or riveting parts together. Moreover, the components made by the SPF process can contain very little residual stress and have no spring-back [44]. The technology for SPF with aluminum alloys has progressed significantly in recent years. More often a non-planar clamp line tooling arrangement comprising an upper and a lower tool is used to replace planar tools to manufacture complex parts with curving edges (see Figure 25).

Perhaps aircraft manufacturing companies may have the most interest in this technique. The first SPF aluminum component fabricated by Boeing for civil aircraft was in 1992 [45], shown in Figure 26. Up to now, the emphasis at Boeing has focused on developing applications for the 5083 alloy because of its non-heat-treatable property, and many parts such as the 737 wing-tip light housing, the 767 moisture shroud, etc. [44] are manufactured in the SPF process, to save both cost and weight. SPF technology is often used along with diffusion bonding (SPF/DB) to further enhance the forming ability. After applying the SPF/DB process in building the F-15E fighter, it is estimated that about 726 parts and 10,000 fasteners are eliminated (Figure 27). The SPF process also made the wide-chord hollow fan blade possible in aero-engines (see Figure 28).

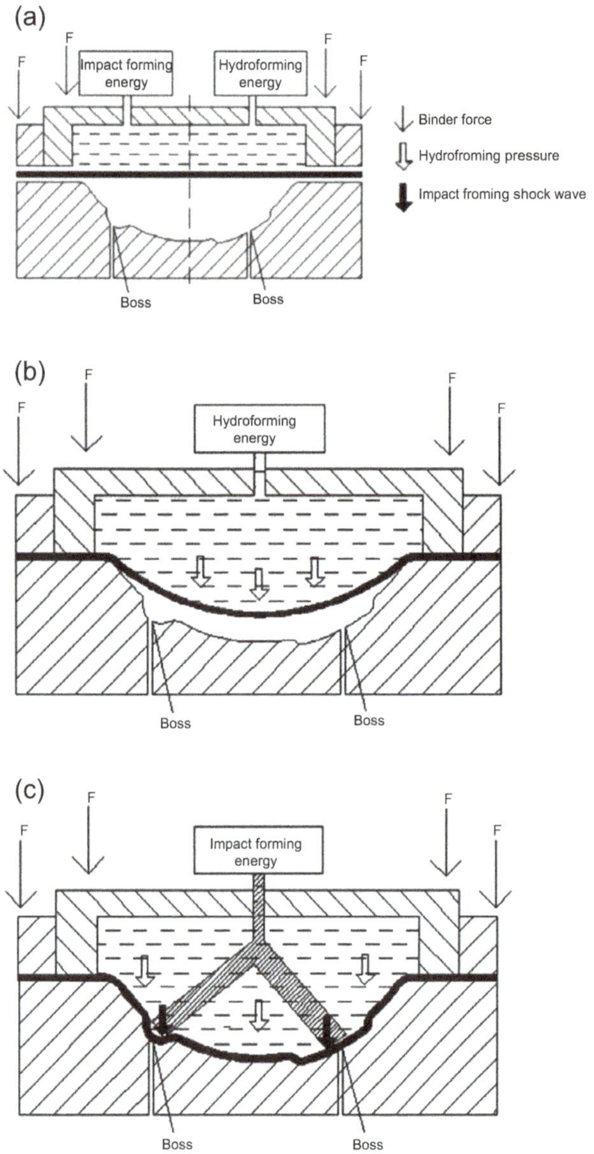

Figure 31. Steps of impact hydroforming technology [50]. (a) Preparation before forming. (b) The hydroforming step. (c) The impact forming step.

Figure 32. The IHF equipment.

Figure 33. Formed part by IHF process.

The promotion of the SPF process is limited in the automobile industry and only niche markets, including specialist vehicles and limited-series sports cars, are suitable for SPF applications [47]. Figure 29 shows the 2006 limited-series reproduction of the 1954 classic Jaguar C-Type in which the SPF process is applied [43]. The reasons why the SPF process is not suitable for Production cars are listed below:

1. The production rate is low and usually for low/medium volume applications.
2. Unique requirements for the materials, which are relatively expensive.
3. The advantage of forming complex part is not highly valued in high-volume models.

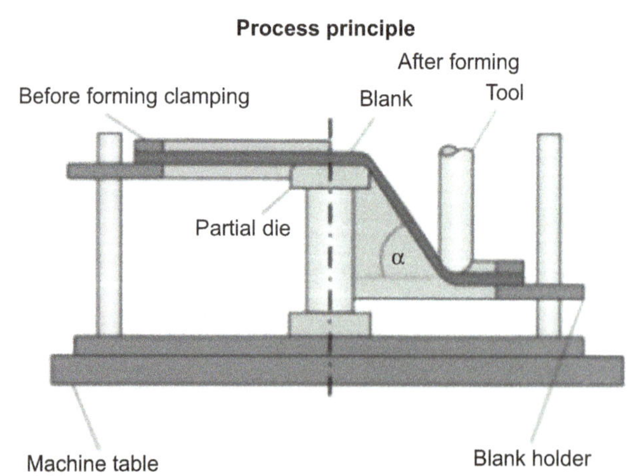

Figure 34. Process principle of AISF process [54].

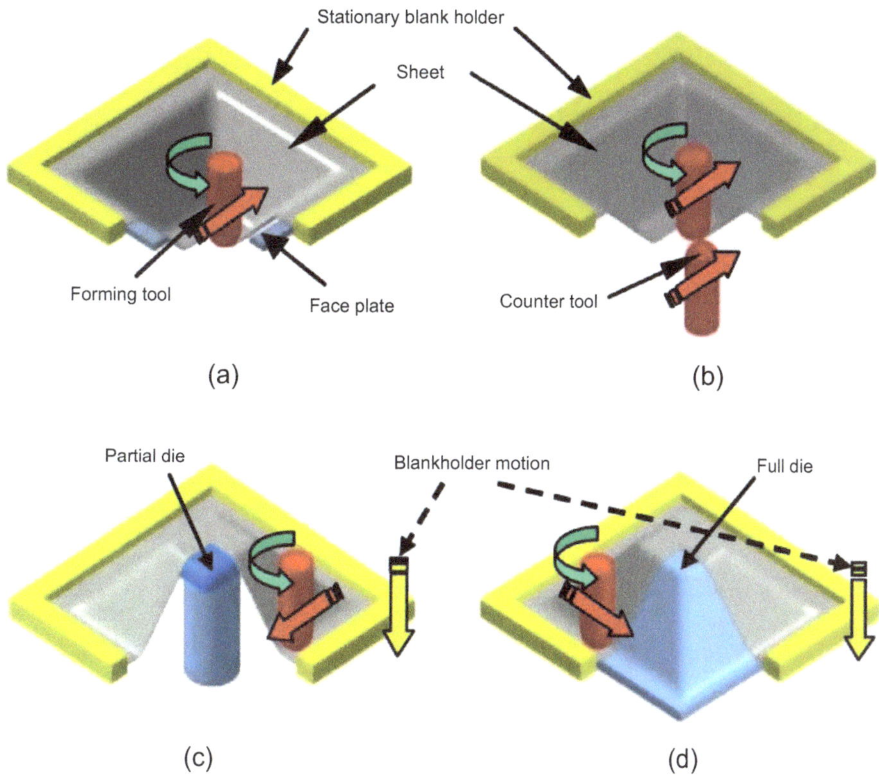

Figure 35. Four categories of AISF process [55]. (a) Single point incremental forming; (b) Incremental forming with counter tools; (c) Two point incremental forming (partial die); (d) Two point incremental forming (full die).

Today the materials used for the SPF process are not limited to metal alloys. Non-metal materials, such as superplastic ceramics [47] (see Figure 30) and glassy metals [49], also demonstrate their ability in the SPF process.

3.5. Impact hydroforming method (IHF)

The innovative method of impact hydroforming (IHF) was proposed by Lang et al. [50] to meet the requirements of forming complex parts with small features which are very common in the automotive and aircraft industries. In the process, most of the geometrical features are formed by hydroforming (see Figure 31b) and some small fillet radius features are rapidly reshaped by high intensity impact energy (see Figure 31c) in a very short time. They also designed an IHF test setup (see Figure 32) to conduct experiments into the process (see Figure 33).

The original idea of impulsive pressure forming can be traced back into the 1960s [50], thus the IPF process can be viewed as a developed method in combination with hydroforming. Moreover, compared with traditional hydroforming methods, impact hydroforming technology can greatly improve the forming speed and the forming accuracy of sheet-metal component. The loading is quite short – less than 10 ms – and strict sealing is not required for the equipment. Moreover, the uniformity of wall thickness of the components can also be assured, as liquid medium fills the whole die cavity and establishes well-distributed pressure [50].

3.6. Asymmetric incremental sheet forming (AISF)

The Asymmetric Incremental Sheet Forming (AISF) process was first described by Mason in 1978 [52] and was developed mainly for the needs of the low volume production of sheet-metal parts. In the AISF process, a metal sheet is placed in the tooling and clamped at the edges, then a tool moves following the required shape in space under CNC control by a succession of "planar" contours or a single "spiral" contour so that the part is obtained as the result of accumulated localized plastic deformations [53]. The AISF process (see Figure 34) is similar to the spinning process but, with the absence of workpiece rotation, asymmetric components can be manufactured by the independent controlling of the *X*, *Y* and *Z* axis movements. One significant advantage of this technology is the ability to form asymmetric shapes at low cost, without the need of expensive dies.

Basically, two main variations of conventional AISF are distinguished: "single point incremental forming" (SPIF) and "two point incremental forming" (TPIF) [54]. Figure 35 shows the different configurations that are included in the group of ASIF techniques. In the SPIF process, the blank edge is clamped without any further support except for the forming tool

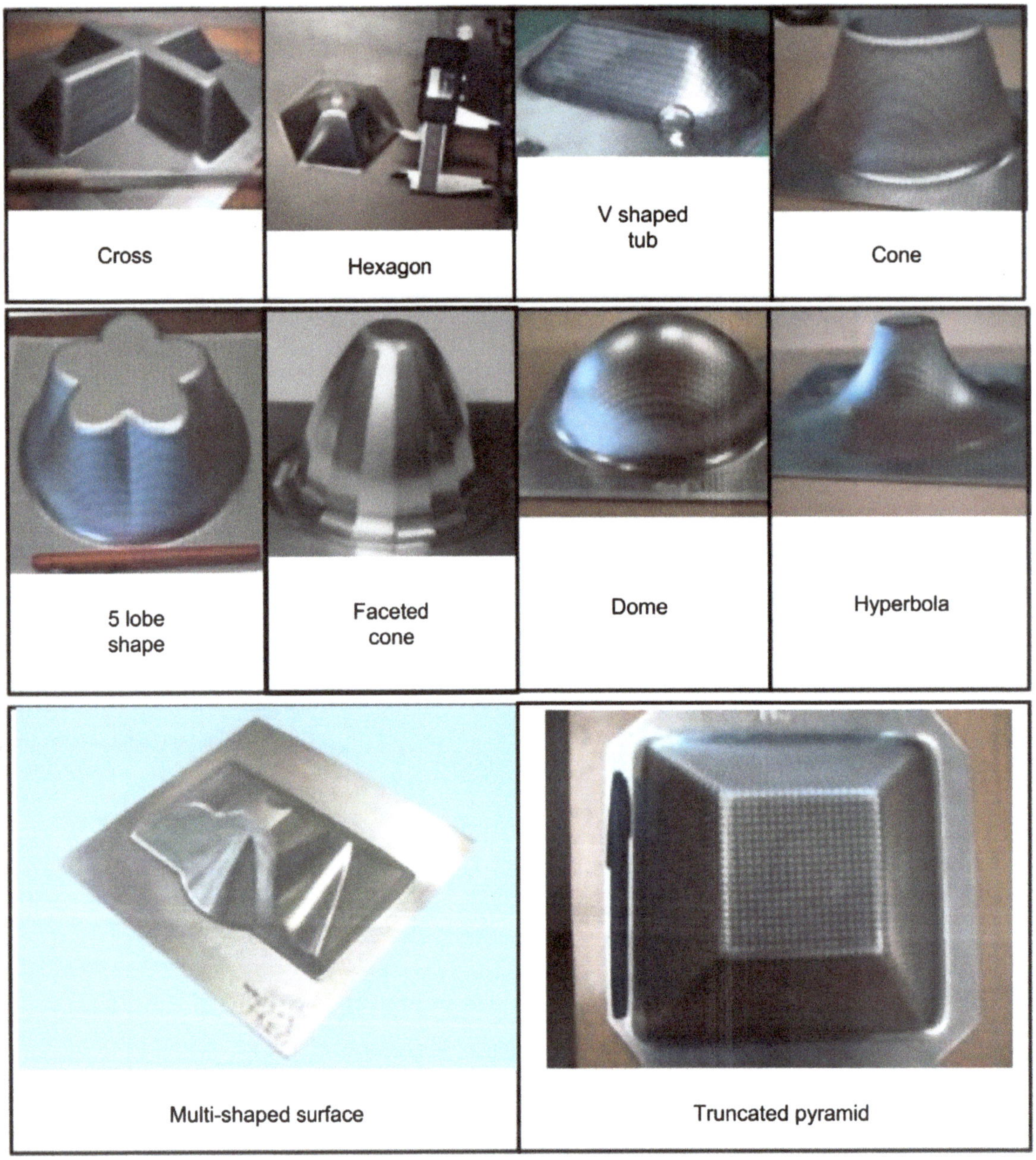

Figure 36. Shapes used to demonstrate the ability of the process and for experiments [60].

(see Figure 35a) [55]; whilst Figure 35b presents a special AISF process with counter tools. Up to now, considerable achievements of this process have been made by Kim and Yang [56], Leach et al. [57], and Felici et al. [58]. In the TPIF process (also called the backward bulge method by Matsubara [59]), there may be a simple support or a partial or full die (see Figures 35c and 35d). Both symmetric and asymmetric shapes can be fabricated using the TPIF or SPIF processes. The configuration shown in Figures 35c and 35d are usually associated with machines dedicated to the incremental forming process, and those in Figures 35a and 35b are usually modified in a CNC milling machine [60].

As mentioned above, the major advantage of asymmetric incremental forming is the property of making asymmetric parts quickly and economically, without using expensive dies. Shapes used to demonstrate the abilities of the process are shown in Figure 36 [60]. Now the AISF process has been used for designs that were made as Rapid Prototypes for the automotive industry [61] (see Figure 37) and the medical industry [62] (see Figure 38).

4. Fundamental research for special sheet and tube forming

In addition to the specific methods of implementing these special sheet and tube forming processes, emphasis has also been paid on the fundamental research concerning those techniques, such as the deformation mechanism, forming limit study, macro-deformation and micro-structure evolution of the

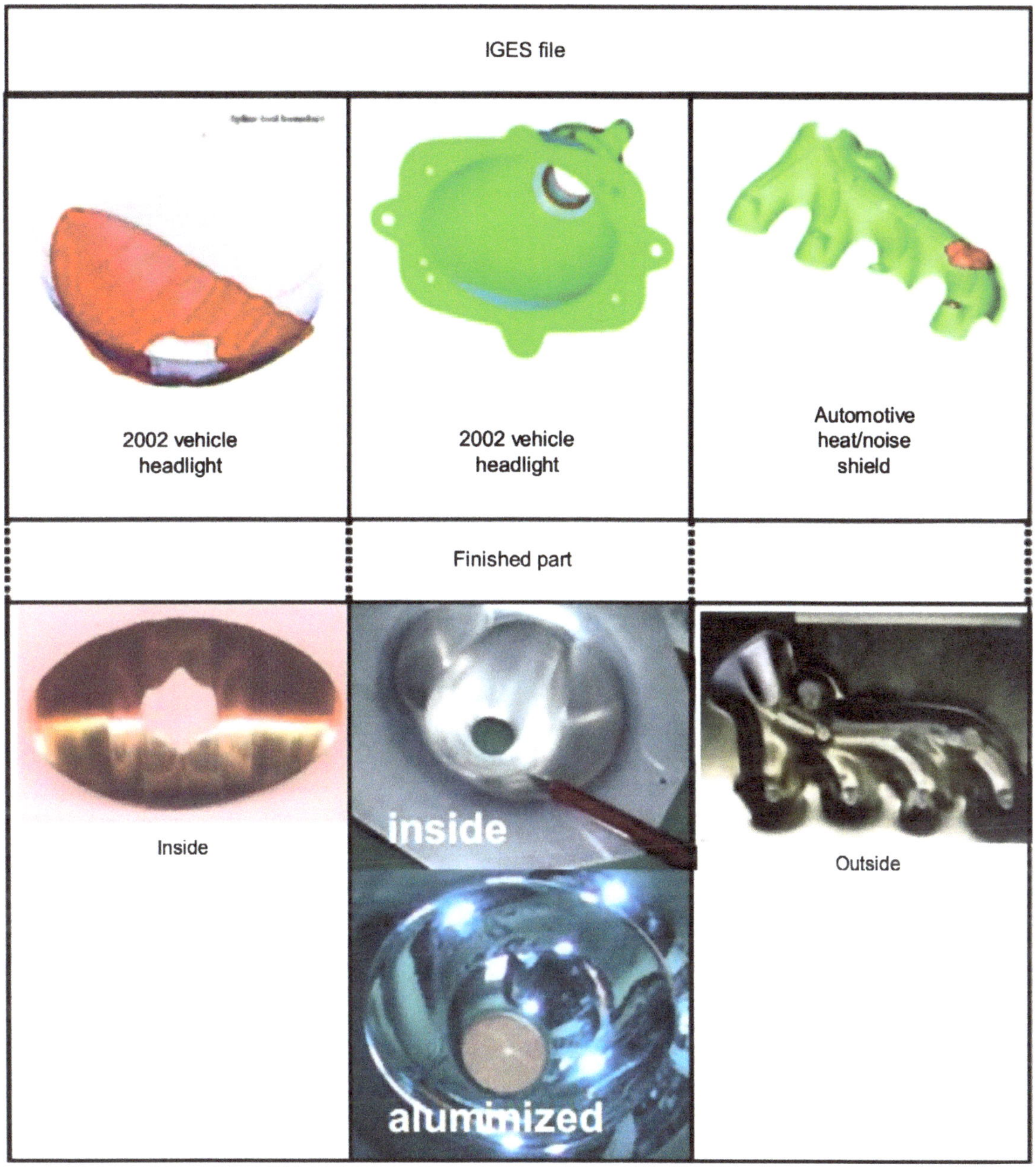

Figure 37. Rapid prototypes for automotive industries [61].

forming process. All the work was done in order to further enhance the formability and help to obtain the optimum processing parameters.

4.1. Mechanism of the HF process

It is widely believed that the main reason why the hydroforming process greatly improves the formability of the material is the introduction of through-thickness normal stress which causes superimposed hydrostatic pressure on the sheet metal.

Wu et al. [63] carried out a detailed analysis of uniaxial tension under superimposed hydrostatic pressure (see Figure 39) and demonstrated that the superimposed hydrostatic pressure lowers the true stress level at yielding by the amount of the superimposed hydrostatic pressure without affecting work-hardening; the increase of sheet metal limit strains for any strain path under superimposed hydrostatic pressure has also been observed.

Fuchs [64] found that the forming limit of tube can be improved remarkably by applying liquid pressure to the inside and outside of the tube simultaneously. Cui et al. [65] designed a special experimental setup (see Figure 40) for double-sided tube hydroforming; the result shows that increase of external pressure has an effect on the fraction of grain boundaries, and on the number and size of the microvoids in the transition zone, and thus increases the critical effective strain in the transition zone. Liu et al. [66] simulated the effect of through-thickness normal stress on the sheet formability using the finite element software based on the ductile fracture criterion.

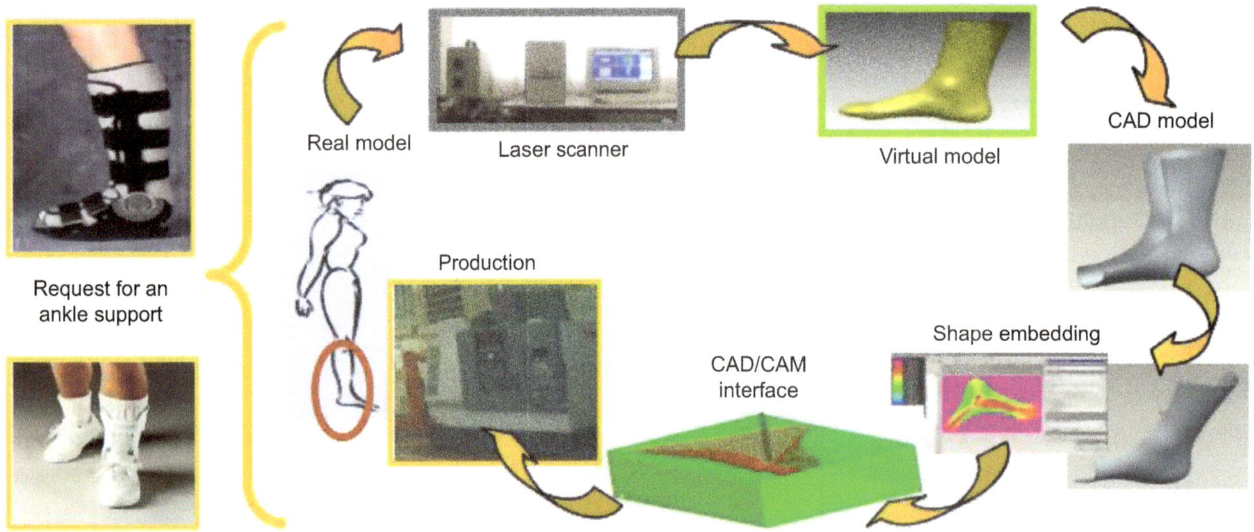

Figure 38. The manufacture of an ankle support [62].

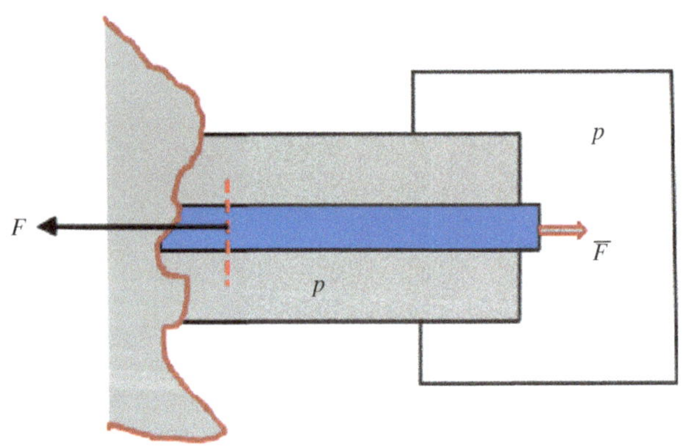

Figure 39. Schematic representation of a tensile sample in a pressure vessel [63].

Assempour et al. [67] studied the effect of normal stress on the prediction of the forming limit diagrams (FLDs) based on the Marciniak and Kuczynski (M-K) model with some modifications on the stress states for consideration of the normal stress effects; as expected, the compressive normal stress shifts the FLD upwards (see Figure 41). Michel and Picart [68] conducted tensile and hydraulic bulging tests (see Figure 42) to investigate the size effects on the constitutive behaviour of brass in sheet metal forming; both experiments showed the same trends, in that the flow curve decreases with a decrease of thickness: a new model considering the size effect was established.

4.2. Mechanism of the SPF process

The deformation mechanism of fine-grain aluminum alloys is generally accepted as grain-boundary-sliding (GBS) creep, which supplies a high strain-rate sensitivity for successful commercial superplastic forming operations [69, 70]. Aluminum alloy 5083 is the most common material in SPF operations and many researches have been focused on it.

Kulas et al. [70] studied the plastic deformation of seven 5083 commercial aluminum materials and found two mechanisms governing plastic deformation in AA5083: grain-boundary-sliding (GBS) creep and solute-drag (SD) creep. Quantitative analysis of stress transients following different strain rates offers conclusive proof that SD creep dominates deformation at high strain rates and low temperature, whereas GBS creep occurs at high temperatures and low strain rates. Furthermore, deformation mechanism maps characterizing deformation in the fine-grained AA5083 materials were also constructed.

Kulas and Green [71] further investigated the failure mechanisms in AA5083 materials under uniaxial tension at elevated temperatures (see Figure 43) and discovered that the deformation dominated by grain-boundary-sliding creep leads to failure controlled by cavitation and the deformation dominated by solute-drag creep leads to failure controlled by flow localization. Further, the cavitation growth rate under SD creep is significantly less than that under GBS creep.

Li et al. [72] studied the superplastic behavior and microstructure of rolled Mg-Gd-Y-Zr alloy sheet and established that the high ductility of Mg-Gd-Y-Zr alloy can be attributed to grain boundary sliding accommodated by dislocation motion assisted by lattice diffusion. Microstructural analysis indicated that the second phases exhibited the significant effect of pinning grain boundaries (see Figure 44).

4.3. Mechanism of the AISF process

The asymmetric single- and two-point incremental sheet forming processes are still in their infancy. Much research work remains to be done and to do this appropriate design of shapes (see Figure 36) need to be developed for the tests of FLDs, springback and roughness.

Jackson and Allwood [73] conducted special tests similar to those designed by Kalpakcioglu [74] using brazed plates in which a grid pattern was machined onto the cross-section of

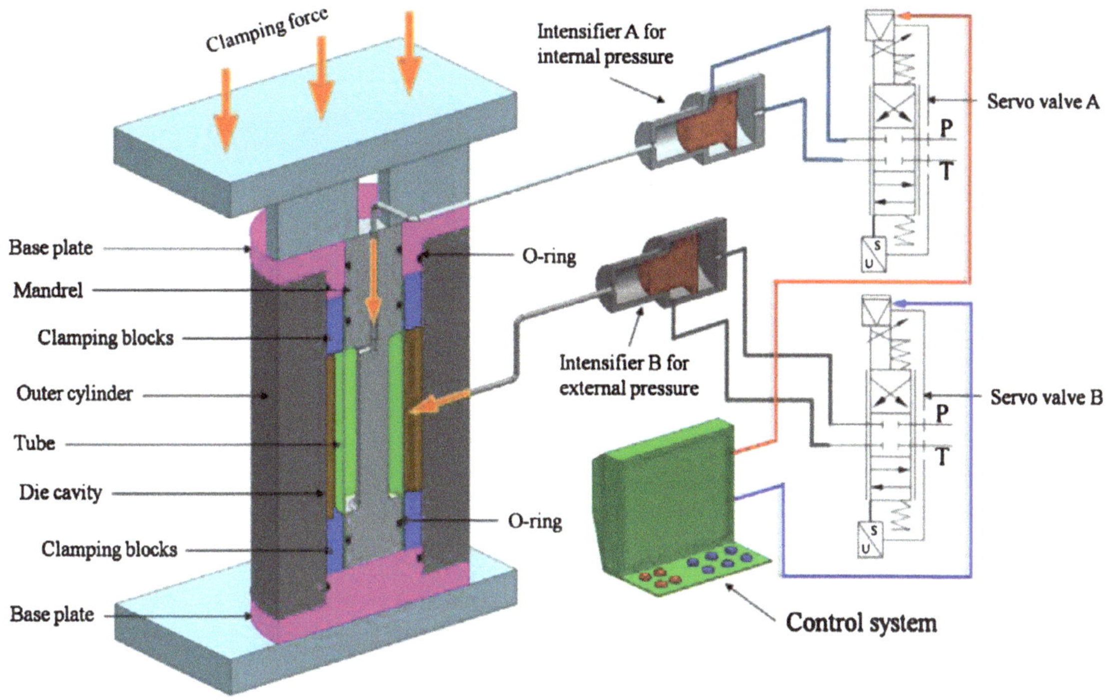

Figure 40. Experimental setup of double-sided tube hydroforming process [65].

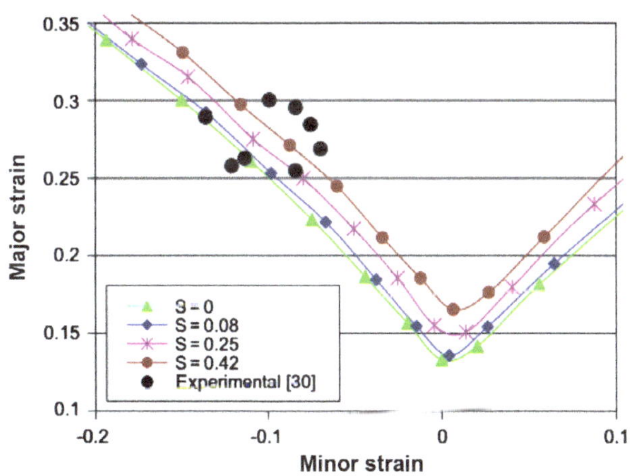

Figure 41. Effect of normal stress on shifting the FLDs for DP600 [67] (S is normalized normal stress).

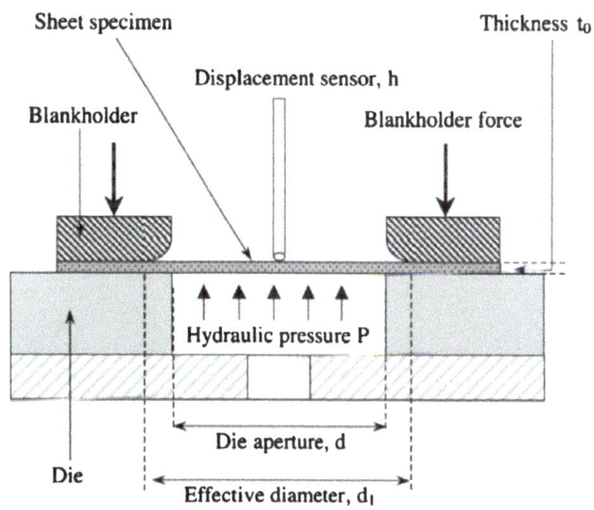

Figure 42. Hydraulic bulging machine [68].

a half-plate (see Figure 45), which was then formed before being separated by heating. The deformation pattern of SPIF, TPIF and pressing was observed and it was revealed that the deformation mechanisms of both SPIF and TPIF are increasing stretching and shear in the radial-axial plane (perpendicular to the tool direction) and shear in the tool direction. Among these, shear in the tool direction, which is likely to be a result of friction between the tool and workpiece, is the most significant strain component. Silva et al. [75] studied the influence of process variables such as the tool radius to propose a new unified view on formability limits in the principal strain space and identified a critical threshold for the ratio between the thickness of the sheet and the radius of the tool that distinguishes between fracture with and without previous necking.

Young and Jeswiet [76] used more than one shape to develop a non-traditional FLD for 1.21 mm thick AA 3003-0 formed by the AISF process. The non-traditional FLD could provide the designer with an additional method of judging if a design can be made in one pass with AISF. (The boundary for safe forming is shown by the dashed line in Figure 46.) Both Filice et al. [77] and Young [76] have found the same tendency of the boundary slope in the composite FLD.

Surface roughness is a major concern in a final product. In AISF the incremental step size, Δz, is the major factor in determining surface roughness [78, 79] which is shown in Figure 47. Junk et al. [79] also studied the effect of pitch Δz and tool radius upon surface roughness and found the interrelationship between these parameters in the AISF process.

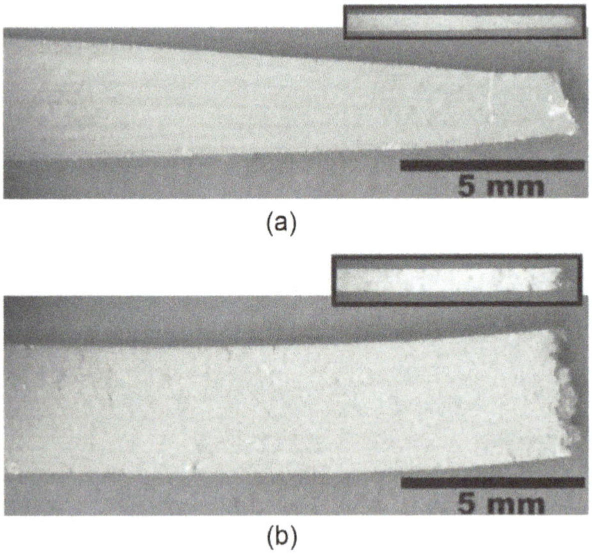

Figure 43. The failure regions of specimens are shown for (a) failure controlled by necking when deformation is by SD creep, (b) failure controlled by cavitation when deformation is by GBS creep [71].

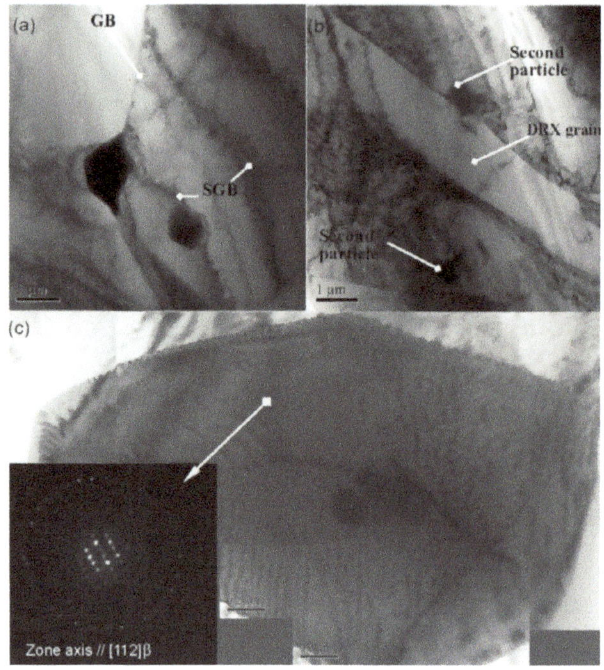

Figure 44. TEM images of the fractured specimen tested at 435 °C and 0.0005 s^{-1}. (a) An example of grain boundary and subgrain boundary, (b) a strip recrystallized grain, (c) dislocation network within the β phase [72].

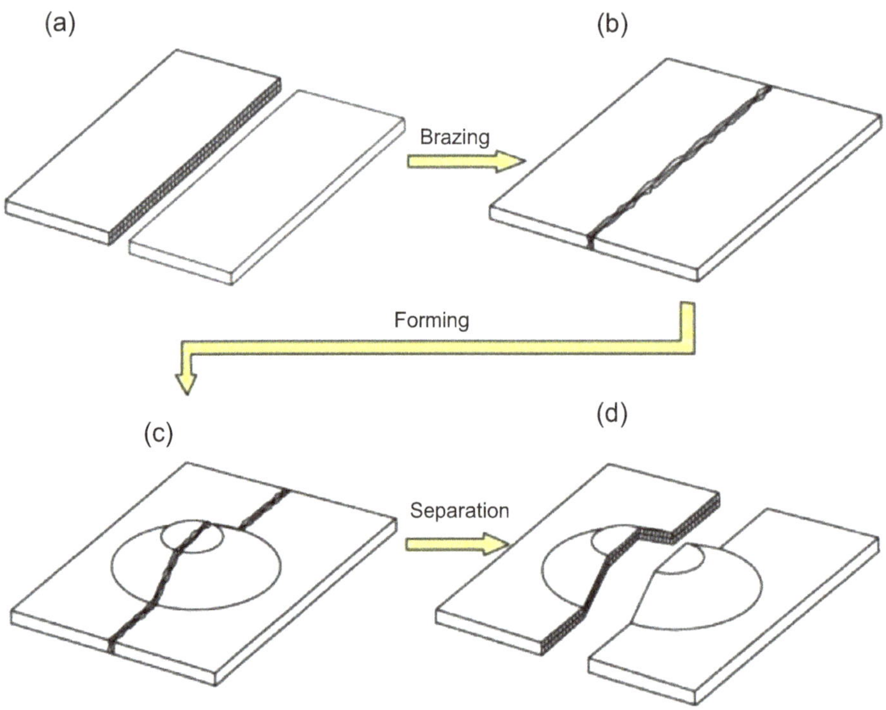

Figure 45. Schematic diagram of method for measuring through-thickness deformation by Jackson et al [73]. (a) Copper plate with gridded cross-section; (b) Plate with brazed joint; (c) Formed plate; (d) Separated plate with distorted grid pattern.

The springback effect is the main cause of the dimensional accuracy problems in the AISF process and has been widely investigated by several authors [80–83] Lasunon and Knight [80] developed a model to compare the accuracy of single-point incremental forming and double-point incremental forming and found the FE model was in good agreement with experimental data (see Figure 48). Several strategies have been proposed to overcome accuracy problems such as systems with flexible programmable back support structures [82] and systems with a localized heating device [83].

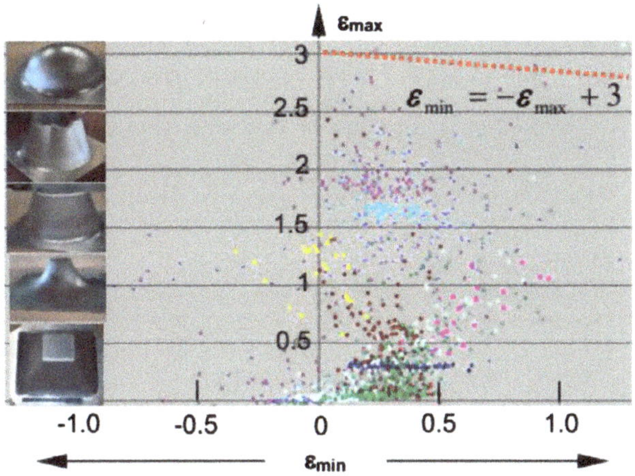

Figure 46. Non-traditional FLD formed by AISF method of 1.21 mm thick AA 3003-0 [76].

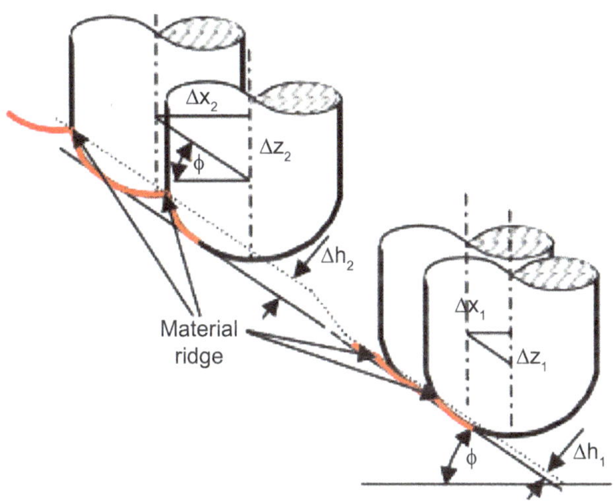

Figure 47. illustration of how the pitch size, Δz, influences roughness [60].

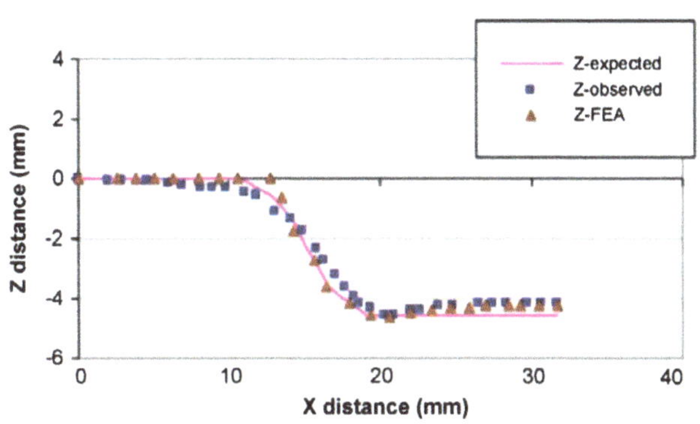

Figure 48. Comparison of profile plots gained from FEM and experimental results [80].

4.4. The issues to be addressed

A large amount of researches on these particular techniques has been undertaken to further enhance the formability of sheet and tube metal, although many problems still remains to be solved or improvements made to the aforementioned processes, among which some issues are listed below:

1. The accurate simulation of forming processes requires in-depth comprehension of the friction mechanism between the blank and the tool, considering the effects of lubrication. With the application of lubrication, formability of metal sheet is increased as a consequence of reduced friction and it is possible to get more homogeneous thickness distributions [84].
2. Some of the special forming methods are conducted under relatively high temperatures, which may affect the microscopic and mechanical properties of the deformed components, so thermal influence should also be identified.
3. The FLD under complex strain paths is very important in forming sheet or tubular components with complex geometries has not yet been investigated. Also the influences of temperature and viscoplastic effect on FLD have attracted great interests from researchers recently [85].
4. Further research targeting onto improved hardware, tool path generation and control strategy development would be helpful in securing industrial opportunities.

5. Conclusions

Lightweight materials such as ultra-high-strength steel, titanium alloys and aluminum alloys are used widely to help to achieve the goal of emissions- and energy-reduction. In this review, recent developments on some particular sheet forming methods and their forming mechanisms (HF, VPF, SGMF, SPF/DB, IHF and AISF) are critically summarized with the objective of evaluating the different research results and indicating the current applications in the aerospace and automotive industries.

In general, some of these techniques have been used extensively to fabricate hard to-form and complex-shaped components, i.e. hydroforming and superplastic forming processes, on the occasions where conventional forming methods cannot fulfill the requirements of the manufacturers. Some other techniques remain at the stage of laboratory research and need to be refined before spreading into modern industries.

References

1. J. Aspacher, Forming hardening concepts, 1st International Conference on Hot Sheet Metal Forming of High Performance Steel, Kassel, Germany, 2008, pp. 77–81.
2. R.R. Boyer, Applications of beta titanium alloys in airframes, Beta Titanium Alloys in the 1990's, Warrendale, PA, TMS, 1993, pp. 335–346.
3. M. Peters, J. Kumpfert, C.H. Ward, C. Leyens, Titanium alloys for aerospace applications, Advanced Engineering Materials 5, 6 (2003) 419–427.
4. K. Martchek, Modelling more sustainable aluminium, The International Journal of Life Cycle Assessment 11, 1 (2006) 34–37.

5. S. Toros, F. Ozturk, I. Kacar, Review of warm forming of aluminum-magnesium alloys, Journal of Materials Processing Technology 207, 1 (2008) 1–12.
6. T. Naka, G. Torikai, R. Hino, F. Yoshida, The effects of temperature and forming speed on the forming limit diagram for type 5083 aluminum-magnesium alloy sheet, Journal of Materials Processing Technology 113, 1 (2001) 648–653.
7. T. Naka, Y. Nakayama, T. Uemori, R. Hino, F. Yoshida, Effects of temperature on yield locus for 5083 aluminum alloy sheet, Journal of Materials Processing Technology 140, 1 (2003) 494–499.
8. D. Li, A. Ghosh, Tensile deformation behavior of aluminum alloys at warm forming temperatures, Materials Science and Engineering: A 352, 1 (2003) 279–286.
9. D. Li, A. Ghosh, Biaxial warm forming behavior of aluminum sheet alloys, Journal of Materials Processing Technology 145, 3 (2004) 281–293.
10. P. Cavaliere, Hot and warm forming of 2618 aluminium alloy, Journal of Light Metals 2, 4 (2002) 247–252.
11. R.R. Boyer, An overview on the use of titanium in the aerospace industry, Materials Science and Engineering: A 213, 1 (1996) 103–114.
12. C. Leyens, M. Peters, Titanium and titanium alloys, Wiley-VCH, Weinheim, 2003.
13. H. Karbasian, A.E. Tekkaya, A review on hot stamping, Journal of Materials Processing Technology 210, 15 (2010) 2103–2118.
14. D. Keith, HSS, AHSS and aluminum jockey for position in the race to cut auto curb weight, American Metal Market Monthly (2010) 1.
15. R. Kuziak, R. Kawalla, S. Waengler, Advanced high strength steels for automotive industry, Archives of Civil and Mechanical Engineering 8, 2 (2008) 103–117.
16. N.P. Lutsey, Review of technical literature and trends related to automobile mass-reduction technology, Institute of Transportation Studies, University of California, Davis, CA, 2010.
17. D. Anderson, New study finds increased use of advanced high-strength steels helps decrease overall vehicle weight, AISI News Release, Detroit, 2009.
18. M. Tolazzi, Hydroforming applications in automotive: a review, International Journal of Material Forming 3, 1 (2010) 307–310.
19. M. Koç, Hydroforming for advanced manufacturing, Woodhead Publishing, UK, 2008.
20. M. Koc, A. Taylan, An overall review of the tube hydroforming (THF) technology, Journal of Materials Processing Technology 108, 3 (2001) 384–393.
21. M. Ahmetoglu, K. Sutter, X.J. Li, T. Altan, Tube hydroforming: current research, applications and need for training, Journal of Materials Processing Technology 98, 2 (2000) 224–231.
22. B. Constantine, R. Roth, J.P. Clark, Substituting tube-hydroformed parts for automotive stampings: an economic model, Journal of Modeling 53, 8 (2001) 33–38.
23. L.H. Lang, Z.R. Wang, D.C. Kang, S.J. Yuan, S.H. Zhang, J. Danckert, K.B. Nielsen, Hydroforming highlights: sheet hydroforming and tube hydroforming, Journal of Materials Processing Technology 151, 1 (2004) 165–177.
24. M. Kleiner, M. Geiger, A. Klaus, Manufacturing of lightweight components by metal forming, CIRP Annals-Manufacturing Technology 52, 2 (2003) 521–542.
25. P.J. Bolt, N.A.P.M. Lamboo, P.J.C.M. Rozier, Feasibility of warm drawing of aluminium products, Journal of Materials Processing Technology 115, 1 (2001) 118–121.
26. L.J. Rhoades, M.L. Rhoades, Die forming metallic sheet materials, US Patent No. 5,085,068, 4 Feb. 1992.
27. Z.J. Wang, J.G. Liu, X.Y. Wang, Z.Y. Hu, B. Guo, Viscous pressure forming (VPF): state-of-the-art and future trends, Journal of Materials Processing Technology 151, 1 (2004) 80–87.
28. Z.J. Wang, Z.R. Wang, H.F. Yang, Experimental investigation for viscous medium drawing of sheet metal under the condition of non-uniform blank holder force, Journal of Plasticity Engineering 6, 2 (1999) 50–52.
29. Z.R. Wang, X.Y. Wang, Z.J. Wang, Numerical simulation and experimental research on the forming of thin-walled complex parts with viscous pressure forming VPF, NUMISHEET, Korea, 2002, pp. 249–252.
30. M. Ahmetoglu, J. Hua, S. Kulukuru, T. Altan, Hydroforming of sheet metal using a viscous pressure medium, Journal of Materials Processing Technology 146, 1 (2004) 97–107.
31. L.B. Shulkin, R.A. Posteraro, M.A. Ahmetoglu, G.L. Kinzel, T. Altan, Blank holder force (BHF) control in viscous pressure forming (VPF) of sheet metal, Journal of Materials Processing Technology 98, 1 (2000) 7–16.
32. X.Y. Wang, J.C. Xia, G.A. Hu, Z.J. Wang, Z.R. Wang, Sheet bulging experiment with a viscous pressure-carrying medium, Journal of Materials Processing Technology 151, 1 (2004) 340–344.
33. J. Liu, Q. Peng, Y. Liu, Z. Wang, Viscous pressure bulging of aluminum alloy sheet at warm temperatures, Journal of Mechanical Science and Technology 21, 10 (2007) 1505–1511.
34. C.C. Zhao, G.J. Dong, H. Xiao, Y.S. Wang, New process of solid granule medium forming, Journal of Mechanical Engineering 8 (2009) 046.
35. G.J. Dong, C.C. Zhao, M.Y. Cao, Flexible-die forming process with solid granule medium on sheet metal, Transactions of Nonferrous Metals Society of China 23, 9 (2013) 2666–2677.
36. M.Y. Cao, C.C. Zhao, G.J. Dong, Numerical simulation on granules medium drawing process parameters of magnesium alloy sheet, The Chinese Journal of Nonferrous Metals 22, 11 (2012) 2992–2999.
37. M. Grüner, M. Merklein, Numerical simulation of hydro forming at elevated temperatures with granular material used as medium compared to the real part geometry, International Journal of Material Forming 3, 1 (2010) 279–282.
38. M. Grünerak, M. Merklein, Influences on the molding in hydroforming using granular material as a medium, AIP Conference Proceedings 1383 (2011) 645–652.
39. J. Bonet, A. Gil, R.D. Wood, R. Said, R.V. Curtis, Simulating superplastic forming, Computer Methods in Applied Mechanics and Engineering 195, 48 (2006) 6580–6603.
40. G.J. Davies, J.W. Edington, C.P. Cutler, K.A. Padmanabhan, Superplasticity: a review, Journal of Materials Science 5, 12 (1970) 1091–1102.
41. W.A. Backofen, I.R. Turner, D.H. Avery, Superplasticity in an Al-Zn alloy, Transactions of ASM 57, 4 (1964) 980–990.
42. J. Bonet, A. Gil, R.D. Wood, R. Said, R.V. Curtis, Simulating superplastic forming, Computer Methods in Applied Mechanics and Engineering 195, 48 (2006) 6580–6603.
43. A.J. Barnes, Superplastic forming 40 years and still growing, Journal of Materials Engineering and Performance 16, 4 (2007) 440–454.
44. L.D. Hefti, Commercial airplane applications of superplastically formed AA5083 aluminum sheet, Journal of Materials Engineering and Performance 16, 2 (2007) 136–141.

45. L.D. Hefti, Using superplastic forming as a means of achieving cost benefits as well as enhancing aircraft performance, High Performance Metallic Materials for Cost Sensitive Applications, John Wiley and Sons, Hoboken, NJ, 2002 pp. 64–72.
46. R. Martin, D. Evans, Reducing costs in aircraft: the metals affordability initiative consortium, Journal of Modeling 52, 3 (2000) 24–28.
47. A.J. Barnes, Superplastic aluminum forming-expanding its techno-economic niche, Materials Science Forum 304 (1999) 785–796.
48. J. Wittenauer, T.G. Nieh, J. Wadsworth, A first report on superplastic gas-pressure forming of ceramic sheet, Scripta Metallurgica et Materialia 26, 4 (1992) 551–556.
49. T. Waniuk, J. Schroers, W.L. Johnson, Timescales of crystallization and viscous flow of the bulk glass-forming Zr-Ti-Ni-Cu-Be alloys, Physical Review B 67, 18 (2003) 184203.
50. L.H. Lang, S.H. Wang, C.L. Yang, Investigation on the innovative impact hydroforming technology, The 11th International Conference on Numerical Methods in Industrial Forming Processes: NUMIFORM 2013 1532, 1 (2013) 791–798.
51. H. Tominga, M. Takamatsu, Hydropunch, a pneumatic-hydraulic-forming machine, International Conference of the Center for High Energy Forming, Estes Park, USA, 1969.
52. B. Mason, Sheet metal forming for small batches, Bachelor thesis, University of Nottingham, May 1978.
53. W.C. Emmens, G. Sebastiani, A.H.V.D. Boogaard, The technology of incremental sheet forming – a brief review of the history, Journal of Materials Processing Technology 210, 8 (2010) 981–997.
54. B.T. Araghi, A. Göttmann, M. Bambach, G. Hirt, G. Bergweiler, J. Diettrich, M. Steiners, A. Saeed-Akbari, Review on the development of a hybrid incremental sheet forming system for small batch sizes and individualized production, Production Engineering 5, 4 (2011) 393–404.
55. M. Bambach, G. Hirt, S. Junk, Modelling and experimental evaluation of the incremental CNC sheet metal forming process, 7th International Conference on Computational Plasticity, Barcelona, Spain, 2003.
56. T.J. Kim, D.Y. Yang, Improvement of formability for the incremental sheet metal forming process, International Journal of Mechanical Sciences 42, 7 (2000) 1271–1286.
57. D. Leach, A.J. Green, A.N. Bramley, A new incremental sheet forming process for small batch and prototype parts, 9th International Conference on Sheet Metal, Leuven, 2001.
58. L. Filice, L. Fratini, F. Micari, Analysis of material formability in incremental forming, CIRP Annals-Manufacturing Technology 51, 1 (2002) 199–202.
59. S. Matsubara, Incremental backward bulge forming of a sheet metal with a hemispherical head tool-a study of a numerical control forming system II, Journal-Japan Society for Technology of Plasticity 35 (1994) 1311–1311.
60. J. Jeswiet, F. Micari, G. Hirt, A. Bramley, J. Duflou, J. Allwood, Asymmetric single point incremental forming of sheet metal, CIRP Annals-Manufacturing Technology 54, 2 (2005) 88–114.
61. J. Jeswiet, E. Hagan, Rapid prototyping of a headlight with sheet metal, Canadian Institute of Mining, Metallurgy and Petroleum, Canada, 2001, pp. 109–114.
62. F. Micari, Single point incremental forming: recent results, Seminar on Incremental Forming, Cambridge University, Cambridge, 22, 2004.
63. P.D. Wu, J.D. Embury, D.J. Lloyd, Y. Huang, K.W. Neale, Effects of superimposed hydrostatic pressure on sheet metal formability, International Journal of Plasticity 25, 9 (2009) 1711–1725.
64. F. Fuchs, Hydrostatic pressure – its role in metal forming, Mechanical Engineering 88, 4 (1966) 34–40.
65. X.L. Cui, X.S. Wang, S.J. Yuan, Deformation analysis of double-sided tube hydroforming in square-section die, Journal of Materials Processing Technology 214, 7 (2014) 1341–1351.
66. J.G. Liu, Z.J. Wang, Q.Y. Meng, Numerical investigations on the influence of superimposed double-sided pressure on the formability of biaxially stretched AA6111-T4 sheet metal, Journal of Materials Engineering and Performance 21, 4 (2012) 429–436.
67. A. Assempour, H.K. Nejadkhaki, R. Hashemi, Forming limit diagrams with the existence of through-thickness normal stress, Computational Materials Science 48, 3 (2010) 504–508.
68. J.F. Michel, P. Picart, Size effects on the constitutive behaviour for brass in sheet metal forming, Journal of Materials Processing Technology 141, 3 (2003) 439–446.
69. T.G. Nieh, J. Wadsworth, O.D. Sherby, Superplasticity in metals and ceramics, Cambridge University Press, Cambridge, 2005.
70. M.A. Kulas, W.P. Green, E.M. Taleff, P.E. Krajewski, T.R. McNelley, Deformation mechanisms in superplastic AA5083 materials, Metallurgical and Materials Transactions A 36, 5 (2005) 1249–1261.
71. M.A. Kulas, W.P. Green, E.M. Taleff, P.E. Krajewski, T.R. McNelley, Failure mechanisms in superplastic AA5083 materials, Metallurgical and Materials Transactions A 37, 3 (2006) 645–655.
72. L. Li, X. Zhang, Y. Deng, C. Tang, Superplasticity and microstructure in Mg-Gd-Y-Zr rolled sheet, Journal of Alloys and Compounds 485, 1 (2009) 295–299.
73. K. Jackson, J. Allwood, The mechanics of incremental sheet forming, Journal of Materials Processing Technology 209, 3 (2009) 1158–1174.
74. S. Kalpakcioglu, On the mechanics of shear spinning, Journal of Engineering for Industry 83, 2 (1961) 125–130.
75. M.B. Silva, P.S. Nielsen, N. Bay, P.A.F. Martins, Failure mechanisms in single-point incremental forming of metals, The International Journal of Advanced Manufacturing Technology 56, 9–12 (2011) 893–903.
76. J. Jeswiet, D. Young, Forming limit diagrams for single-point incremental forming of aluminium sheet, Proceedings of the Institution of Mechanical Engineers, Part B: Journal of Engineering Manufacture 219, 4 (2005) 359–364.
77. L. Filice, L. Fratini, F. Micari, Analysis of material formability in incremental forming, CIRP Annals-Manufacturing Technology 51, 1 (2002) 199–202.
78. E. Hagan, J. Jeswiet, Analysis of surface roughness for parts formed by computer numerical controlled incremental forming, Proceedings of the Institution of Mechanical Engineers, Part B: Journal of Engineering Manufacture 218, 10 (2004) 1307–1312.
79. S. Junk, G. Hirt, I. Chouvalova, Forming strategies and tools in incremental sheet forming, Proceedings of the 10th International Conference on Sheet Metal, Belfast, 2003.
80. O. Lasunon, W.A. Knight, Comparative investigation of single-point and double-point incremental sheet metal forming processes, Proceedings of the Institution of Mechanical Engineers, Part B: Journal of Engineering Manufacture 221, 12 (2007) 1725–1732.

81. S. Junk, G. Hirt, I. Chouvalova, Forming strategies and process modelling for CNC incremental sheet forming, CIRP Annals-Manufacturing Technology 53, 1 (2004) 203–206.
82. H. Meier, V. Smukala, O. Dewald, J. Zhang, Two point incremental forming with two moving forming tools, Key Engineering Materials 344 (2007) 599–605.
83. J.R. Duflou, B. Callebaut, J. Verbert, H.D. Baerdemaeker, Improved SPIF performance through dynamic local heating, International Journal of Machine Tools and Manufacture 48, 5 (2008) 543–549.
84. H.Y. Wu, C.H. Chiu, J.Y. Wang, S. Lee, Effect of lubrication on deformation characteristics of a superplastic 5083 Al alloy during bi-axial deformation, Materials Science and Engineering: A 427, 1 (2006) 268–273.
85. M. Serenelli, M. Bertinetti, P. Turner, J. Signorelli, A theoretical study on forming limit diagram predictions using viscoplastic polycrystalline plasticity models, Key Engineering Materials 473 (2011) 327–334.

17

Effect of post-welding heat treatment on wear resistance of cast-steel die with surfacing layer

Wujiao Xu*, Xin Ma, Nongjie Tang, Li Zhu, Wuhua Li, and Yongfeng Ding

College of Material Science and Engineering, Chongqing University, Chongqing 400044, PR China

Abstract – The wear resistance capability of die surfacing layer under different Post-Welding Heat Treatments (PWHT) was analysed by Finite Element (FE) simulation and experiments. Taking hot forging process of a crankshaft as an example, a wear model of the hot forging die coated with surfacing layer was established using FE software DEFORM-3D. The simulation results indicated that the wear resistance capability of the die surfacing layer is optimal when tempering temperature and holding time are 550 °C and 4 h respectively. To verify the wear computational results, 16 groups of PWHT orthogonal wear tests were performed at a temperature of 400 °C, which is a similar temperature to that occurs in an actual hot forging die. The wear-test result showed a good agreement with the FE simulation. SEM observation of the wear debris on 16 specimens showed that oxidative wear is dominant when the temperature was in 400 °C. Furthermore, when tempering temperature and holding time were 550 °C and 4 h respectively, the carbide alloy dispersively distributes in the metallographic structure, which helps to improve the wear resistance of the surfacing layer.

Key words: Post-welding heat treatments, Surfacing layer, Wear resistance capability, Finite Element (FE) simulation

1. Introduction

In a hot forging process, complex loading conditions often lead to several types of die failures such as wear, plastic deformation, fatigue, etc., among which wear accounts for 70% of all the hot-forging die failures [1]. The lifespan of hot forging dies is one of the most important concerns in hot forging. From a economic point of view, accurate prediction of the die service life and extending it as much as possible are of significant importance to forging industry.

At present, closed-die forging processes are being developed towards large-scale, high-precision and high quality production. For heavy forgings, the corresponding hot-forging dies are normally very large. Traditionally, a hot-forging die may be made into an integral part made of a hot working steel. For this type of the die, the strength of die-cavity surfaces may be improved through PVD, PACVD, CVD coatings and diffusion processes such as Nitriding, in order to extend the service life of the die [2–5]. However, the cost per unit workpiece for such a die remains very high due to the high cost on the die material and on heat treatment.

For the production of heavy forgings, it has been proposed to apply the Functionally Graded Material (FGM) for manufacturing dies [6]. Considering the requirement of a hot forging process and the need to reduce the die cost, die construction with casting-steel substrate coated with the surfacing layer was adopted, as illustrated in Figure 1.

Since wear is the predominant factor affecting the die life in hot forging, the concept of the hot forging cast-steel die coated with the surfacing layer could be a useful solution. To effectively support the development of such a die, in-depth understanding of the die-wear would be needed.

Die-wear related topics such as wear mechanisms, wear modelling, prediction of die life, wear resistance, etc., have been subjected to intensive research in the past and recently. Archard wear model has been applied to the prediction of tool life in hot forging process, as a general purpose model [7]. Painter et al. developed a numerical simulation approach to predict wear and then optimized the process [8]. Schaefer predicted the tool wear with consideration of the changes of the hardness in the tool surface-layer [9]. Taking the deviation of the temperatures in preheated billets into consideration, Kim and Kang used a finite element method to study wear and plastic deformations of the dies [10]. Abachi et al. analysed the die wear in hot forging by a finite volume method, based on which a value of wear coefficient was recommended [11].

*e-mail: xuwujiao_cq@163.com

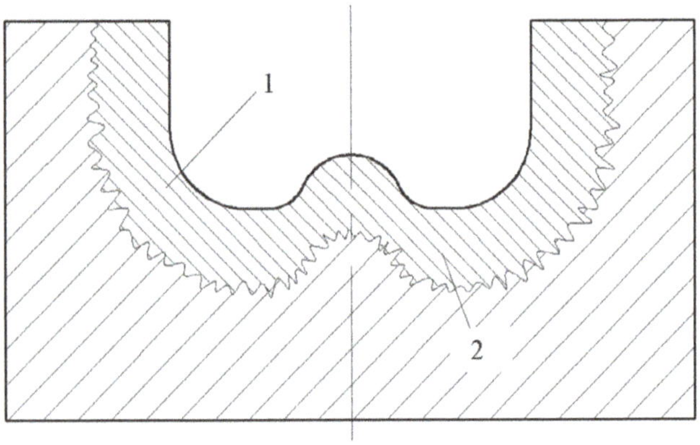

Figure 1. The schematic of a hot-forging cast-steel die coated with the surfacing layer: 1 – the surfacing layer, 2 – the casting-steel.

In this paper, besides the analysis of die-wear through FE modelling, analysis of the wear resistance capability of the die surfacing layer subjected different Post-Welding Heat Treatments (PWHT), are reported.

2. The die material

To study the wear resistance of a hot forging cast-steel die coated with surfacing layer, gas metal arc welding with 545E flux cored wires was used in surfacing welding. The cast-steel substrate was made of ZG25MnCrNiMo. 545E flux cored wires were utilized in the surfacing process. The dimensions of the die-block were 600 mm × 380 mm × 190 mm.

To ensure reliability of the forging die, thickness of the surfacing layer should not be less than 20 mm. At least 10 layers were created in the welding process, in which the thickness of each layer was about 2 mm. Oxide skin and welding slag on the surface were removed after every surfacing layer was added. The welding current and voltage were 130 A and 26 V respectively. The welding velocity was set to be 180 mm min^{-1} and the welding inter-pass temperature was chosen to be 300 ± 20 °C. Before welding the substrate was preheated to 350 ± 25 °C. Chemical compositions for casting-steel substrate (ZG25MnCrNiMo) and surfacing layer (545E) are shown in Table 1 and the mechanical properties listed in Table 2.

The post-welding heat treatment, i.e. tempering process was carried out in an induction heating furnace. For 16 groups of orthogonal experiment, selected tempering temperatures were 420 °C, 480 °C, 550 °C, 600 °C and holding times 1 h, 2 h, 4 h and 6 h respectively.

To predict the wear of the cast-steel dies coated with surfacing layers under different PWHT, it is essential to establish constitutive relationships of the surfacing layers under various PWHT. To gain the constitutive relationships for the 16 groups of the experiment, a series of high temperature (250 °C, 300 °C, 350 °C, 400 °C, 450 °C) compression tests were carried out on a Gleeble thermo-mechanical simulator (Gleeble 3500, heating rate: 10 °C/s, holding time: 150 s and amount of compression: 8% and strain rate: 0.0001 s) which are shown

Table 1. Chemical composition of the casting-steel substrate and surfacing layer (wt.%).

Chemical composition	C	Si	Mn	P	S	Cr	Ni	Mo	Cu	Fe
ZG25MnCrNiMo (casting steel substrate)	0.27	0.53	1.37	0.02	0.01	0.55	0.43	0.27	0.10	97.57
545E (surfacing layer)	0.10	0.77	0.83	0.008	0.007	11.32	2.12	2.43	–	82.42

Table 2. The mechanical properties of the casting-steel substrate and surfacing layer.

Die material	Young modulus E (GPa)	Poisson ratio v	Hardness (HRC)	Coefficient of friction	Yield stress $\sigma_{0.2}$ (MPa)
ZG25MnCrNiMo (casting steel substrate)	42.5 ± 1.6	0.3 ± 0.016	24 ± 0.7	–	512 ± 12.1
545E (surfacing layer)	27.6 ± 0.9	0.3 ± 0.021	48 ± 1.2	0.2 ± 0.019	588 ± 15.3

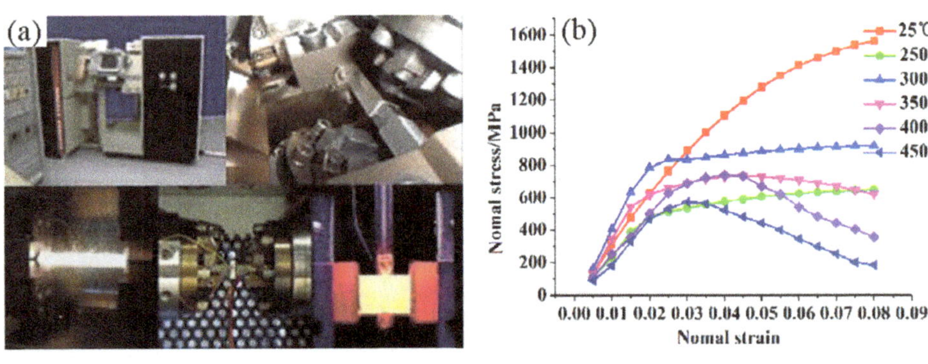

Figure 2. (a) The Gleeble thermo-mechanical simulator; (b) the constitutive relationship of the surfacing layer in different temperatures when the PWHT was carried out at 550 °C and 4 h.

Table 3. The main parameters used in FE simulation.

Parameters	Value
Billet material	AISI-5140
Billet initial forging temperature	1150 °C
Die initial-temperature	250 °C
Coefficient of heat convection	0.02 N/s mm °C
Radiation coefficient	0.3
Contact-surface heat transfer coefficient	5
Machine-ram speed speed	280 mm/s
Displacement increment	1 mm

in Figure 2a. One group of the constitutive relationships whose tempering temperature and holding time were 550 °C and 4 h respectively are shown in Figure 2b. This was then, late, applied to the establishment of the wear model through using DEFORM-3D analysis software.

3. Wear computation and experimental verification

3.1. The wear model established through DEFORM-3D

To study the wear behaviour of the hot forging cast-steel die coated with surfacing layer, the hot forging of a crankshaft was analysed as a study case. For this crankshaft, there were two operations involved in the hot forging, namely the pre-forging and the finish forging. Using DEFORM-3D, a coupled thermo-mechanical simulation of the crankshaft forging was carried out. The main parameters of the simulation are listed in Table 3.

The Archard wear model, expressed in equation (1), was adopted to simulate the wear behaviour of the surfacing layer:

$$dw_i = \int K \frac{P_i^a v_i^b}{H^c} \mathrm{d}t \tag{1}$$

where dw is the wear depth of the die cavity, P and v the contact pressure and relative sliding velocity respectively. Generally a and b are set as 1, and c equals to 2 for the simulation of the tool steels as the tool-material [7]. H is hardness of the tool material and dt the time increment. K is wear coefficient and the subscript i refers to the surface node on the forging die. The effect of the different PWHT on the constitutive relationship of the surfacing layer was considered when calculating the wear depth.

The simulated finished forging and the die model shown in Figure 3a. After the wear calculation was completed, the wear-depth distribution in the forging-die cavity was displayed (Figure 3b). It can be observed that the area having severe wear in the die cavity appears at the cavity for the forming of the crankshaft front-end (the wear corresponding to the flash land was not considered in the analysis reported in this paper). The field distributions of the crankshaft can be seen in Figure 4. It can be seen that the severe wears occur in those zones where the contact pressure, the velocity and the temperature were higher.

Due to that the severely worn area was at the die-cavity section where forming the crankshaft front-end took place, the section A-A was chosen to be focused for the study of the effect of PWHT on the wear depth of the surfacing layer. The circumferential coordinate system defined to illustrate the wear depth in the section A-A is shown in Figure 3c.

Figure 5 shows the computed wear-depth along the section A-A in the forging die with different PWHTs (including different tempering temperatures and holding time). Among the 16 combinations of the PWHT conditions, the optimal tempering temperature and holding time were found to be 550 °C and 4 h respectively. Under this PWHT condition the die wear-depth peak (shown in the circumferential coordinate system) in section A-A, in a single forging cycle, was 9.92×10^{-5} mm, which is shown in Figure 7a.

3.2. Wear test

The wear-resistant plate material and pressure-head material used in the wear test are 545E and AISI-5140 receptively, which are the same kinds of the materials for the surfacing

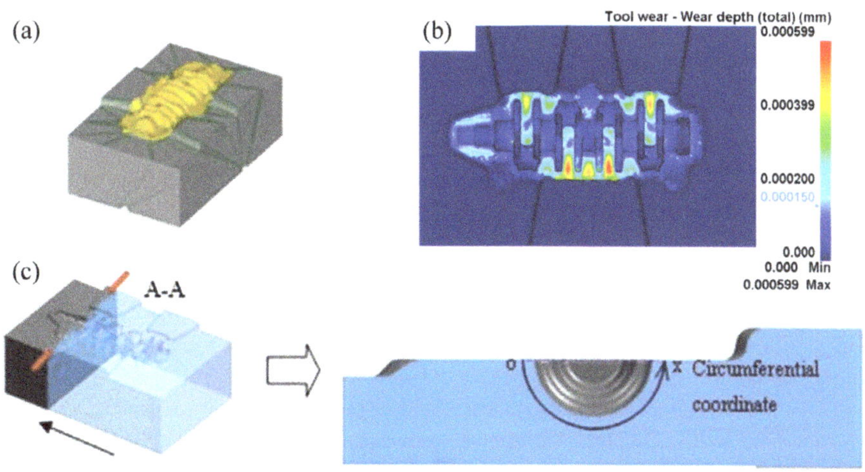

Figure 3. (a) The finish forging of the crankshaft; (b) wear depth distribution of the crankshaft; (c) circumferential coordinate system of the section A-A.

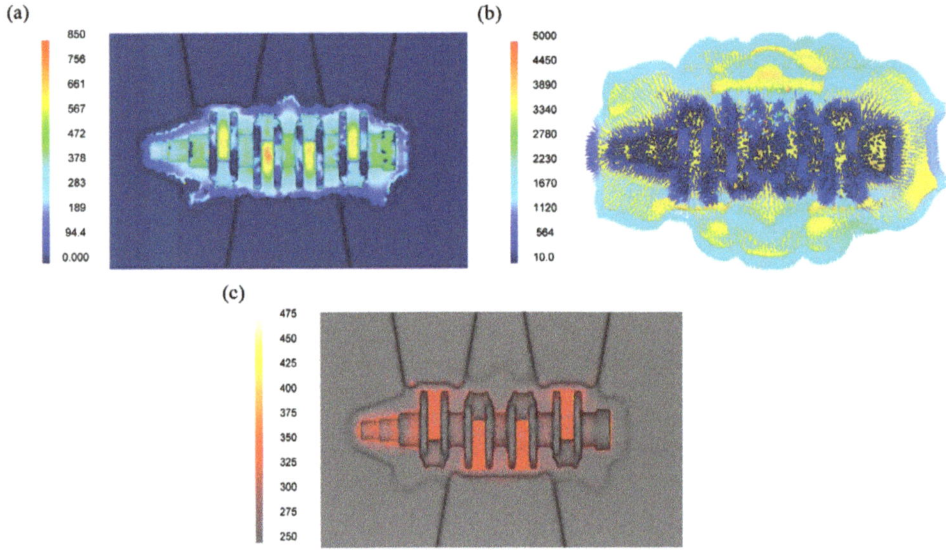

Figure 4. Results of the FE Simulation of crankshaft forging: (a) Contact pressure distribution (MPa); (b) Velocity distribution (mm/s); (c) Temperature distribution (°C).

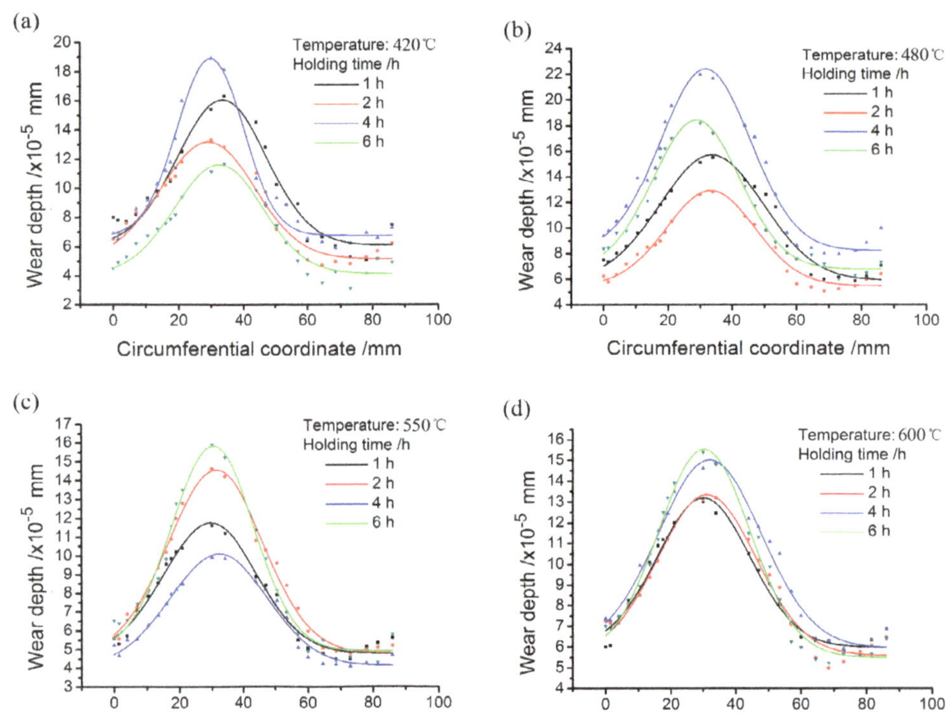

Figure 5. The computed wear-depth of the section A-A in the forging die with the tempering temperature of: (a) 420 °C, (b) 480 °C, (c) 550 °C and (d) 600 °C.

layer and the billet that were used in the forging test respectively. The applied contact force and relative sliding velocity used in the wear test were similar to that obtained from the FEM simulation of crankshaft hot forging. The wear test was carried out on an HSR-2 M wear-testing machine (temperature: 400 °C; force: 50 N, speed: 400 r/min; time: 3 min; and length: 5 mm). Figure 6 shows the HSR-2 M wear-testing machine and the friction pairs used in the wear test.

To verify the wear calculation result presented in Figure 7a, a series of wear tests were conducted. Figure 7b shown the wear-test result corresponding to 100 forging cycles. It can be seen that the optimal tempering temperature and holding time are also 550 °C and 4 h respectively, and the wear depth of wear-resistance plate, 0.00987 mm, is minimum, comparing it to that corresponding to all the PWHT conditions. The wear test results show a close agreement with the wear calculation result.

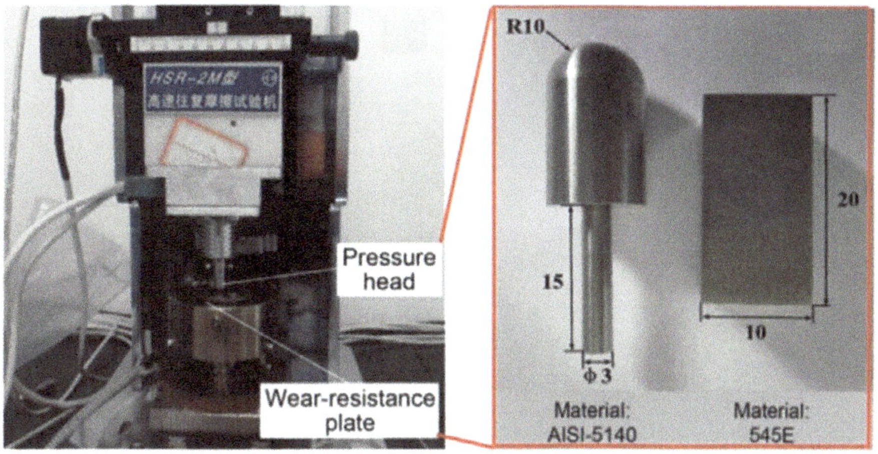

Figure 6. The HSR-2M wear-test machine and the friction pairs used in the wear test

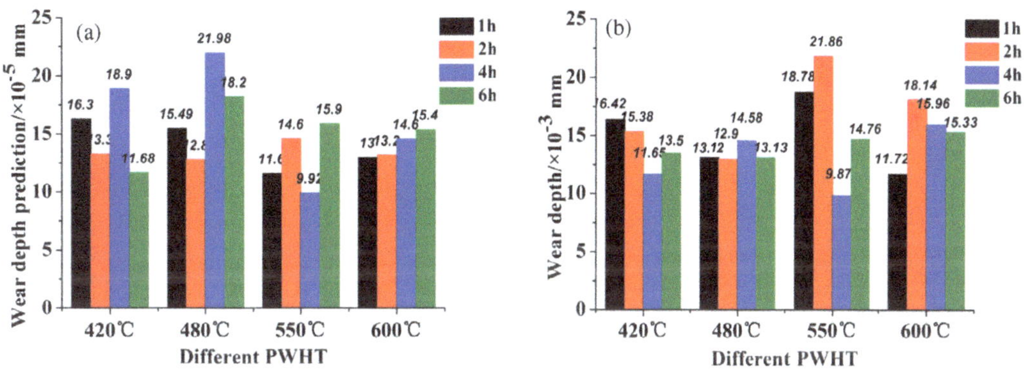

Figure 7. (a) Computed wear-depths for different PWHT conditions (single cycle simulation); (b) measured wear depths (after 100 forging cycles).

3.3. Forging of the crankshaft

In order to further verify FE simulation results, a serious of preparatory work was carried out. Firstly, surfacing welding was performed on a hot-forging cast-steel die, and the surfacing welding parameters were mentioned in the section of the die material used. After the weld surfacing, the cast-steel die was air cooled to about 300 °C. Then a post-welding heat treatment was carried out on the die for stress relief (corresponding to an optimal condition, i.e. tempering temperature 550 °C, holding time 4 h). These will then be followed by crankshaft forging. A 6300T hot forging press was used for the forging of the crankshafts.

After 10,390 forging cycles, the die cavity was examined using an CMM machine. Figure 8a is the point cloud of the worn surfacing layer, and the Figure 8b shows a comparison between the point cloud tested and the designed model of the finish forging die, from which the wear depth of surfacing layer of the die can be obtained. Figure 8c illustrates the finish forging die which is coated with surfacing layer. Comparison of the die wear depth at the section A-A between the simulation result and forging is illustrated in Figure 8d, which shows a good agreement with each other. It indicates the FEM simulation of the hot forging die wear is valid.

4. Wear mechanism and strengthening mechanism

4.1. SEM observation and wear mechanism

In order to gain insight into the wear mechanism, the wear debris were examined using SEM. The wear debris images corresponding to the optimal PWHT are illustrated in Figure 9. SEM examinations identified that there are three main types of wear during the forging process, i.e. abrasive wear, adhesive wear and oxidative wear. The adhesive wear debris corresponded to a rough surface, an uneven thickness and little oxygen. The shape of abrasive wear debris is banded. The oxidative wear debris corresponded to a smooth surface, an even thickness, and they contain a lot of oxygen. The proportion of the three types of wear debris was calculated using MATLAB, and it can be seen that the proportion of oxidative wear debris increases gradually as the holding time increases.

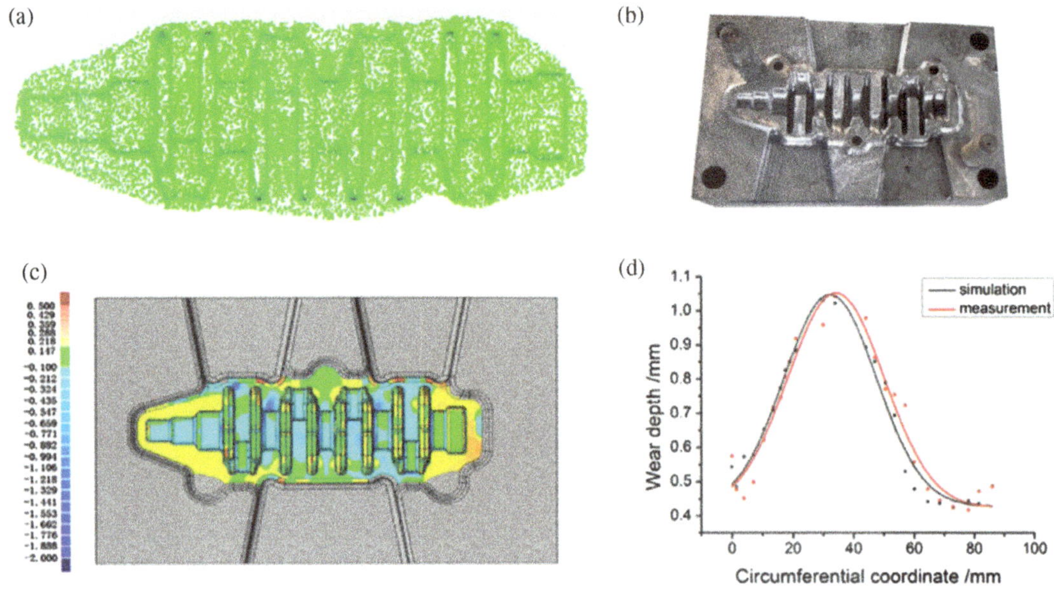

Figure 8. (a) The point cloud of the worn surfacing layer, (b) comparison of the point cloud tested and the design model of finish forging die, (c) finish forging die coated with surfacing layer and (d) comparison of the die wear depth at section A-A between simulation and the practical production.

Figure 9. SEM images of wear debris when tempering temperature was 550 °C and holding time of: (a) 1 h, (b) 2 h, (c) 4 h and (d) 6 h.

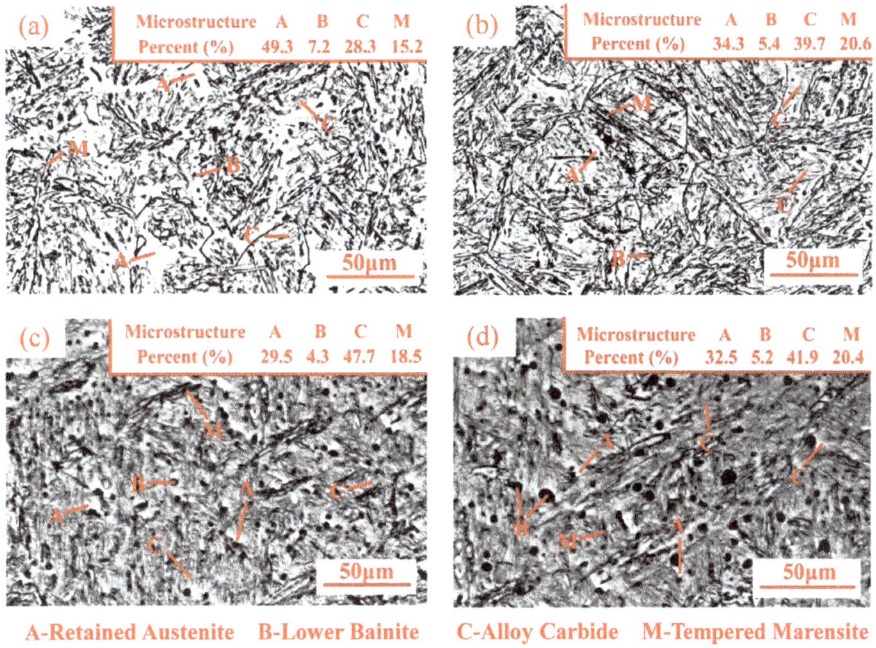

Figure 10. Microstructure of the surfacing layer with the tempering temperature is 550 °C and holding time of: (a) 1 h, (b) 2 h, (c) 4 h and (d) 6 h.

Therefore, oxidative wear mechanism plays a dominant role among all the wear mechanisms.

4.2. Microstructure investigation and strengthening mechanism

To understand the mechanism of strengthening due to PWHT on the wear resistance of the surfacing layer, the microstructure investigations were conducted. Specimens from the surfacing layers were etched with 4% nitric acid solution for 15 s. Then, a quantitative metallographic analysis system was used to calculate the percentage of microstructures and the result can be seen in Figure 10.

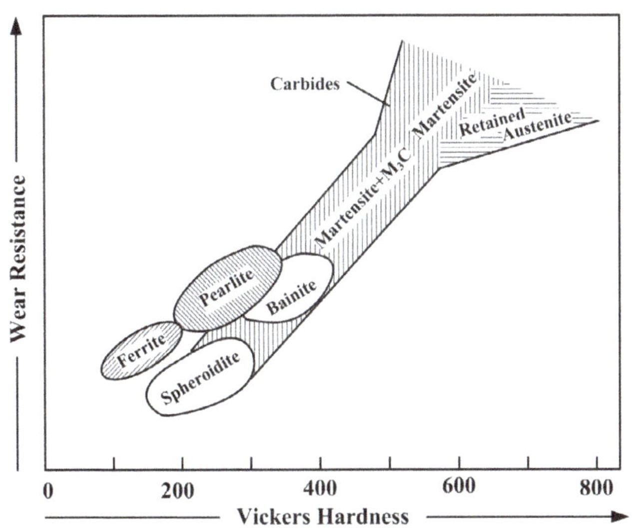

Figure 11. The influence of microstructures on wear resistance.

It can be seen from Figure 10a that the microstructure of the surfacing layer is composed of tempered martensite, lower bainite, retained austenite and few undissolved alloy carbide when the tempering temperature was 550 °C and holding time 1 h. When the tempering temperature was kept as 550 °C and the holding time extended to 2 h, the retained austenite dissolved partly and the alloy carbide precipitation slightly increased, shown in Figure 10b. It is obvious, as shown in Figure 10c, that the retained austenite transformed almost completely and the alloy carbide was dispersively distributed in the metallographic structure when the holding time was 4 h, which is closely related to the wear resistance improvement of the surfacing layer. While the holding time was prolonged to 6 h, the alloy carbide decreased, as illustrated in Figure 10d. The decrease in dispersivity of alloy carbide resulted in the reduced wear-resistance of the surfacing layer, comparing to that with holding time of 4 h. Glaeser [12] studied the relative wear resistance of different microstructures for steel, from which it can be known that the wear resistance of alloy carbide is better when it dispersively distributes in the metallographic structure. The study reported in this paper, once again, confirms such a finding.

5. Conclusions

For large hot forging, replacing a single piece die made of the hot working steel with the die construction consisting of casting-steel substrate coated with surfacing layer is beneficial, in terms of reducing the die cost and prolonging the service life of the die. The study reported in this paper is focused on the effect of Post Welding Heat Treatments (PWHT) on the wear resistance of the surfacing layer, based on which the following conclusions are drawn:

1. The FE results showed that wear depth of the surfacing layer is determined significantly by the PWHT, and the optimal tempering temperature and holding time are 550 °C and 4 h respectively. The wear test and forging test results show a good agreement with the FE simulation.
2. Similar to normal hot-forging dies, SEM examinations identified three main types of wear debris existing when a surfacing layer structure was used for the die, and the oxidative wear debris were dominant.
3. Microstructure investigations on the surfacing layer with different PWHT conditions showed that the alloy carbide dispersively distributing in the metallographic structure could improve the wear resistance of the surfacing layer effectively, when the tempering temperature and holding time are 550 °C and 4 h respectively.

Acknowledgements. This study was financially supported by the National Natural Science Foundation of China through a funded project (No. 51275543) and the Fundamental Research Funds for the Central Universities of China (No. CDJZR14130005 and No. CDJZR13130087).

References

1. K. Lange, L. Cesr, M. Geiger, Crip. Ann-manuf. Techn. 41 (1992) 667–675.
2. L. Dobrzanski, M. Polok, J. Mater. Process. Technol. 155 (2004) 1995–2001.
3. L. Dobrzanski, D. Pakula, A. Křiž, J. Mater. Process. Technol. 175 (2006) 179–185.
4. K.T. Rie, A. Gebauer, J. Wöhle, Surf. Coat. Technol. 74 (1995) 375–381.
5. H. Hoche, J. Schmit, Surf. Coat. Technol. 205 (2011) 145–150.
6. V. Birman, L.W. Byrd, Appl. Mech. Rev. 60 (2007) 195–216.
7. J.F. Archard, J. Appl. Phys. 24 (1953) 981–988.
8. B. Painter, R. Shivpuri, T. Altan, J. Mater. Process. Technol. 59 (1996) 132–143.
9. B.A. Behrens, F. Schaefer, J. Mater. Process. Technol. 167 (2005) 309–315.
10. D.H. Kim, B.M. Kim, C.G. Kang, Finite. Elem. Anal. Des. 41 (2005) 1225–1269.
11. S. Abachi, M. Akkok, I.M. Gokler, Tribol. Int. 43 (2010) 467–473.
12. W.A. Glaeser, Materials for Tribology, Elsevier Science Ltd, Amsterdam, The Netherlands, 1992.

18

Effect of welding parameters on mechanical properties of GTAW of UNS S31803 and UNS S32750 weldments

Prabhu Paulraj* and Rajnish Garg

University of Petroleum & Energy Studies, Dehradun 248007, Uttarakhand, India

Abstract – Duplex Stainless Steel (DSS) and Super Duplex Stainless Steel (SDSS) pipes were welded by Gas Tungsten Arc Welding (GTAW) process. The effect of welding parameters such as heat input, cooling rate, shielding/purging gas composition and interpass temperature on tensile strength, hardness and impact toughness were studied. The microstructure analysis revealed presence of intermetallic phases at root region of the weldments. All mechanical properties were improved at lower heat input and high cooling rate due to grain refinement and balanced microstructure [ferrite and austenite]. All weldments exhibited higher strength than base materials. Weld root region was harder than centre and cap region. SDSS is more susceptible to sigma phase formation due to higher alloying elements and weld thermal cycles, which lead to considerable loss of toughness. Higher nitrogen contents in shielding and purging gas resulted strengthening of austenite phase and restriction of dislocations, which ultimately improved mechanical properties. Higher interpass temperature caused reduction in strength and toughness because of grain coarsening and secondary phase precipitation.

Key words: DSS, SDSS, Tensile strength, Hardness, Impact toughness, Heat input, Shielding gas, Interpass temperature

1. Introduction

Duplex stainless steels (DSS) and Super Duplex Stainless Steels (SDSS) are dual phased steels comprising ferrite and austenite theoretically in equal proportions. They have exceptional mechanical and corrosion properties. Super duplex stainless steels have high chromium and molybdenum content which makes them highly corrosion resistive and higher mechanical strength. DSS and SDSS have wide range of applications in offshore, chemical, paper and pulp industries [1–3].

Welding is a key fabrication technique in applications of duplex and super duplex stainless steels. Welding of DSS and SDSS grades are challenging operation due to their complex microstructure. Improper welding cycles may lead to destroy material properties such as drastic reduction in toughness, strength and corrosion resistance [4].

It is important to obtain balanced microstructure (50% Austenite and 50% Ferrite) after welding. The high arc energy processes (high heat input) such as Submerged Arc Welding (SAW), Gas Metal Arc Welding (GMAW) and Gas Tungsten Arc Welding (GTAW) induce slower cooling rates. This leads to formation of higher amount of austenite as well as wide heat affected zones (HAZ). On the other hand, low arc energy processes like Electron Beam Welding (EBW), Laser Beam Welding (LBW) cause faster cooling and hence very less austenite reformation. The faster cooling may even lead to precipitation of chromium nitrides [5].

The intermetallic phase precipitation is a major issue in DSS/SDSS fabrication. It has been found that toughness is the most sensitive property which gets affected due to secondary phase formation [6]. Hence prolonged exposure between 600 and 1000 °C is not favourable for DSS/SDSS applications [7].

During welding of duplex stainless steels and super duplex stainless steel, it is important to prevent oxidation of the weldments in order to avoid loss of corrosion resistance. Hence, inert gas such as argon or helium gas purging technique are commonly used to prevent oxidation. Two to five percent nitrogen addition in argon gas [8] improves and assist phase balance.

From the literature review, it was found that there has not been a thorough research work done by considering all the welding parameters of GTAW of DSS and SDSS. The main aim of this work is to characterize mechanical properties of

*e-mail: paulrajprabhul@gmail.com

Table 1. Base metal chemical composition.

Material grade	Cr	Mo	Ni	N	C	PREN	Remarks
UNS S31803	22.9	3.03	7.92	0.15	0.017	35.15	DSS-Low PREN
UNS S31803	22.9	3.04	7.63	0.17	0.019	36.30	DSS-High PREN
UNS S32750	25.1	3.71	8.9	0.2	0.016	40.36	SDSS-Low PREN
UNS S32750	25.1	3.75	8.86	0.21	0.028	41.40	SDSS-High PREN

Table 2. Base material properties.

Grade	Tensile strength (MPa)	Hardness (VHN)	Impact toughness [J] at −46 °C
DSS-Low PREN	750	255	127
DSS-High PREN	762	260	109
SDSS-Low PREN	806	301	140
SDSS-High PREN	830	310	119

the DSS and SDSS materials welded by GTAW technique, which is most commonly used welding process for DSS and SDSS. This paper focuses on effect of welding parameters on mechanical properties of DSS and SDSS weld. We have studied various mechanical properties such as Tensile Strength, Ductility, Impact toughness and Hardness with respect to the welding variables heat input, cooling rate, shielding gas, purging gas and inter-pass temperature.

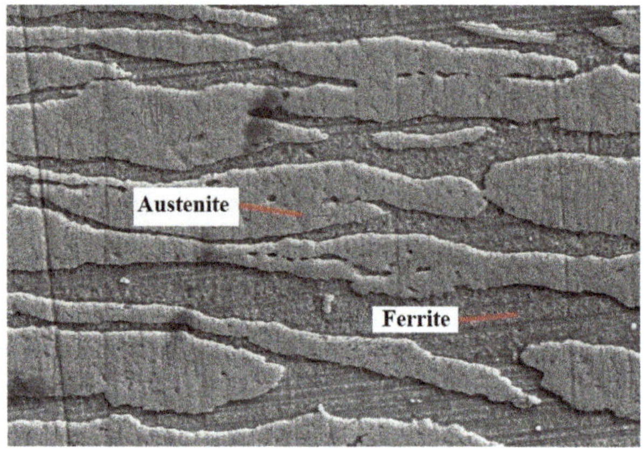

Figure 1. Typical base material microstructure.

2. Experimental details

2.1. Research materials

DSS and SDSS pipes with different PREN were used for this study. Table 1 shows the chemical composition and PREN value of base materials. The pipes were of 2"diameter, 5.54 mm thickness (Sch 80) and 150 mm long. Table 2 shows mechanical properties of base materials. Figure 1 shows typical microstructure of base material comprising dark ferrite phase and bright austenite phase.

The filler metal used for this work was a Sandvik manufactured filler metal of 2 mm size. The chemical composition of filler metal is given in Table 3.

2.2. Welding

DSS and SDSS pipes were welded with Gas Tungsten Arc Welding (GTAW) process. Ar+2%N and Ar+5%N gas mixture was used as shielding gas and back purging gas (refer Tables 5 and 6). Welding experiments were carried out by varying material grade, welding heat input, cooling rate and by varying shielding gas, purging gas and inter-pass temperature in order to study the effect of welding parameters on mechanical properties of DSS and SDSS welds. The welding parameters ranges are tabulated in Table 4.

During welding experiments, one of the above parameter was varied and other parameters were kept constant. Tables 5 and 6 show experimental details of this section.

After welding, welded pipes were liquid penetrant examined to ensure no surface defects and followed by radiographic examined to ensure no defects though out the thickness. Subsequently, mechanical test samples were prepared as per ASME Sec-IX.

2.3. Metallography

Metallographic studies were performed using optical microscopy. First the specimen were polished on emery sheets up to 1200 grit fineness. Then cloth polishing was done using alumina powder of 0.05 μm size. The weldments etched with 20% NaOH solution and they were examined under optical microscopes. The ferrite content measurements were done by point count method in accordance with E562 standard.

2.4. Mechanical testing

After successful completion of welding, the specimen were subjected to mechanical testing. The weldments were characterised with tensile test, hardness test and impact test. For tensile test, ASME IX standard followed. The material thickness and width were 5.30 mm and 13.20 mm respectively. Three sets of tests were performed and average value was taken to this study. Fractured surfaces were examined with SEM images. Charpy V-notch impact tests were performed on a pendulum type impact tester as per ASTM A370 standard. The specimen dimensions used for impact tests was $5 \times 10 \times 55$ mm. All the impact tests were carried out at −46 °C.

Hardness measurements were taken on transverse section of weldments where hardness values were measured at weld metal, HAZ and base metal. Also hardness was measured along thickness of weldments. The ASTM E92 standard was followed with 10 kg of test load.

Table 3. Filler metal chemical composition.

Base metal	Filler grade	C	Si	Mn	P	S	Cr	Ni	Mo	N
DSS	22.8.3.L	≤0.02	0.5	1.6	≤0.02	≤0.015	23	9	3.2	0.16
SDSS	25.10.4.L	≤0.02	0.3	0.4	≤0.02	≤0.015	25	9.5	4	0.25

Table 4. Welding specifications.

Welding position	5G
Groove design	Single V groove 65–75° groove angle, 1–1.5 mm root face, 2.5–4 mm root gap
Welding current (A)	80–150
Arc voltage (V)	10–12
Welding speed (mm/min)	40–80
Number of weld layers	4–5
Inter pass temperature (°C)	100–140
Gas flow rate (L/min)	13–18

Table 5. Welding parameters to study effect of shielding/purging gas and interpass temperature.

Specimen	PREN	Heat input (kJ/mm)	Shielding gas	Purging gas	Interpass temperature (°C)
1	35.55	1.05	Ar+2%N	Ar+2%N	120
2	35.55	1.05	Ar+5%N	Ar+2%N	120
3	35.55	1.05	Ar+2%N	Ar+5%N	120
4	35.55	1.05	Ar+2%N	Ar+2%N	160

Table 6. Welding parameters to study effect of shielding/purging gas and interpass temperature for SDSS weldments.

Exp. no.	PREN	Heat input (kJ/mm)	Shielding gas	Purging gas	Interpass temperature (°C)
1	41.4	1.05	Ar+2%N	Ar+2%N	120
2	41.4	1.05	Ar+5%N	Ar+2%N	120
3	41.4	1.05	Ar+2%N	Ar+5%N	120
4	41.4	1.05	Ar+2%N	Ar+2%N	160

Table 7. Results of mechanical tests for DSS joints.

Exp. no.	Specimen	Heat input (kJ/mm)	Tensile strength (MPa)	Impact toughness at −46 °C (J)	Hardness (VHN)
1	DSS- Low	1.05	795	168	266
2	PREN	1.10	780	140	264
3		1.15	767	123	259
4		1.20	755	85	255
5	DSS- High	1.0	805	115	275
6	PREN	1.05	802	104	270
7		1.1	785	97	266
8		1.15	777	94	261

Table 8. Results of mechanical tests for SDSS joints.

Exp. no.	Specimen	Heat input (kJ/mm)	Tensile strength (MPa)	Impact toughness at −46 °C (J)	Hardness (VHN)
1	SDSS- Low	0.95	862	140	319
2	PREN	1.05	844	130	312
3		1.15	827	126	298
4		1.25	805	107	288
5	SDSS-	0.75	896	90	326
6	High	1.0	880	81	320
7	PREN	1.1	855	72	315
8		1.2	835	64	302

Table 9. Results of mechanical tests by varying shielding/purging gas and interpass temperature for DSS.

Experiment No.	Heat input (kJ/mm)	Tensile strength (MPa)	Hardness (VHN)	Impact toughness at −46 °C (J)
1	1.05	805	275	112
2	1.05	820	281	131
3	1.05	818	277	135
4	1.05	784	290	86

Table 10. Results of mechanical tests by varying shielding/purging gas and interpass temperature for SDSS.

Experiment No.	Heat input (kJ/mm)	Tensile strength (MPa)	Hardness (VHN)	Impact toughness at −46 °C (J)
1	1.05	880	312	130
2	1.05	895	322	146
3	1.05	888	317	143
4	1.05	868	319	104

3. Results and discussion

The results of mechanical tests for first part of this work (i.e. to study the effect of heat input on weldments) are tabulated in Tables 7 and 8. The Experiment no. 1–4 were done on low PREN DSS/SDSS grades and Experiment no. 5–8 are done on high PREN DSS/SDSS grades.

Effect of shielding/purging gas composition and interpass temperature on mechanical properties on DSS and SDSS welds are shown in Tables 9 and 10.

3.1. Microstructure of weldments

A typical microstructure of weldments is shown in Figures 2 and 3. The weld region microstructure differed across the thickness of the pipe. It can be divided into cap and root regions. Due to constant reheating effect in the root region, secondary austenite was found to be precipitated at the root region. The microstructure at cap region cosist of

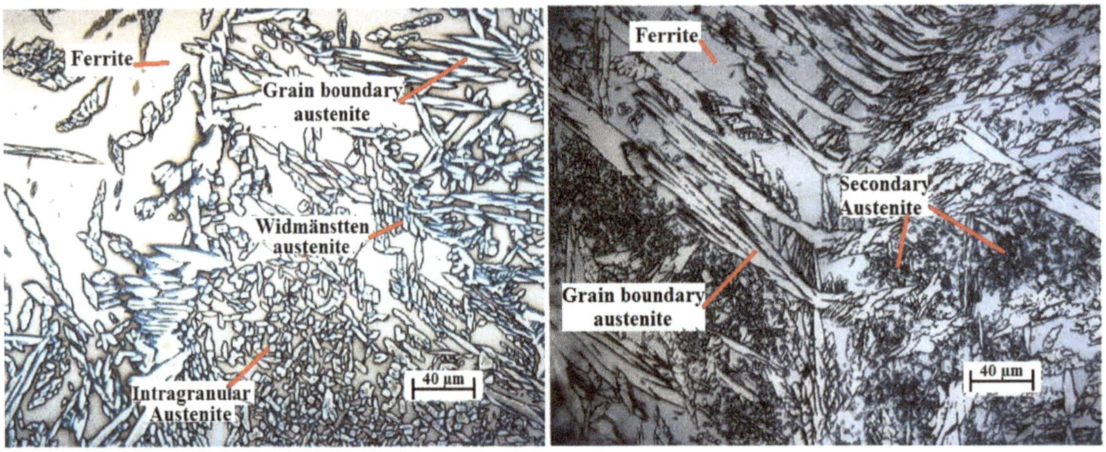

(a) (b)

Figure 2. (a) Microstructure of weld cap region, (b) microstructure of weld root region.

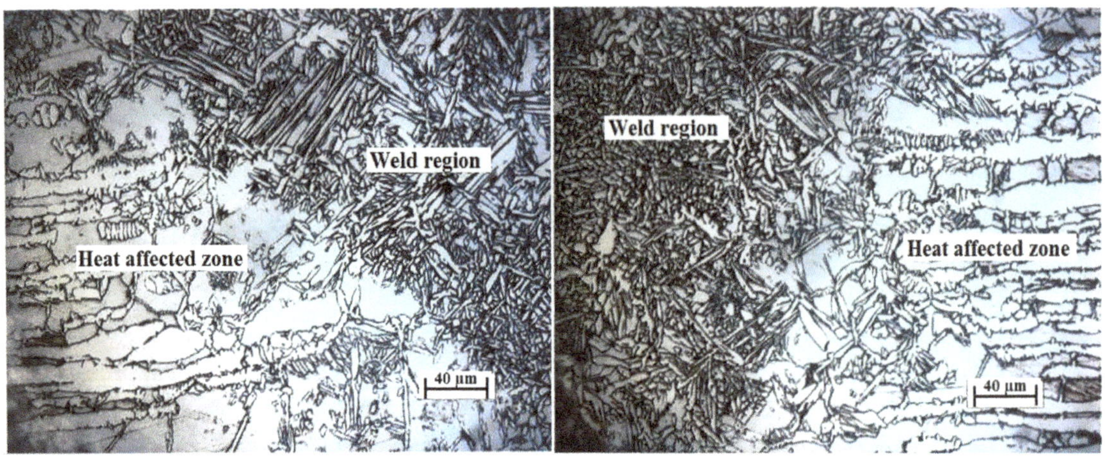

(a) (b)

Figure 3. Microstructure of Heat affected zone (a) left hand side, (b) right hand side.

Table 11. Ferrite content (%) measurements DSS weldments with variable heat input.

Specimen	Heat input (kJ/mm)	Cap	Root	HAZ-cap	HAZ-root
DSS- Low PREN	1.05	54	38	54	48
	1.10	46	43	52	43
	1.15	39	36	55	51
	1.20	35	34	50	48
DSS- High PREN	1.05	55	35	65	60
	1.0	45	39	51	50
	1.1	44	42	50	44
	1.15	38	35	44	43

intergranualar, intragranualar and widmanstten austenite structure formed inside ferrite matrix. There was no evidence of secondary phases in weld cap region. It was also observed that ferrite content in weld cap was more than that of weld root region.

In this study, we noticed HAZ comprised of coarse grains of ferrite and austenite. There were no intermetallic phase formation observed in HAZ region and no evidence of chromium nitrides formation in the weldments.

The ferrite content has been measured using point count method. The test results are tabulated in Tables 11 and 12. It was observed that at higher heat input, lesser ferrite was formed due slower cooling rates which indirectly facilitated higher austenite formation (Tables 13 and 14).

3.2. Tensile strength tests

Tables 6 and 7 show tensile test results. All the samples exhibited excellent tensile strength. All the weldments had higher strength value than the base metal. Hence the welds were fractured outside the weld region. The weld strength was found to be more 5–10% as compared to base material. This is because of higher tensile properties of the filler wire. The highest strength obtained was 805 MPa and 896 MPa for DSS and SDSS respectively. The fractography is shown in Figure 4 which indicates ductile mode of fracture where micro/macro void coalescence (i.e. dimples) are visible.

Table 12. Ferrite content (%) measurements SDSS weldments with variable heat input.

Specimen	Heat input (kJ/mm)	Cap	Root	HAZ-cap	HAZ-root
SDSS- Low PREN	0.95	56	39	42	39
	1.05	48	46	49	48
	1.15	48	43	49	47
	1.25	46	38	48	45
SDSS- High PREN	0.75	64	46	61	56
	1.0	58	49	56	55
	1.1	55	51	51	48
	1.2	52	45	48	43

Table 13. Ferrite content measurements (study effect of shielding/purging gas and interpass temperature for DSS weldments).

Exp. no.	Cap	Root	HAZ-cap	HAZ-root
1	54	38	54	48
2	51	37	50	46
3	49	34	48	46
4	68	46	61	43

Table 14. Ferrite content measurements (study effect of shielding/purging gas and interpass temperature for SDSS weldments).

Exp. no.	Cap	Root	HAZ-cap	HAZ-root
1	56	39	42	39
2	51	37	41	37
3	49	34	41	39
4	70	46	64	44

It was observed that low heat input gives higher tensile strength as compared to high heat inputs. This is due to higher cooling rates induced in low heat input welding which results in finer grains and more ferrite formation. Improved strength of the weldments is due to higher ferrite formation and fine grain size [1]. The graph showing variation of tensile strength with heat input is shown in Figure 5.

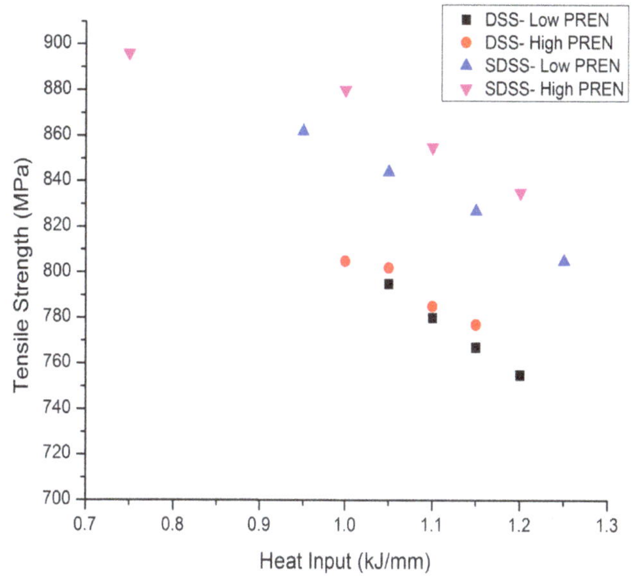

Figure 5. Variation of tensile strength with heat input.

The effect of shielding gas and purging gas on mechanical properties was analysed as shown in Tables 9 and 10. For both DSS and SDSS welds, it was found that increase in Nitrogen content in shielding and purging gas had a positive effect on mechanical properties of weldments. Nitrogen being an austenite stabilizer, affects the weld microstructure. It dissolves mostly in austenite and gives solution strengthening effect especially to austenite phase. Hence there was a slight increase in strength of the weldments [9]. On the other hand increase in interpass temperature results in decrease in tensile strength. This is due to high temperature which leads to coarsening of grains and loss of phase balance. Possibilities for intermetallic phases formation due to slow cooling could have reduced tensile strength of the materials.

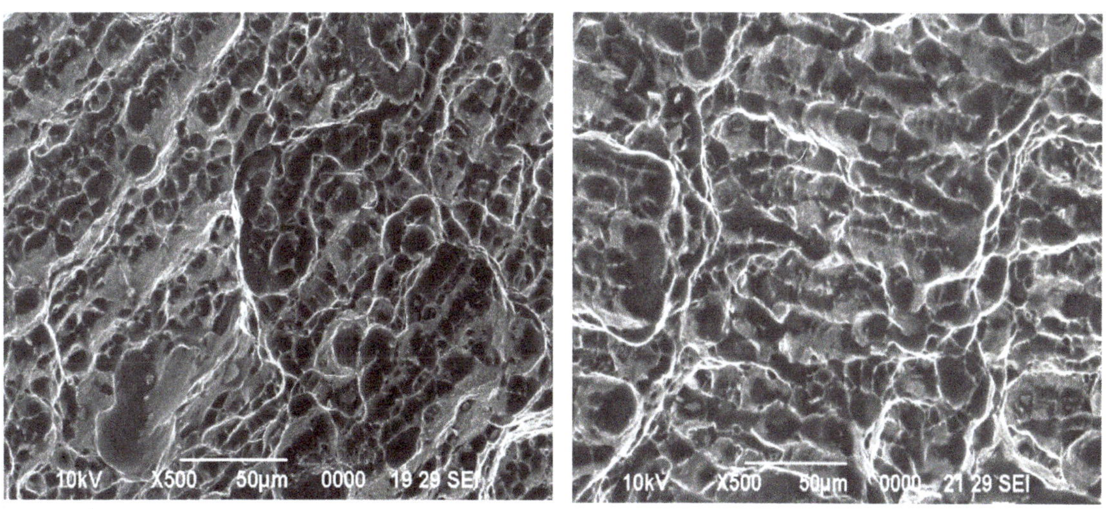

(a) (b)

Figure 4. SEM images of fractured surfaces in tensile tests (a) DSS, (b) SDSS.

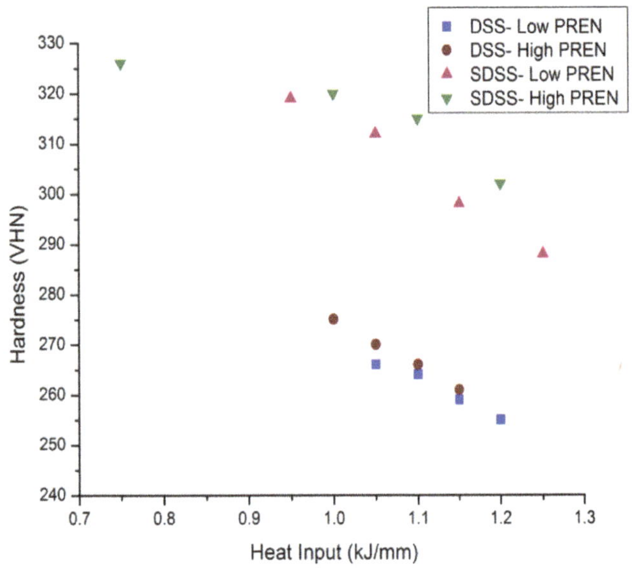

Figure 6. Variation of hardness with heat input.

3.3. Hardness

The results of hardness measurements taken on Vickers hardness machine are shown in Tables 7 and 8. The hardness value of weld metals was found to be higher than the base material. This is due to higher alloying element filler wire and repeated heating and cooling cycle. Effect of heat input on hardness value was analysed and developed scattered diagram as shown in Figure 6. It was observed that with low heat input (i.e. higher cooling rate), high hardness values were noted due to grain refinement.

The hardness values were analysed across the transverse section of the weldments. It was noted that weld region exhibited higher hardness than HAZ due to higher alloying element filler wire and grains refinement. Figure 7 shows hardness variation along the length the joint.

The hardness values were also measured along a vertical line of weld region i.e. weld cap, weld centre, weld root. There was a difference between hardness values at each region. A typical distribution of hardness along the thickness of the weldments is shown in Figure 8. The weld root region exhibited higher hardness values due to continuous repeated heating and cooling cycles and grain refinement. Another possibility is the presence of brittle intermetallic phases could have increased hardness value at weld root [10]. Weld centre had lower hardness than weld cap region. The weld cap hardness was slightly lower than root but higher than weld centre because of faster cooling rate on the cap surface than inside the weld. From this, we can conclude that temper bead technique will minimize the hardness value at the surface if there is an application limitation for surface hardness value.

There was a slight increase in hardness of DSS and SDSS weldments when high nitrogen content is used in shielding and purging gas owing to hardening of austenite phase. It was observed that an increase in inter-pass temperature had increased hardness value of the weldments and HAZ. This could be because of increase in inter-pass temperature may lead for formation of hard and brittle intermetallic phases due to slow cooling.

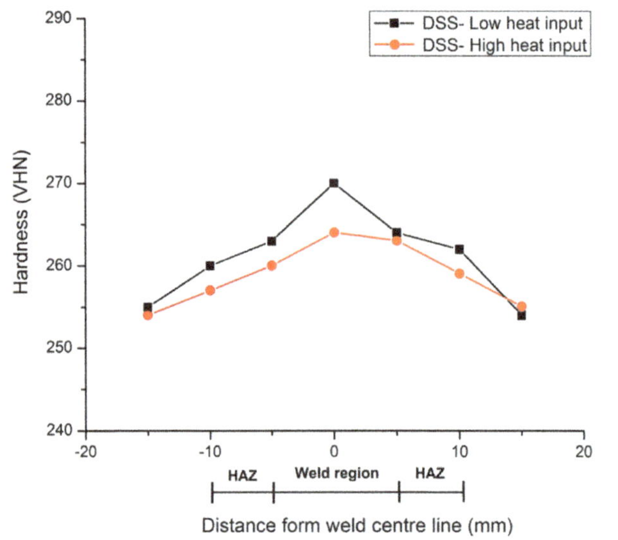

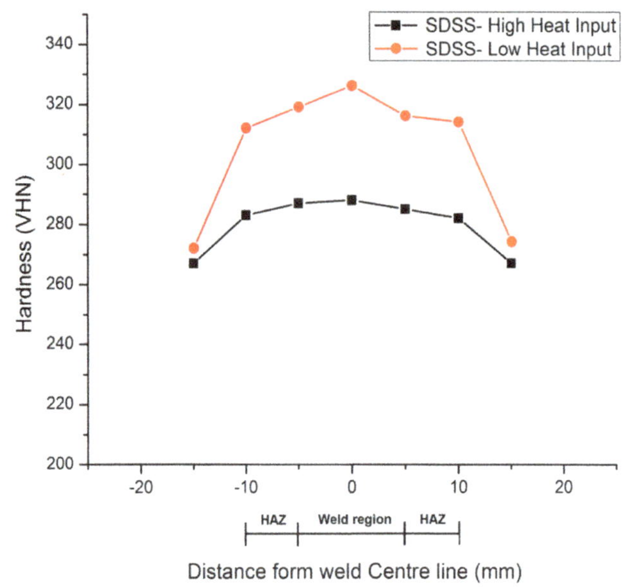

Figure 7. Hardness measurements along transverse section (a) DSS weldments, (b) SDSS weldments.

3.4. Impact toughness

Impact toughness of duplex stainless steel welds depends on many factors such as grain size, austenite formation, secondary austenite and intermetallic phase precipitation. The effect of heat input on impact toughness of weld metal is shown in Figure 9. It was observed that high heat input reduced toughness value of the joints. This is because of formation of coarse grains due to slow cooling because of high heat input. Although formation of austenite was facilitated by high heat input, precipitation of intermetallic phases (i.e. sigma phase and secondary austenite) is unavoidable at higher heat input which may leads to reduce impact toughness properties of the weld. This observation is in line with the literature [6, 7].

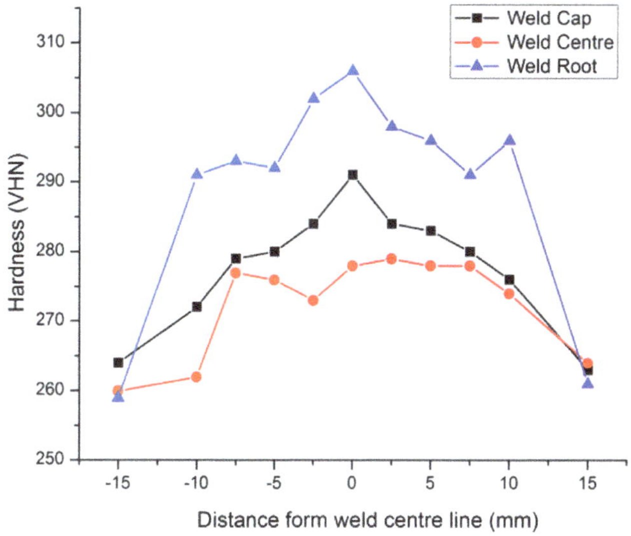

Figure 8. Typical hardness distribution in weld region.

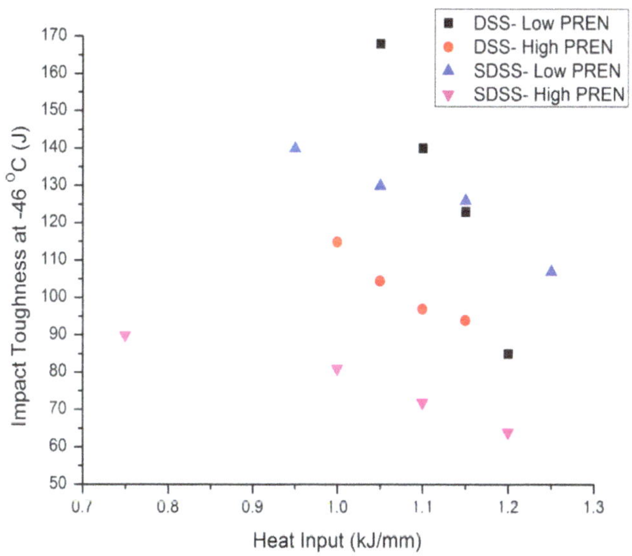

Figure 9. Variation of impact toughness (at −46 °C) with heat input.

SDSS base material has higher toughness values because of high content of Cr and Mo alloying elements. From literature, we noted that these higher alloying elements may promote formation of intermetallic phases at high temperatures. Hence, it is very important to control weld parameters in order to avoid intermetallic phases formation as SDSS is more prone to sigma and secondary austenite phase formation than DSS. Intermetallic phases will drastically reduce toughness value of the weldments.

Figure 10 shows fractography images of the joints taken on SEM. It can be seen that both materials showed signs of ductile fracture as dimples were obserevd. Fractography for DSS/SDSS specimen showed more dimples and small average dimple size which suggests ductile mode of fracture was dominant in fractured welded samples [11]. Also, we observed some cleavages (i.e. flat surafces) in both specimens which indicates brittle phase because of intermetallic components.

Impact value will vary from weld centre line to the base metal. Fusion line, HAZ and weld will have different impact toughness properties. In order to understand the impact toughness values through weld line, V-notch were made at weld metal, fusion line, 2 mm from fusion line and 5 mm from fusion line. Figure 11 shows impact toughness variation on weld centre, fusion line, 2 mm from FL and 5 mm from FL. Impact toughness value of fusion line was observed as low when compare to weld centre, FL+2 and FL+5. It is obvious that grain growth end at fusion line and it becomes solidification line during welding, which may reduce impact value. At V-notch positions 2 mm and 5 mm from fusion line, we observed higher toughness value as it is nothing but base metal with little exposure to temperature gradients.

Higher nitrogen content in shielding and backing gas mixture has assisted to improve impact toughness value on both DSS and SDSS materials. This could be due to the fact that nitrogen will assist to obtain austenite and ferrite phase balance. Increase in nitrogen content will broaden separation between two dislocation planes, which restricts the dislocations to their own slip planes [9].

An increase in interpass temperature had a negative effect on impact toughness due to prolonged exposure to sensitive temperature range and slow cooling rate, which promotes

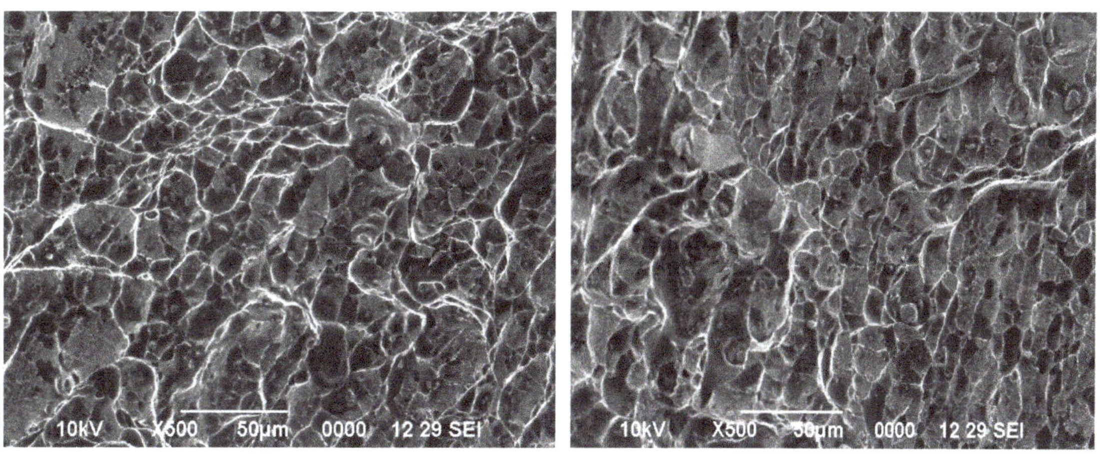

(a) (b)

Figure 10. SEM images of fractured samples during Charpy V-notch impact test (a) DSS, (b) SDSS.

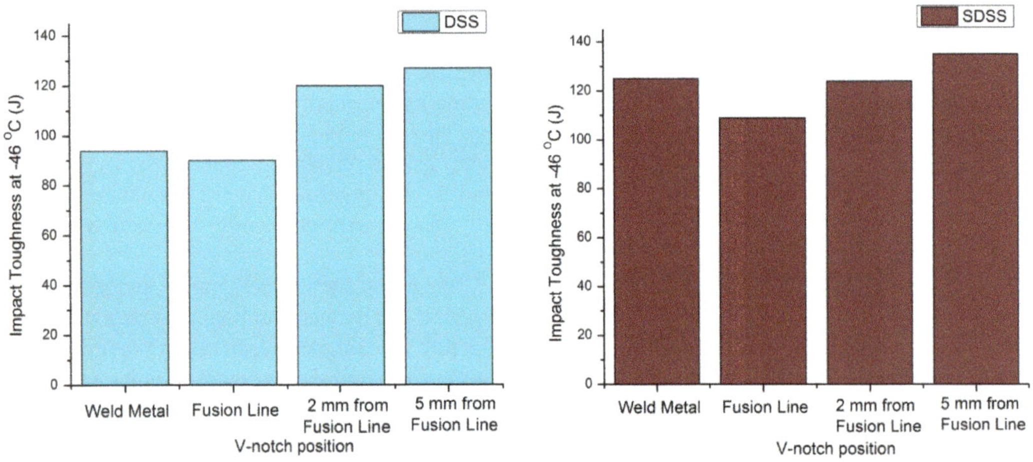

Figure 11. Impact toughness values at different V-notch positions.

formation of sigma phase in weld zone. This brittle sigma phase is the main cause for reduction of impact toughness values in weldments and HAZ.

4. Conclusions

- GTAW of DSS and SDSS was successfully done and effect of welding parameters on mechanical properties were studied. For DSS, heat input of 1.0–1.1 kJ/mm and for SDSS, heat input of 0.75–1.05 kJ/mm were found to give excellent mechanical properties.
- Tensile strength of all weldments was 5–10% higher than that of base material. The highest strength values were 805 MPa for DSS and 862 MPa for SDSS weldments. There was a decrease in strength of weldments with increase in heat input due to grain coarsening and less ferrite content.
- Hardness values decreased with higher heat input for both DSS and SDSS. At weld root area, high hardness values were observed due to presence of hard and brittle intermetallic phases and varying weld thermal cycles.
- Impact toughness decreased with increase in heat input because of intermetallic phase formation, secondary austenite formation and increase in grain size at higher heat input. The highest toughness values were 168 J for DSS and 140 J for SDSS at −46 °C. There was a significant loss of toughness in SDSS samples as they are more prone to intermetallic/secondary phases formation.
- Higher nitrogen contents in shielding and purging gas (i.e. Ar+5%N) slightly improved tensile strength and impact toughness due to strengthening of austenite phase than Ar+2%N gas mixture but Ar+2%N has also given equally comparable results.
- An increase in interpass temperature from 120 °C to 160 °C reduced strength and toughness value due to coarse grain formation and sigma phase formation.
- Based on the experiments and test results, improved mechanical properties were obtained with low heat input, low inter-pass temperature and Ar+5%N shielding/back purging gas. Due to higher alloying elements in SDSS, welding variables shall be properly controlled in order to avoid intermetallic and secondary austenite phase formation. Following weld parameters were recommended to obtain better mechanical properties of DSS and SDSS material. (a) Heat input (0.75–1.1 kJ/mm), (b) interpass temperature (<120 °C), (c) stringer bead technique, (d) shielding/purging gas – Ar+2%N.

References

1. I. Alvarez-Armas, S. Degallaix-Moreuil (Eds.), Duplex stainless steels, John Wiley & Sons, New Jersey, USA, 2013.
2. M. Mohammed Asif, K.A. Shrikrishna, P. Sathiya, S. Goel, The impact of heat input on the strength, toughness, microhardness, microstructure and corrosion aspects of friction welded duplex stainless steel joints, J. Manuf. Process. 18 (2015) 92–106.
3. J. Charles, Duplex stainless steels – a review after DSS'07 held in Grado, Steel Research International 79 (2008) 455–465.
4. R. Cervo, P. Ferro, A. Tiziani, Annealing temperature effects on super duplex stainless steel UNS s32750 welded joints. I: Microstructure and partitioning of elements, J. Mater. Sci. 45 (2010) 4369–4377.
5. A.V. Jebaraj, L. Ajaykumar, Influence of microstructural changes on impact toughness of weldment and base metal of duplex stainless steel AISI 2205 for low temperature applications, Procedia Eng. 64 (2013) 456–466.
6. G. Fargas, M. Anglada, A. Mateo, Effect of the annealing temperature on the mechanical properties, formability and corrosion resistance of hot-rolled duplex stainless steel, J. Mater. Process. Technol. 209 (2009) 1770–1782.
7. S. Topolska, J. Łabanowski, Effect of microstructure on impact toughness of duplex and superduplex stainless steels, J. Achieve. Mater. Manuf. Eng. 36 (2009) 142–149.
8. E. Taban, E. Kaluc, T.S. Aykan, Effect of the purging gas on properties of 304H GTA welds, Welding Journal 93 (2014) 124s–130s.
9. K. Migiakis, G.D. Papadimitriou, Effect of nitrogen and nickel on the microstructure and mechanical properties of plasma

welded UNS S32760 super-duplex stainless steels, J. Mater. Sci. 44 (2009) 6372–6383.

10. M. Martins, L.C. Casteletti, Effect of heat treatment on the mechanical properties of ASTM A 890 Gr6A super duplex stainless steel, J. ASTM Int. 2 (2005) 1–14.

11. S.S.M. Tavares, J.M. Pardal, L.D. Lima, I.N. Bastos, A.M. Nascimento, J.A. de Souza, Characterization of microstructure, chemical composition, corrosion resistance and toughness of a multipass weld joint of super duplex stainless steel UNS S32750, Mater. Charact. 58 (2007) 610–616.

19

Hot stamping of AA6082 tailor welded blanks: experiment and FE simulation

Jun Liu, Haoxiang Gao, Omer El Fakir, Liliang Wang*, and Jianguo Lin

Department of Mechanical Engineering, Imperial College London, London SW7 2AZ, UK

Abstract – An advanced forming technology, solution Heat treatment, Forming and in-die Quenching (HFQ®), has been employed to form AA6082 tailor welded blanks (TWBs). In comparison with conventional stamping of TWBs, the mechanical properties and formability of AA6082 laser TWBs could be improved under the HFQ® forming condition. The TWB was divided into three physical zones, i.e. base metal, heat affected zone (HAZ) and weld zone, based on the hardness distribution. It was found that the degraded hardness of the weldment can be restored after HFQ® forming. TWBs of AA6082 with different thickness ratios of 2 (2–1 mm), 1.3 (2–1.5 mm) and 1 (1.5–1.5 mm) were used to study the TWB thickness ratio effects on the forming behaviour. Hemispherical punch dome tests on the TWBs with varying thickness ratios demonstrated different formabilities, and indicated increased displacement of the weld line with increasing thickness ratio. Finite element (FE) modelling was adopted to analyse the weld line movement and strain distributions during HFQ® forming.

Key words: Aluminium, Tailor welded blank, Hot stamping, Solution heat treatment, Artificial ageing

1. Introduction

In recent decades, aluminium alloys have attracted the attention of many researchers, engineers and designers as promising structural materials for the automotive industry applications. As a commercial alloy with high stiffness to density and high strength to density ratios, AA6082 is suitable for structural sheet-metal fabrications.

Tailor welded blanks (TWBs) consisting of multiple sheet materials joined with welds has been widely used in lightweight structures. The market for aluminium alloys is expanding into various sectors, most significantly in automotive industries where the use of tailor welded blanks in aluminium is developing rapidly. Due to the development of advanced forming technologies, it is now possible to form tailor welded blanks to produce body panels with complex structure and reduced weight [1]. The most commonly used welding technology for manufacturing the AA6082 tailor welded blanks is laser beam welding. Laser welding has various advantages, including high productivity, high weld quality in terms of controllable heat affected zone (HAZ) and weld zone, low distortion, flexibility due to the movable heat source, reliability and precision [2].

Many studies have concentrated on the mechanical properties of TWBs [3], the movement of weld line [4], forming process design [5], finite element simulation [6], forming limit (formability) [7] and failure prediction for TWBs [8]. Generally, the overall formability of TWBs is significantly affected by the welding parameters [9], especially for heat-treatable aluminium alloys. Miles et al. [10] have reported that the aluminium friction stir welded blanks showed similar formability as the base material in plan-strain, but had rapidly decreasing formability as biaxial strain conditions were approached. The weld area in aluminium TWBs is typically weaker than the base material, which adds difficulty in forming such TWBs with conventional methods. Additionally, the other parameters influencing the formability of TWBs may include the thickness ratio and the orientation of the weld line to the loading direction. It is well accepted that the limit dome height and strain distribution during stretch forming were greatly influenced by the thickness variation [3].

Foster et al. [11] have developed a novel process, solution Heat treatment, Forming and in-die Quenching (HFQ®[1]), which was able to overcome the major problems in forming AA6082 TWBs, such as low formability and microstructure variations. Heat treatment is required for the heat-treatable aluminium alloys to achieve desired microstructure and thus specified mechanical properties. Previous studies have shown that heating a material to its solution heat treatment temperature,

*Corresponding author: liliang.wang@imperial.ac.uk

[1] HFQ® is a registered trademark of Impression Technologies Ltd.

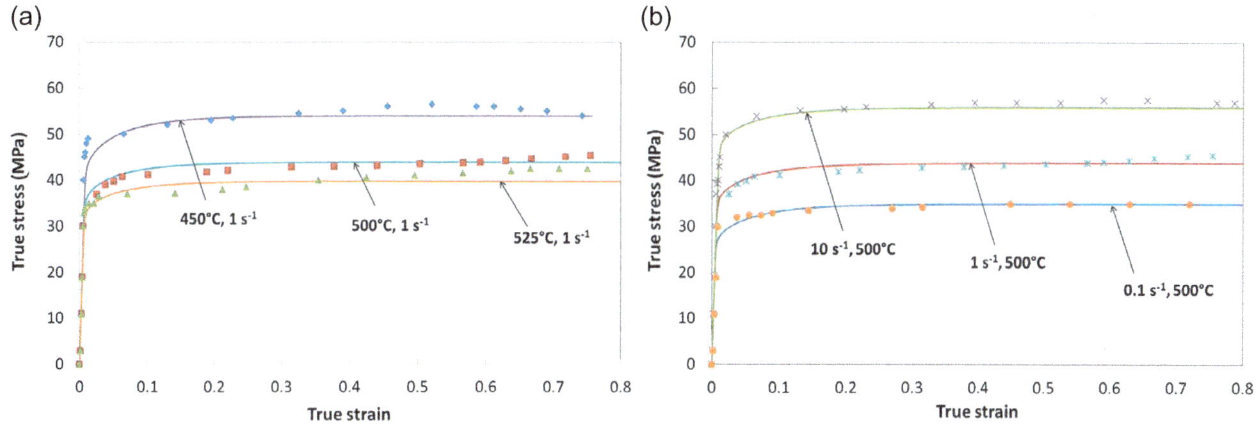

Figure 1. Uniaxial tensile test results of AA6082 at (a) different temperatures and (b) various strain rates.

e.g. 450–550 °C, significantly increases its formability, due to the complete dissolution of alloying elements and/or precipitates into the material matrix resulting in a single phase, which reduces the obstructions to dislocation motion and enhances material flow [12]. This material characteristic has been utilised for the stamping of complex-shaped high-strength aluminium panel components, and has led to applications of the HFQ® process.

The aim of this work is to apply the novel HFQ® process into the forming of laser welded AA6082 TWBs. The degraded hardness of the TWB can be restored after HFQ® forming and artificial ageing. AA6082 TWBs with different thickness combinations of 2–1, 2–1.5 and 1.5–1.5 mm were used to study the TWB thickness ratio effects on the forming behaviour. Finite element simulations were performed for evaluating the strain distributions and weld line movement during deformation.

2. Experimental details

2.1. Material

The base material used in this study is aluminium alloy AA6082-T6. Uniaxial tensile tests were performed on AA6082 using a Gleeble 3800 thermo-mechanical testing system. Figure 1 shows the stress-strain curves of AA6082 at various temperatures and strain rates. The symbols are the results from experimental determinations. The experimental results were used to calibrate a set of dislocation-based hardening constitutive equations [13] interpreting the unified viscoplastic behaviour at warm/hot forming conditions. Good agreements between the experiment (symbols) and modelling (solid curves) have been obtained. The calibrated material model was applied in PAM-STAMP simulations to predict the HFQ® forming process.

The blanks (initial size of 600×300 mm^2) with thicknesses of 1, 1.5 and 2 mm were initially cut along the length (rolling) direction, and then welded together parallel to the rolling direction into three different thickness combinations, i.e. 2–1, 2–1.5 and 1.5–1.5 mm TWBs. Laser welding was carried out on the blanks using a Nd:YAG source through a 0.2 mm fibre with a power of 2.1 kW. The welding speed was 25 mm/s. In this study, the investigations were focused only on butt welds with fully penetrated joints.

2.2 HFQ® dome test

The HFQ® process mainly consists of four phases, i.e. solution heat treatment (SHT), transfer, cold die forming and quenching, and post-forming heat treatment (artificial ageing). All the TWBs were initially heated to 525 °C (SHT temperature) and maintained at the temperature for 1 min to ensure complete dissolution of second phases during solution heat treatment. The hot TWB was then rapidly transferred to the cold die which was immediately closed to form the dome shaped component. The transfer was completed within 10 s, during which the blank temperature decreased to around 450 °C before the forming started. The HFQ® process may comprise a post-forming heat treatment stage for heat-treatable aluminium alloy components which includes heating the formed component to an artificial ageing temperature and holding it at that temperature to allow the precipitation hardening to occur. Typical temperature for AA6082 is 190 °C, and the time is 9 h in this study so as to achieve a complete T6 temper condition after artificial ageing.

Hemispherical dome tests were conducted to investigate the formability of AA6082 TWBs at speeds of 75, 250 and 400 mm/s under HFQ® conditions. The tool used for the HFQ® forming test is shown in Figure 2. It comprises of the top punch pole, 40 mm diameter hemispherical punch, top/bottom blank holders, and the TWB. During testing, the punch was fixed, while the top blank holder was moved down at the desired rate to first clamp the TWB between the top and bottom blank holders, and then to deform the blank over the punch to the limit dome height of the TWB. Graphite grease (Omega 35) was used as the lubricant and was evenly applied to the punch surface to reduce the friction during forming. Two gas springs were utilised to maintain the blank holding force at a constant value of 20 kN. The top blank holder was specially manufactured so as to accommodate the TWBs with different thickness combinations.

For a sheet metal forming process, localised necking may appear at a specific forming stage just before fracture happens,

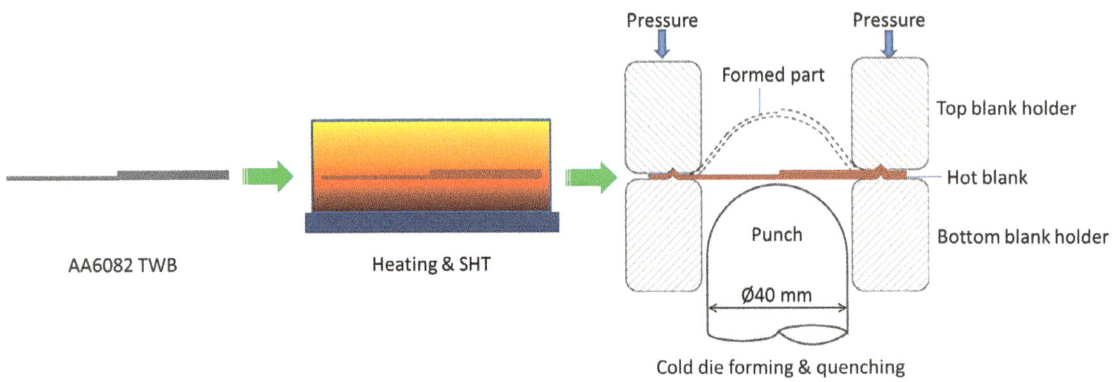

Figure 2. Schematic diagram of the HFQ® forming process.

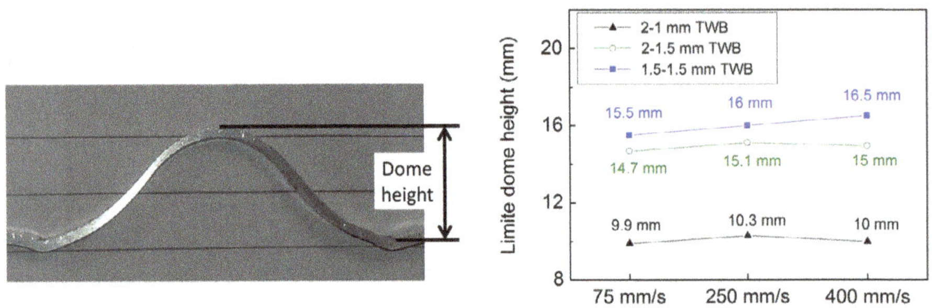

Figure 3. Limit dome height of HFQ® formed tailor welded blanks.

which is an indication of localised plastic deformation [14]. As a stroke controlled test, the stroke of the punch was always set to a value at which localised necking was estimated to initiate, and then refined through multiple trial runs to the correct value. Due to the test piece being quite sensitive to changes in the stroke, even a small adjustment could cause fracture in the formed part. Three tests for each condition have been performed so as to ensure a good repeatability of the forming results.

To determine the mechanical properties of the TWB before and after HFQ®, hardness values were measured along the mid thickness of the blank using a Vickers Microhardness tester with a load of 200 g and a dwell time of 12 s.

3. Results and discussion

3.1. HFQ® dome test

3.1.1. Limit dome height

The limit dome height (LDH) is the position at which localised necking starts to occur on the formed component, and was used to compare the formability of the TWBs under HFQ® conditions. In general, the LDH increases with the decreasing blank thickness ratio and increases as the forming speed increases from 75 to 400 mm/s, although the increasing trend for the 2–1 mm TWBs is not as significant as the 1.5–1.5 mm TWBs. The premature fracture occurred at higher blank thickness ratios, because the localised strain primarily occurred in the thinner blank at higher circumferential stress levels. The increase in the LDH with increasing forming speed is due to the enhanced strain rate hardening effect at higher forming speed. In addition, the blank temperature was higher at fast forming speed due to less heat loss into the cold punch and air, which is beneficial for the formability improvement (Figure 3).

3.1.2 Deformation characteristics

Figure 4 shows the different features of localised necking (or cracking) for the TWBs with various thickness combinations. It can be seen from the 2–1 mm TWB in Figure 4a that the most severe plastic deformation occurred in the 1 mm blank parallel to the weld line. In this case, the failure is found at a small distance away from the weld line, which is referred to as parallel failure in the study.

For the 2–1.5 mm TWB shown in Figure 4b, two localised thinning regions are found in a mixture mode combining the parallel necking and the circumferential cracking. The cold die quenching was not effective due to the rough surface topography around the weld line, hence a higher blank temperature in this region has resulted in lower strength and thus localised necking took place along the weld line. The circumferential necking in the thin blank was due to the stress concentration in the thinner blank, as the circumferential stress within the blank is proportional to the ratio between the punch radius and blank thickness (R/t).

Figure 4c shows the circumferential necking occurring approximately halfway between the base and apex of the formed 1.5–1.5 mm TWB part. Since the two part of the TWB have similar thickness, the weld line stayed in the centre after forming and the stain distribution on either side of the weld line is

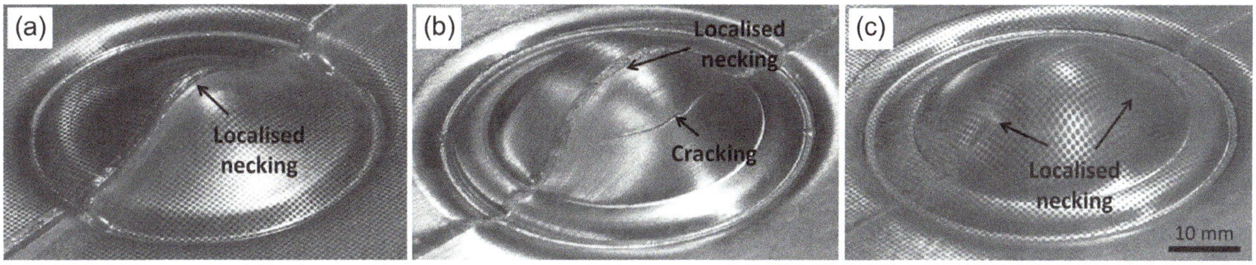

Figure 4. HFQ® formed parts of (a) 2–1 mm TWB, (b) 2–1.5 mm TWB and (c) 1.5–1.5 mm TWB.

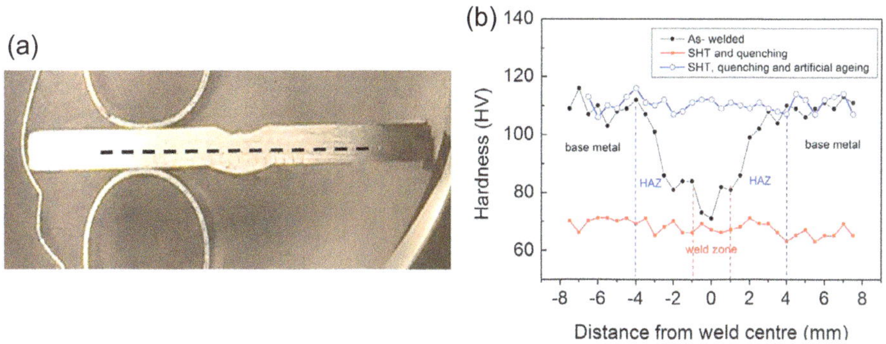

Figure 5. (a) Cross-section of a 1.5–1.5 mm TWB and (b) hardness profiles of the TWB at different heat treatment conditions.

symmetrical. The strain distribution will be detailed in Section 3.2.3.

3.1.3. Microhardness

Microhardness values of the as-welded specimen show a typical V-shape, as plotted in Figure 5, where the lowest value is located in the centre of the weld zone. After solution heat treatment at 525 °C and the subsequent water quenching, all the hardness decreased significantly but were at a consistent level at different measured locations. An artificial ageing treatment (190 °C for 9 h) strengthened the material, and as a result, achieved the full hardness (HV > 110) of the alloy.

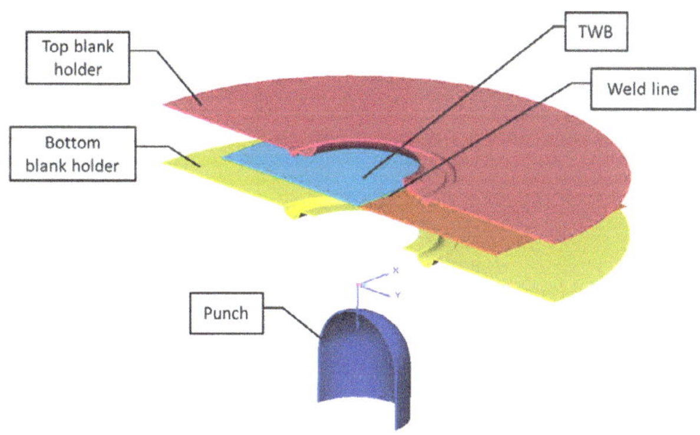

Figure 6. PAM-STAMP simulation model (cross-sectional view).

3.2. Finite element simulation

3.2.1. Simulation model

Finite element (FE) simulations of the hemispherical punch forming at HFQ® conditions were conducted using the commercial software PAM-STAMP and a developed temperature and strain rate dependent material model. As shown in Figure 6, the simulation model comprises of a punch, TWB, and the blank holder sets. The tools (i.e. blank holders and punch) were modelled using rigid elements. The TWB was defined using the built-in feature provided by PAM-STAMP. Since the HFQ® forming process could eliminate the welding effects and restore the mechanical properties of the weld zone, no special consideration was taken for the weld zone and the weld section mesh was set to the same as that of base material. The initial blank temperature was set to be 450 °C, which is identical to the experimental condition. A friction coefficient of 0.35 was chosen to account for surface interaction between the sliding sheet and the die assembly.

3.2.2. Temperature evolution

The HFQ® process is a hot sheet forming operation that incorporates part of the thermal tempering process required for heat-treatable aluminium alloys [15]. The hot blank is formed and held between cold dies, during which a rapid quenching is performed. The temperature of the deforming sheet is not uniformly distributed, which makes the deformation more complicated.

Figure 7 shows the temperature history during HFQ® at a forming speed of 75 mm/s. Temperature data at two points of the formed component was recorded using a data logger at a frequency of 6 Hz. The blank temperature was at 450 °C when the stamping started. The temperature at Point 1 dropped

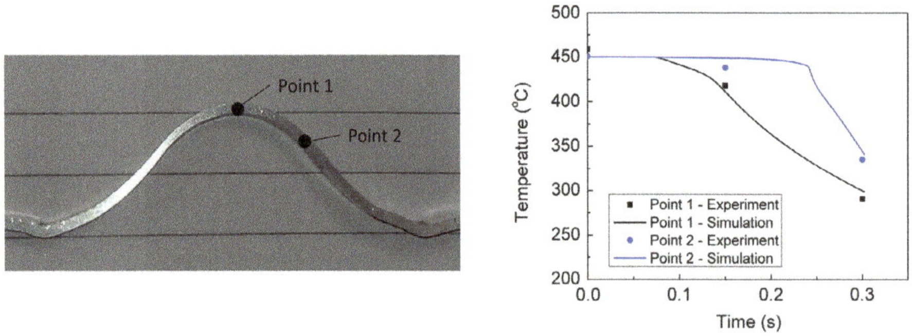

Figure 7. Temperature history during HFQ® forming at 75 mm/s between experiment and simulation.

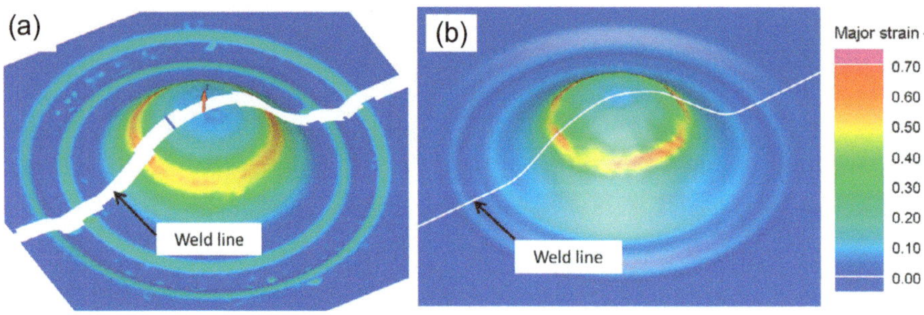

Figure 8. Comparison of strain distributions in the 1.5–1.5 mm TWB between (a) experiment and (b) simulation.

rapidly when the punch firstly contacted the blank at the dome apex, but there was some delayed temperature drop at Point 2 due to the later contact between this point and the punch. The predicted temperature evolution from simulation, see Figure 7, has a good agreement with the experimental measurements.

3.2.3. Strain distribution

Figure 8 shows the strain comparison of the 1.5–1.5 mm TWB between experiment and simulation at the onset of localised necking. The strain distributions are able to depict deformation characteristics of this thickness combination. As shown in Figure 4c, the failure type was circumferential necking occurring approximately halfway between the base and apex of the formed part. As the stroke increased, the punch stretched the sheet material to a higher strain level, and the eventual failure was initiated by necking from excessive strain in a circumferential location. The good agreement between experiment and simulation also indicates the feasibility of the FE simulation in prediction of the deformation behaviour, e.g. localised necking, weld line movement, etc. for other TWBs.

3.2.4. Weld line movement

Due to blank thickness mismatch, the weld line shifts during the dome tests of TWBs. The weld line moves towards the thicker blank side because the thinner blank section has been stretched more by the moving punch than the thicker blank section. Table 1 lists the maximum displacement of the weld in the HFQ® formed TWBs with different thickness combinations. For each TWB type, the stamping stroke was maintained at the same level. It was found that the weld line displacement was nearly independent of the forming speed. Thus, only the displacement of the weld line at a forming speed of 250 mm/s was presented here.

Table 1. Weld line displacement at the dome apex (unit: mm).

	Experiment	Simulation
2–1 mm TWB	1.0	1.04
2–1.5 mm TWB	0.6	0.576
1.5–1.5 mm TWB	0	0

FE simulation is a useful method to evaluate the displacement of a whole weld line. As can be observed from Figure 9, the amount of weld line movement on the punch surface tends to increase as the thickness ratio is increased from 2–1.5 to 2–1. The maximum movement for the 2–1 mm TWB is found in the centre of the weld line at the dome apex. For the 2–1.5 mm TWB, the maximum values are found along the weld line in the centre area. In addition, there is localised necking occurring parallel to the weld line for the dissimilar thickness TWBs, as shown in Figure 4. This is because most of the stretching is concentrated on the thinner blank section with lower strength. The deformation is not uniformly distributed and the weld line shifts towards the thicker side, finally leading to parallel localised necking in the thinner blank section.

The blank thickness ratio in a TWB is an important factor impacting the movement of weld line. A lower thickness ratio (less thickness variation within the TWB) could have a more uniform deformation during forming. In real applications, the orientation and the position of the weld line can be adjusted according to the FE simulation. Clamping force may be

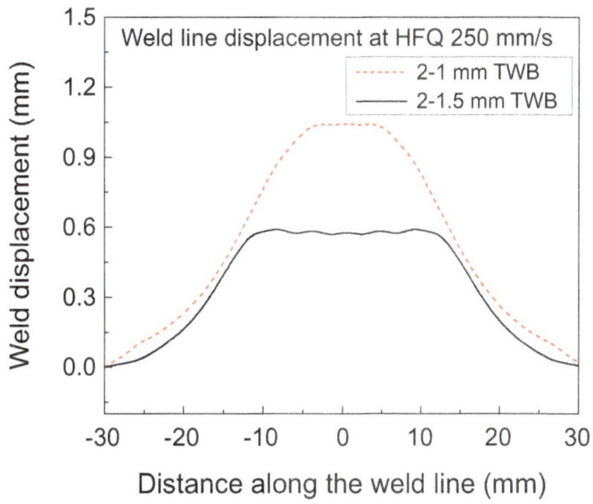

Figure 9. Predicted displacement of weld line for HFQ® formed TWBs with different thickness ratios.

applied on the weld line to eliminate the tearing failure (or localised necking) and produce a more uniformly distributed strain distribution in the thicker and thinner gauges of TWBs [6].

Conclusions

AA6082 tailor welded blanks with various thickness combinations have been used in the HFQ® forming process. The mechanical properties of the TWB have been restored after solution heat treatment, subsequent quenching and artificial ageing, indicating the advantages of the HFQ® process. The HFQ® formed TWBs exhibited different failure modes: localised necking parallel to the weld line for the 2–1 mm TWB, circumferential necking for the 1.5–1.5 mm TWB, and mixture of the parallel and circumferential necking for the 2–1.5 mm TWB. Thickness ratio has an important effect on the extent of the weld line movement and it was found that the weld line shift increases as the thickness ratio is increased. The non-uniform deformation may result in localised failure in the TWBs. Therefore, it is very important to identify the failure location, especially for the forming of complex-shaped components. The simulation of the HFQ® process could be further developed to predict the failure occurrence as well as to facilitate the forming process design.

Acknowledgements. The financial support from Innovate UK, Ultra-light Car Bodies (UlCab, reference 101568), is gratefully appreciated.

References

1. R.W. Davies, H.E. Oliver, M.T. Smith, G.J. Grant, JOM 51 (1999) 46.
2. X. Cao, W. Wallace, C. Poon, J.P. Immarigeon, Mater. Manuf. Process. 18 (2003) 1.
3. S.K. Panda, D.R. Kumar, H. Kumar, A.K. Nath, J. Mater. Process. Tech. 183 (2007) 321.
4. Y. Heo, Y. Choi, H.Y. Kim, D. Seo, J. Mater. Process. Tech. 111 (2001) 164.
5. M.A. Ahmetoglu, D. Brouwers, L. Shulkin, L. Taupin, G.L. Kinzel, T. Altan, J. Mater. Process. Tech. 53 (1995) 684.
6. B. Kinsey, Z. Liu, J. Cao, J. Mater. Process. Tech. 99 (2000) 145.
7. R.W. Davies, G.J. Grant, M.A. Khaleel, M.T. Smith, H.E. Oliver, Metall. Mater. Trans. A 32 (2001) 275.
8. A.A. Zadpoor, J. Sinke, R. Benedictus, Mater. Design. 30 (2009) 1423.
9. A.A. Zadpoor, J. Sinke, R. Benedictus, Key Eng. Mater. 344 (2007) 373.
10. M.P. Miles, T.W. Nelson, B.J. Decker, Metall. Mater. Trans. A 35 (2004) 3461.
11. A. Foster, T.A. Dean, J. Lin, Process for forming aluminium alloy sheet components, 2010, European Patent EP2324137B1.
12. R.P. Garrett, J. Lin, T.A. Dean, Int. J. Plasticity 21 (2005) 1640.
13. J. Lin, T.A. Dean, J. Mater. Process. Tech. 167 (2005) 354.
14. W.F. Hosford, Mechanical Behavior of Materials, Cambridge University Press, Cambridge, UK, 2005.
15. A.D. Foster, M.S. Mohamed, J. Lin, T.A. Dean, Steel Res. Int. 79 (2009) 133.

20

Structuring by electromagnetic forming and by forming with an elastomer punch as a tool for component optimisation regarding mechanical stiffness and acoustic performance

Verena Psyk[1,*], Petr Kurka[1], Simon Kimme[1], Markus Werner[1], Dirk Landgrebe[1], Andreas Ebert[2], and Mathias Schwarzendahl[2]

[1] Fraunhofer Institute for Machine Tools and Forming Technology, Reichenhainer Strasse 88, 09126 Chemnitz, Germany
[2] Westfalia Presstechnik GmbH & Co. KG, Gewerbering 26, 08451 Crimmitschau, Germany

Abstract – Ecological and economic reasons are forcing industry to improve efficiency and to save energy and resources. Especially for the automotive and transport industry – but also for all other industrial branches dealing with accelerated masses – lightweight design is a key issue for achieving this goal. In current product designs, insufficient geometric stiffness of the part often prohibits exploiting the full potential of weight reduction offered by modern materials. Ideally adapting the geometry to the load profile by implementing appropriate structures frequently allows wall thickness and weight reduction. At the same time, the acoustic properties can be improved in many cases, because the natural frequency of the modified structure is higher compared to a conventionally-designed structure, thus leading to a reduction of awkward low frequency noise (humming). To enable a target-oriented design, the influence of structures manufactured by elastomer tools and by electromagnetic forming (EMF) was analysed systematically. Knowledge about the influence of the structuring technology and the process parameters on the resulting structure properties and the part performance allows conclusions regarding the optimisation of the part functionality via a suitable selection and design of the process.

Key words: Lightweight design, Electromagnetic forming, Elastomer tool, Structure, Stiffness, Acoustic Performance, Sequential forming, Incremental forming

1. Introduction and motivation

Due to climate change and global warming, industry is being forced to take ecological aspects into consideration when developing new products, such as the reduction of the carbon-footprint. Over the last 20 years, Europe's transport sector has been responsible for approximately 30% of Europe's energy consumption making it a key player in climate and resource aspects [1]. For this reason, consequent implementation of lightweight design strategies is an important approach to increasing efficiency especially for the automotive and transport sector industry, but also for other industrial branches dealing with accelerated masses. In addition to the direct saving of resources and energy due to reduced material usage, over the complete product life cycle significant amounts of energy can be saved via reduction of moving masses [2]. For example, decreasing the weight of a standard passenger car by about 100 kg can result in fuel savings of about 300–800 L over the vehicle's lifetime and to a reduction of the CO_2 emission by 9 g/km [3]. Developments as e.g. the multi-material concept implemented in the Audi A3 [4] already succeeded in reducing vehicle weight, so that since 2000, a trend of decreasing greenhouse gas emissions can be observed in Europe (compared to 1990) [1]. Nevertheless, according to [5] almost all car manufacturers will miss the European carbon emission reduction targets for 2020, so that further cutting of vehicle weight is indispensable.

*e-mail: verena.psyk@iwu.fraunhofer.de

2. State of the art

2.1. Lightweight construction

It is well known that maximum weight savings can only be achieved if all potentials from the disciplines of material sciences, construction design, and systems engineering are considered in a holistic view and thus the approaches of material-based, form- and structure-based, and conditional lightweight design are followed up in parallel [6, 7]. In addition to the mere substitution of conventional materials by typical lightweight ones such as aluminium and magnesium alloys, high-strength steels, and plastics, also the component design offers enormous weight saving potential [7–9]. A brief review of weight reduction via adapted part design starting from early and more theoretical approaches [10] up to current work [11] can be found in reference [12].

Precisely, the implementation of additional structuring elements can increase the stiffness of components, so that the wall thickness can be significantly reduced if the material strength is sufficiently high. Currently, insufficient geometric stiffness of the part frequently prohibits exploiting the full potential of wall thickness and weight reduction offered by modern high-strength materials. At the same time, the acoustic properties can frequently be improved by implementing stiffening elements, because the natural frequency of the modified structure is usually higher compared to the conventionally designed one (see e.g. [7]). As a result of this, the awkward low frequency noise, which is typical for thin-walled parts, such as e.g. seat components or panels, is significantly reduced. At the same time, the structured material features damping properties so that less structure-borne noise is emitted.

However, the optimum exploitation of these effects requires a sophisticated adaptation of the structuring elements to the part specific load profile [8] as well as a sophisticated choice and design of the structuring process. Frequently, the resulting increase in component complexity cannot be realised by using conventional manufacturing techniques any longer [13].

2.2. Electromagnetic forming

Electromagnetic forming is a high-speed forming process using the energy density of pulsed electromagnetic fields to apply Lorentz forces to electrically conductive workpieces without mechanical contact and to form them accordingly [14]. In comparison to conventional forming processes as e.g. deep drawing, one element of the set of tools (usually the punch) is replaced by the so-called tool coil or inductor, and only the counterpart (the die) remains as form defining tool in the process [15]. Depending on geometry and alignment of workpiece and inductor, tubes or hollow profiles can be compressed or expended and flat or curved sheet metal can be formed [16]. In order to allow easier comparison to other forming processes, the physically correct volume forces are frequently mathematically transformed to surface forces – the so-called magnetic pressure [17].

The typical pulse duration in electromagnetic forming (EMF) is in the range of 15–100 μs. Within this time the magnetic pressure rises to maximum values of up to several hundreds of megapascale and decreases again [18]. The workpiece is accelerated to velocities of up to several hundreds of meters per second and reaches strain rates in the range of $10^3\ s^{-1}$ to $10^4\ s^{-1}$ [19].

Compared to conventional forming techniques EMF excels due to

- the process-specific high velocities and strain rates, which allow
  - higher formability of many materials without material failure as necking or cracking (e.g. [20, 21]),
  - very accurate forming results since springback effects are significantly reduced [22],
  - extremely short duration of the deformation process itself.
- contact-free force application, which allows
  - flexibly using one and the same inductor for manufacturing different geometry variants [23]. Therefore, the economic efficiency of the process is very good especially for applications with high geometric variety and small lot size,
  - a very surface-friendly forming process without mechanical wear at the workpiece surface facing the inductor. This allows processing of sensitive and even pre-coated surfaces [24], provided that the coating can withstand the occurring strains,
  - processing without lubricants. This reduces cleaning effort during production and makes the process environmentally friendly [25].

A brief review of the process principle and its variants as well as process specific advantages and recommendations for practice is summarized in reference [26]. More comprehensive information regarding these points as well as concerning material behaviour at high velocities, tools, and applications of electromagnetic forming can be found in reference [16].

In addition to the application as a stand-alone process EMF, also features high potential for extension of forming limits if it is applied in combination with conventional forming techniques. Ideally, the process combinations are composed in such a manner that the particular advantages of the individual techniques complement each other. Examples for such process combinations are

- combined curving, electromagnetic compression, and hydroforming [27],
- combined extrusion and electromagnetic compression [28, 29],
- combined joining by electromagnetic compression and subsequent hydroforming [30],
- deep drawing with integrated electromagnetic forming [31–33],
- combined bending and electromagnetic calibration [22], and
- combined roll forming and electromagnetic forming [34].

All publications concordantly proof the feasibility of the process combinations and the advantages compared to the individual forming processes alone.

2.3. Forming with elastomer tools

Similar to EMF, also in case of forming with elastomer tools one active part of the set of tools is replaced by an elastomer block while the counterpart remains as form defining tool. According to DIN8584 it can be differentiated between deep drawing with flexible punch and deep drawing with flexible die. In both cases a pressure is applied to the workpiece by the elastomer and consequently it aligns to the form defining tool. In addition to sheet metal forming processes, forming with elastomer tools also allows the expansion of tubes [35]. In international literature, elastomer tools are frequently considered as working medium, because the forming shows some similarities to hydroforming based technologies. However, as shown in reference [36], forming with elastomer tools differs from working media based processes with fluid or gaseous media.

Forming and cutting by forming with elastomer tools was patented by Guerin [37, 38]. The tool setup in the so-called Guerin process consists of a flexible cushion, which draws the workpiece over the die. A solid container serves as housing of the elastomer and prevents bulging at the sides. Contrary, the Marform process uses an additional blankholder usually made of steel in order to prevent wrinkling and allow higher drawing depths. Moreover, process variants applying the elastomer tool as a thick-walled membrane have been developed (e.g. Verson-Wheelon-process or Demarest-process) [41].

Originally, natural rubber was used as tool material, but the development of new plastic materials led to comprehensive investigations regarding suitable tool elastomers. Thiruvarudchelvan [35] summarises that due to the relatively high temperature stability (up to 150 °C), the high lifetime (up to 50,000 forming processes), and the wide possibilities to adjust the hardness, polyurethane is most appropriate. It turned out, that for forming workpiece materials of higher strength also higher strength of the elastomer should be chosen.

Thiruvarudchelvan [35] and Sala [41] point out that in comparison to conventional deep drawing, forming with elastomer shows the following important main benefits, which are quite similar to the advantages of EMF:

- the stress distribution in the workpiece is more uniform, thus delaying necking and cracking and allowing higher formability,
- the elastomer can be flexibly applied for different workpiece materials, thicknesses, and geometries. This is especially beneficial with regard to cost efficiency when forming a high number of variants in small lot sizes,
- the low tribological load allows achieving high surface qualities without lubricants.

Due to the possibilities of economically forming a high number of variants in small lot sizes, initially especially the aerospace industry pushed the developments in the field of forming with elastomers. However, the increasing demand for mass customised products raises interest in the technology in other sectors, too.

2.4. Resume and objective

As shown in the previous sections, the smart application of innovative processes as e.g. forming with elastomer and electromagnetic forming enables an extension of existing process limits. This allows higher design freedom and improves economic efficiency in the production at the same time. The advantages of the technologies suggest that they are predestined for manufacturing (local) structuring elements in order to improve the component with regard to lightweight design aspects.

However, lack of knowledge regarding the precise correlations between manufacturing technology and structuring parameters, on the one hand, and properties and functionality of the component, on the other hand, have been preventing the full industrial exploitation of this potential. To enable an efficient and requirement dependant product and production planning, the influence of different process parameters on the resulting structure properties and the related part performance was systematically analysed and the results provide the basis for an according application oriented design (compare Figure 1).

3. Description of the structuring processes

3.1. Structuring task and form defining tools

Within the analysis, aluminium sheets made of EN AW-5754, an alloy, which is typically applied in the automotive industry for manufacturing of structural parts without specific requirements to the surface quality, were considered. The structuring area was chosen as 220 × 220 mm and regular hexagonal honeycomb structures of three different sizes were regarded. The comb size (i.e. the width between the opposing sides) was set to 39 mm, 50 mm, and 70 mm, respectively. The according dies were designed as conventional steel tools. The form defining surface was realised as a partial structure, meaning that the dies consist of 5 mm thick vertical division walls between the individual combs and flat bottoms. The depth of the die cavities was 20 mm, i.e. sufficiently high to allow free forming of the comb domes according to the chosen process parameters without any interaction with the die bottom (compare Figure 2). The edges of the division walls were rounded with a radius of 2.5 m to avoid shearing of the workpiece, here.

3.2. Structuring with an elastomer punch

The principle setup used for structuring with an elastomer punch is illustrated in Figure 3 together with an exemplary structured part. At the beginning of the process, the flat sheet metal workpiece is clamped between elastomer and die. During the process the plunger moves down and the elastomer block and the workpiece are deformed together up to the maximum plunger stroke. As soon as the plunger moves back, the pressure between workpiece and elastomer block is relieved. The deformation of the elastomer is completely reversed and the sheet workpiece is subject to elastic springback.

In principle, the plunger stroke is the significant control parameter for the deformation and the resulting structure height increases with increasing plunger stroke. However, the resulting structure depth frequently does not match this value. Depending on the deformation of the elastomer and the springback it can be both, higher or lower than the plunger stroke as explained in the following.

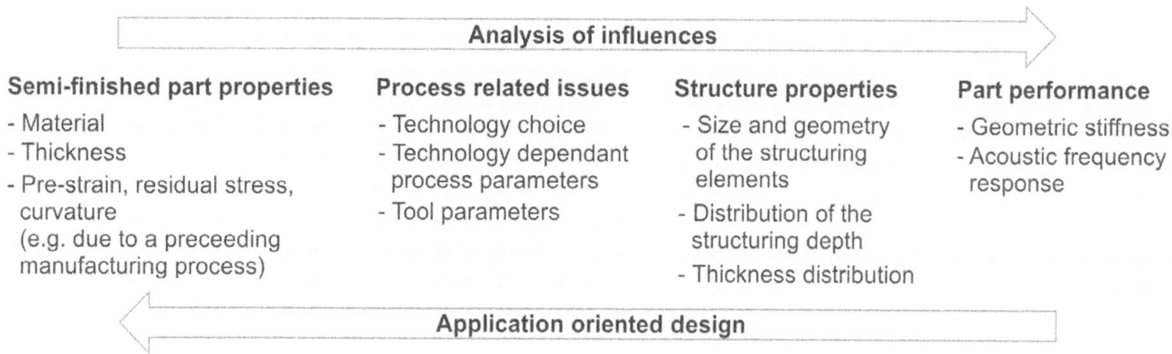

Figure 1. Analysis and application oriented design of the structuring process [39].

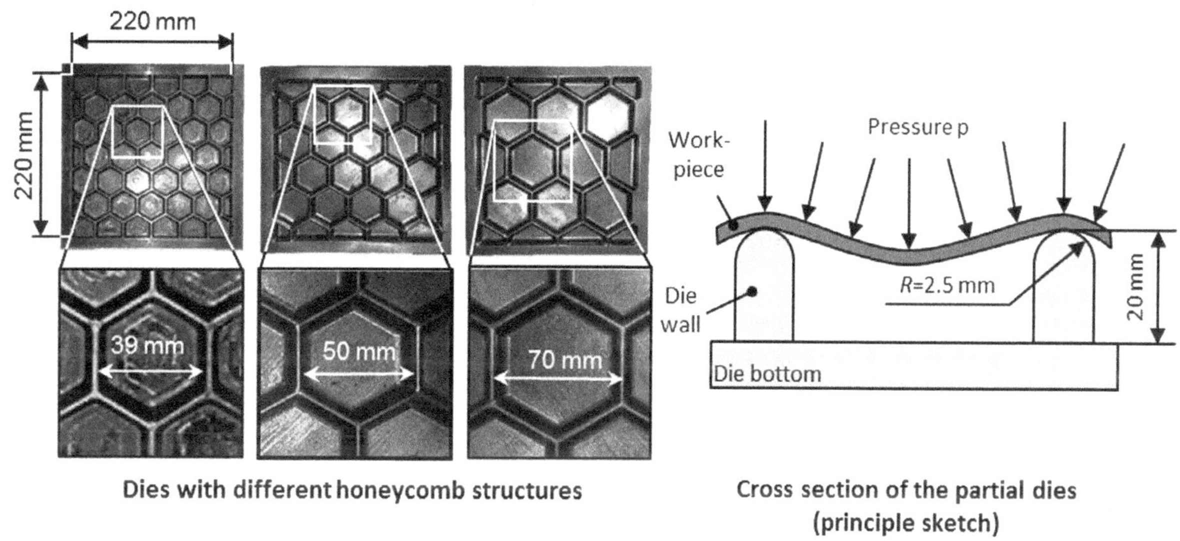

Figure 2. Partial dies for forming of different regular hexagonal structures.

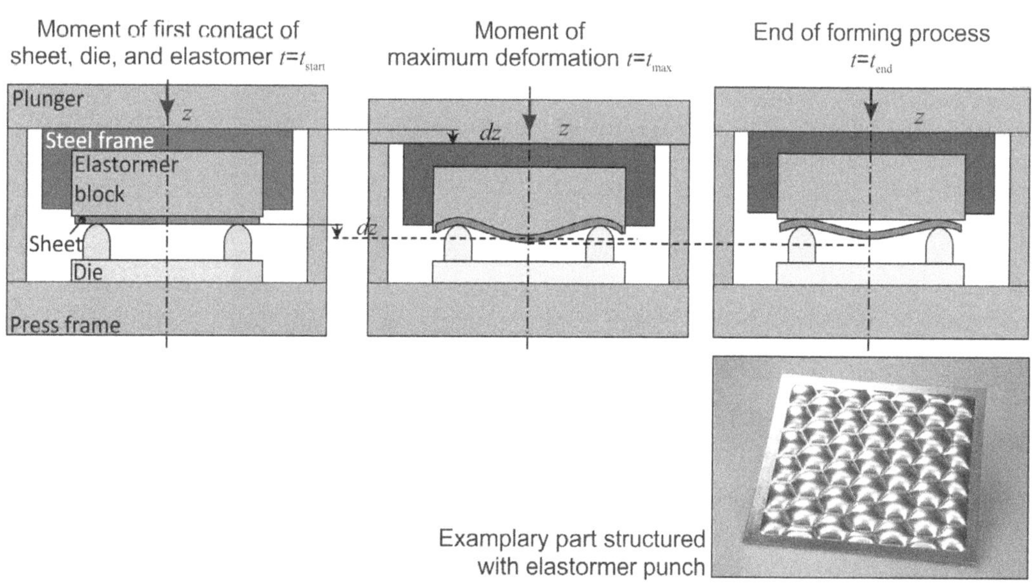

Figure 3. Forming with elastomer punch.

Since some areas of the workpiece are already in contact with the solid die walls at the beginning of the process, the displacement is locally prevented. However, the elastomer can be regarded as an incompressible body, so that its volume has to be constant. Consequently, higher displacements occur in the structure domes as indicated by the first dotted line in Figure 3. This effect can be

- undercompensated by the elastic springback resulting in structure heights that are higher compared to the plunger stroke,
- fully compensated by the elastic springback resulting in structure heights are equal to the plunger stroke, and
- overcompensated by the elastic springback resulting in structure heights that are lower compared to the plunger stroke.

High structure heights in relation to the according plunger stroke were observed especially in case of large structuring elements and small punch strokes.

3.3. Structuring by electromagnetic forming

For the investigation of structuring by EMF, an innovative process variant – the so-called sequential electromagnetic forming – has been applied. Conventional electromagnetic sheet metal forming approaches, aim at processing the complete workpiece or forming section within one single discharging process. However, with increasing size the electrical energy and voltage required for forming rise, resulting in higher loading of all machine and tool components. Consequently, the development of durable machines and tools becomes more complex and finally impossible for large forming areas. For this reason, application examples of EMF have been limited to small and medium sized parts and part areas up to now. However, recent developments at the Fraunhofer IWU [39] and in China [40] focus on extending this process limit via enhancing the conventional EMF by a sequential – i.e. a stepwise – manufacturing approach.

For applying this process variant to the current structuring task a tool coil with rectangular winding geometry was designed, realised, and tested with the partial structuring dies presented in Section 3.1. The geometry of the coil winding is shown in Figure 4a. In principle, the pressure distribution applied for structuring should be uniform, but in EMF only the areas of the workpiece close to the coil winding are pressurised, so that the coil considered here features a non-pressurised area in the centre as it is known from spirally wound tool coils (see e.g. [16]). This deficit can be overcome by the sequential forming approach. For this purpose, the workpiece is moved relative to the coil, after the first EMF sequence, so that the zone initially positioned in the non-pressurised centre area of the coil is shifted to the pressurised area in the second forming sequence (see Figure 4b). As shown in Figure 4c the workpiece features undeformed regions after the first forming sequence, which are structured during the forming sequence 2. This sequential forming approach can be used for stepwisely structuring even large flat sheets or flat sections of hollow bodies (if accessible for the coil). Various shapes of the structuring elements can be realised using one and the same coil.

4. Characterisation of the structure properties

To characterise the resulting structures and identify differences between the parts structured with an elastomer and those structured by sequential EMF, the structuring depth d – i.e. the depth of the comb domes – and the wall thickness distribution was measured.

4.1. Achieved structure depth

Figure 5 summarises the influence of the characteristic load parameter of the process, i.e. the press stroke in case of structuring with elastomer punch and the maximum magnetic pressure applied during structuring by EMF, on the structure depth. As expected, for one and the same sheet thickness and comb size of the structure higher press strokes as well as higher magnetic pressures lead to an increasing structure depth until finally the process specific forming limit of the material is achieved and cracking occurs at the edges of the hexahedrons.

However, comparing the forming limit of the two processes quantitatively, it can be seen that structuring by EMF allows significantly higher structure depths compared to structuring with an elastomer punch. The different forming limits are indicated in the diagram, but the principle effect can be seen clearly in the photo comparing a part structured by EMF to one loaded with an elastomer tool up to material failure, too. Since no cracking occurred in the left front edge of the failed part the domes in this area clearly represent the maximum achievable depth and it is obvious that they are less deep compared to the sheet structured by EMF. This indicates that in case of structuring by EMF the forming limit of the material can be increased compared to alternative and more conventional techniques, an effect that is well known from previous publications as e.g. [9].

On the other hand, the components structured with an elastomer punch feature a much more homogeneous depth of the dome heights within one and the same structured part. In the diagram this can be seen via the vertical bars, indicating the minimum and the maximum occurring comb height. This effect is probably owed to the specific pressure distribution and to the sequential forming approach. Due to this forming strategy, there are workpiece areas which are pressurised in the first or second forming sequence only, while other zones are slightly preformed after the first forming sequence and re-pressurised in the second step. In the latter areas, the slight preforming after the first forming sequence causes an increasing gap width between tool coil and workpiece in the second forming sequence. In general, increasing gap widths result in lower magnetic pressure, because the magnetic field is expanded to a bigger gap volume between coil and workpiece [16]. At the same time, the preformed material will feature higher strength compared to the original sheet due to strain

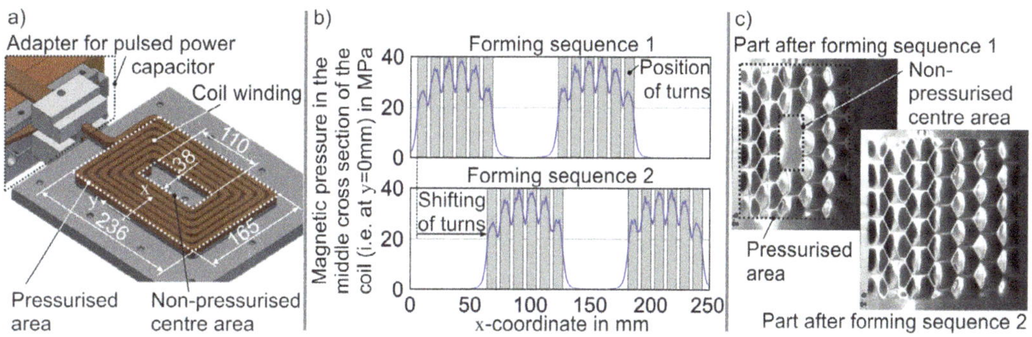

Figure 4. (a) Tool coil for structuring by EMF; (b) position of turns and according magnetic pressure distribution during forming sequences 1 and 2; (c) structured part after forming sequences 1 and 2 [39].

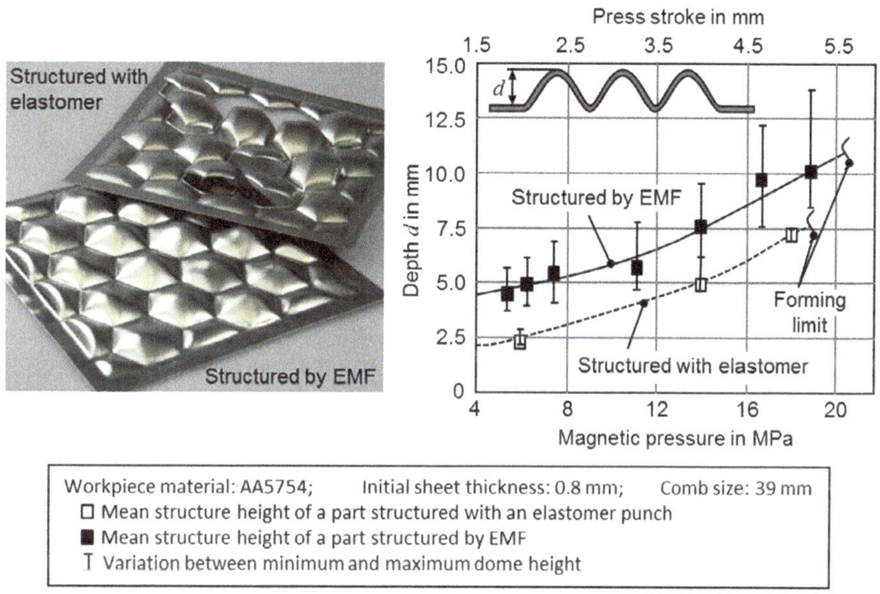

Figure 5. Achievable structure depth depending on the press stroke and on the magnetic pressure, respectively.

hardening effects. Both influences lead to relatively lower structure depth in the areas that are pressurised twice.

Complementary to the influence of the press stroke and the magnetic pressure, Figure 6 illustrates the effect of the comb size on the structure height. It can clearly be seen that if the same characteristic load parameter, i.e. press stroke and magnetic pressure, respectively, is applied to the same material and sheet thickness, an increasing size of the combs leads to higher structure depths. During deformation the comb edges can be considered as a kind of fixing, because neither in forming with an elastomer punch nor during EMF any material flow over these edges can be observed. This means that in case of larger comb sizes the distance between these restraint areas is longer leading to lower geometric stiffness and accordingly lower resistance against deformation. Consequently, deeper structures are achieved for one and the same load.

4.2. Wall thickness distribution

The measured wall thickness distributions of parts structured with an elastomer tool and such structured by EMF is exemplarily shown in Figure 7. The significant process parameters plunger stroke and magnetic pressure were adapted in such manner that the resulting structure depth of the compared parts is approximately the same. Different measurement points were chosen in relation to the structure geometry. Thereby, measurement points were set on the comb edges, in the centre of the comb domes and two equidistant points in-between. This means that for different dome sizes the distance between individual measurement points varies. However, the qualitative wall-thickness distribution shown in Figure 7 is representative for all cases regarded in this study. It clearly shows, local minima of the wall thickness distribution at the comb edges (i.e. at the position of the die walls) for both, parts structured by EMF and such structured with an elastomer punch. Moreover, the comparison clearly shows that for similar structuring depths the parts structured with an elastomer feature significantly more pronounced thinning in these regions. The reason for this might be that in case of an elastomer punch, the pressure forces are more focused to the edge areas compared to structuring by EMF. Consequently, in case of structuring by EMF localised thinning occurs at higher structure

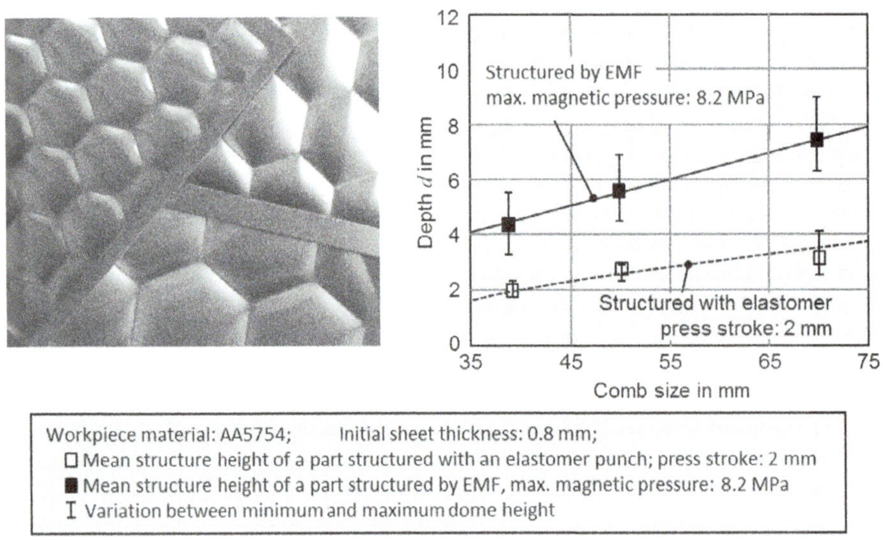

Figure 6. Achievable structure depth for different comb sizes.

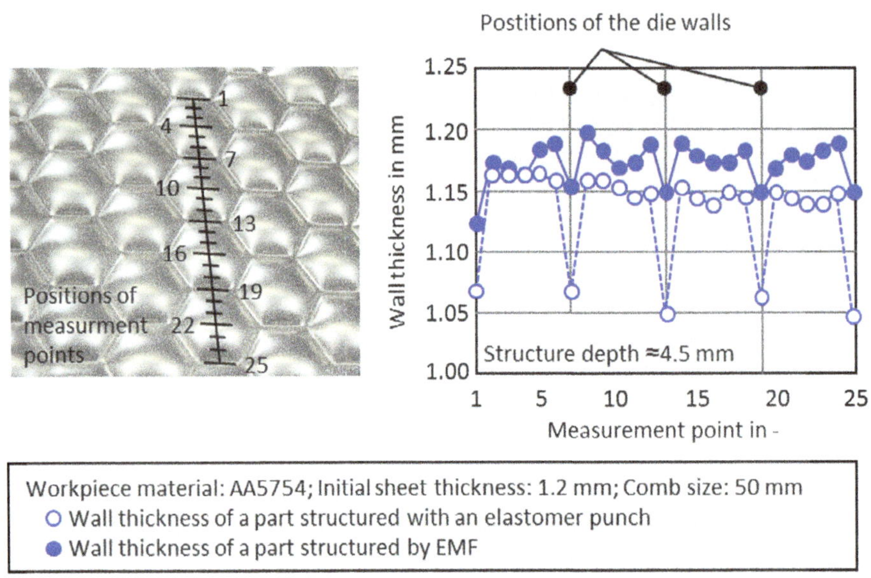

Figure 7. Wall thickness distribution of parts structured with an elastomer punch and via EMF.

depths only, explaining the higher forming limit of the material observed in Figure 5.

5. Performance of structured parts

To investigate the performance of the structured sheets, bending tests and measurements of the acoustic frequency response were carried out. As a reference, also unstructured sheets (i.e. structuring depth $d = 0$ mm) were included in the measurement.

5.1. Geometric stiffness of the structured parts

The setup of the bending tests is shown in Figure 8. The force was applied in the centre cross section of the part at those positions where the bending punch contacts a comb edge. Force-displacement-curves were recorded during bending. As also shown in Figure 8 on the examples of flat sheets of different thicknesses, these curves initially rise linearly, indicating an elastic deformation of the sheet before a further non-linear rise indicates plastic deformation.

The slope of the linear section was chosen to describe the workpiece stiffness via a so-called stiffness indicator. Figure 9 compares this stiffness indicator for flat sheets and parts structured with various parameters. Different structure depths for one and the same comb size and sheet thickness have been realised by applying different press strokes in forming with elastomer tools and different magnetic pressure pulses in structuring by EMF (compare Figure 5). It is obvious, that the stiffness value is higher in case of higher sheet thicknesses and rises with increasing structure depth. The influence of the

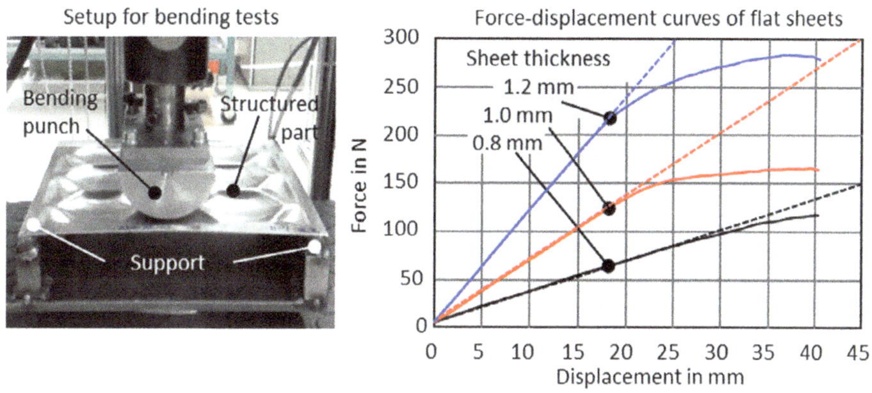

Figure 8. Setup for bending tests and measured force-displacement curves for flat sheets of different thicknesses.

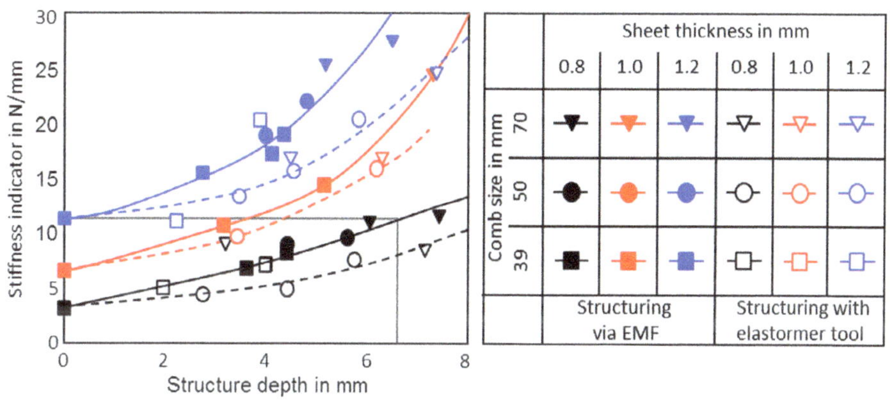

Figure 9. Geometric stiffness indicators of flat sheets and specimens structured using various parameters.

comb size seems to be secondary compared to the structure depth. Considering specimens structured via EMF, the comparison of the stiffness indicators shows that a flat sheet with a wall thickness of 1.2 mm can be replaced by a structured sheet with a thickness of 0.8 mm and a structuring depth of approximately 6.5 mm without losing stiffness. This allows weight savings of more than 30%.

Comparing the two technologies shows by trend that the stiffness indicator determined for specimens structured with elastomer tools is slightly lower compared to those structured by EMF, provided that the same structuring depth is considered. A possible explanation for this effect is given by the more significant thinning in case of parts structured with elastomer tools (compare Sect. 4.2). Due to the locally reduced wall thickness, the overall stiffness of the part is reduced. Consequently, higher structuring depths are required to achieve the same stiffness and if increasing the structure depth is not possible any longer due to the process specific forming limit, higher wall thicknesses have to be applied in order to achieve a specific required stiffness. This means that parts structured with elastomer tools still offer high weight saving potential compared to conventional flat sheets. However, due to the relatively lower stiffness indicator for the same structure depth and due to the lower process specific forming limit the weight saving potential is not as high as in case of parts structured via EMF.

5.2. Acoustic performance of the structured parts

For the measurement of the acoustic part properties, the structure was excited by noise of increasing frequencies coupled in by an electro dynamic shaker. The frequency response function from exciting force to sound pressure was measured in a defined distance from the structure. Natural frequencies of the system are indicated as local maxima of the response function. The experimental setup and exemplary measured curves are shown in Figure 10.

With regard to the application of sheet metal parts in an environment arising vibrations of different frequencies as e.g. a driving car or a milling machine, especially low frequent natural frequencies are of interest. These lead to low frequent noise (humming), which is often experienced as very awkward and consequently should be avoided. Therefore, special attention was paid to the first natural frequency and the influences on this parameter. Figure 10 clearly proves that for flat sheets the first natural frequency increases with increasing sheet thickness.

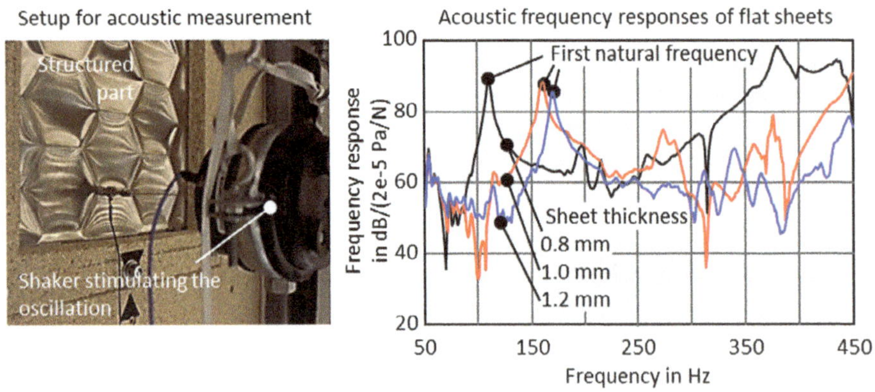

Figure 10. Setup for acoustic measurement and measured acoustic frequency responses for flat sheets of different thicknesses.

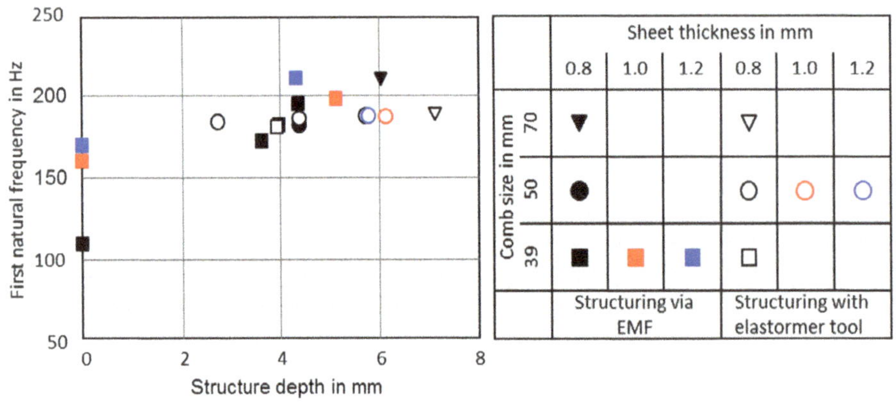

Figure 11. First natural frequencies of flat sheets and specimens structured using various parameters.

The influence of the structuring on the first natural frequency was determined via the same measurement. Figure 11 compares the first natural frequencies of flat sheets and structured parts with various parameters and both technologies regarded in this study. Similar to the investigation of the influences on the stiffness indicator, also here different structure depths have been realised for the same combination of sheet thickness and comb size by applying different press strokes and magnetic pressures, respectively (compare Figure 5).

The measurement results clearly show that in general the structuring leads to a shifting of the first natural frequency to higher values. Furthermore, the measurement results of the parts structured with by EMF suggest that higher structure depth leads to higher first natural frequency, while in case of the parts structured with elastomer tool no clear influence of the structure depth on the first natural frequency can be detected. However, it has to be mentioned that for the parts structured by EMF the first natural frequency could be identified less accurately compared to the flat sheets and the parts structured with an elastomer tool. The most probable reason for this is that in case of structuring via EMF in one and the same part the height of the individual domes varies stronger (compare Figure 5). Nevertheless, the experimental results clearly prove that especially for thin sheets structuring with an elastomer and via EMF is a proper tool in order to shift the first natural frequency to higher values, thus reducing awkward, low frequency noise in practise.

6. Summarising comparison of the technologies

To provide fundamental knowledge for a target oriented design of structures, the influences of different parameters on the resulting structure properties and the according part performance was investigated for structuring using an elastomer punch and by EMF, respectively. The high weight saving potential (here up to more than 30%) was exemplarily proved.

Furthermore, similarities and differences of parts structured with elastomer tools and via electromagnetic forming were identified. It was shown that both structuring by EMF or with elastomer tools

- clearly increases the geometric stiffness of the part,
- allows thickness and weight reduction (if the material strength is sufficient),
- increases the first natural frequency of the acoustic frequency response, and thus
- reduces awkward low frequency noise.

In comparison to structuring with elastomer tools structuring by EMF

- leads to less localised thinning and consequently,
- allows deeper structures without material failure, and
- leads to a slightly higher stiffness of the structure.

On the other hand, parts structured with elastomer tools show

- more homogeneous structure depth within one and the same part and
- clearer first natural frequency of the acoustic frequency response.

On the whole, structuring by EMF offers a slightly higher weight saving potential, while the homogeneity of the parts is better in case of structuring with elastomer tools.

7. Implications and influences

The performed analysis gives first quantitative information about the correlations of material, process, and tools on the resulting structure and the related part performance. Thus, it builds the basis for a target oriented design of the structure and the according structuring process including the selection of the most suitable forming technology, appropriate process parameters, and eligible tool properties. Future work will focus on developing a methodology using numerical simulation technologies, which allows performing this design process in a systematic and efficient way. Thereby, the implementation of structures is not limited to regular geometries, but it is also possible to apply arbitrary designs including bionic ones that are inspired e.g. by cobwebs or the wings of a dragonfly. This will enable dimensioning application and load optimised local structures in sheet metal components leading to weight reduction and improved acoustic performance of the product.

Additionally, it was already successfully proved that both technologies regarded in this study, EMF and forming with elastomer tools can successfully be implemented in classical forming tools. Many examples have shown that frequently such combined processes feature significant benefits compared the standard technology. Thus, it can be assumed that the conclusions regarding the manufacturing of structures gained here can be exploited for local structuring of real part geometries. The implementation can be realised e.g. via deep drawing with integrated EMF or integrated forming with elastomer tools in a series production chain. Thus, lightweight potentials, which are currently known on a more theoretical level only, will be made exploitable for industrial production.

Acknowledgements. The published research results are the outcome of collaborative work of the Fraunhofer Institute for Machine Tools and Forming Technology – IWU and WESTFALIA Presstechnik GmbH & Co, an automotive supplier. They were partly achieved within the project "Lokal versteifte Leichtbaukomponenten" (locally reinforced lightweight components), funded by the Sächsische Aufbaubank SAB with financial means of the Europäischer Fonds für Regionale Entwicklung EFRE.

References

1. European Union (editor), EU energy in figures – statistical pocketbook 2013, DOI: 10.2833/15026.
2. G. Reinhart, S. Teufelhart, M. Ott, J. Schilp, Potentials of generative manufactured components for gaining resource efficiency of production facilities, in: Sustainable production for resource efficiency and ecomobility. International Chemnitz manufacturing colloquium. Reports from the IWU, vol. 54, Verlag Wissenschaftliche Scripten, Auerbach, 2010, pp. 703–710.
3. H. Helms, U. Lambrecht, et al., Energy savings by lightweighting, Final Report for the International Aluminium Institute, Institut für Energie- und Umweltforschung GmbH, Heidelberg, 2003.
4. AUDI AG (editor), Workshop Audi A3, 2012, https://www.audi-mediaservices.com
5. T. Lawrence, Developing vehicles to meet carbon emissions reduction targets, 2015. Online available at http://www.paconsulting.com/our-thinking/developing-vehicles-to-meet-carbon-emissions-reduction-targets/
6. M. Kleiner, M. Geiger, A. Klaus, Manufacturing of lightweight components by metal forming, Annals of the CIRP "Manufacturing Technology", 53rd General Assembly of CIRP 52 (2003) 521–542.
7. L. Kroll, P. Blau, M. Wabner, U. Frieß, J. Eulitz, M. Klärner, Lightweight components for energy-efficient machine tools, CIRP Journal of Manufacturing Science and Technology 4 (2011) 148–160.
8. L. Zhao, J. Ma, T. Wang, D. Xing, Lightweight design of mechanical structures based on structural bionic methodology, Journal of Bionic Engineering 7 (2010) 224–231.
9. S. Golovashchenko, Material formability and coil design in electromagnetic forming, Journal of Materials Engineering and Performance 16 (2007) 314–320.
10. A.G.M. Michell, The limits of economy of material in frame-structures, Philosophical Magazine 8 (1904) 589–597.
11. A. Wanner, Minimum-weight materials selection for limited available space, Materials and Design 31 (2010) 2834–2839.
12. M.A. Carruth, J.M. Allwood, M.C. Moynihan, The technical potential for reducing metal requirements through lightweight product design, Resources, Conservation and Recycling 57 (2011) 48–60.
13. G. Reinhart, S. Teufelhart, Load-adapted design of generative manufactured lattice structures, Physics Procedia 12 (2011) 385–392.
14. G.W. Harvey, D.F. Brower, Metal forming device and method, US patent 2,976,907, 1961.
15. M. Plum, Electromagnetic forming, in: ASM Handbook "Forming and Forging" vol. 14, 1988, pp. 644–653.
16. V. Psyk, D. Risch, B.L. Kinsey, A.E. Tekkaya, M. Kleiner, Electromagnetic forming – a review, Journal of Materials Processing Technology 211 (2011) 787–829.

17. M. Kleiner, C. Beerwald, W. Homberg, Analysis of process parameters and forming mechanisms within the electromagnetic forming process, CIRP Annals – Manufacturing Technology 54 (2005) 225–228.
18. J.F. Daube, Die Abhängigkeit der Durchmesseränderung elektromagnetisch umgeformter metallischer Werkstücke von Anlagen- und Werkstückparametern, Dr.-Ing.-Dissertation, Akademie der Wissenschaften der DDR, 1973.
19. C. Beerwald, Grundlagen der Prozessauslegung und –gestaltung bei der elektromagnetischen Umformung, Dr.-Ing.-Dissertation, Technische Universität Dortmund, 2005.
20. E. El-Magd, M. Abouridane, High speed forming of the lightweight wrought alloys, in: Proceedings of the 1st International Conference on High Speed Forming ICHSF 2004, Dortmund, 2004, pp. 3–12. Online available at https://eldorado.tu-dortmund.de/bitstream/2003/27055/1/01-el-Magd_etal-040320a.pdf
21. S.B. Kim, H. Huh, H.H. Bok, M.B. Moon, Forming limit diagram of auto body steel sheets for high-speed sheet metal forming, Journal of Materials Processing Technology 211 (2011) 851–862.
22. S.F. Golovashchenko, Springback calibration using pulsed electromagnetic field, AIP Conference Proceedings 778 (2005) 284.
23. I.V. Belyy, S.M. Fertik, L.T. Khimenko, Spravochnik Po Magnitno-Impul' Snoy Obrabotke Metallov [Electromagnetic metal forming handbook], 1977. English translation by M.M. Altynova, online available at http://www.mse.eng.ohio-state.edu/~Daehn/metalforminghb/index.html
24. W. Bertholdi, J. Daube, Die elektrohydraulische und die elektromagnetische Umformung von Metallen. Urania – Gesellschaft zur Verbreitung wissenschaftlicher Kenntnisse, 1966.
25. J. Erdösi, M. Meinel, Elektrische Hochgeschwindigkeitsbearbeitung, Fertigungstechnik und Betrieb 34 (1984) 600–601.
26. R. Neugebauer, V. Psyk, Take the fast route with electromagnetic forming. Metalworking equipment news, May–June 2012, pp. 92–96.
27. V. Psyk, Prozesskette Krümmen – Elektromagnetisch Komprimieren – Innenhochdruckumformen für Rohre und profilförmige Bauteile, Dr.-Ing. Dissertation, Technische Universität-Dortmund, 2010, ISBN: 978-3-8322-9026-9.0
28. A. Jäger, D. Risch, A.E. Tekkaya, Verfahren und Vorrichtung zum Strangpressen und nachfolgender elektromagnetischer Umformung, German patent, DE 102009039759.0, 2009.
29. A. Jäger, D. Risch, A.E. Tekkaya, Thermo-mechanical processing of aluminum profiles by integrated electromagnetic compression subsequent to hot extrusion, 2010, DOI: 10.1016/j.jmatprotec.2010.06.016.
30. V. Psyk, T. Lieber, P. Kurka, W.-G. Drossel, Electromagnetic joining of hybrid tubes for hydroforming, Procedia CIRP 23 (2014) 1–6.
31. V. Vohnout, A hybrid quasi-static/dynamic process for forming large sheet metal parts from aluminum alloys, Ph.D. Dissertation, Ohio State University, 1998.
32. V. Psyk, C. Beerwald, A. Henselek, W. Homberg, A. Brosius, M. Kleiner, Integration of Electromagnetic Calibration into a Deep Drawing Process of an Industrial Demonstrator Part, Key Engineering Materials, 344 (2007) 435–442.
33. F. Taebi, O.K. Demir, M. Stiemer, V. Psyk, L. Kwiatkowski, A. Brosius, H. Blum, A.E. Tekkaya, Dynamic forming limits and numerical optimization of combined quasi-static and impulse metal forming, Computational Materials Science 54 (2012) 293–302.
34. I. Eguia, A. Mangas, R. Iturbe, M.A. Gutiérrez, Electromagnetic forming of longitudinal strengthening ribs in roll formed automotive profiles, in: Proceedings of the 4th International Conference on High Speed Forming – ICHSF 2010, Columbus, pp. 198–207, DOI: 10.17877/DE290R-12593.
35. S. Thiruvarudchelvan, The potential role of flexible tools in metal forming, Journal of Materials Processing Technology 122 (2002) 293–300.
36. M. Fukuda, K. Yamaguchi, An Analysis for Deep Drawing of Cylindrical Shell with Rubber Die, Bulletin of The Japan Society of Mechanical Engineers 14 (1971) 1971.
37. H.E. Guerin, Method and apparatus for cutting sheet metal, Patentschrift US 2,055,077, 1936.
38. H.E. Guerin, Method and apparatus for forming sheet metal, Patentschrift US 2,190,659, 1940.
39. V. Psyk, P. Kurka, S. Kimme, M. Werner, D. Landgrebe, A. Ebert, M. Schwarzendahl, Optimisation of component performance via structuring, in: 4th International Conference on New Forming Technology – ICNFT2015, Glasgow, 2015, MATEC Web of Conferences, vol. 21, 2015, p. 11001, DOI: 10.1051/matecconf/20152111001.
40. X.H. Cui, J.H. Mo, J.J. Li, J. Zhao, Y. Zhu, L. Huang, Z.W. Li, K. Zhong, Electromagnetic incremental forming (EMIF): a novel aluminum alloy sheet and tube forming technology, Journal of Materials Processing Technology 214 (2014) 409–427.
41. G. Sala, A numerical and experimental approach to optimize sheet stamping technologies: part II – aluminium alloys rubber-forming, Materials and Design 22 (2001) 299–315.

21

Low temperature high density plasma nitriding of stainless steel molds for stamping of oxide glasses

Tatsuhiko Aizawa[1,*], Tatsuya Fukuda[2], and Hiroshi Morita[3]

[1] Department of Engineering and Design, Shibaura Institute of Technology, 3-9-14 Shibaura, Minato-City, Tokyo 108-8548, Japan
[2] Research Center, Kimura Precision, Co. Ltd., Kyoto 615-8245, Japan
[3] Nano-Coat Laboratory, LLC., Tokyo 144-0045, Japan

Abstract – Various kinds of stainless steels have been widely utilized as a die for mold- and direct-stamping processes of optical oxide glasses. Since they suffered from high temperature transients and thermal cycles in practice, they must be surface-treated by dry and wet coatings, or, by plasma nitriding. Martensitic stainless steel mold was first wet plated by the nickel phosphate (NiP), which was unstable at the high temperature stamping condition; and, was easy to crystalize or to fracture by itself. This issue of nuisance significantly lowered the productivity in fabrication of optical oxide-glass elements. In the present paper, the stainless steel mold was surface-treated by the low-temperature plasma nitriding. The nitrided layer by this surface modification had higher nitrogen solute content than 4 mass%; the maximum solid-solubility of nitrogen is usually 0.1 mass% in the equilibrium phase diagram. Owing to this solid-solution with high nitrogen concentration, the nitrided layer had high hardness over 1400 HV within its thickness of 50 μm without any formation of nitrides after plasma nitriding at 693 K for 14.4 ks. This plasma-nitrided mold was utilized for mold-stamping of two colored oxide glass plates at 833 K; these plates were successfully deformed and joined into a single glass plate by this stamping without adhesion or galling of oxide glasses onto the nitrided mold surface.

Key words: Plasma nitriding, High nitrogen solute concentration, Martensitic stainless steel, Solid solution hardening, Mold, Mold-stamping

1. Introduction

Various kinds of stainless steels have been widely utilized as a mold-die for injection molding of plastics and for mold- and direct-stamping processes of optical oxide glasses. For examples, the case-hardened AISI420 martensitic stainless steels, graded by STAVAX in commercial, is used as a mold for injection molding [1]. AISI316L austenitic stainless steels are also employed as a mold-die for direct stamping of oxide glasses [2]. In the former, heat treatment or case-hardening processes can afford to provide the sufficient hardness to molds, working around 500 K. In the latter, since the glass-transition temperature (T_g) of oxide glasses is much higher than plastics, most of molds must be coated or plasma nitrided to build up the hardened layer with heat resistance and endurance against galling [3].

In addition, these stainless steel mold substrates are often wet-plated by the nickel phosphate (NiP) [4]. Then, fine micro-textures are machined into this amorphous NiP layer; this original micro-patterns are duplicated onto the oxide glasses by mold-stamping as an optical element. However, this NiP is unstable at the high temperature stamping condition; it is easy to crystalize or to fracture by itself. This issue of nuisance significantly lowers the productivity in fabrication of optical elements at present [5].

The authors [6–8] have developed the high density plasma nitriding process for hardening of Fe-Cr binary alloys and stainless steel molds. In particular, low temperature RF-DC plasma nitriding is effective to make solid solution hardening of martensitic and austenitic stainless steels [8, 9].

In the present paper, the stainless steel mold is surface-treated by the low-temperature plasma nitriding at 693 K for 7.2 ks and 14.4 ks, respectively. First, SEM-EDX analysis is made to describe the micro-structure change; the nitrided layer by this surface modification has higher nitrogen solute content than 4 mass%. Compared to the maximum solid-solubility of nitrogen of 0.1 mass% in the equilibrium phase diagram, this high nitrogen distribution characterizes the present nitrided

*e-mail: taizawa@sic.shibaura-it.ac.jp

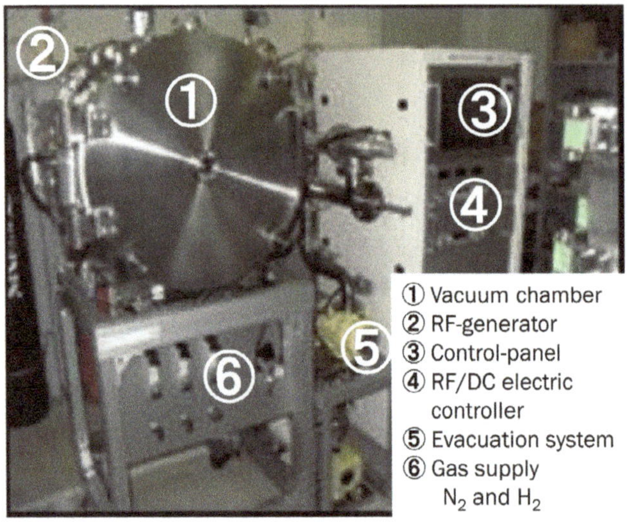

Figure 1. RF-DC plasma nitriding system, working even in the low temperature.

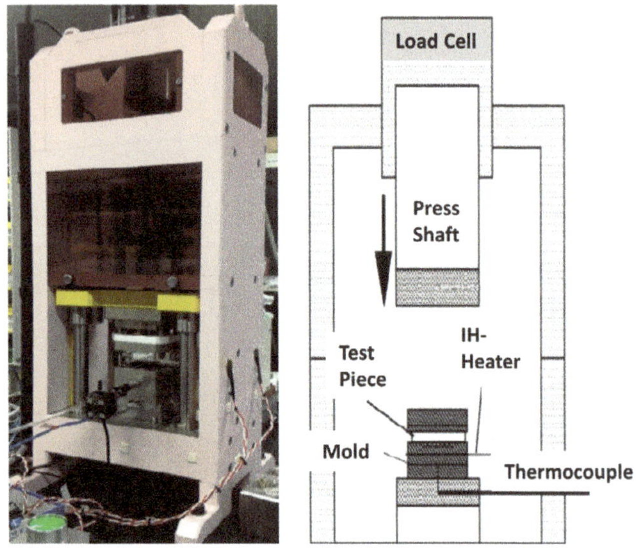

Figure 2. A mold stamping system for bet-shaping and joining of oxide glasses.

stainless steels. In second, XRD and XPS analyses reveal that no nitrides are synthesized as precipitates by the present plasma nitriding and that nitrogen interstitials occupy the vacancy sites in the α'-supercells. Since this occupation process accompanies with straining in crystals, the nitrided layer is hardened by this solid solution. In fact, the micro-Vickers testing proves that the nitrided layer has high hardness over 1400 HV within its thickness of 50 μm. This nitrogen solid-solution treated stainless steel has thermal resistivity even at the mold-stamping conditions up to 900 K. Two colored oxide glass plates are deformed and joined with success into an optical element by mold-stamping.

2. Experimental procedure

2.1. Plasma nitriding system

In this nitriding system, the input-and-out power matching takes place automatically around 2 MHz with the prompt duration of 1 ms at most. This high controllability enables to make full use of meso-plasma pressure range over 50 Pa. Figure 1 illustrates the present system. In this chamber, both RF- and DC-plasmas are independently ignited and controlled in the meso-pressure range. Since both the nitrogen atoms and NH-radicals work as a main activated species, the stainless steels can be nitrided even below 700 K. Heating unit is located under this DC-biased cathode plate. In the following nitriding experiments, the specimens are located on the cathode table and nitrided under the nitrogen and hydrogen mixed gas condition after pre-sputtering for 1.8 ks at 693 K. The flow-rate ratio of this mixed gas has influence on the yield of NH-radicals; after preliminary experiments [7, 8]. N_2 gas flow is controlled by 100 mL/min, and, H_2, by 20 mL/min, respectively. RF-voltage, DC-bias and pressure are constant by 250 V, −300 V and 70 Pa, respectively. After nitriding at 693 K for 7.2 ks or 14.4 ks, the nitrided specimens and molds are picked up after cooling down to 473 K.

2.2. Mold-stamping system

Mold-stamping test was performed by the CNC table-top stamping system, as shown in Figure 2. Both the applied load and stroke are numerically controlled up to 50 kN and 100 mm, respectively. Since each step in the stamping advances by 0.1 μm, fine micro-patterns can be imprinted onto the work materials by using a precise mother tool [10]. The ultrahigh frequency induction heating device, working around 2 MHz, is utilized to make rapid heating and quenching. The upper and lower molds are directly heated to sustain the holding temperature during mold-stamping. The sequential control of input power enables to suppress the temperature deviation within the tolerance of 1 K. This accurate thermal control results in high qualification in mold-stamping just above the glass transition temperature. To be free from severe oxidation, the vacuum cassette mold system is constructed so that the mold-stamping takes place under 100 Pa.

2.3. Specimen and mold

Martensitic stainless steel AISI420 specimens and mold-substrates were prepared to have either surface polished. The specimen was shaped by $25 \times 25 \times 5$ mm^3; the mold substrate, the circular disc with the diameter of 25 mm and the thickness of 5 mm.

2.4. Observation and measurement

Scanning Electron Microscope (SEM) as well as Electron Diffraction X-ray (EDX) were utilized to make micro-structure analysis and to make nitrogen element mapping on the cross-section of nitrided specimens. Micro-Vickers hardness testing was performed to describe the surface hardness distribution as well as the hardness depth profile. The applied load was constant by 0.25 N (or 25 g). XPS (X-ray photoelectric

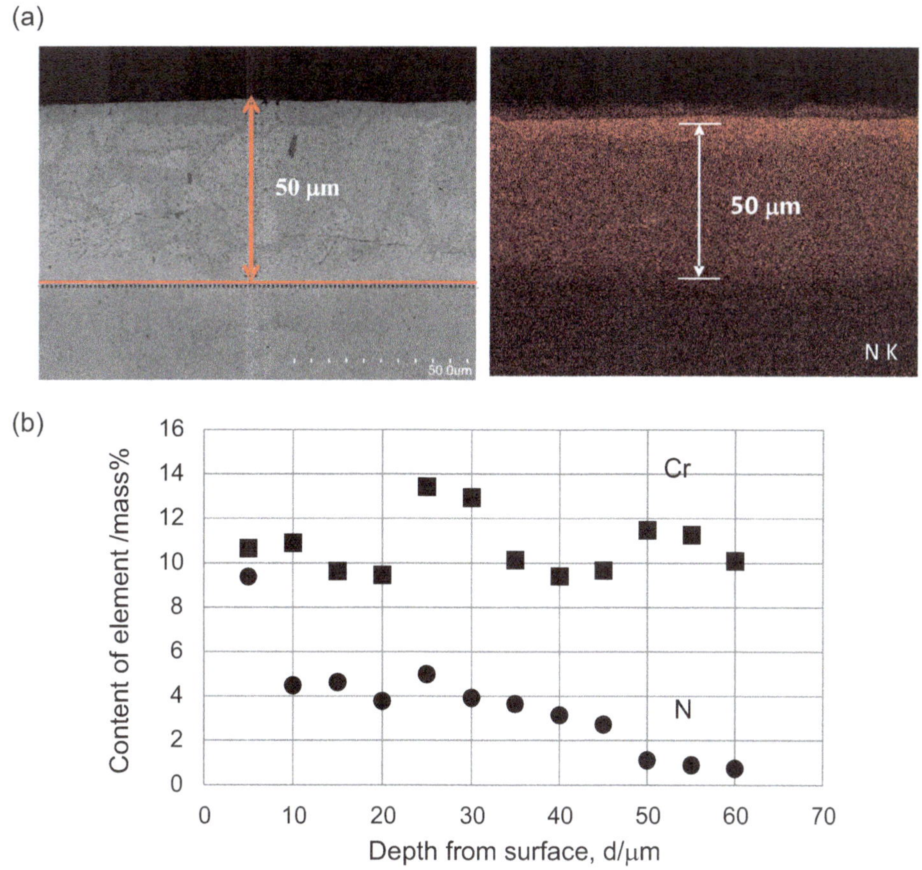

Figure 3. (a) SEM image and nitrogen mapping by EDX on the cross-section of nitrided layer. (b) Distributions of chromium and nitrogen solute in the depth from the surface of nitrided at 693 K for 14.4 ks.

spectroscopy) was also used to investigate the intrinsic chemical shift to occupation of nitrogen solutes at the vacancy site in the α'-supercells. In this measurement, the monochromic Al-K_α by 25 W was utilized with the probe diameter of 100 μm. The scanning width of energy was 0.1 eV.

3. Experimental results and discussion

3.1. Inner nitriding process

The nitrided AISI420 specimen at 693 K for 14.4 ks was halved by wire-cutting to investigate the microstructure and nitrogen solute map on its cross-section. SEM and EDX were utilized to describe the nitrogen concentration in the depth of nitrided layer. Figure 3a depicts the SEM image and the nitrogen mapping by EDX in the cross-section of nitrided martensitic stainless steel specimen, respectively. Only after nitriding for 14.4 ks, the nitrided layer thickness reaches to 50 μm.

EDX was also utilized to measure the chromium and diffusing nitrogen contents from the surface to the depth in each step by 5 μm. Figure 3b depicts the depth profiles of both element contents. At the vicinity of surface, the nitrogen solute content becomes 9 mass% in average. The maximum solid solubility of nitrogen into irons was only 0.1 mass% after the phase diagram of Fe-N system [11]. The highest surface nitrogen contents of plasma nitrided tool steels and stainless steels by using the commercial process at 823 K was reported to be 0.1 to 0.2 mass% [12]. This high surface nitrogen content significantly characterizes the present inner nitriding process at low temperature.

The nitrogen content decreases from this high concentration at the surface down to 4 mass% in the middle range of thickness and goes to nearly zero near the nitriding front end. In particular, the plateau with constant nitrogen content around 4 mass% is formed in the range of 10–40 μm in depth. As reported in [12, 13], since every inner nitriding process in the high temperature plasma nitriding is governed by the nitrogen diffusion mechanism, the nitrogen content exponentially decreases from the surface to the depth toward the nitriding front end. This nitrogen content depth profile in the nitrided stainless steels is intrinsic to the present low temperature plasma nitriding.

On the other hand, little change is detected in the chromium concentration along the depth of nitrided layer. In fact, the average of chromium content is 12 mass%, which is the same as the initial content of AISI420 stainless steels, with the deviation of 1.5 mass%, as shown in Figure 3b. This implies that no chromium is used for precipitation reaction of CrN during plasma nitriding in different from the high temperature plasma nitriding, where large amount of chromium constituents in the matrix was used to synthesize CrN [12, 13]. Hence, the thermal and corrosion resistance, intrinsic

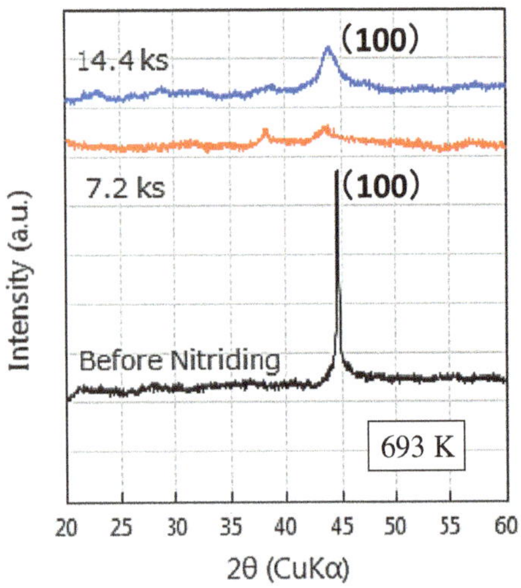

Figure 4. XRD diagram of nitrided AISI420 martensitic stainless steels at 693 K for 7.2 ks and 14.4 ks.

to original stainless steels, must be never lost even after the present plasma nitriding.

XRD analysis was also performed to investigate whether the nitride precipitates were synthesized by plasma nitriding at 693 K for 7.2 ks and 14.4 ks. As shown in Figure 4, no significant peaks for nitrides were seen in the range of $30° < 2\theta < 40°$; e.g., little significant peaks other than a main peak at $2\theta = 44°$ were seen in the nitrided specimen for 14.4 ks. The detected small peak detected at $2\theta = 38°$ in the nitrided specimen for 7.2 ks might be a transient phase by solid solution process with high nitrogen content at the surface [14, 15].

Comparing the main peaks for (1 0 0)-plane of α'-Fe lattice before and after nitriding, the original peak at $2\theta = 44.7°$ for AISI420 stainless steels is shifted in the lower 2θ side down to $2\theta = 44°$. This implies that the α'-Fe lattice structure expands itself by site occupation of nitrogen solute into vacancy sites. Hence, the presence of nitrogen solutes in the α'-supercell with 4–9 mass% in the nitrided layer accompanies with straining or distortion of original α'-lattice [16].

3.2. Solid-solution process

When the nitrogen interstitial atoms are present as a solute in the α'-supercell, its binding energy is different from the original nitrogen atom in solids with weak bonding and from the bound nitrogen in CrN. XPS is employed to analyze the binding energy of nitrogen in the nitrided layer at 693 K for 14.4 ks. First, the argon sputtering was used to clean the surface by the rate of 3 nm/min. Then, the measured energy was standardized by the carbon energy of 284.8 eV. Figure 5 depicts the XPS diagram of N1s on the surface of nitrided layer, at its center and below the nitrided layer, respectively. Fine energy profiles are detected both on the surface and in the nitrided layer. No peaks are detected below the nitrided layer.

The measured N1s spectra have the peak at 399.6 eV. After [17], the bound N1s in CrN has the binding energy of 396.2 eV; the unbound N1s or much weakly bound N1s in the polyimide, 400.2 eV after [18]. This reveals that the nitrogen atoms present in the nitrided layer are not bound in CrN but relatively weak-bound in the α'-supercell. Further precise XPS analysis and discussion is needed to define the binding state of nitrogen solutes in the nitrided layer.

The above EDX and XPS analyses suggest that the nitrogen solute is structured into α'-supercell through its occupation of vacancies and have weaker bonding than nitrogen in CrN. After the first principle calculation on the occupation of nitrogen atoms in the α'-supercell in [19], let us consider this occupation process during the present plasma nitriding.

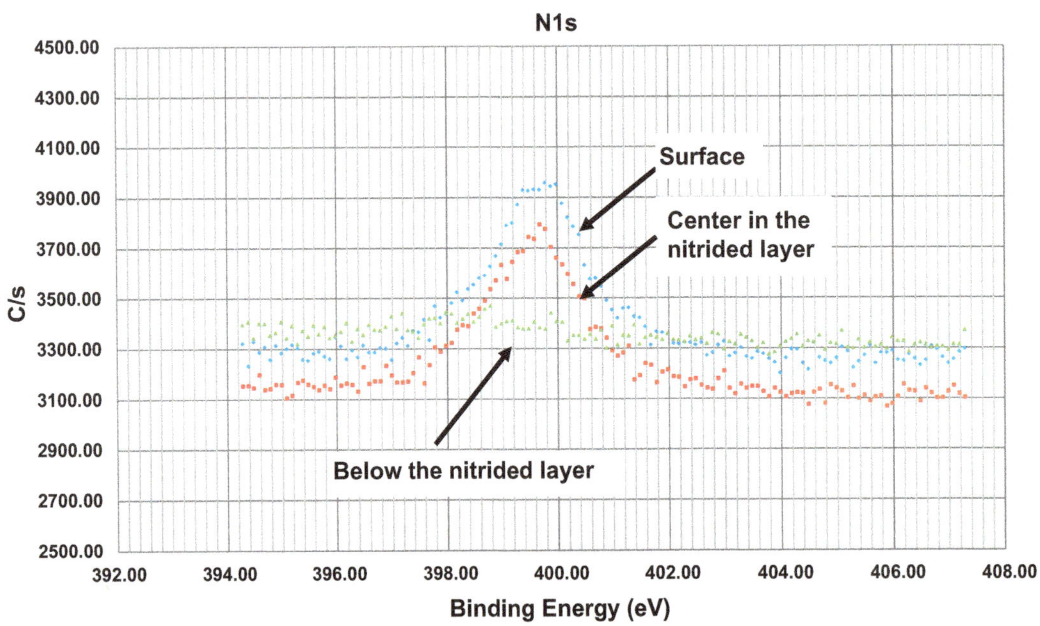

Figure 5. XPS spectra measured at the surface of the nitrided layer, at its center and below the nitrided layer.

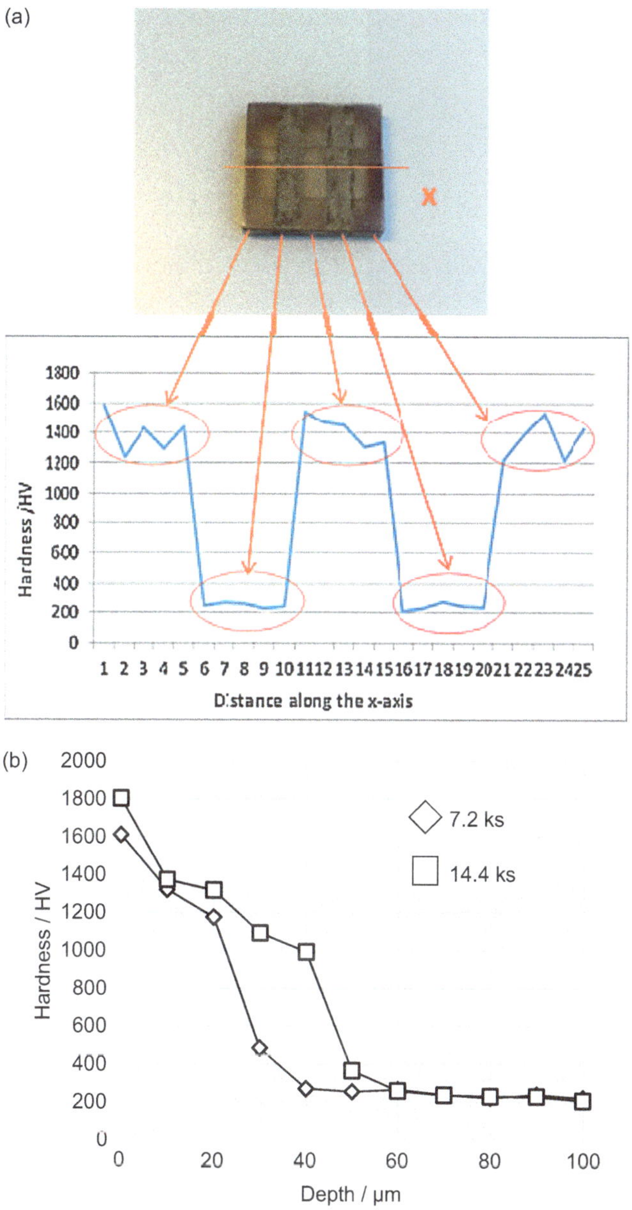

Figure 6. (a) Variation of the surface hardness distribution measured across two masking stripes on the nitrided specimen at 693 K for 7.2 ks. (b) Hardness depth profiles of the nitrided region in the two-stripe masked specimens at 693 K for 7.2 ks and 14.4 ks, respectively.

From the nitrogen mapping by EDX, N/(N + Fe) ratio is 27 to 28 at.%, corresponding to 9 mass%, at the vicinity of surface and that N/(N + Fe) ratio is 15–17 at.%, corresponding to 4 mass%, in the middle of the nitrided layer except for the vicinity of nitriding front end. Assuming this occupation process is driven by nitrogen concentration from the surface, both the octahedral and tetrahedral vacancies, or, V_{oct} and V_{tra} are expected to be gradually occupied by nitrogen interstitials from the surface. At the vicinity of surface with higher nitrogen content around 9 mass%, two pairs of V_{oct} and V_{tra} are occupied by nitrogen solutes. One pair of V_{oct} and V_{tra} are occupied to invoke two nitrogen solutes into the α'-supercell in the middle of the nitrided layer.

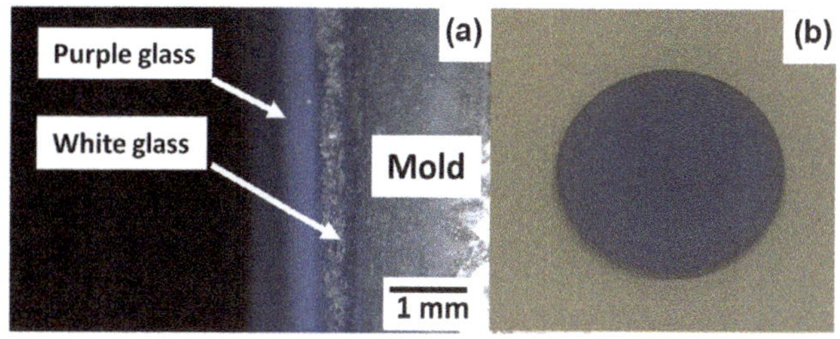

Figure 7. Hot mold stamping to join two colored oxide glass plates to one optical element at 833 K. (a) Two color oxide glass sheets on the molds. (b) Deformed and joined two-color oxide sheet.

3.3. Solid solution hardening

ASIS420 specimen with two stripe masks on its surface was prepared and nitrided at 693 K for 7.2 ks to investigate the hardness distribution across these two stripe masks along the center-line on the specimen surface as shown in Figure 6a. Since the masked regions are free from nitriding and remain as they were before nitriding, the measured hardness is equivalent to the AISI420 matrix hardness of 250 HV. On the other hand, the unmasked or nitrided regions have much higher hardness than 1400 HV even after plasma nitriding for 7.2 ks.

Figure 6b depicts the hardness depth profile of nitrided region in the two-stripe masked specimen at 693 K for 7.2 ks and 14.4 ks, respectively. As seen in Figure 6a, high surface hardness is attained to be 1600 HV for 7.2 ks and 1800 HV for 14.4 ks, respectively. This high hardness monotonically decreases in the depth of nitrided layer toward the nitriding front end. Comparing this hardness depth profile for 14.4 ks to the nitrogen content depth profile in Figure 3b, high nitrogen concentration from the surface to the depth of 50 μm corresponds to high hardness layer in the nitrided region in depth. This proves that solid solution with high nitrogen contents drives the hardening process of stainless steels.

3.4. Hot mold-stamping test

The nitrogen solid-solution hardened AISI420 stainless steel has dense microstructure with nitrogen solute occupation into α'-supercell. Its surface property as well its engineering endurance are improved from the martensitic stainless steels before nitriding. In order to demonstrate this possibility, the nitrided AISI420 substrate at 693 K for 14.4 ks was shaped to fabricate a flat mold for hot mold-stamping of oxide glasses.

In this experiment, two colored oxide glass plates with different glass transition temperature (T_g) and thickness were mold-stamped at 833 K for 3.6 ks in vacuum atmosphere under 100 Pa by using this nitride mold. Using the programmed thermal cycling schedule, the stamped glasses are free from fracture. Figure 7a depicted the cross-section of two oxide glass plates and nitrided mold. Under the reduction of thickness by 8%, two oxide glasses were uniformly compressed and joined together with each other. Figure 7b shows the jointed glass plate after demolding from the nitrided mold. A joined two-color glass product is successfully fabricated to have no defects and cracks.

This success in hot mold-stamping implies that the α'-supercell structure occupied by nitrogen solutes at the vacancy sites could promote the heat resistance of original AISI420 stainless steels in vacuum. Further research is needed to prove the improvement of thermal endurance of this plasma nitrided molds by the hot hardness testing.

4. Conclusion

The solid-solution hardened stainless steels by low temperature plasma nitriding work as a heat resistant mold for hot stamping and injection molding of oxide glasses and high-T_g plastics. This mold with the nitrided layer thickness over 50 μm and the surface hardness over 1400 HV by nitriding at 693 K for 14.4 ks or 4 h, is attractive for mold-stamping of optical elements since it is free from hot galling and severe abrasive wear. In particular, the average nitrogen solute content of 4 mass% or 15 at.% is useful to prevent the hot-stamping mold from galling with oxide glasses in the hot and vacuum conditions. Further studies are planned to utilize this nitrided AISI420 material for shaping and machining to have well-defined geometry to optical element and to use it as a mold to duplicate its geometry into the high-T_g oxide glasses with sufficient accuracy in dimension.

5. Implications and influences

The present paper has contribution to surface treatment for dies and molds. In particular, low temperature plasma nitriding is used to make solid solution hardening of stainless steel mold. This mold is utilized for mold-stamping of oxide glasses. Two colored oxide glass sheets are uniformly deformed and joined into a single optical glass.

Acknowledgements. The authors would like to express their gratitude to Mr. S. Kurozumi and Mr. H. Yamagata (Shibaura Institute of Technology) for their help in mold stamping experiments of oxide glasses. This study is financially supported by MEXT with the contract of #411419.

References

1. T. Aizawa, F.-J. Hoe, S. Yukawa, Res. Rep. SIT 55 (2011) 34–43.
2. A. Ikeda, T. Fukuda, MITI-supporting industry report on the surface treatment of molds for oxide glass stamtry, 2013-SIR-05 (2013) 113–116.
3. K. Dohda, T. Aizawa, Manuf. Lett. 2 (2014) 82–85.
4. http://www/nacro.co.jp (cited in 12/24/2015)
5. K. Okada, Y. Hirata, K. Nitta, S. Inazawa, SEI-Technical Review 167 (2005) 92–95.
6. T. Aizawa, Y. Sugita, Prof. 6th SEATUC Conference, Bangkok, 2012.
7. T. Aizawa, T. Fukuda, K. Itoh, Int. Steel Res. 87 (2012) 215–220.
8. T. Aizawa, Y. Sugita, Res. Rep. SIT 57 (2013) 1–10.
9. M. Anzai (Ed.), Surface treatment for high qualification of dies and molds, Nikkan Kogyo Shimbun, Japan, 2009, pp. 43–61.
10. T. Aizawa, K. Itoh, T. Inohara, Proc. 6th ICOMM (2010) 77–82
11. Japan Institute of Metals (Ed.), Metals Data Book (Maruzen, Tokyo, 1974), p. 25.
12. Y. Hiroka, K. Inoue, Denki-Seiko 80 (2010) 15–24.
13. N. Granito, H. Kuwahara, T. Aizawa, J. Mater. Sci. 37 (2002) 835–844.
14. D. Santojojo, T. Aizawa, M. Shinji, H. Morita, Proc. 8th ICOMM 90 (2014) 1–8.
15. Abdelrahman, T. Aizawa, Proc. 10th SEATUC, Tokyo (2016) (CD-ROM).
16. D. Manova, et al., Nucl. Instrum. Methods Phys. Res. B 242 (2006) 285–288.
17. F. Esaka, et al., Surf. Sci. 16 (1997) 42–47.
18. G.M. Rignasese, et al., Phys. Rev. Lett. 79 (1997) 5174–5180.
19. C. Domain, et al., Phys. Rev. B 69 (2004) 114112.

Continuous extrusion and rolling forming of copper strips

Xinbing Yun*, Mo Zhou, Tian Tian, and Ying Zhao

Engineering Research Center of Continuous Extrusion, Ministry of Education, Dalian Jiaotong University, Dalian 116028, P.R. China

Abstract – Continuous extrusion and rolling technology was proposed as a new strip production technology. It conducts a hot rolling process using waste heat after continuous extrusion. The continuous extrusion and rolling forming was simulated with DEFORM-3D^{T}. Influences of extrusion wheel velocity and rolling reduction on the continuous extrusion and rolling forming were analyzed. It was shown that as extrusion wheel velocity increases, torque of extrusion wheel, chamber force and rolling force, will drop; temperature of the billet in the area of abutment which is highest will increase. As the rolling reduction is increased, torque of the extrusion wheel and force acting on the chamber decrease, while torque and force of the rolls increase. The experimental results showed that a homogeneously distributed and equiaxed grains microstructure can be formed in copper strip billets with an average grain size of about 80 μm, after continuous extrusion. Grains of the copper strips are stretched clearly, during rolling, along the rolling direction, to form a stable orientation. Nevertheless, the grain boundaries are still relatively clear to see.

Key words: Continuous extrusion and rolling, Copper strip, Numerical simulation, Rolling reduction

1. Introduction

Continuous extrusion and rolling forming technology is a burgeoning and environmental manufacturing technology. It is combined with continuous extrusion process and rolling process in a row. The principle of continuous extrusion and rolling process is that oxygen-free copper plates or strips is pressed in the groove of the extrusion wheel by coining wheel, and moves with the friction of extrusion wheel rotating, then is obstructed by the abutment, turns to the chamber. The billet fills the chamber until extruded from the die. The temperature of the strips from continuous extrusion is about 400–600 °C in the exit of die [1], there is a distance between continuous extrusion machine and rolling mill, and strips supplied by continuous extrusion could be rolled down until reaching a certain length with the temperature about 360–400 °C in the entrance of rolls.

Compared to traditional processes of cold rolling after continuous extrusion, the continuous extrusion and rolling process method has the following advantages [2–5]: (1) In this process, the frictional resistance for energy dissipation is transferred into the deformation driving force, and deformation heat and friction heat are used to reach the thermal deformation temperature of billet, so the heating process is eliminated; and (2) both the continuous extrusion and rolling completely achieve the thermal deformation of the billet, during which the dynamic recrystallization occurs. Therefore, the grain is significantly refined and the material property is improved; however, the size of the billet is nearly the same as that of the finished product, thus reducing the workforce of the finishing process.

Yun et al. proposed the principle of continuous extrusion and rolling technology, based on theoretical calculation and numerical simulation [6]. But it did not consider the metal flow regularity in the process with variation of the technological parameter.

In this work, the finite element model of continuous extrusion and rolling process was established by using DEFORM-3D software platform in order to simulate the whole process, and the metal flow regularity was analyzed in the continuous extrusion and rolling process with variation of the technological parameter, such as the influence of extrusion wheel velocity, rolling reduction and strip size.

2. Numerical simulation of continuous extrusion and rolling

The DEFORM-3D software was used for the numerical simulation of continuous extrusion and rolling process of copper strips. The influence of extrusion wheel velocity and rolling reduction on the continuous extrusion and rolling forming process were analyzed. The temperature of billet, effective-stress of billet, torques of extrusion wheel and rolls, force of chamber, rolling force were compared and studied during forming.

*Corresponding author: yunxb@djtu.edu.cn

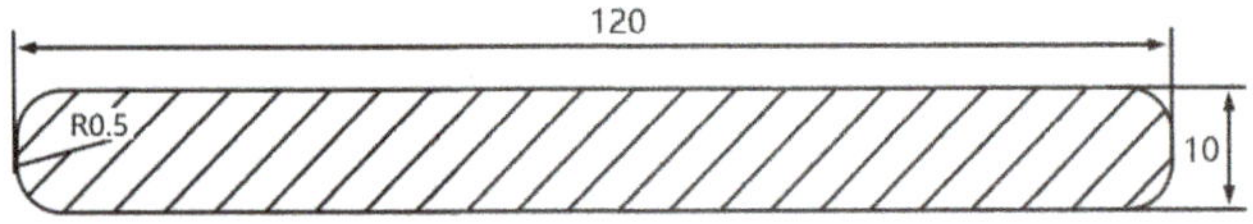

Figure 1. Cross-sectional shape of the rolled strip.

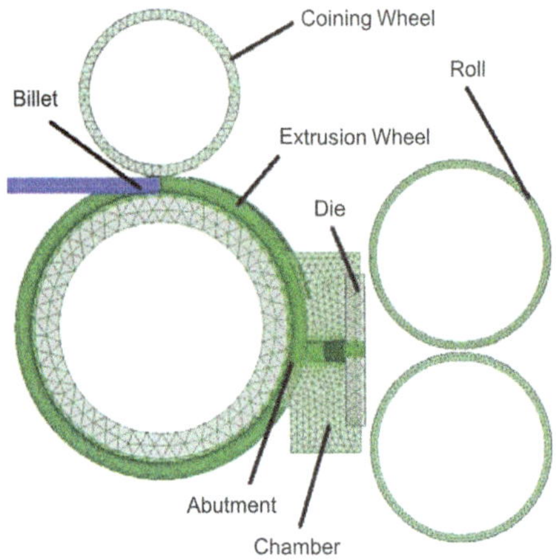

Figure 2. Finite element geometric model.

The size of a copper strip produced by continuous extrusion process is depicted in Figure 1.

The finite element simulation model which was established by using TLJ400 continuous extrusion machine and the roll with diameter of 250 mm as the model is depicted in Figure 2, where the representative geometry parameters are indicated. To save time and reduce the computer storage capacity, 1/2 of the integral model was used for the simulations based on the symmetry [7, 8].

In this work, the rigid viscoplastic finite element method was used. The billet was set as the rigid-plastic body, whereas the die and other parts were set as rigid bodies. The shear friction driving model was used for the friction between the contact surfaces of the billet and the die. All of the initial parameters in numerical simulation are shown in Table 1 [9–12].

Table 1. The initial parameters of numerical simulation.

Condition	Value
Material of billet	Copper
Material of tool and die	AISI-H13
Diameter of billet (mm)	Φ20
Initial temperature of billet (°C)	20
Initial temperature of tool and die (°C)	450
Rotation velocity of extrusion wheel (rpm)	6, 9, 12
Friction factor of billet and extrusion wheel	0.95
Friction factor of billet and guide roller	0.05
Friction factor of billet and chamber and die	0.3
Friction factor of billet and roller	0.35
Conductivity of billet (N/s K)	30
Heat capacity of billet (N m^2/K)	0.3

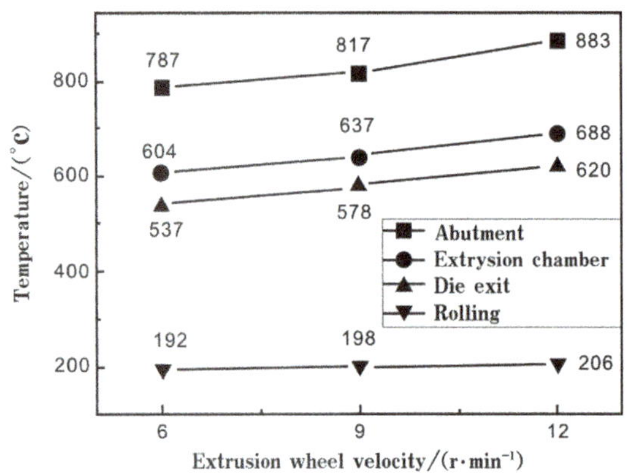

Figure 3. Effects of velocities of extrusion wheel on temperature.

3. Simulation results and analysis

3.1 Influences of the extrusion wheel velocity on the continuous extrusion and rolling forming

Extrusion wheel velocity was set to 6 r/min, 9 r/min, 12 r/min for the numerical simulations. Comparison of temperature, effective-stress, torque of extrusion wheel and rolling, force of chamber, and rolling force are analysed in different extrusion wheel velocity.

It can be observed that the temperature lines in the continuous extrusion and rolling process are escalating trend with extrusion wheel velocity increased in Figure 3. The temperature of billet in the area of abutment is hottest which was increased from 787 °C to 883 °C, temperature of billet in the area of chamber is ascended from 604 °C to 688 °C, temperature of billet in the area of exit of die is risen from 537 °C to 620 °C, and temperature of billet in the area of rolls is from 192 °C up to 206 °C. What makes increment of temperature of billet is that speed difference between billet and extrusion wheel become bigger with greater velocity of extrusion wheel, work from extrusion wheel to qualities of heat of billet is increased, extent of plastic transformation of billet is severer with extrusion wheel velocity increased. Due to temperature of billet in the continuous extrusion process is risen, the temperature in the rolling process is also increased, and the abutment and chamber will be suffered more abraded. As a result, it's not advisable to choose overlarge velocity of extrusion wheel for continuous extrusion and rolling process.

Figure 4 is depicted that effective-stress of billet is risen slightly since extrusion wheel velocity increased. As extrusion wheel velocity increased, temperature of billet is higher, flow velocity of billet is rapider, effective-stress is also ascended.

From Figure 5, it can be seen that the torque of extrusion wheel will be dropped from 5.44 N mm to 4.80 N mm when extrusion wheel is increased from 6 r/min to 12 r/min, which because increase of temperature of billet makes extent of plastic transformation of billet severer, work of billet to plastic transformation needed from extrusion wheel is reduced.

Because of resistance of strip suffered by rolls decreasing, torque of rolls is also dropped from 16.08 N mm to 12.96 N mm with extrusion wheel increased 6 r/min to 12 r/min (Figure 6).

It can be figured out, from Figures 7 and 8, that the values of force of chamber and rolling force lines are all falling trend with extrusion wheel velocity increased. The value of force of

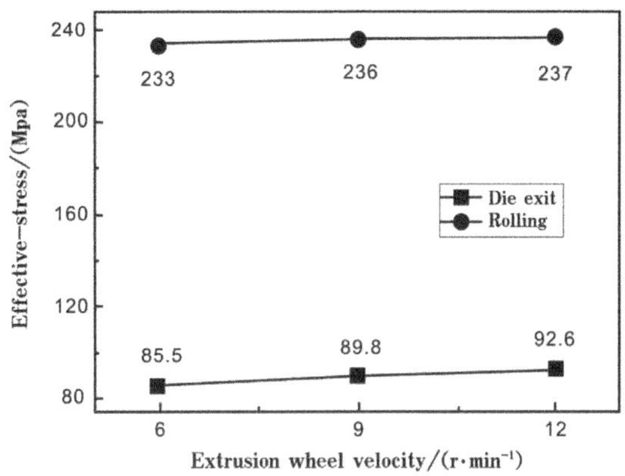

Figure 4. Effects of velocities of extrusion wheel on effective-stress.

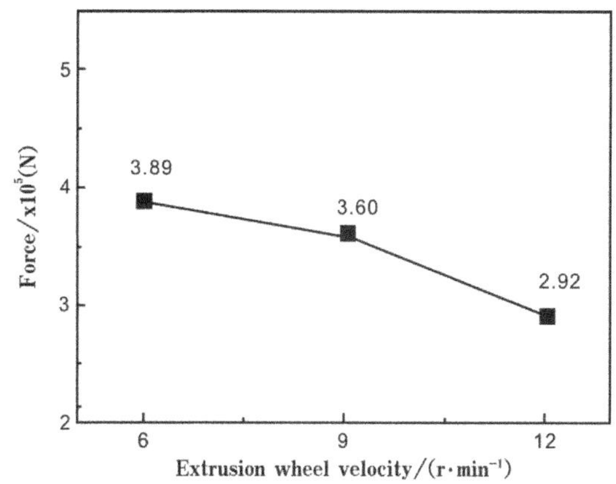

Figure 7. Effects of velocities of extrusion wheel on force of chamber.

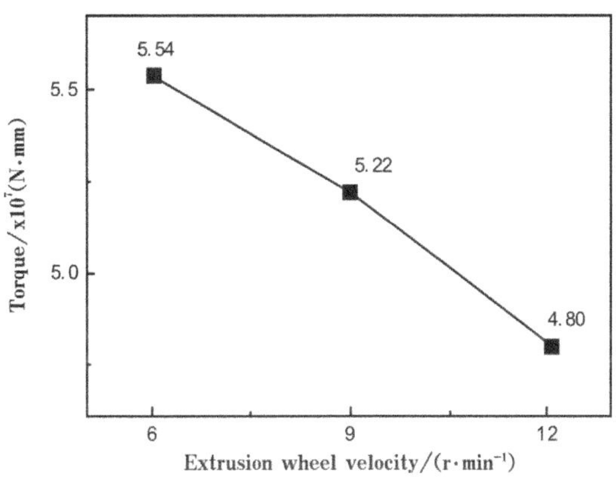

Figure 5. Effects of velocities of extrusion wheel on torque of extrusion wheel.

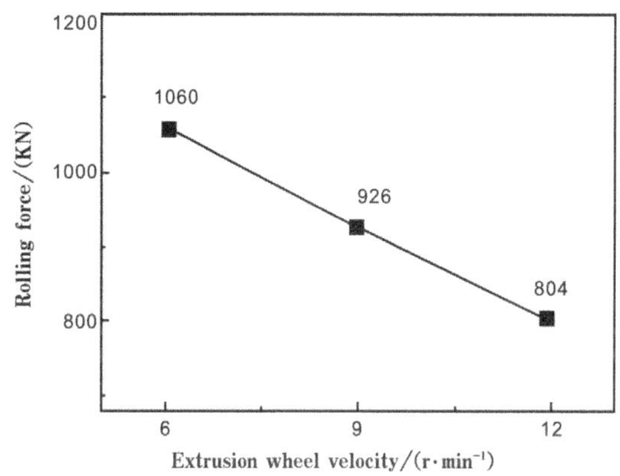

Figure 8. Effects of velocities of extrusion wheel on rolling force.

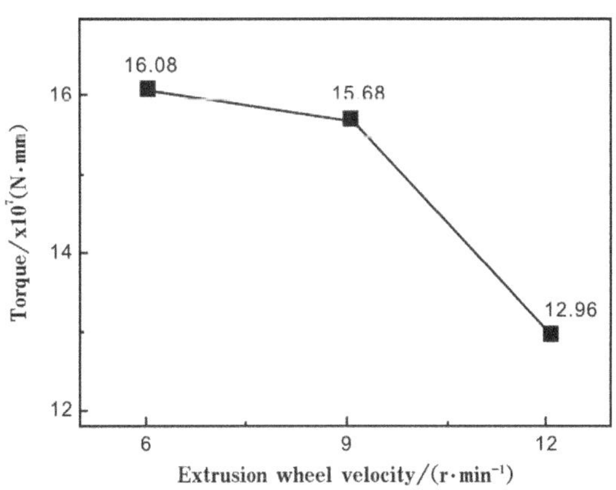

Figure 6. Effects of velocities of extrusion wheel on torque of roll.

chamber is dropped from 3.89×10^5 N to 2.92×10^5 N, which as a result of resistance power of plastic transformation of billet is reduced as temperature of billet risen. And value of rolling force is decreased from 1060 kN to 804 kN. We can draw a conclusion that rolling process will become easy with rapider velocity of extrusion wheel, but the rapider velocity of extrusion wheel, the higher temperature of billet in the process, the higher temperature of abutment, the less working life of the moulds.

3.2. Influences of the rolling reduction on the continuous extrusion and rolling forming

The value of rolling reduction is one of the most important parameters in the process of continuous extrusion and rolling. In this part, transformation regularity of the size of strips will be discussed which are presented by continuous extrusion and the size is 120×10 mm^2. The rolling reduction respectively are 2 mm, 4 mm, 6 mm, 8 mm.

Figure 9 depicts temperature of billet is decreased when rolling reduction is increased. The temperature of billet in the area of abutment is hottest which is dropped from 822 °C to 813 °C, temperature of billet in the area of chamber is fell from 650 °C to 629 °C, temperature of billet in the area of exit of die is descended from 585 °C to 563 °C, and temperature of billet in the area of rolls is from 206 °C down to 187 °C.

Owing to temperature of billet in the continuous extrusion decreased, effective-stress of billet in the area of exit of die is increased in Figure 10. However effective-stress of billet in the area of rolls is dropped, because resistance power from rolls to

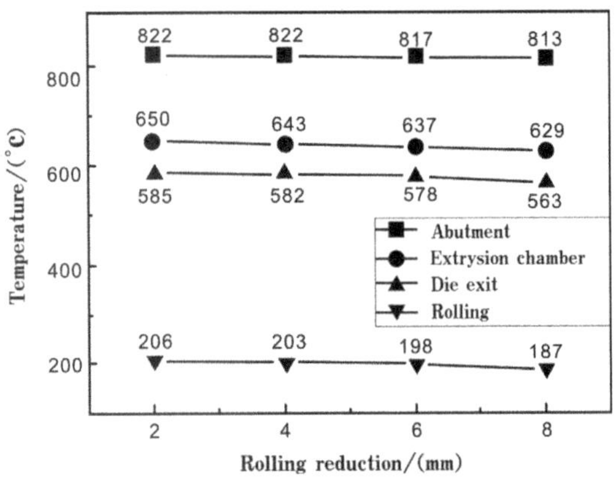

Figure 9. Effects of rolling reduction on temperature.

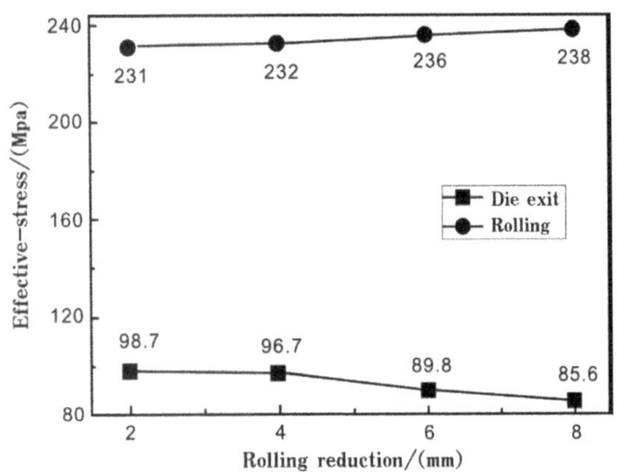

Figure 10. Effects of rolling reduction on effective-stress.

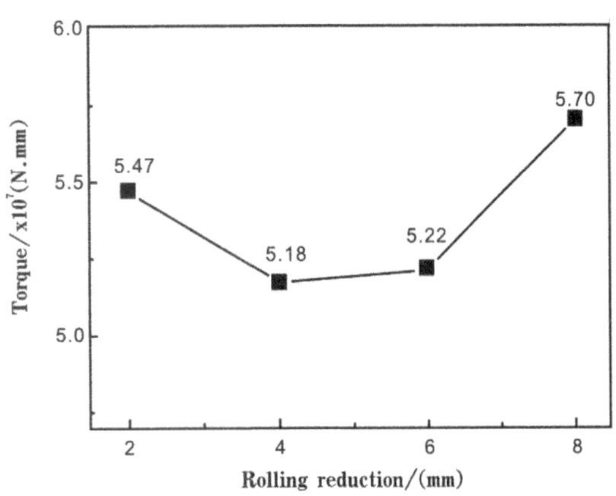

Figure 11. Effects of rolling reduction on torque of extrusion wheel.

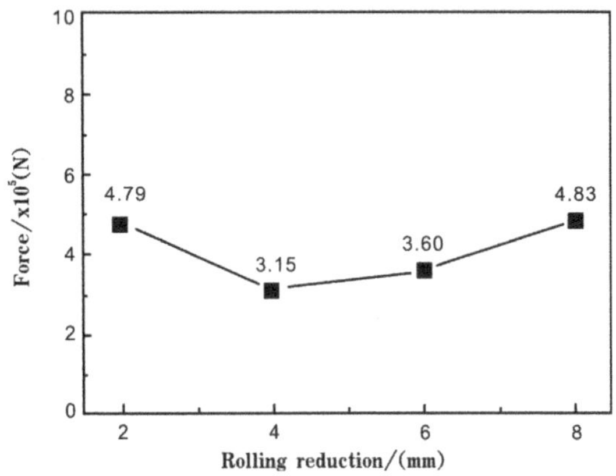

Figure 12. Effects of rolling reduction on force of chamber.

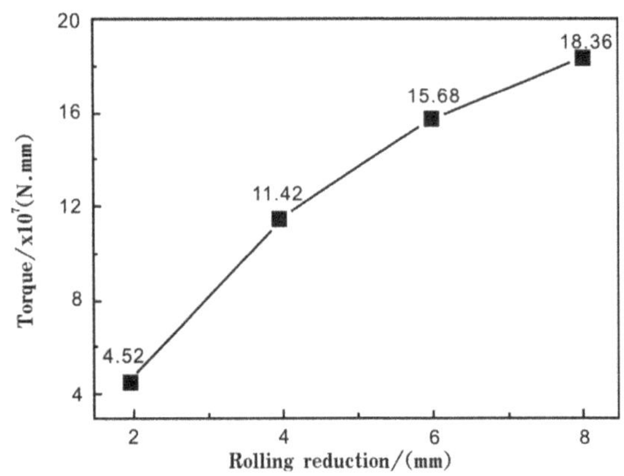

Figure 13. Effects of rolling reduction on torque of roll.

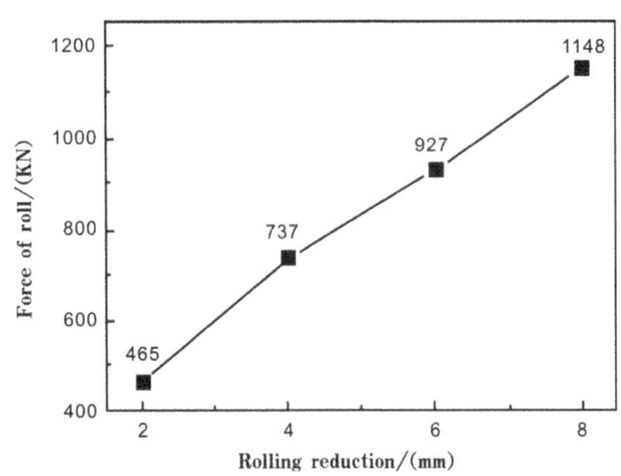

Figure 14. Effects of rolling reduction on rolling force.

billet is reduced as reduction decreased, which makes billet flowing harder and more slowly.

It can be obviously observed that value of torque of extrusion wheel and force of chamber is declined at first, then ascended following in Figures 11 and 12. When rolling reduction is from 2 mm up to 4 mm, torque of extrusion wheel is descended from 5.47 N mm to 5.18 N mm, and force of chamber is from 4.79×10^5 N to 3.15×10^5 N. When rolling reduction is from 4 mm up to 8 mm, torque of extrusion wheel is increased from 5.18 N mm to 5.70 N mm, and force of chamber is from 3.15×10^5 N to 4.83×10^5 N.

While value of torque and stress of rolls are risen accompanied by rolling reduction increased. Torque of rolls in Figure 13 is increased from 4.52 N mm to 18.36 N mm, as rolling reduction increasing from 2 mm to 8 mm. At the same time, force of rolls in Figure 14 are risen from 465 kN to 1148 kN. In other words, it is not suitable for rolling process to take overlarge rolling reduction, if not, service life of rolls will be influenced.

Figure 15. Experiment of continuous extrusion and rolling.

Figure 16. The billet.

Figure 17. Copper strips of continuous extrusion.

Figure 18. Microstructure of a copper strip.

4. Experiment

4.1. Experimental procedures

In the experiment, TLJ400 continuous extrusion machine and rolling mill with a diameter of 250 mm were used for continuous extrusion and rolling process, (Figure 15). The wheel speed was 5 rpm, the size of extrusion slab was 120 mm × 10 mm and the rolling reduction was 4 mm.

The specimens in extrusion wheel groove, deformation chamber, after extrusion and after rolling were taken out and prepared by wire cutting along the longitudinal direction. Metallographic specimens were corroded by compositing acid ($FeCl3+HCl$) after furbishing and polishing. Microstructures were observed under the OLYMPUS BX41M metalloscope.

4.2. Results and discussion

The billet of continuous extrusion and rolling exhibited typical casting structure of the upward as-cast copper rod with diameter $\Phi 20$ mm. The average size of grain was over 1 mm, the microstructure at the middle and adge is small equiaxed grains and oversize column crystal respectively (Figure 16).

Figure 17 shows the microstructure of copper after continuous extrusion forming. A homogeneously distributed and equiaxed grains microstructure can be found in the copper strip billet with a grain size of about 80 μm.

Within the grains, twins can be observed clearly. These twins were annealing twins instead of deformation twins. This was because that copper is a face-centered cubic metal which symmetry is high. Copper has more slip systems and is easy to slip. It is difficult for copper to occur twinning in the deformation to produce deformation twins, except at very low temperature which is extremely hard to slip. The annealing twins is a kind of growing twins formed during grain growth. When the grains are grown by grain boundary migration, atomic layer stacking order of the grain boundary corner accidental wrong heap, there will be coherent twin boundaries, followed by the formation of annealing twins in the grain boundary corner. Stacking fault energy of copper is low and it is easy to form annealing twins [13].

The grains of the copper strip after rolling were stretched obviously along the rolling direction to form a stable orientation, but the boundaries were still relatively clear, as shown in Figure 18.

5. Conclusions

Based on the research conducted, the following conclusions may be drawn:

1. As the extrusion wheel velocity increases from 6 r/min to 12 r/min, torque of the extrusion wheel will be

reduced from 5.44 N mm to 4.80 N mm, force on the chamber dropped from 3.89 × 105 N to 2.92 × 105 N, and rolling force decreased from 1060 kN to 804 kN. Temperature of the billet in the area of abutment where the temperature is highest increases from 787 °C to 883 °C.
2. As the rolling reduction is increased from 2 mm to 8 mm, torque of the extrusion wheel descends from 5.47 N mm to 5.18 and force of the chamber from 4.79 × 105 N to 3.15 × 105 N, while torque of the rolls increases from 4.52 N mm to 18.36 N mm, and the force of rolls from 465 kN to 1148 kN.
3. A homogeneously distributed and equiaxed grains microstructure can be formed in copper strip billets with an average grain size of about 80 μm after continuous extrusion. Grains of the copper strip are stretched during rolling obviously, along the rolling direction, to form a stable orientation, and the boundaries are still relatively clear to see.

High-precision copper strips can be widely used in the fields like electronic materials, connectors, car terminals, coaxial cables, transformers, lead frames, and so on. Since continuous extrusion and rolling has obvious merits such as a short process chain, lower energy-consumption, higher quality, environmentally beneficial, and so on, it is certain that it will be used to replace traditional processes for producing copper belts. Therefore, it has excellent application prospects.

Acknowledgements. The work reported in this paper was supported by the National Natural Science Foundation of China (No. 51175055) and the Doctoral Program of the Ministry of Education of China (No. 20132124110003).

References

1. C.D. Chu, et al., Mechanical Science and Technology 20 (2001) 556–557.
2. Y.L. Cheng, et al., Journal of Plasticity Engineering 21 (2014) 77–81.
3. H.Y. Zhang, et al., Casting Forging Welding 23 (2010) 213–215.
4. X.B. Yun, et al., Journal of Plasticity Engineering 18 (2011) 2–5.
5. P.Y. Wu, et al., Transactions of Nonferrous Metals Society of China 17 (2007) 280–286.
6. X.B. Yun, et al., Transactions of Nonferrous Metals Society of China 23 (2013) 1108–1113.
7. X. Chen, et al., Journal of Plasticity Engineering 17 (2010) 68–72.
8. X. Cao, et al., Forging & Stamping Technology 38 (2013) 91–95.
9. B. Li, et al., AIP Conference Proceedings 1532 (2013) 918–923.
10. Y.H. Kim, et al., Journal of Materials Processing Technology 97 (2000) 153–157.
11. Z.X. Fan, et al., Journal of Plasticity Engineering 15 (2008) 136–141.
12. Y.H. Kim, et al., Journal of Materials Processing Technology 80–81 (1998) 671–675.
13. S. Xia, et al., Chinese Journal of Nature 32 (2010) 94–100.

Tribological impact of the finishing technology onto meso- and microforming processes

Stephan F. Jahn[1], Benedikt Müller[1,*], Felix S. Hilfinger[1], and Andreas Schubert[1,2]

[1] Professorship Micromanufacturing Technology, Technische Universität Chemnitz, 09107 Chemnitz, Germany
[2] Fraunhofer Institute for Machine Tools and Forming Technology, Reichenhainer Str. 88, 09126 Chemnitz, Germany

Abstract – Tribological conditions between workpiece and tool have a major impact onto forming processes. Besides lubrication, temperature, and material, they are mainly influenced by the tool surface characteristics, especially by its topography, which is a result of finishing technology. Thereby formed lubricant pockets are of prime importance for the tribosystem due to its ability to separate the tribopartners by storing lubricant. The influence of four different tool finishing processes and their according surface topographies onto frictional behaviour have been investigated and quantified by the barrel compression test, in dry and lubricated tribosystems. The tool steel surfaces were stochastically microstructured by grinding and spark erosion as well as deterministic structured with micro calottes, applied by jet electrochemical machining and compared to a polished reference tool. The allocation of open and closed lubricant pockets over tool surfaces before and after the forming process was analysed by a self-made MATLAB tool.

Key words: Meso- and microforming, Tool finishing, Lubricant pockets, Friction factor, Jet-ECM, EDM, Barrel compression test, MATLAB tool

1. Introduction

Microproducts such as downsized mechanical or electrical systems or microsystems sensors and actuators are of growing demand in many branches due to their great benefits which are compact size, low weight and power consumption as well as low manufacturing costs. Since micro-manufacturing technologies gains more and more importance in producing microparts, a well-founded knowledge of their frictional behaviour is required for reproducible production processes of high quality. Especially in meso- and microforming processes tribological conditions are of major concern in terms of final shape, wear and surface quality [1]. So called size effects need to be considered as well while scaling down dimensions into sub-millimetre range [2]. Tiny tool structures which are required to form tiny microstructures could not always be finished by polishing. Hence, the surface of these tool structures is defined by the manufacturing technology, e.g. milling, spark erosion (electro discharge machining [EDM]), electrochemical machining (ECM), or laser ablation. Forming processes are highly influenced by the friction between tool and workpiece resulting of the mechanical contact caused by process related high loads. Friction has a massive impact in terms of forming resistance and therefore onto the total forming force needed. It effects the material flow and thus the final shape of the workpiece. Surface topography of the forming tool is a major characteristic of the tribosystem. Roughness peaks interfere with the workpiece surface during relative motion which increases the friction force and promotes abrasive wear. In general, by lowering the roughness of the tool surface friction force is decreasing. But very smooth surfaces, e.g. produced by polishing, could act friction-increasing because of increasing adhesion forces. Liquid lubricants, applied to separate the tribopartners and lower shear forces, could be squeezed out in case of a very smooth surfaces. Microstructured surfaces could provide the ideal tool topography to reach minimal friction forces [3]. Surfaces that are characterized by a high number of small roughness valleys can store lubricant in lubricant pockets. Valleys that are not connected to the edge of the contact area are called closed lubricant pockets and lower friction forces by developing a selective hydrostatic pressure under the applied forming load. Open lubricant pockets are linked to the edge of the contact area. They are unable to build up a hydrostatic pressure and therefore have only a little friction-lowering impact. Surface characterisation in terms of proportion of material and open lubricant pockets and closed lubricant pockets can be described by the ratio of the closed void area as a function of the material penetration and its maximum α_{clm}. By integration of the function, the normalised closed void volume V_{cl} can be calculated, which describes the percentage of

*e-mail: benedikt.mueller@mb.tu-chemnitz.de

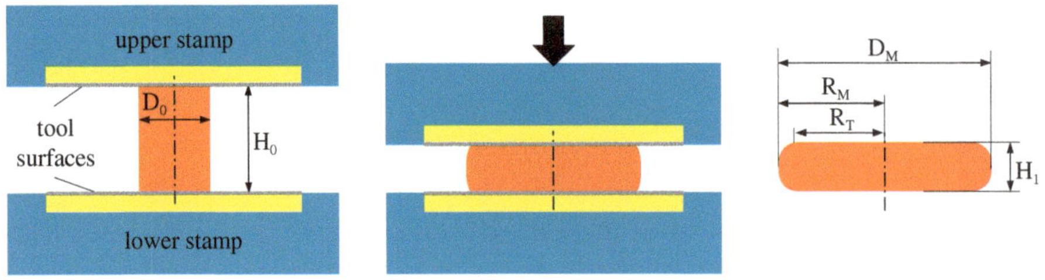

Figure 1. Principle and geometric parameters of the barrel compression test.

the lubricant store volume of a 3D-surface [4]. Depending on the manufacturing technologies, tool surfaces can differ in terms of surface roughness as well as shape and orientation of the created microstructure. In case of cutting processes such as milling or grinding, a characteristic kinematic roughness will describe the topography of the tool surface. In case of abrasive techniques like Jet-ECM [5], spark erosion (electro discharge machining [EDM]) or laser ablation, a stochastic or deterministic allocation of positive or negative asperities in the form of elevations or cavities will define the surface structure. A major advantage of deterministic structures is the possibility to create closed lubricant pockets in a defined shape and arrangement over the contact area. Jet electrochemical machining (Jet-ECM) is a technology which allows a fast generation of surface structures in the micrometer range. Like all electrochemical processes, Jet-ECM works independently from material hardness. By applying an electric current to a free electrolyte jet, a defined and localized volume of material can be removed based upon anodic dissolution [6]. Within this study, Jet-ECM is used to generate precise deterministically distributed calottes over the tool surface which should work as closed lubricant pockets.

Friction conditions in forming processes can be determined and evaluated by friction tests like the pin extrusion test [7], the double cup extrusion test [1], the slide-compression test [8], or the barrel compression test [9]. In this paper, the barrel compression test was chosen to quantify the tribological impact of four differently finished tool surfaces. The barrel compression test provides a simple way to calculate the friction factor out of geometrical parameters, measured before and after the test (Figure 1). The parameters measured for calculation are the initial height H_0, the initial diameter D_0, the height after compression H_1 and the maximum sample diameter D_m, which depends on the friction-dependent barrelling of the cylinder while being compressed. The friction factor describes the relation between friction shear stress τ and flow stress k_f:

$$m = \sqrt{3} \times \frac{\tau}{k_f} \tag{1}$$

According to [9] the friction factor m is calculated by:

$$m = \frac{\frac{R}{H} \times b}{\frac{4}{\sqrt{3}} - \frac{2b}{3\sqrt{3}}} \tag{2}$$

The average radius of the cylinder after deformation R, the top radius R_T and the barrel parameter b can be calculated easily from the geometric parameters H_0, D_0, H_1 and D_M [9].

$$R = \frac{D_0}{2}\sqrt{\frac{H_0}{H_1}} \tag{3}$$

$$b = 4\frac{R_M - R_T}{R} \times \frac{H_1}{H_0 - H_1} \tag{4}$$

$$R_T = \sqrt{3\frac{H_0}{H_1} \times R_0^2 - 2R_M^2} \tag{5}$$

2. Experimental procedure

Four different pairs of upper and lower tool stamps were made of hardened 1.3343 tool steel (64 HRC). All stamps were polished. One polished pair of stamps was kept as reference tool set (Figure 2a). Two pairs of stamps were stochastically microstructured by creating grinding marks through outside peripheral grinding (Figure 2b) and by spark erosion (Figure 2c). The fourth pair was deterministically structured by using Jet-ECM (Figure 2d). Micro calottes with diameters of 200 μm and depths of 11 μm were generated. The cylinder samples (Figure 3a) were machined on a precision-turning machine from one bar of AlMg3 and Cu-ETP, respectively. Figure 2e shows the resulting sample surface, due to the kinematic roughness of the turning process. The samples were heat-treated after machining at 380 °C in a 10 mbar nitrogen atmosphere. Each sample was measured twice in height and diameter by a micrometre before and after the compression. The cylinder size was measured with $D_0 = 4.999 \pm 0.005$ mm and $H_0 = 7.498 \pm 0.008$ mm. Images from tool and sample surfaces were taken by an optical microscope Nikon MM-400 and by a 3D-Laser scanning microscope Keyence VK-9710. 3D-Laser images were used to calculate the areal roughness values over 213×284 μm^2 (Figure 2f). 3D-Laser images over 1000×1000 μm^2 were taken to characterize the stamps' surfaces concerning the ratios of material area as well as open and closed void areas. Therefore, a MATLAB tool was developed.

Tool stamps (Figure 3b) were integrated into a servo precision forming press from Hegewald & Peschke (Figure 3c) which is located in a clean room of class 100,000. Before each compression test, the samples were cleaned in an ethanol ultrasonic bath for 5 min and tool surfaces were wiped with ethanol. For lubricated experiments, one drop of forming oil was applied manually on the tool surface as well as on the top side

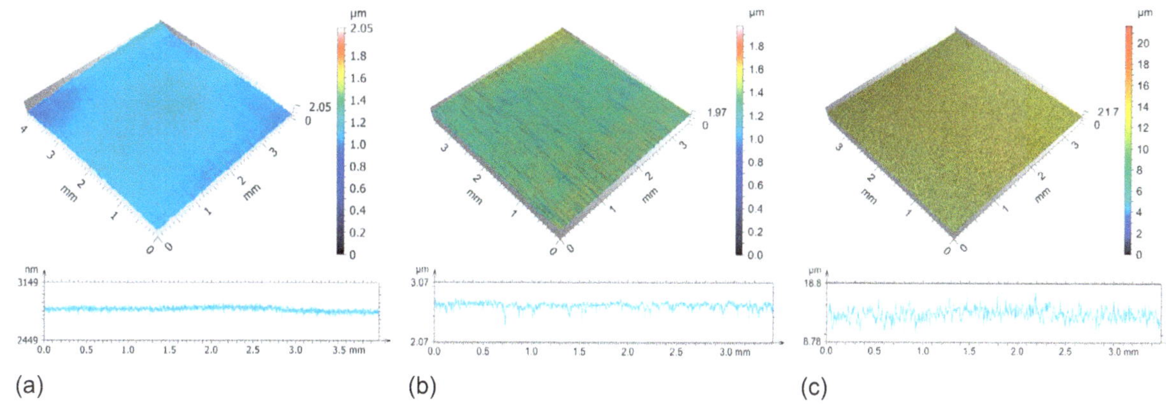

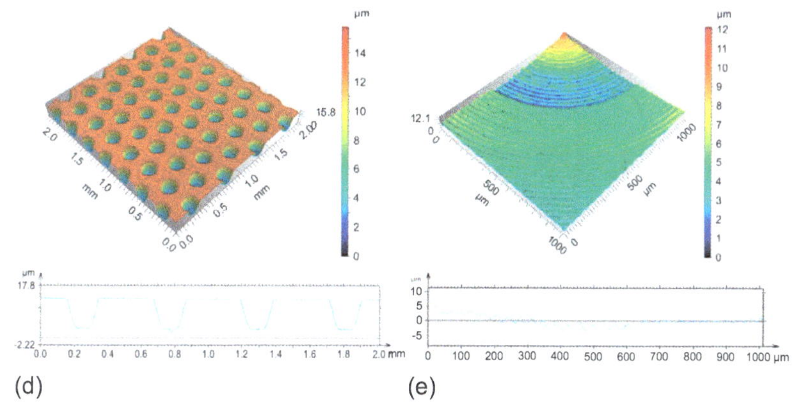

surface	Sa [μm]	Sq [μm]
polished	0.032	0.041
grinding marks	0.053	0.071
erosion craters	0.891	1.118
micro calottes	2.936	3.542
AlMg3 samples	0.586	0.704
Cu-ETP samples	0.561	0.698

(f)

Figure 2. 3D images, 2D profiles and surface roughness of tool and sample surfaces. (a) Polished, (b) grinding marks, (c) erosion craters, (d) micro calottes, (e) sample surface, (f) surface roughness.

Figure 3. AlMg3 cylindrical sample (a), sample on a polished lower tool stamp (b), precision forming press (c).

Table 1. Investigated parameters.

Parameter	Values
Specimen material	AlMg3/Cu-ETP
Forming degree φ (Eq. (4))	−0.17/−0.34/−0.51/−0.69/−0.88/ −1.10/−1.30
Stamp structure	polished/grinding marks/erosion craters/micro calottes
Lubrication	no lubrication/forming oil (Plantoform MBO 2797)

of the cylinder after placing it into the lubricant drop. In case of dry tests, the sample was placed directly on the plane tool surface. Each parameter combination according to Table 1 was conducted twice under consideration of the machine elastic recovery to ensure the desired forming degrees. Forming degree φ was calculated by

$$\varphi = \ln\frac{H_1}{H_0} \tag{6}$$

During the tests, forming temperature and stamp velocity were held constantly at 20 °C and 3 mm/min. The mean values of the normal pressure load p which were calculated by the measured maximum forming force F and the forming degree:

$$p = \frac{F}{A} \tag{7}$$

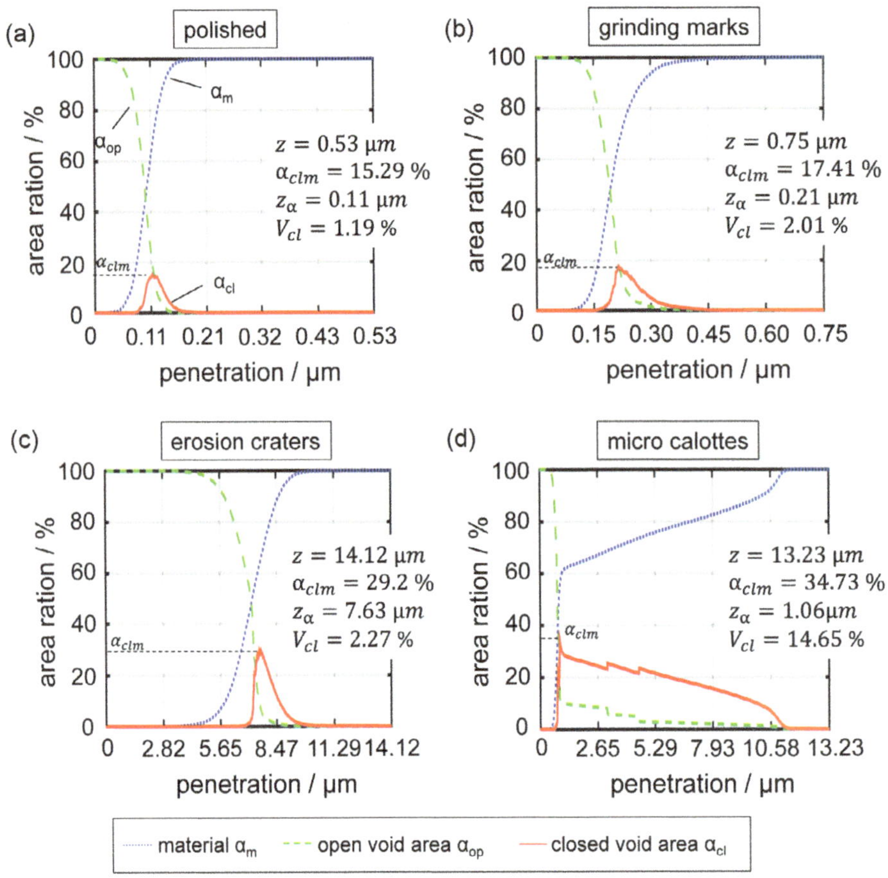

Figure 4. Area ratio diagrams with profile height z, closed void area α_{clm}, penetration depth of closed void area z_α and closed void volume V_{cl}.

whereas F is the maximum force applied to the sample. The area A of the sample is the top surface area of the sample after the compression test. Because of the barrelling of the sample the diameter of the top D_{T} is not the same as the maximum diameter D_{M}, which is measured after the test using a micrometre. Ebrahimi and Najafizadeh [9] published an equation (8) for the barrelling test, where the top radius R_{T} was calculated out of the initial radius R_0 and the maximum radius R_{M} of the sample:

$$R_{\text{T}} = \sqrt{3\frac{H_0}{H_1}R_0^2 - 2R_{\text{M}}^2} \tag{8}$$

Equation (8) can then be modified for the top diameter D_{T}:

$$D_{\text{T}} = 2 \times \sqrt{\frac{3}{4}\frac{H_0}{H_1}D_0^2 - \frac{D_{\text{M}}^2}{2}} \tag{9}$$

With equation (7) the normal pressure p was calculated by (10) and (11):

$$p = \frac{F}{\frac{\pi}{4} \times {D_{\text{T}}}^2} \tag{10}$$

$$p = \frac{4 \times F \times H_1}{\pi \times \left(3 \times D_0^2 \times H_0 - 2 \times D_{\text{M}}^2 \times H_1\right)} \tag{11}$$

3. Results and discussion

Figure 4 shows the area ratio diagrams for the four different tool surfaces. Dotted lines represent material, dashed lines open void areas and solid lines closed void areas with its maximum at α_{clm}. The polished and ground surfaces show no significant difference in terms of void area ratios, penetration depth of α_{clm} and the closed void volume V_{cl}. The eroded surface exhibits a similar spreading of area ratios but with a higher closed void area. However, Figure 4c obviously indicates that the penetration depth of α_{clm} is very high. Therefore no closed lubrication pockets can be created by sealing the closed void area after first contact of roughness peaks of tool and workpiece surfaces. Compared to that, the surfaces with micro calottes show high values for α_{clm}, too, but at a lower penetration depth, i.e., the closed lubricant pockets are sealed and effective at a moderate contact pressure between tool and workpiece. Furthermore, V_{cl} is more than five times higher compared to all the other tool surfaces, which means that a high amount of lubricant can be stored within the micro calottes.

Figure 5 shows the friction factor for experiments with AlMg3 and Cu-ETP samples in dependency on the forming degree. Because normal pressure plays an important role in friction conditions, the diagrams in Figure 6 show the friction factor m as a function of the mean normal pressure p. Continuous lines represent dry experiments and dotted lines

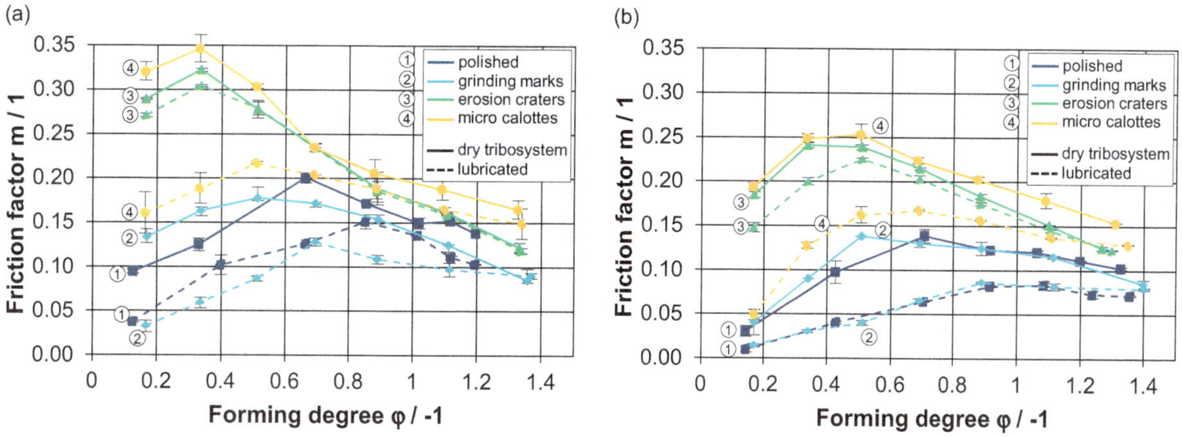

Figure 5. Friction factor over forming degree for AlMg3 (a) and Cu-ETP (b).

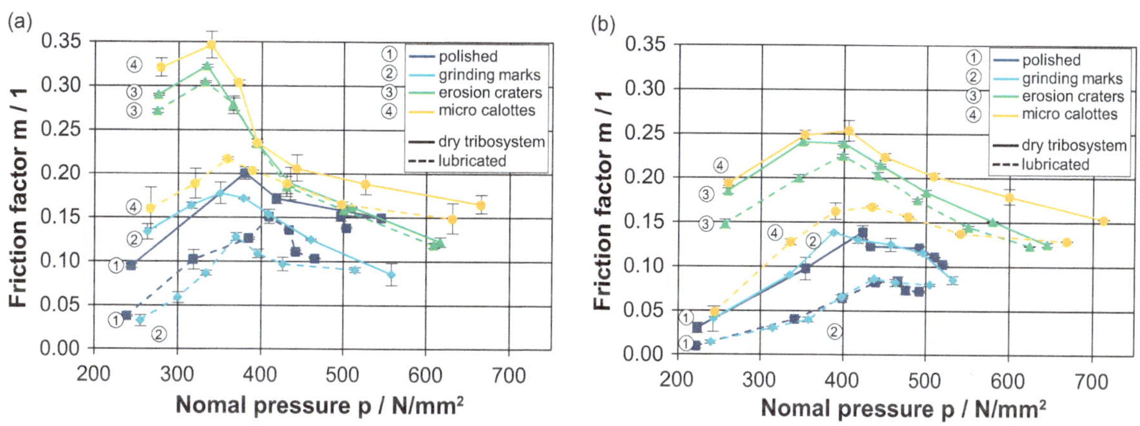

Figure 6. Friction factor over normal pressure for AlMg3 (a) and Cu-ETP (b).

lubricated tribosystems. All tool and material combinations show a related course of friction factor with increasing values to a maximum, followed by a slight decay. The rising part of the curve can be explained by the increasing friction shear stress τ due to an increasing real contact area resulting of smoothed roughness peaks on the workpiece surface. Therefore, according to equation (1), the friction factor is increasing with rising forming degree. In case of lubricated tribosystems, with increasing normal pressure, the lubricant gets more and more squeezed out of the lubrication pockets. This leads to a less hydrostatic support, which means, τ and therefore the friction factor is increasing. The drop of friction factor can be explained by its dependency on the flow stress k_f and friction shear stress. While τ is levelling out to a constant value with increasing real contact area at high normal pressure, the friction factor decreases due to a rising flow stress as a result of strain hardening.

Both materials show a similar spreading of friction factors for forming degrees over 0.8 with almost comparable values. At forming degrees below 0.8, aluminium leads to 1.5–2 times higher friction factors than copper. In general it can be said, that a higher tool surface roughness leads to increasing friction forces. The quite low and very similar roughness values for tools with grinding marks and the polished reference result

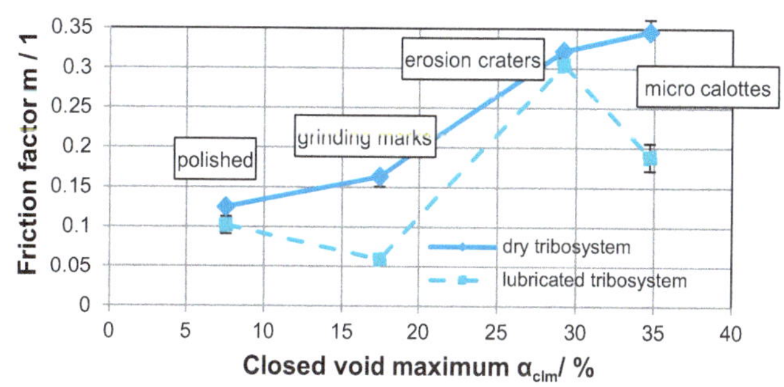

Figure 7. Friction factor over closed void maximum α_{clm} for AlMg3 samples at $\varphi = -0.51$.

in relatively low friction factors. Contrary to that, the high roughness values of surfaces with erosion craters and micro calottes result in up to two times higher friction factors, especially in unlubricated tests. This matches with the fact that rough tool surfaces interfere with the relative motion because the softer workpiece material is pressed into the roughness valleys [10]. At high normal pressures (550 N/mm^2 for aluminium and 600 N/mm^2 for Cu-ETP) the friction factor reducing effect is still enabled for the micro calottes-shaped tool configuration. Figures 5 and 6 also show the friction-decreasing

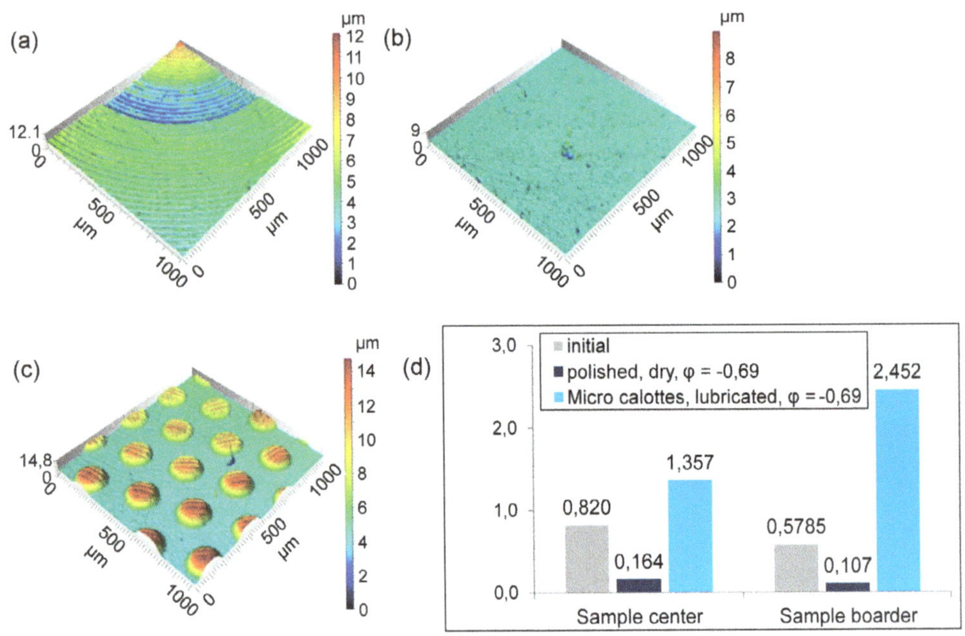

Figure 8. AlMg3 sample face side surface of initial (a), polished, dry, $\varphi = -0.69$ (b), micro calottes, lubricated, $\varphi = -0.69$ (c) and root mean square roughness Sq of sample centre and boarder (d).

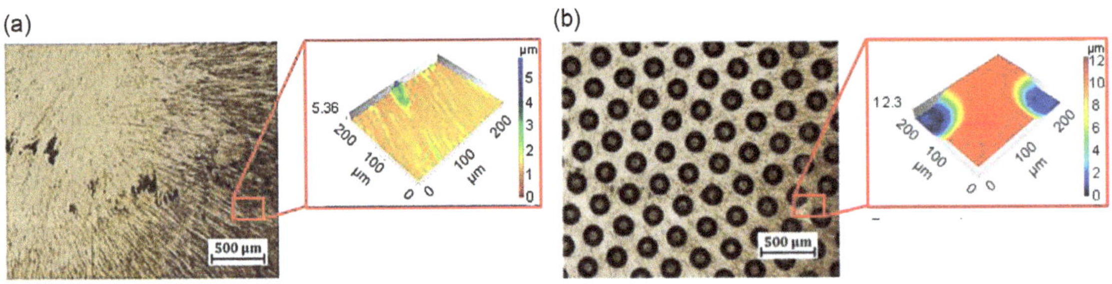

After experiments: Sa = 0.188 µm; Sq = 0.349 µm
Initial: Sa = 0.032 µm; Sq = 0.041 µm

Between calottes:
After experiments: Sa = 0.061 µm; Sq = 0.085 µm
Initial: Sa = 0.055 µm; Sq = 0.072 µm

Figure 9. Pick-up at tool surfaces after last compression, polished (a) and micro calottes (b).

effect of lubrication in case of the polished reference tool and for tools with grinding marks and micro calottes. Tool surfaces with erosion craters form an exception: lubrication does not lead to a significant decrease of friction. The diagram in Figure 7 plots the friction factor in dependency on the closed void area for aluminium samples at a forming degree of -0.51 for dry and lubricated tribosystems. Despite the fact that the ratio of closed void area α_{clm} and therefore the possible area of hydrostatic pressure is even higher compared to polished tools and tools with grinding marks the difference of friction factors for the both values for erosion craters is negligible. The most likely reason for this phenomenon is the fact, that the maximum of closed area ratio lies in a profile depth of about 7.63 µm. The sample surface must be flattened and pressed into the tool surface in a high level before the lubricant can be trapped in the resulting closed lubricant pockets. Until this point mainly less effective open lubricant pockets are responsible for taking external load by building a hydrodynamic pressure. But with increasing forming degrees and increasing normal pressure the lubricant gets easily squeezed out of the open lubrication pockets, due to their connection to the boarder of the contact zone. Therefore the friction lowering effect is vanishing for tools microstructured with erosion craters.

Figure 8 illustrates the resulting sample face side surface of dry compression with the polished reference tool (Figure 8b) and lubricated compression with micro calottes-tool (Figure 8c) at a forming degree of -0.69 in comparison to the initial face side surface (Figure 8a). Both compressed samples reached a friction factor of -0.2 but resulting in strongly different surface qualities, despite their similar tribological conditions. By comparing the shift of root mean square roughness Sq from initial to resulting surfaces at sample centre and sample boarder (Figure 8c) it is evident, that the polished reference tools cause a strong flattening of the sample surface and a decreasing surface roughness, whereas the microstructured tool increases the resulting surface roughness by up to four times. The consequence of using the structured forming tool is a generated negative form of the micro calottes in the shape of small hills on the sample's surface.

Figure 9 shows the difference between polished and microstructured stamp surfaces after the experiments. The image of the polished surface in Figure 9a indicates clearly a high pick-up due to strong flattening impact onto the compression samples because of the very low surface roughness of the initial polished tool surface. In case of dry compression, this results in a higher real contact area without a separation of tool and workpiece surface by any lubrication and increases the appearance of adhesion wear. The 3D-scan in Figure 9a shows workpiece material remaining on the polished tool surface, which leads to an drastically increase of surface roughness from initially Sq = 0.041 μm up to resulting 0.349 μm. Compared to that, the tool stamps covered with micro calottes reveal no significant increase of surface roughness. The initial root mean square roughness between the calottes is only increasing from Sq = 0.072 μm to resulting 0.085 μm.

4. Summary

By using the barrel compression test, the friction factor was determined for four different tool topographies, with two different workpiece materials, in dry and lubricated tribosystems, at forming degrees from −0.17 to −1.3. The tool topographies were results of possible finishing technologies polishing, grinding and spark erosion as well as a deterministic microstructuring by Jet-ECM. The friction factor increases rapidly with higher surface roughness for forming degrees below $|\varphi| = 0.8$ for both tested workpiece materials, whereas higher forming degrees are less dependent on the tool surface topography. By using a lubricated tribosystem, the friction was lowered in case of polished, ground and with micro calottes structured tools. Tools with erosion craters show no significant impact under the influence of lubrication due to the absence of closed lubricant pockets within the interacting tool surface profile. Besides tribological conditions, workpiece and tool surfaces are affected drastically by the tool surface condition. Dry compressions with polished tools result in smooth workpiece surfaces but increase the appearance of pick-up on the tool system. Whereas tools structured with micro calottes indicate just minimal pick up in form of higher surface roughness but develop a not intended negative form of micro calottes on the workpiece surface.

But in fact, calottes are a promising technology. A strong effect of friction factor reduction was noticeable. In further researches this effect will be examined by varying the calottes geometry in terms of depth, diameter and distance to each other.

Acknowledgements. This research is supported by the Deutsche Forschungsgemeinschaft (DFG) in context of the Collaborative Research Centre/Transregio 39 PT-PIESA, subproject A02.

References

1. U. Engel, Wear 260 (2006) 265–273.
2. F. Vollertsen, D. Biermann, H. Hansen, I. Jawahir, K. Kuzman, CIRP Ann. 58 (2009) 566–587.
3. E. Brinksmeier, O. Riemer, S. Twardy, Int. J. Mach. Tools Manuf. 50 (2010) 425–430.
4. S. Weidel, U. Engel, Int. J. Adv. Manuf. Technol. 33 (2007) 130–136.
5. M. Hackert-Oschätzchen, G. Meichsner, H. Zeidler, M. Zinecker, A. Schubert, Proceedings of ESAFORM 2011 1353 (2011) 1337–1343.
6. M. Hackert-Oschätzchen, R. Paul, A. Martin, G. Meichsner, N. Lehnert, A. Schubert, JMPT 223 (2015) 240–251.
7. B. Eichenhüller, U. Engel, in Proceedings of 3rd ICOMM'08, Carnegie Mellon University Pittsburgh, Pennsylvania, USA, 2008, pp. 207–210.
8. P. Groche, J. Stahlmann, J. Hartel, M. Köhler, Tribol. Int. 42 (2009) 1173–1179.
9. R. Ebrahimi, A. Najafizadeh, J. Mater. Process. Technol. 152 (2004) 136–143.
10. K. Steinhoff, W. Rasp, O. Pawelski, Tribologie und Schmierungstechnik 2 (1998) 31–36.

24

Investigation into initiation and propagation of cracks in the coated surfaces of spur gears with submodelling and irreversible cohesive-zone modelling techniques

Jiling Feng[1,*], Yi Qin[2], and Hanshan Dong[3]

[1] Division of Mechanical Engineering, School of Engineering, Manchester Metropolitan University, Manchester M1 5GD, UK
[2] Department of Design, Manufacturing and Engineering Management, University of Strathclyde, Glasgow G1 1XJ, UK
[3] Surface Engineering, University of Birmingham, Birmingham B15 2TT, UK

Abstract – Spur gears are one of the commonly used transmission parts in industry, due to its simplicity in structures and low cost in manufacturing. Due to a performance requirement, a spur gear may be coated with a specific coating arrangement. Under working conditions, the coating on the teeth may be damaged due to contact fatigue, in the forms such as micro-pitting and/or delamination. The failure mechanism of the coated surface under the gearing contact loading has been investigated intensively through experiment. A comprehensive computational model, which could be used to investigate the propagation of cracks in the coated surfaces, is still lacking. In the research reported in this paper, several finite element modelling techniques, including that for submodelling and irreversible cohesive zone modelling (CZM), have been developed to investigate the failure mechanisms of the coated surfaces of gears under the gearing contact fatigue loading. These techniques not only allow the localized stresses distribution and deformation in the interested locations in the coating and the substrate to be investigated in detail but also enable visual observation on the development of fatigue damages in the coating.

Key words: Coated surface, Spur gear, Fatigue damage, FE modelling, Submodelling, Cohesive-zone modelling

1. Introduction

Thin, hard PVD (physical vapour deposition) and CVD (Chemical Vapour Deposition) coatings have been widely used in a variety of industrial applications, due to the enhanced hardness and wear resistance of the contact surface [1]. Thin hard coatings have positive influences on the fatigue and corrosion fatigue behaviours of the mechanical components, which justifies their uses in some critical systems such as automotive systems [2] and aeronautic systems [3]. Prior to the initiation of the research reported in this paper, it was found that there was a need to develop a more reliable, theoretical and numerical model for the analysis of coating surfaces in the gears, although some effort had been made on the analysis of the fatigue damages of the coated gears [4, 5]. To-date, analysis of the coatings in engineering surfaces still largely relies on the uses of empirical formulas or is based on experience. Clearly, a more accurate computational model, allowing for the specific working conditions to be taken into account, is much needed from the point of view of the coating design and analysis. Most recently, an innovative modelling method, which combined with the parameterized modelling, submodelling technique and cohesive zone model (CZM), was developed to determine the load-bearing capacity for the coated products [6, 7]. This comprehensive modelling method allows for many key factors, including geometry, material properties and pre-heated treatment, to be considered in the coating design and analysis. This method and model have been validated by a series of micro-indentation tests and gearing contact tests [6, 8]. In this study, the innovative modelling method was developed further by implementing irreversible cohesive law with unloading-reloading hysteresis, with which the failure mechanisms of the coated spur gear under the contact fatigue loading has been investigated.

*Corresponding author: j.feng@mmu.ac.uk

2. Localizations of the damage on the tooth flank

Based on a conventional standard procedure for gear design, pinions are more apt to pit than gears [9]. This is because the sliding motion of the pinion tends to pull metal

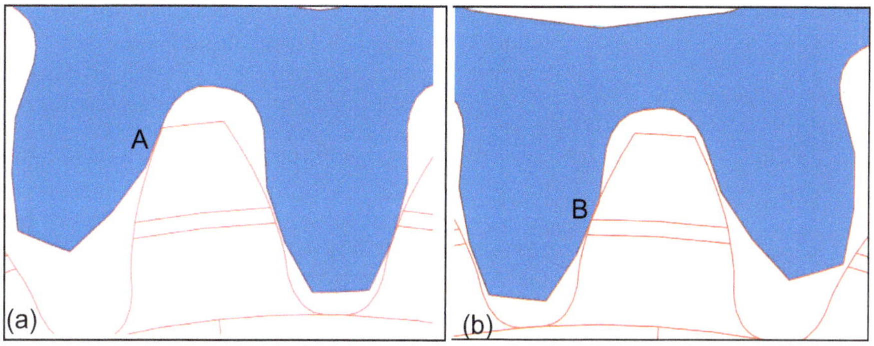

Figure 1. (a) Point A represents the location where the pinion starts to engage with the wheel; (b) Point B represents the location where only one pair of teeth is subjected to the load.

away from the pitch line due to its role as a driver. In addition, the pinion has more cycles of operation than the gear wheel and therefore more apt to fail due to fatigue damage. In this study, investigation of damage behaviour only focuses on the teeth flank of the pinion.

In the teeth flank of pinion, there are two locations where micro-pitting is most likely to occur due to the severe contact loading conditions. These two locations are based on the two extremely points (point A and point B in Figure 1). The point A refers to the location, where the wheel tip touches pinion root and engagement begins. As the pinion has small number of teeth, the radius of curvature of involute profile at the root tends to reduce toward zero, which induces extremely high localized stresses, and hence gives rise to pitting away of the metal. Point B is a typical location, where the load is only supported by one pair of teeth, and hence the highest contact stress is expected to occur. Therefore, Point B is considered as the area where the micro-pitting is easily to occur. In the following sections, the damage performance such as localized contact stresses, plastic deformation and cracks at these locations under the cyclic loading, will be investigated using the developed methodology.

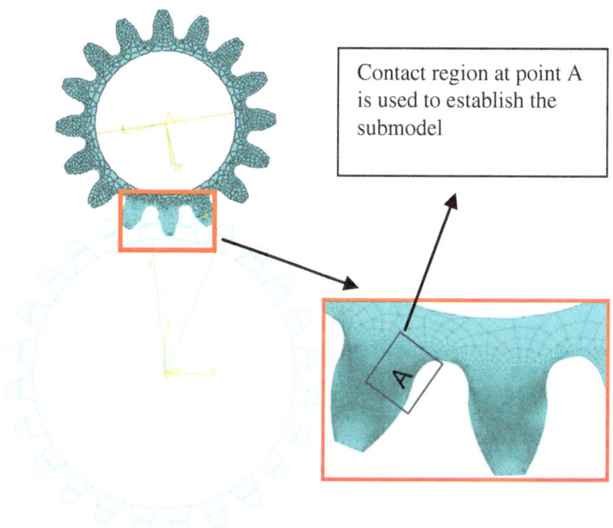

Figure 2. Engagement of the pinion and wheel, where the gear starts to touch root of tooth of pinion (point A) and the engagement starts. The sub-model, representing the contact around the point A, is created, in order to analyze localized contact stresses and crack propagation at this area.

3. Methodology

In this paper, the fatigue damage of the coated surface of a pair of spur gears was investigated using the newly developed modelling procedure, which combines several modelling techniques, including the parameterised FE modelling, irreversible CZM and sub-modelling. Parameterised modelling allows for materials and geometrical parameters as well as loading conditions to be described using parameters, enabling many key factors to be incorporated in the analysis as a whole. CZM enables the reproduction of the initiation and propagation of cracks in the coating-layer under complex gear/pinion contact loading conditions. The sub-modelling helps to scale down the geometry of the global model from the macro-scale to the micro/nano-scale. The sub-modelling is a powerful tool for developing understandings of the performance of coating structures and its failure mechanism in a great detail. The case examined in this study was line contact between a pair of gears, where pinion and wheel have 16 and 24 teeth, respectively. The central distance between gears is 91.1 mm, where addendum diameter for pinion is 82.405 mm and for wheel is 118.417 mm. The surface of the gears will be deposited a very thin coating layer with the thickness of 2 micron. Evidently, there is significant difference existed in the geometrical dimensions of gears and thickness of coatings, which causes difficulty in the establishing of a unified model to consider different length scales. In order to address the significant difference of dimensions between the wheel/pinion and coating thickness, a two-step analytical procedure was used: (1) determining the contact conditions (pressure and friction) in the contact area using a global model; and (2) generating a representative sub-model within the contact region so that coating performance can be investigated at the micron-scale level. The techniques described above was shown in Figure 2, where a finite element global model is used to capture the features of geometrical shape and engagement of gears and a sub-model is used to predict the detailed performance of the coated surface. It is scaled down to a micron level where cohesive elements

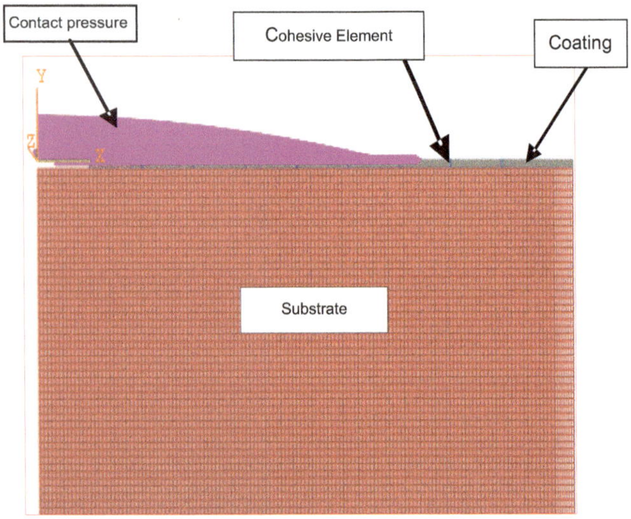

Figure 3. A sub-model, focusing on the contact region at the surface of the pinion in the global model, where each cohesive-zone element was inserted between each pair of the continuum elements within the coating.

were introduced in order to predict the performance of the coating system and its damage micro mechanism.

4. Cohesive zone model

The cohesive law, specifying the relationship of the traction and separation across the two virtual adjacent surfaces, provides a phenomenological description for the progressive damage and eventual fracture of the material [10]. Implementation of a cohesive zone model between the continuum elements has been proved to be an efficient technique for the simulation of nucleation and propagation of cracks in the coating surface under monotonic loading [1]. An example of incorporating cohesive elements in a sub-model is shown in Figure 3.

As well known, under the fatigue loading, deterioration of the coating material and cohesive strength are expected to be related largely to the loading history. Accordingly, CZM with unloading-reloading hysteresis would need to be implemented in a model, allowing for the fatigue failure mechanisms under the contact cyclic loading to be investigated.

4.1. Cohesive-zone model under the monotonically loading

An irreversible bilinear cohesive-zone modelling approach under the monotonically loading was incorporated with the other modelling techniques including parameterized modelling technique and sub-modelling technique to investigate the crack initiation within the coating layer. The bilinear cohesive-zone law, characterised as the linear function of softening relationship of traction-separation of material degradation, can be written as:

$$G_c = \frac{1}{2} T_{max} \delta_c \tag{1}$$

where: G_c (45 J/m^2) is the critical energy-release rate governing the damage evolution and T_{max} indicates the maximum traction; and δ_c (1%) is the characteristic cohesive-zone length to which the separation reaches when the crack surfaces will be generated. A crack in the coating layer may be initiated when the separation reaches the critical value of, δ_0, at which point the traction achieves its maximum value, T_{max}. The parameters of cohesive properties, including the energy-release rate, are crucial to the accuracy of simulation of the cracking within the coating and delaminating of a bonding interface. In this study, the adhesive energy at the atomistic level, was calculated using CASTEP software, which provides an alternative way to determine the value of the cohesive properties. It also provides an approach of linking micro- and nano-mechanical modelling. By means of the cohesive-zone modelling, the critical load in terms of the initiation of the first crack could be predicted. More details on the bilinear cohesive constitutive law, the implementation and applications can be found in author's previous work [6].

4.2. Cohesive-zone model under cyclic loading

The fatigue can be considered as one of the major causes that induce the failures in many coated surface-systems. Under a fatigue loading condition, the material deteriorates with the time and the prevailing cohesive strength must be related to the loading history. The cohesive constitutive equation described in equation (1) has been approved an efficient way to simulate the crack within the coating and the delamination of the interface between coating and substrate under the monotonic loading [6]. This equation, however, may results in infinite life under cyclic loading since the dissipation mechanism during the process unloading and re-loading is not taken into account. To develop a model that considers cyclic loading, an irreversible cohesive law with an unloading-reloading hysteresis function, as shown in Figure 4, is employed to simulate the mechanical response of the cohesive zone. This cohesive constitutive law, through introduction of damage variable, D, enables the degraded cohesive properties [11]. The modified constitutive equation incorporating the damage variable, D, is written as:

$$\begin{Bmatrix} t_n \\ t_s \\ t_t \end{Bmatrix} = \begin{bmatrix} K_{nn}(1-D) & & \\ & K_{ss}(1-D) & \\ & & K_{tt}(1-D) \end{bmatrix} \begin{Bmatrix} \varepsilon_n \\ \varepsilon_s \\ \varepsilon_t \end{Bmatrix} \tag{2}$$

With the condition of cyclic loading, the modified cohesive properties, as described above, are considered in the developed FE model by use of an UMAT subroutine in ABAQUS 11. Each cohesive element possesses four nodes and two integration points. Two types of unloading process, described in equations (1) and (2), respectively, were considered in the subroutine programme. The degraded stiffness of cohesive element during the unloading/reloading process is determined by the current state of displacement, where both of the damage mechanism under monotonic and cyclic loading are considered.

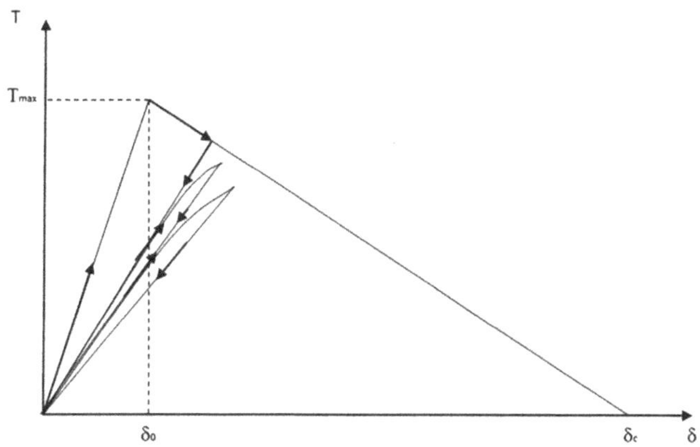

Figure 4. Cohesive traction-separation law under cyclic loading.

Table 1. Material properties of coating and substrate.

Materials	TiN/CrN coating	Substrate
Thickness (micron)	2	–
Hardness (HV)	1680	300
Young's Modulus (GPa)	240	120
Poisson's ratio	0.3	0.3

5. Results

5.1. Configuration of the coating surface of gears

Crack initiation and propagation analysis had been carried out for a 42CrMo4 spur gear with deposited multilayer TiN/CrN coatings. Two different surface systems were investigated in this study and they are defined as the following: (1) Model I: a model PVD TiN/CrN coating on a T-conditioned 42CrMo4, where the substrate was in the simple quenched-and-tempered state; (2) Model II: a model PVD TiN/CrN coating on a nitride 42CrMo4 substrate.

The spur gears were manufactured in 42CrMo4 steel and then the gears were super-finished. For the Model I, the steel substrate of spur gear was directly coated with TiN/CrN coating. For the Model II, prior to the coating deposition, the part of steel substrates was subjected to active-screen plasma nitriding, whereas the other part was in the simple quenched-and-tempered state. This type of surface system was referred as nitrided 42CrMo4 substrate. Coating deposition was performed with an industrial cathodic arc evaporation PVD system. The configurations of the two types of gears are summarizes as:

- Model I: PVD TiN/CrN coating on a T-condition 42CrMo4,
- Model II: PVD TiN/CrN coating on a nitride 42CrMo4 substrate.

In both FE global and the sub-model, the coating and substrate are characterised as homogenous elastic plastic material. The depth-dependent hardness of the hardened layer (after case-hardening) was converted into yield strength, which was obtained by a least square fitting-technique, and the strength values were defined with a field method in the FE model.

The main material properties for both substrate and coating are given in Table 1.

Table 2. Critical loads for two types of surface systems.

Critical load for crack initiation	Model I		Model II	
	Point A	Point B	Point A	Point B
Maximum pressure (MPa)	1438	1534	3768	3245
Average pressure (MPa)	1321	1279	3331	2918
Normal force (N)	1644	5014	4883	10,588
Torque (Nm)	170		358	

5.2. Computational and experimental results

To determine the critical load at which a crack was initiated, different levels of loading (Toque) were applied on the global model (pinion/gear system). The pressure distribution at the contact region (point A and point B) from the global model was then transferred to the sub-model as a boundary condition. The critical loads predicted through the FE simulation for two types of configurations of the gears are shown in Table 2. An example of crack propagation in the surface of coating as the loading cycles increases is shown in Figure 5.

The results from the FE model in this study were compared with the experimental results obtained from DGMK-FZG (Deutsche Wissenschaftliche Gesellschaft für Erdöl, Erdgas und Kohle, Forschungsstelle für Zahnräder and Getriebebau der TU München) short micro-pitting test. In such tests, different levels of load (Torque) were applied on the pinion/gear system. During the experiment, the critical loads were defined as the loads under which surface cracks nucleate, where the cracks were observed with scanning electron microscopy (SEM) (Figure 6).

6. Discussion

As shown in Table 2, average pressure at point A was greater than that at point B for both types of gears although the normal force acted on this region was much smaller than that in the region around point B. The reason for that was the small radius of curvature of involute profile at the root leading to the high localized stresses and eventually giving rise to early nucleation of a crack in this area. It was found from both experimental and computational results that pre-treatment with plasma nitriding improved the surface properties dramatically, where the critical load associated with initiation of the first crack for the Model II system is nearly twice higher than that for the Model I system. The greater critical load with the TiN/CrN on the nitrided 42CrMo4 substrate is largely due to the improved strength in the hardened case. As a consequence, the load-bearing capacity is increased for such a system.

The critical loads predicted from the simulation results in both conditions are greater than from experimental test, for instance, the normal force predicted in point B with Model I coated teeth is 1644 N, which is about 20% greater than from

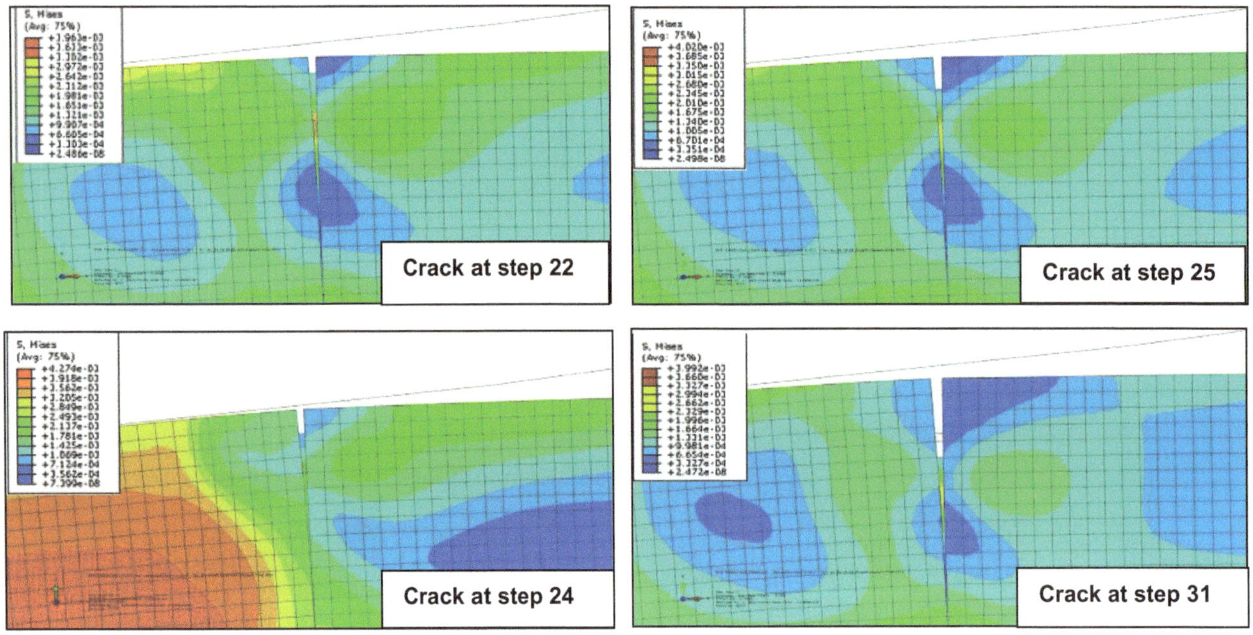

Figure 5. Crack propagation with an increase of the loading cycle.

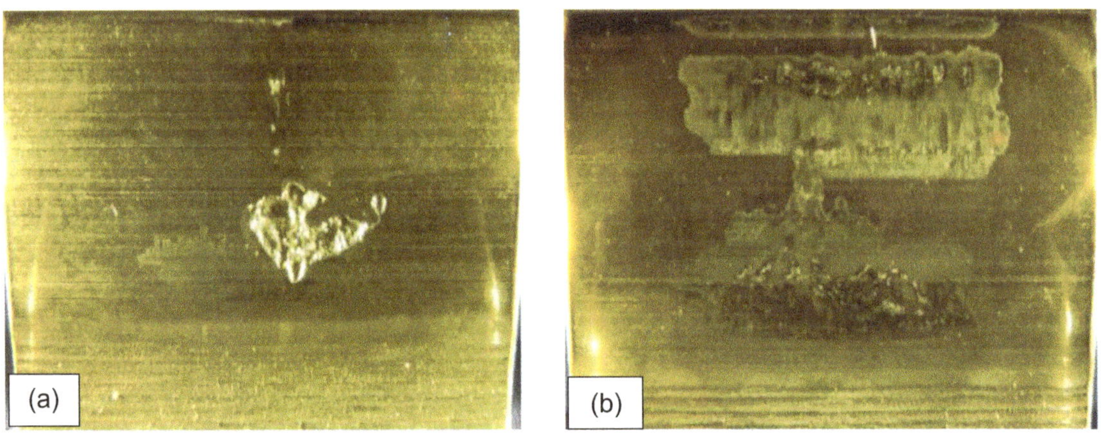

Figure 6. Photographs of the damages in the coated tooth (Model I and Model II). (a) Model I failed at the key loading stage 6 (normal force is 1333 N in point A and 3999 N in point B); and (b) Model II failed at the key loading stage 8 (normal force is 2357 N in point A and 7072 N in point B).

the experimental test. This discrepancy existed between the experimental and simulation results may be caused by the reason that the lubrication and effect of the roughness of surface were not considered in the FE modelling.

The simulation results also show that the strength of the coated layer degraded with increase of the loading cycles, which led to damages of the coated surface once the crack was nucleated. With an increase of the contact loading cycles, the cracks propagate rapidly to the depth of the coating, which will eventually result in the debonding of the coating from the substrate.

7. Conclusions

Several numerical modelling techniques were combined to develop a comprehensive model for the prediction of the failures of the coated surfaces on engineering components, and contact spur-gears were used as a test case in this study. The developed model not only allows localized stress distributions and deformations in the interested locations in coatings and substrate materials to be investigated in detail but also enables visual observation on the development of fatigue damages in the coating, due to the inclusion of a numerical capability of simulating multi-cycle loading in cohesive-zone modelling. The experimental observation on the damages of the coated tooth-surfaces of the gears suggests that similar phenomenon can be predicted with the numerical model developed. Further work is being carried out for improving accuracy of the predicted adhesion energy in coatings and for optimizing the arrangement of cohesive elements in order to further improve accuracy and efficiency of numerical modelling for predicting failures of the coated layers in engineering surface systems.

Acknowledgements. The research reported in this paper was supported by the EU FP7 project "Multiscale Modelling for Multilayered Surface Systems (M3-2S)", Grant No. CP-FP213600-2 M3-2S. Beneficial discussion with and support from the project partners are acknowledged.

References

1. S. Baragetti, F. Villa, The Open Materials Science Journal 8 (2014) 87–98.
2. K. Bobzin, N. Bagcivan, N. Goebbels, et al., Surface and Coatings Technology 204 (2009) 1097–1101.
3. A. Feuerstein, A. Kleyman, Surface and Coatings Technology 204 (2009) 1092–1096.
4. S. Baragetti, F. Tordini, Engineering with Computers 27 (2011) 127–137.
5. M. Sraml, J. Falsker, The International Journal of Advanced Manufacturing Technology 31 (2007) 1066–1075.
6. J. Feng, Y. Qin, Q. Zeng, E. Alamandoz, G. Fuente, H. Dong, R. Raghavan, J. Michler, Journal of Multiscale Modelling 1 (2011), 1–22.
7. J. Feng, Y. Qin, Applied Mechanics and Materials 446 (2014) 491.
8. J. Feng, Y. Qin, et al., Steel Research International, Proceeding of Metal Forming (2012), 1059–1062.
9. D.W. Dudley, Handbook of Practical Gear Design, Dudley Engineering Co. McGraw-Hill Book Company, US, 1984.
10. X.P. Xu, A. Needleman, Journal of the Mechanics and Physics of Solids 1397 (1994) 42.
11. A. Abdul-Baqi, P.J.G. Schreurs, M.G.D. Geers, International Journal of Solid and Structure 42 (2005) 927.

25

Development of micro rotary swaging tools of graded tool steel via co-spray forming

Chengsong Cui[1,*], Alwin Schulz[1], Matthias Steinbacher[1], Eric Moumi[2], Bernd Kuhfuss[2], Florian Böhmermann[3], and Oltmann Riemer[3]

[1] Foundation Institute for Materials Science, Badgasteiner Str. 3, 28359 Bremen, Germany
[2] Bremen Institute for Mechanical Engineering, University of Bremen, Badgasteiner Str. 1, 28359 Bremen, Germany
[3] Laboratory for Precision Machining (LFM), University of Bremen, Badgasteiner Str. 2, 28359 Bremen, Germany

Abstract – In order to meet the requirements of micro rotary swaging, the local properties of the tools should be adjusted properly with respect to abrasive and adhesive wear, compressive strength, and toughness. These properties can be optimally combined by using different materials in specific regions of the tools, with a gradual transition in between to reduce critical stresses at the interface during heat treatment and in the rotary swaging process. In this study, a newly developed co-spray forming process was used to produce graded tool material in the form of a flat product. The graded deposit was subsequently hot rolled and heat treated to achieve an optimal microstructure and advanced properties. Micro plunge rotary swaging tools with fine geometrical structures were machined from the hot rolled material. The new forming tools were successfully applied in the micro plunge rotary swaging of wires of stainless steel.

Key words: Spray forming, Graded tool steel, Micro rotary swaging

1. Introduction

The selection of tool materials for micro rotary swaging processes is an important issue. Adapted material properties are required as the tools are loaded differently in the various functional areas. For example, the tool surface should be hard and resistant to wear and the tool body should have sufficient toughness [1, 2]. The varying properties can be optimally combined by using adapted materials in the specific regions of the tools. The combination of materials can be achieved by the use of a gradual transition in between to reduce critical stresses at the interface during heat treatment and in the swaging process. Such graded materials have much lower tendency to delaminate than surface-coated materials.

On the other hand, tool materials for micro rotary swaging should be precisely machinable by micro-machining because the required geometrical structures of the tool surfaces are significantly smaller than 1 mm in more than one dimension [2]. To meet this requirement, the tool materials should exhibit fine and homogenous microstructures since the precision machining of the tool structures may be disturbed by large carbides or significant segregation [3].

A newly developed co-spray forming process is supposed to produce such graded tool materials [4]. In this process, two melts of different tool steels are co-sprayed and deposited on a flat substrate, resulting in a graded flat product when the two sprays partially overlap. By means of combining two tool steels in a single deposit, different microstructure and properties can be combined [5, 6]. In addition, a fine and homogeneous microstructure can be achieved in the spray-formed materials due to rapid solidification [7, 8].

In this study, a flat deposit of graded tool steel (HS6-5-3C/X110CrMoV8-2) was produced by this novel co-spray forming process. The graded deposit was subsequently hot rolled and heat treated to achieve optimal microstructure and material properties. Micro plunge rotary swaging tools were machined from the graded steel and tested in the deformation of wires of stainless steel 1.4301 (AISI 304).

2. Production of tool material

A high speed steel HS6-5-3C (AISI M3:2) was applied in the tool surface area and a cold working steel X110CrMoV8-2 in the tool body. A large number of carbides in HS6-5-3C are responsible for high hardness and excellent wear resistance. X110CrMoV8-2 shows significantly higher toughness compared to the high speed steel. The co-spray forming of the graded steel was carried out on the spray forming facility SK1+ at the University of Bremen (see Figure 1). The steels

*e-mail: cscui@iwt.uni-bremen.de

Figure 1. Co-spray forming of a graded steel deposit (X110CrMoV8-2/HS6-5-3C).

were melted in separate crucibles, and atomized by means of nitrogen gas jets generated by two free-fall gas atomizers. A flat ceramic substrate, moving from left to right, was used to collect the atomized droplets. HS6-5-3C was first spray-deposited on the substrate to form the lower layer of the deposit, followed by the deposition of X110CrMoV8-2 and the forming of the upper layer of the deposit. Since the two spray cones overlapped partially, a gradient zone was generated between the two layers. The superheating temperature of both melts was 180 °C. The pouring nozzle diameter was 4.5 mm for HS6-5-3C and 6.5 mm for X110CrMoV8-2. The melt mass flow rates were 0.22 kg/s and 0.47 kg/s, respectively. The gas atomizers scanned in the transverse direction of the substrate. The scanning angle of the atomizers was kept at 10.5°, and the scanning frequency at 5 Hz. The atomization gas pressure was 0.2 MPa for the HS6-5-3C melt and 0.5 MPa for the X110CrMoV8-2 melt, corresponding to a gas mass flow rate of 0.17 kg/s and 0.33 kg/s, respectively. The substrate was preheated to 600 °C and moved at a translational speed of 6 mm/s. The vertical spray distance from the atomizers to the substrate was 580 mm. The central distance between the atomizers was 340 mm. A distance of 156 mm between the axes of the two sprays at the substrate surface was determined as the tilting angles of both atomizers were 9°. The spray-formed deposit was in the form of a flat product, with the dimensions of 250 mm width, 500–600 mm length, and 55–60 mm thickness.

The spray-formed deposit was soft annealed to reduce hardness and avoid cracking. It was heated at a rate of 50 K/h to 840 °C and held at 840 °C for 4 h, followed by a controlled cooling at a rate of 20 K/h to 500 °C and a subsequent uncontrolled furnace cooling to RT. Samples were taken from the deposit for chemical composition testing, porosity measurement and metallographic study.

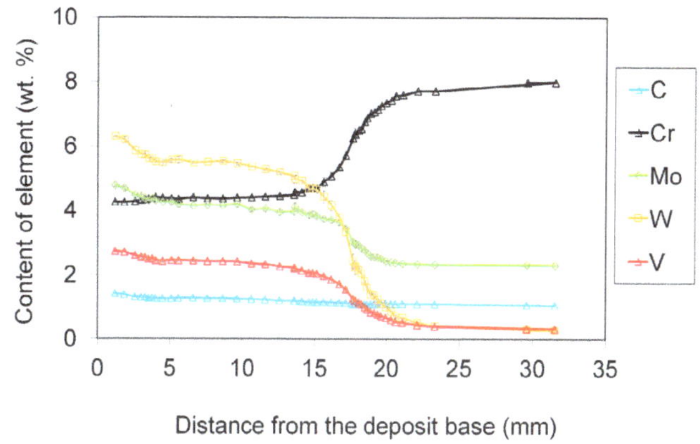

Figure 2. Main element distributions of the graded deposit (X110CrMoV8-2/HS6-5-3C, after soft annealing).

The spray-formed graded deposit was subsequently hot rolled on a lab rolling mill at the University of Bremen to eliminate porosity and break up the carbide network. Rolling samples with a thickness of 44–49 mm, a width of 32 mm and a length of 200 mm were machined from the graded deposit. The samples were preheated in a furnace to 1100 °C and rolled in three passes to a final thickness of approximately 24 mm. For each rolling pass, the true strain and the strain rate of the sample were about 0.2–0.3 and 1 s^{-1}, respectively. The total strain of the sample was 0.6–0.7.

Microscopic samples and semi-finished tools were machined from the hot rolled material. They were hardened and tempered prior to final grinding or milling: (1) they were austenitized in a vacuum furnace at 1080 °C for 20 min, followed by gas quenching with 6 bar nitrogen; (2) they were subsequently tempered three times at 550 °C for 2 h and cooled in air. The hardness of the X110CrMoV8-2 was about 65 HRC and about 66 HRC for the HS6-5-3C.

2.1. Element distributions in the graded deposit

The chemical composition of the graded deposit was analyzed by spark optical emission spectroscopy (S-OES, ARL3460). The sample was ground at intervals and the S-OES measurement was conducted at the newly ground surfaces (starting from the base of the sample). In this way, the element distributions in the thickness of the graded deposit were determined. The distributions of the alloying elements C, Cr, Mo, W and V in the spray-formed deposit are presented in Figure 2. The gradient zone starts at the position about 15 mm from the base of the deposit, and it ends at the position about 20 mm from the base of the deposit. It has been proved that the slope of the gradient of the elements in the deposit strongly depends on the distance between the spray axes [4, 5]. Partial overlapping of the spray cones occurred when both spray cones were tilted by 9°, and the droplets of the two different steels mixed in the overlapped spray as well as in the mushy layer of the deposit. Further homogenization of the composition in the gradient zone might take place since some of the elements might diffuse over a short distance at high temperature due to the concentration difference between the steels.

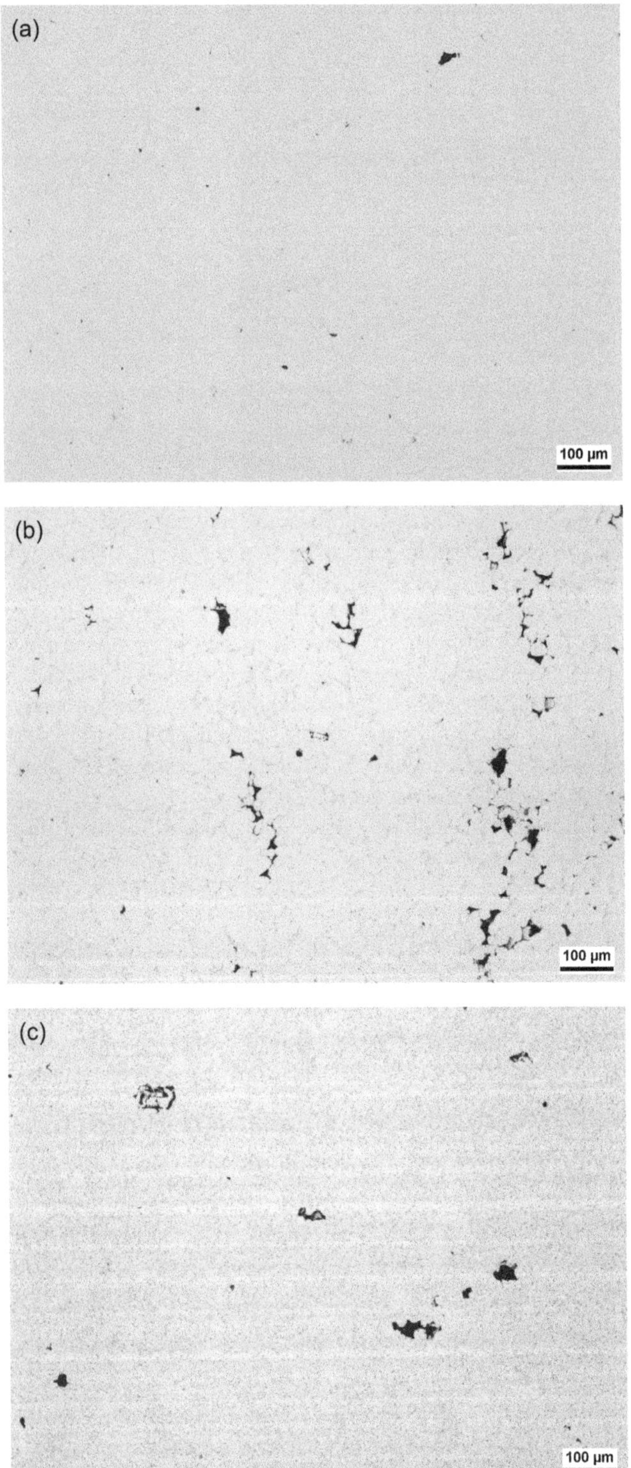

Figure 3. Porosity of the graded deposit: (a) X110CrMoV8-2; (b) X110CrMoV8-2+HS6-5-3C; and (c) HS6-5-3C.

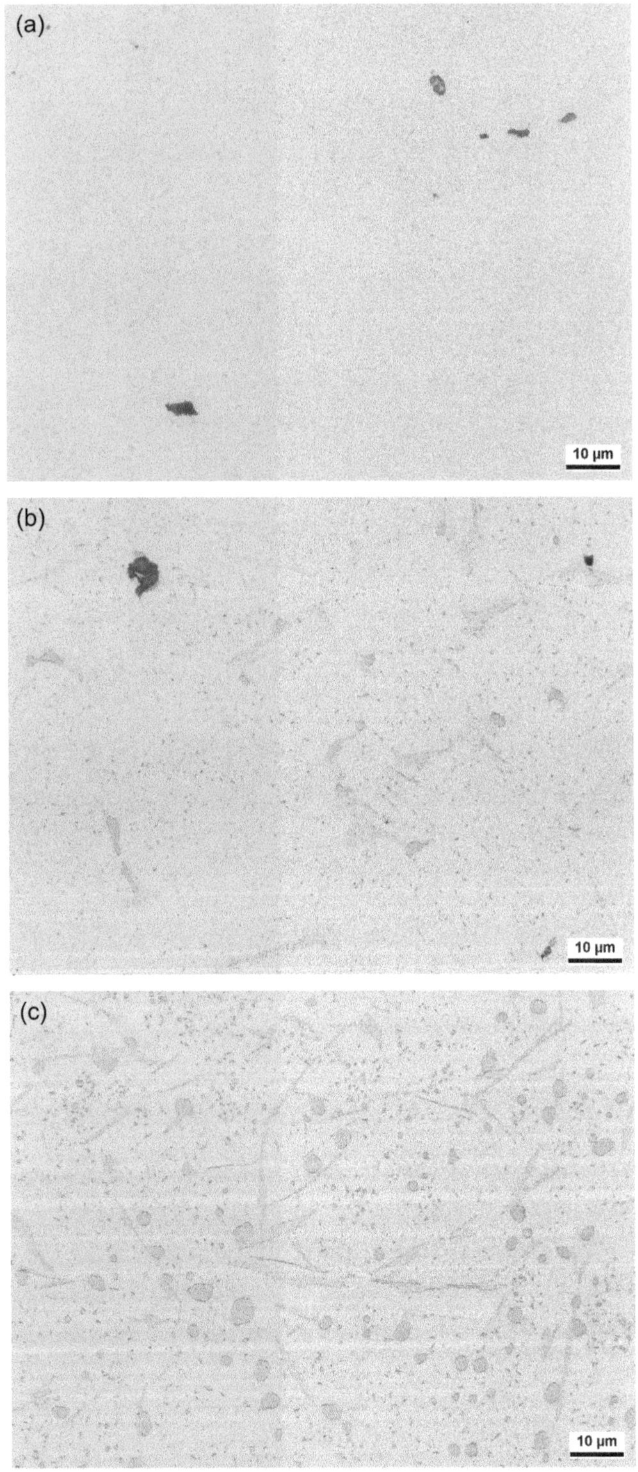

Figure 4. Polished microsections of the graded deposit: (a) X110CrMoV8-2; (b) X110CrMoV8-2+HS6-5-3C; and (c) HS6-5-3C.

2.2. Material characterisation of the graded deposit

Figure 3 shows the porosity of the graded deposit. Small pores are seen in the deposit, particularly in the gradient zone. The irregular shape of the pores in the gradient zone indicates relatively cold spray condition. During the spray forming of multilayer materials, a porous zone is frequently formed at the interface between the layers of the deposits [9, 10]. When the first layer is spray-formed, the upper surface of the layer often shows high amount of porosity since the surface temperature is relatively low and the solidification shrinkage as well

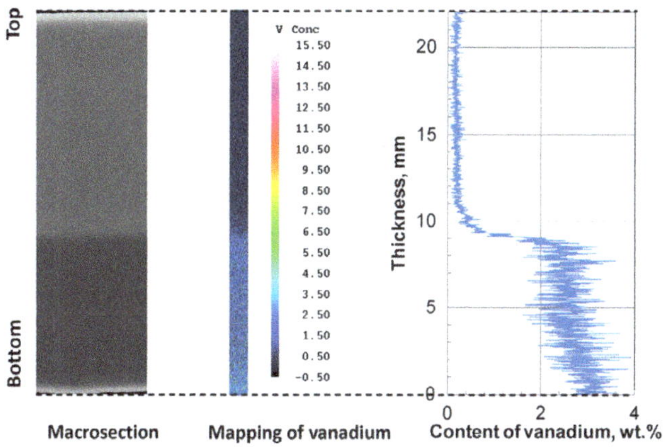

Figure 5. Element distribution across the thickness of the graded tool steel (X110CrMoV8-2/HS6-5-3C) after hot rolling.

as the interstices between the droplet fragments and solid particles in the surface area cannot be refilled by sufficient liquid phase. If the second spray is close to the first spray, an overlapping of the sprays contributes to higher mass input and enthalpy input to the interfacial area of the deposit. This helps reduce the porosity in this area.

The graded deposit after soft annealing is mainly composed of ferrite and carbides. Some relatively long and continuous carbide structures (carbide network) are located at the prior austenite grain boundary (it is more clearly seen after etching [5]). The carbides in the interior of the prior austenite grains in the upper layer (X110CrMoV8-2) of the deposit are very fine. In the lower layer (HS6-5-3C) of the deposit, some granular carbide particles (V-rich MC type) are observed. In the gradient zone a mixed microstructure can be seen (Figure 4).

2.3. Element distributions in the hot rolled material

The distributions of the main chemical elements in the hot rolled material were analyzed by means of electron probe microanalysis (model: JEOL JXA-8200). A typical element distribution (vanadium) across the thickness of the hot rolled steel is shown in Figure 5. The blue bar is the mapping of the element (about 1 mm width). The measurement interval is 11 μm in both directions. The curve on the right side was obtained by calculating the mean values of the element content from the mapping of the element. A picture of a macro-etched sample is combined with the element distribution in the figure. The material of the upper part is X110CrMoV8-2 (0.5 wt.% V), and the material of the lower part is HS6-5-3C (3 wt.% V). The gradient zone is in the range of 8–10 mm from the base of the hot rolled sample.

2.4. Microstructure of the hot rolled graded steel

Representative microstructures of the hot rolled steel (after hardening and tempering) are shown in Figures 6 and 7 (optical microscopy, Model: Axiophot, ZEISS). The etchant Nital

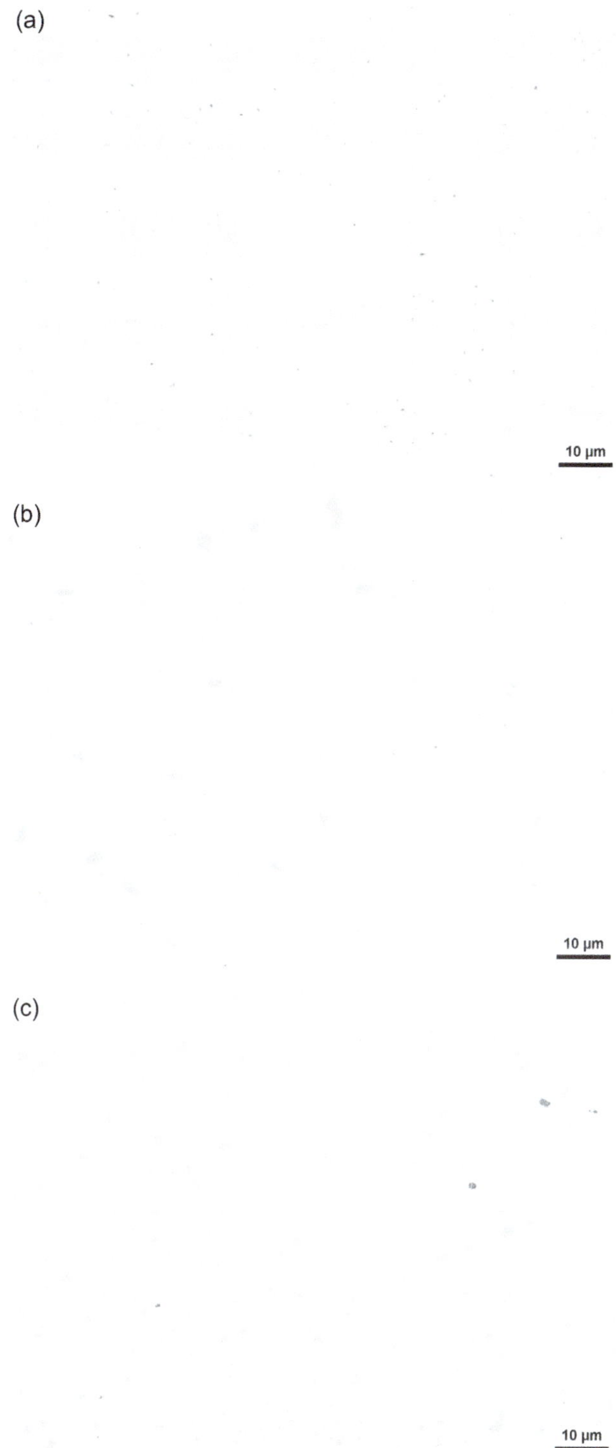

Figure 6. Polished microsections of the graded tool steel processed by co-spray forming, hot rolling, hardening and tempering: (a) X110CrMoV8-2; (b) X110CrMoV8-2+HS6-5-3C; and (c) HS6-5-3C.

(3% alcoholic HNO_3 at RT) was used to reveal the microstructure of the samples. The porosity in the spray-formed deposit was essentially eliminated by the hot rolling, although the deformation ratio was rather low. The hot rolled material

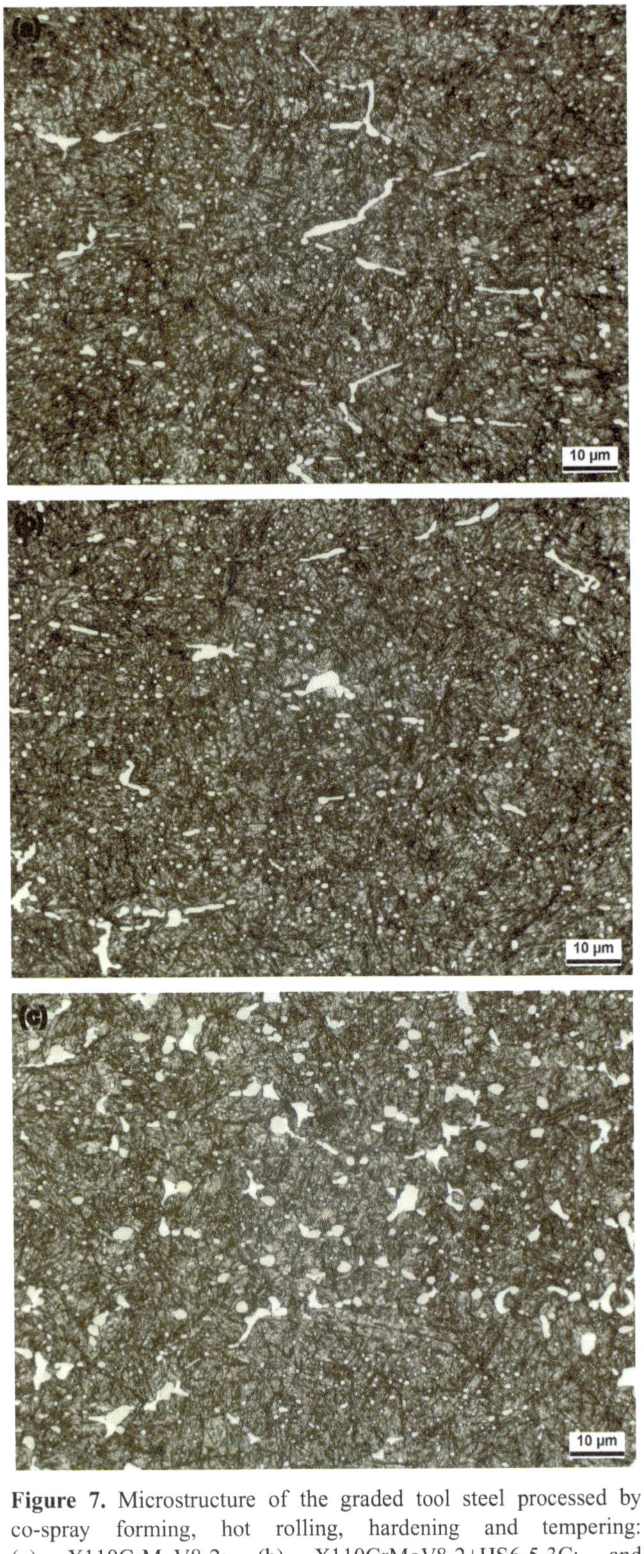

Figure 7. Microstructure of the graded tool steel processed by co-spray forming, hot rolling, hardening and tempering: (a) X110CrMoV8-2; (b) X110CrMoV8-2+HS6-5-3C; and (c) HS6-5-3C.

showed a relatively fine and homogeneous microstructure. The carbide network of the deposit was also broken up to some extent due to hot rolling. There are remarkably more and larger carbides in the lower layer (HS6-5-3C) than in the upper layer

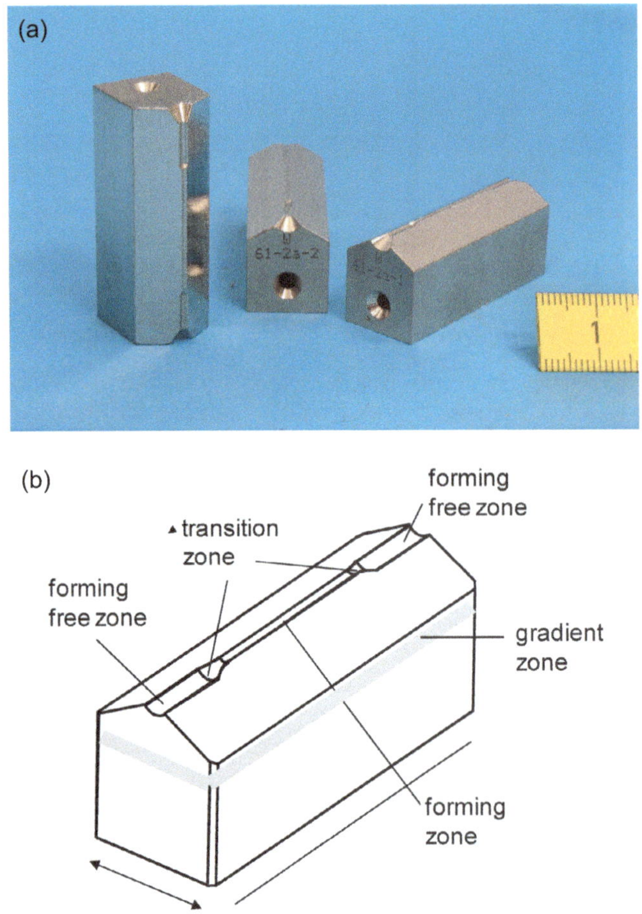

Figure 8. Micro plunge rotary swaging tools of graded tool steel (X110CrMoV8-2/HS6-5-3C): (a) set of swaging tools; and (b) micro swaging tool with indicated different functional areas.

(X110CrMoV8-2) of the graded steel. A gradual transition of the microstructure is obviously seen in the graded steel.

3. Machining of micro rotary swaging tools

A set of micro plunge rotary swaging tools, for deforming wires to a diameter of 0.8 mm, were manufactured in the hardened state by micro milling. The tool surface consisted of HS6-5-3C, the tool body of X110CrMoV8-2, with the gradient zone in between (see Figure 8). The machining was carried out on a DMG Sauer Ultrasonic 20 linear machine tool. The machining operation comprised multiple, consecutive roughing and finishing steps. CAD/CAM-programming was used for tool path generation. Hard coated tungsten carbide ball-end mills of 1.0 mm for roughing and 0.5 mm diameter for finishing were applied. Material allowance for the finishing cut was 10 µm. The depth of cut and the width of cut for finishing were set to a_p = 5 µm and a_e = 5 µm, respectively. The spindle speed and the feed velocity were kept constant at n = 40,000 min^{-1} and v_f = 1,000 mm/min.

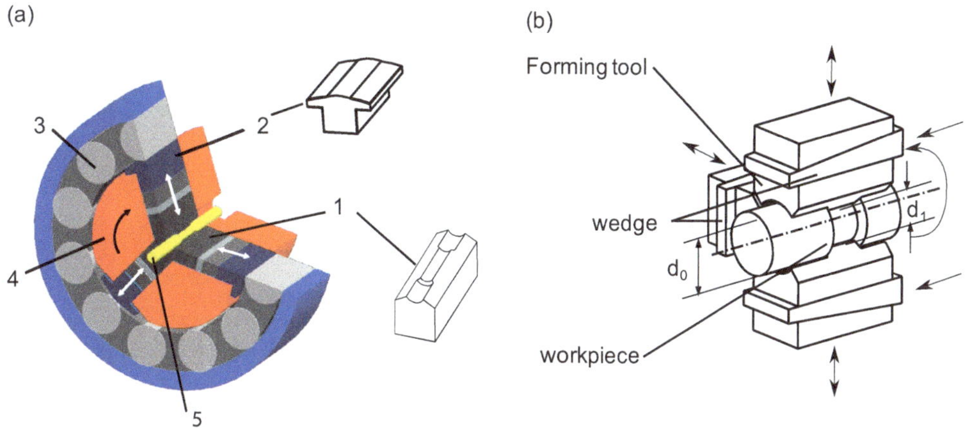

Figure 9. (a) Assembly of rotary swaging head and (b) principle of plunge rotary swaging.

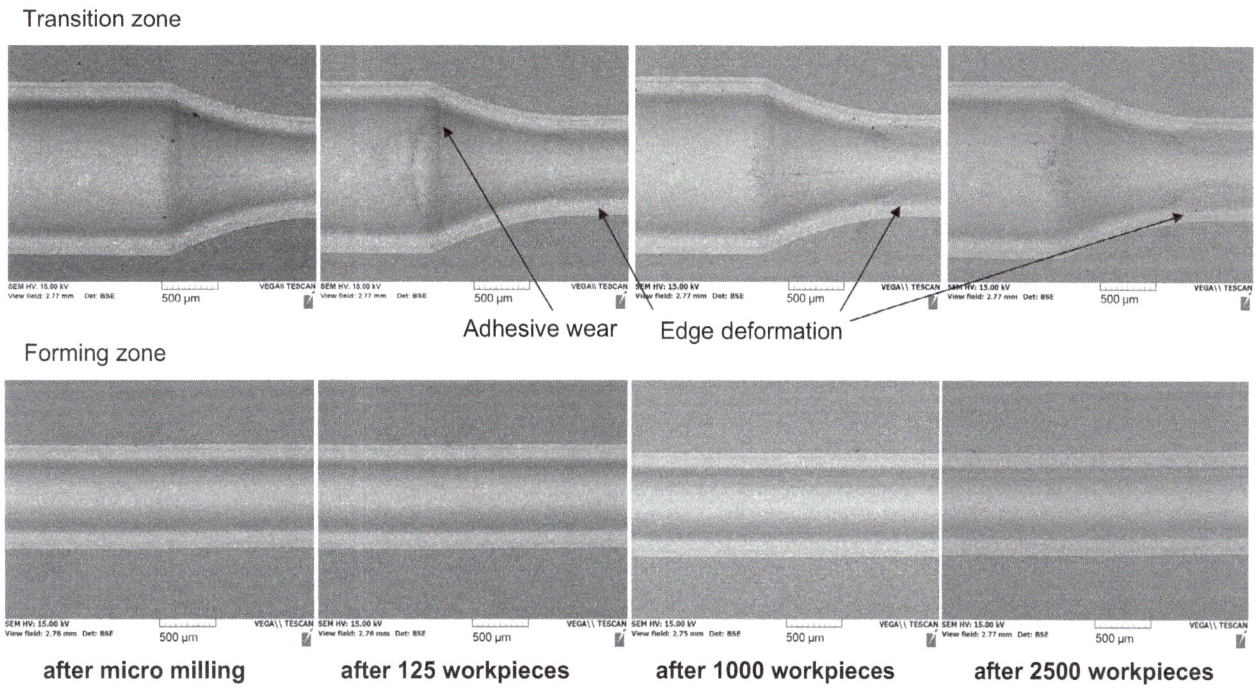

Figure 10. Tool surfaces of the micro plunge rotary swaging tools (forming zone Ø = 0.8 mm).

4. Micro rotary swaging of steel wires

Rotary swaging is an incremental forming process for tubes, bars and wires. Forming takes place in the rotary swaging head (Figure 9a) by a quick oscillating movement of the forming tools (1). Due to the rotation of the driven shaft (4) when the base jaws come across a cylinder roller (3) the forming tools are pushed inwards and the workpiece (5) is deformed [11].

There are two main variations of the process: the infeed rotary swaging during which the diameter of the workpiece is reduced over the complete feed length, and the plunge rotary swaging which allows local diameter reductions. In this study, the plunge rotary swaging was applied and Figure 9b shows the principle of this variation. In plunge rotary swaging, additionally to the oscillating movement, the forming tools achieve a radial displacement towards the workpiece due to the axial displacement of the wedges.

The graded tools were applied in a rotary swaging machine (Type Felss HE 3/DE) to reduce the diameter of wires of stainless steel 1.4301 (AISI 304) from the initial diameter of 1.0 mm to the final diameter of 0.8 mm. The wires were in a soft annealed state (surface hardness 300 HV0.1). The tool oscillated with a frequency of 100 Hz and the radial displacement was set to 35 μm/s. The total strain of the wires was 0.45. The gradient zone between the two steels (X110CrMoV8-2/HS6-5-3C) did not show signs of fracture or cracking after the test (up to 2500 workpieces).

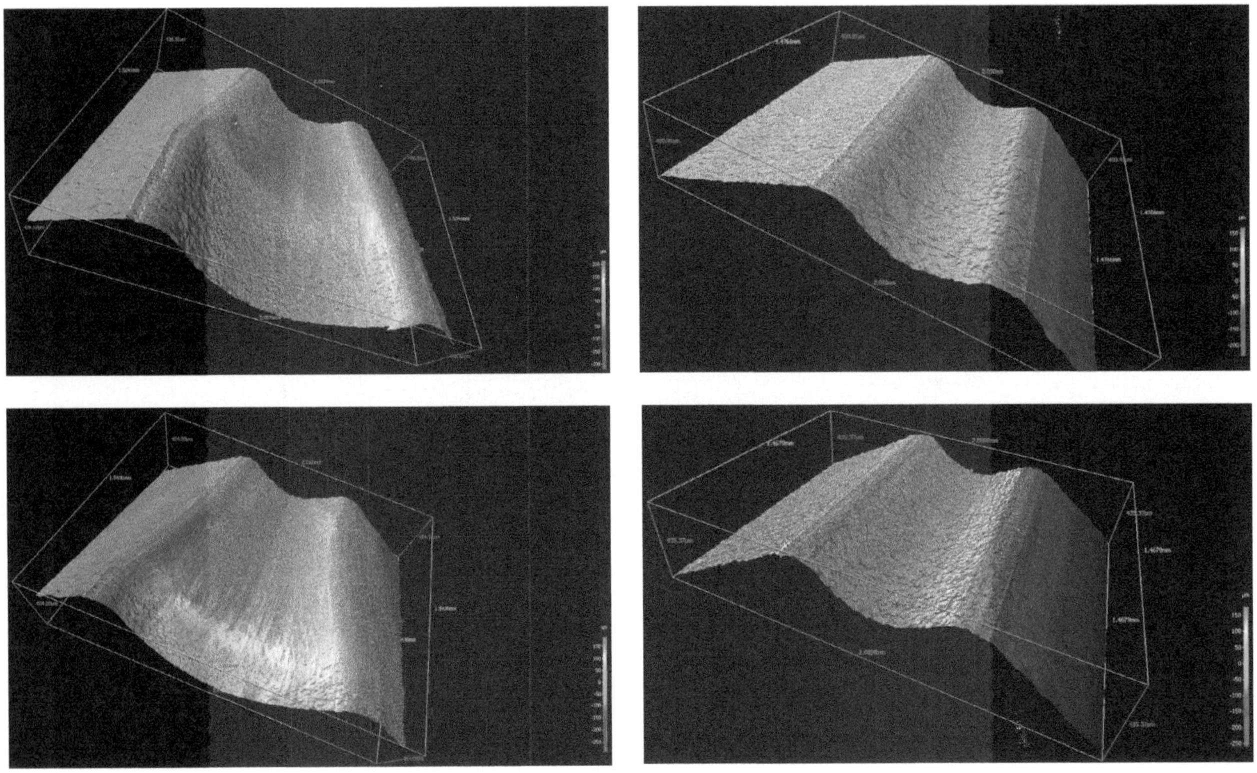

Figure 11. Topography of the tool surfaces: (a) transition zone and (b) forming zone after micro milling, (c) transition zone and (d) forming zone after micro plunge rotary swaging.

The tool surfaces and the steel wires before and after the rotary swaging were investigated by means of scanning electron microscope (model: CAMSCAN CS44, Elektronen-Optik-Service GmbH). Stereoscopic images were used to retrieve 3D information of the tool surfaces (using MeX software from Alicona). It showed that the tool surfaces were smooth and the fine geometrical structures of the tools were precisely machined by micro milling. Only slight traces of micro milling were observed at the tool surfaces. The excellent micro-machinability of the graded steel was guaranteed due to its relatively fine and homogeneous microstructure.

The tool surfaces before and after micro rotary swaging are shown in Figure 10. By comparing the tool surfaces, it is seen that slight adhesion of the workpiece on the tool surfaces and slight deformation/abrasive wear in the transition zone appeared. The forming zone of the tools was almost unchanged after deforming 2500 workpieces. The topography derived from the surface measurement in SEM with Alicona MeX shows clearly the three-dimensional morphology of the tool surfaces (see Figure 11). The steel wires deformed by the swaging tools are shown in Figure 12. The shape, dimensional accuracy, and surface quality of the deformed wires are satisfactory.

It was noticed that the degree of deformation/abrasive wear at the entrance of the forming zone increased with increasing number of workpieces (Figure 10, upper pictures). The damage could be more serious if the tools are applied to deform more workpieces. This might lead to poor surface roughness, low size and shape accuracy of the workpieces and even complete failure of the deformation (fracture of the workpieces). In order to reduce deformation/wear of the swaging tools, several strategies can be considered:

1. applying stronger tool materials at the tool surfaces (increasing carbide content and hardness, improving tool properties and performance by means of adaptive heat treatment, etc.);
2. reducing the deformation ratio of the wires;
3. reducing the temperature of the tools by coolant and improving the lubrication of the tools to reduce the friction between the wire and the tools.

5. Summary and outlook

The graded steel (X110CrMoV8-2/HS6-5-3C) was produced by means of co-spray forming and hot rolling, showing a relatively fine and homogeneous microstructure. After hardening and tempering, the hardness of the steels X110CrMoV8-2 and HS6-5-3C was about 65 HRC and 66 HRC, respectively. The as-hardened graded steel could be machined precisely to micro plunge rotary swaging tools with smooth tool surfaces and fine geometrical structures. The new forming tools were successfully applied in the micro plunge rotary swaging of wires of stainless steel. The deformed wires showed satisfactory shape, size and surface quality. Damages of the tools are supposed to be reduced by applying stronger tool materials at the tool surface, reducing deformation ratio

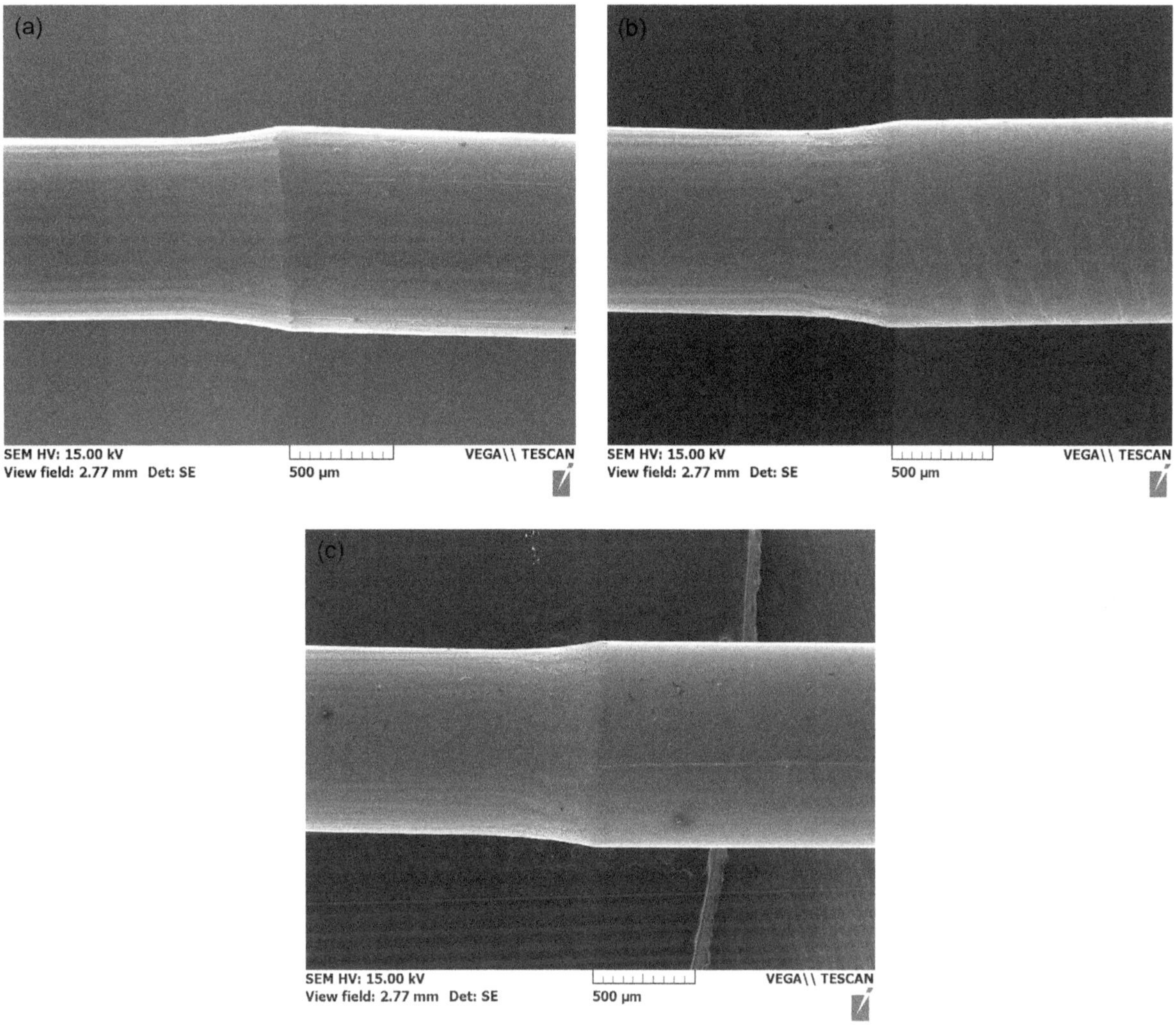

Figure 12. AISI 304 steel wires deformed from Ø = 1 mm to Ø = 0.8 mm by micro plunge rotary swaging: (a) workpiece no. 125; (b) workpiece no. 1000; and (c) workpiece no. 2500.

of the workpieces, and reducing the tool temperature and the friction between the wire and the tools. These suggestions are topics for further investigations.

Acknowledgements. The authors gratefully acknowledge the financial support by Deutsche Forschungsgemeinschaft (DFG, German Research Foundation) for the Subprojects A4, C2, and C6 within the SFB 747 (Collaborative Research Center) "Mikrokaltumformen – Prozesse, Charakterisierung, Optimierung".

References

1. A. Schulz, in: Micro metal forming, edited by F. Vollertsen, Springer-Verlag, Berlin, 2013.
2. F. Vollertsen, H. Schulze Niehoff, Z. Hu, Int. J. Mach. Tools Manuf. 46 (2006) 1172.
3. H. Flosky, F. Vollertsen, Key Eng. Mater. 549 (2013) 511.
4. C. Cui, A. Schulz, Metall. Mater. Trans. 44B (2013) 1030.
5. C. Cui, A. Schulz, V. Uhlenwinkel, Steel Res. Int. 84 (2013) 1075.
6. C. Cui, A. Schulz, V. Uhlenwinkel, Mater. Sci. Eng. Technol. 45 (2014) 652.
7. U. Fritsching, Spray simulation – modelling and numerical simulation of sprayforming metals, Cambridge University Press, Cambridge, 2004.
8. E.J. Lavernia, Y. Wu, Spray atomization and deposition, John Wiley, New York, 1996.
9. C. Meyer, V. Uhlenwinkel, R. Ristau, P. Jahn, H. Müller, P. Krug, W. Trojahn, in: Proceedings of 4th International Conference on Spray Deposition and Melt Atomization (SDMA 2009), Universität Bremen – Staats-und Universitätsbibliothek (Verlag), CD version, 2009.
10. C. Meyer, N. Ellendt, L. Mädler, H. R. Müller, F. Reimer, V. UhlenwinkelMat.-wiss. u. Werkstofftech. 45/8 (2014) 642.
11. B. Kuhfuß, E. Moumi, V. Piwek, Microsystem. Technol. 14 (2008) 1995.

Permissions

All chapters in this book were first published in MFR, by EDP Sciences; hereby published with permission under the Creative Commons Attribution License or equivalent. Every chapter published in this book has been scrutinized by our experts. Their significance has been extensively debated. The topics covered herein carry significant findings which will fuel the growth of the discipline. They may even be implemented as practical applications or may be referred to as a beginning point for another development.

The contributors of this book come from diverse backgrounds, making this book a truly international effort. This book will bring forth new frontiers with its revolutionizing research information and detailed analysis of the nascent developments around the world.

We would like to thank all the contributing authors for lending their expertise to make the book truly unique. They have played a crucial role in the development of this book. Without their invaluable contributions this book wouldn't have been possible. They have made vital efforts to compile up to date information on the varied aspects of this subject to make this book a valuable addition to the collection of many professionals and students.

This book was conceptualized with the vision of imparting up-to-date information and advanced data in this field. To ensure the same, a matchless editorial board was set up. Every individual on the board went through rigorous rounds of assessment to prove their worth. After which they invested a large part of their time researching and compiling the most relevant data for our readers.

The editorial board has been involved in producing this book since its inception. They have spent rigorous hours researching and exploring the diverse topics which have resulted in the successful publishing of this book. They have passed on their knowledge of decades through this book. To expedite this challenging task, the publisher supported the team at every step. A small team of assistant editors was also appointed to further simplify the editing procedure and attain best results for the readers.

Apart from the editorial board, the designing team has also invested a significant amount of their time in understanding the subject and creating the most relevant covers. They scrutinized every image to scout for the most suitable representation of the subject and create an appropriate cover for the book.

The publishing team has been an ardent support to the editorial, designing and production team. Their endless efforts to recruit the best for this project, has resulted in the accomplishment of this book. They are a veteran in the field of academics and their pool of knowledge is as vast as their experience in printing. Their expertise and guidance has proved useful at every step. Their uncompromising quality standards have made this book an exceptional effort. Their encouragement from time to time has been an inspiration for everyone.

The publisher and the editorial board hope that this book will prove to be a valuable piece of knowledge for researchers, students, practitioners and scholars across the globe.

List of Contributors

Majid Khalili
Department of Industrial Engineering, Islamic Azad University, Karaj Branch, Karaj, Iran

Ugochukwu C. Okonkwo, Imhade P. Okokpujie and Jude E. Sinebe
Department of Mechanical Engineering, Nnamdi Azikiwe University, PMB 5025 Awka, Nigeria

Chinedu A.K. Ezugwu
Department of Mechanical Engineering, Landmark University, PMB 1001 Omu-Aran, Kwara, Nigeria

Deqiang Yin
School of Manufacturing Science and Engineering, Sichuan University, Chengdu, P.R. China
Department of Design, Manufacture and Engineering Management, University of Strathclyde, Glasgow G11XQ, Scotland, UK

Zhenhai Xu and Yi Qin
Department of Design, Manufacture and Engineering Management, University of Strathclyde, Glasgow G11XQ, Scotland, UK

Jiling Feng
School of Engineering, Manchester Metropolitan University, Manchester, M15 6BH, UK

Ersyzario Edo Yunata
Department of Regional Environment, Shibaura Institute of Technology, 3-9-14 Shibaura, Minato-City, Tokyo 108-8548, Japan

Tatsuhiko Aizawa
Department of Engineering and Design, Shibaura Institute of Technology, 3-9-14 Shibaura, Minato-City, Tokyo 108-8548, Japan

Ken-ichiro Mori, Tomoyoshi Maeno and Takuya Suganami
Department of Mechanical Engineering, Toyohashi University of Technology, Toyohashi, Aichi 441-8580, Japan

Purwo Kadarno, Ken-Ichiro Mori, Yohei Abe and Tatsuro Abe
Department of Mechanical Engineering, Toyohashi University of Technology, Toyohashi, Aichi 441-8580, Japan

Luis M. Alves, Carlos M.A. Silva and Paulo A.F. Martins
IDMEC, Instituto Superior Técnico, Universidade de Lisboa, Av. Rovisco Pais, 1049-001 Lisboa, Portugal

Yong Jin, Tetsuhide Shimizu and and Ming Yang
Graduate School of System Design, Tokyo Metropolitan University, Hino 191-0065, Japan

Shigeo Yasuhara
Japan Advanced Chemicals, 3131-4 Tana, Chuo-ku, Segamihara-shi, Kanagawa 252-0244, Japan

Aniruddha V. Muley
Netaji Subhas Institute of Technology (NSIT), Dwarka, New Delhi 110078, India

S. Aravindan and I.P. Singh
Indian Institute of Technology Delhi, Hauz Khas, New Delhi 110016, India

Colin S. Harrison
Mechanical and Electrical Engineering, School of Engineering, Glasgow Caledonian University, Cowcaddens Road, Glasgow G4 0BA, UK

Xibing Gong and Kevin Chou
Mechanical Engineering Department, The University of Alabama, Tuscaloosa, AL 35487, USA

Ted Anderson
Advanced Manufacturing Team, Marshall Space Flight Center, Huntsville, AL 35811, USA

Aaron C.L. Lam, Zhusheng Shi, Yo-Lun Yang and Jianguo Lin
Department of Mechanical Engineering, Imperial College London, London SW7 2AZ, UK

Xia Huang and Yuansong Zeng
AVIC Beijing Aeronautical Manufacturing Technology Research Institute, Beijing 100024, PR China

Kazushi Koshimizu, Qiu Zheng, Tetsuhide Shimizu and Ming Yang
Tokyo Metropolitan University, 6-6 Asahigaoka, Hino-shi, Tokyo 191-0065, Japan

Jürgen Leopold
TBZ-PARIV GmbH, 09126 Chemnitz, Germany

Steven Peters, Jung-Hoon Chun and Gisela Lanza
Institute of Production Science (wbk), Karlsruhe Institute of Technology (KIT), Karlsruhe, 76131 Germany
Laboratory for Manufacturing and Productivity, Massachusetts Institute of Technology (MIT), Cambridge, MA, 02139 USA

Lang Lihui, Liu Kangning, Gaoshen Cai, Xiying Yang, Chan Guo and Guoliang Bu
School of Mechanical Engineering and Automation, Beihang University, Beijing 100191, China

Wujiao Xu, Xin Ma, Nongjie Tang, Li Zhu, Wuhua Li and Yongfeng Ding
College of Material Science and Engineering, Chongqing University, Chongqing 400044, PR China

Prabhu Paulraj and Rajnish Garg
University of Petroleum & Energy Studies, Dehradun 248007, Uttarakhand, India

Jun Liu, Haoxiang Gao, Omer El Fakir, Liliang Wang and Jianguo Lin
Department of Mechanical Engineering, Imperial College London, London SW7 2AZ, UK

Verena Psyk, Petr Kurka, Simon Kimme, Markus Werner, Dirk Landgrebe
Fraunhofer Institute for Machine Tools and Forming Technology, Reichenhainer Strasse 88, 09126 Chemnitz, Germany

Andreas Ebert and Mathias Schwarzendahl
Westfalia Presstechnik GmbH & Co. KG, Gewerbering 08451 Crimmitschau, Germany

Tatsuhiko Aizawa
Department of Engineering and Design, Shibaura Institute of Technology, 3-9-14 Shibaura, Minato-City, Tokyo 108-8548, Japan

Tatsuya Fukuda
Research Center, Kimura Precision, Co. Ltd., Kyoto 615-8245, Japan

Hiroshi Morita
Nano-Coat Laboratory, LLC., Tokyo 144-0045, Japan

Xinbing Yun, Mo Zhou, Tian Tian and Ying Zhao
Engineering Research Center of Continuous Extrusion, Ministry of Education, Dalian Jiaotong University, Dalian 116028, P.R. China

Stephan F. Jahn, Benedikt Müller and Felix S. Hilfinger
Professorship Micromanufacturing Technology, Technische Universität Chemnitz, 09107 Chemnitz, Germany

Andreas Schubert
Fraunhofer Institute for Machine Tools and Forming Technology, Reichenhainer Str. 88, 09126 Chemnitz, Germany

Jiling Feng
Division of Mechanical Engineering, School of Engineering, Manchester Metropolitan University, Manchester M1 5GD, UK

Yi Qin
Department of Design, Manufacturing and Engineering Management, University of Strathclyde, Glasgow G1 1XJ, UK

Hanshan Dong
Surface Engineering, University of Birmingham, Birmingham B15 2TT, UK

Chengsong Cui, Alwin Schulz and Matthias Steinbacher
Foundation Institute for Materials Science, Badgasteiner Str. 3, 28359 Bremen, Germany

Eric Moumi and Bernd Kuhfuss
Bremen Institute for Mechanical Engineering, University of Bremen, Badgasteiner Str. 1, 28359 Bremen, Germany

Florian Böhmermann and Oltmann Riemer
Laboratory for Precision Machining (LFM), University of Bremen, Badgasteiner Str. 2, 28359 Bremen, Germany

Index